1st International Seminar on Smart Molecule of Natural Resources (ISSMART 2019)

Smart Molecules of Natural Resources for Biotechnology Perspective in Industrial Revolution 4.0

Journal of Physics: Conference Series Volume 1374

Malang, Indonesia
11 - 12 July 2019

Editors:

Nia Kurnianingsih
Rista Nikmatu Rohmah
Hazna Noor Meidinna
Lidwina Faraline Triprisila

ISBN: 978-1-7138-1131-2
ISSN: 1742-6588

Printed from e-media with permission by:

Curran Associates, Inc.
57 Morehouse Lane
Red Hook, NY 12571

Some format issues inherent in the e-media version may also appear in this print version.

Institute of Physics
Dirac House, Temple Back
Bristol BS1 6BE UK

Phone: 44 1 17 929 7481
Fax: 44 1 17 920 0979

techtracking@iop.org

Additional copies of this publication are available from:

Curran Associates, Inc.
57 Morehouse Lane
Red Hook, NY 12571 USA
Phone: 845-758-0400
Fax: 845-758-2633
Email: curran@proceedings.com
Web: www.proceedings.com

TABLE OF CONTENTS

BIOLOGICAL PHYSICS

SEPARATION OF FLAVONOIDS IN THE EXTRACT POLYALTHIA LONGIFOLIA (SONN.) THW. LEAVES FROM INDONESIA AND THE PHILIPPINES 1
D S Nahari, S Prasetyawan, M A G Beltran, A Aulanni'am

THE EFFECT OF SNAKE FRUIT (SALACCA EDULIS REINW.) SEED EXTRACT ON BODY WEIGHT IN FEMALE RATS MODEL WITH IRON DEFICIENCY ANEMIA 8
D Melati, D G Tamtomo, D Indarto

EFFECTS OF 25%, 37.5% AND 50% MUSA ACUMINATA EXTRACT AS A DENTURE CLEANSER ON THE FLEXURAL STRENGTH AND SURFACE ROUGHNESS OF ACRYLIC RESIN 15
D Puspitasari, A Setiawan, D F Annisa, S R Pramitha, M L Apriasari

EFFECT OF ACTIVE COMPOUND OF PASAK BUMI ROOT (EURICOMA LONGIFOLIA, JACK) AS AN INHIBITOR OF CDK2 METHYLATION: IN SILICO STUDY 22
E Y Rahman, M Ali, B B Purnomo, N Kania, D H Utomo, P W Nurikhwan

PRODUCTION AND OPTIMIZATION OF LIPASE BY ASPERGILLUS NIGER USING COCONUT PULP WASTE IN SOLID STATE FERMENTATION 28
E Suyanto, E S Soetarto, M N Cahyanto

ANTIOXIDANT AND ANTI-INFLAMMATORY ACTIVITY OF CHLOROGENIC ACID ON LEAD-INDUCED FIBROBLAST CELLS 34
E Girsang, I N E Lister, C N Ginting, S L Nasution, S Suhartina, U Z Munshy, R Rizal, W Widowati

BIOSORPTION OF SYNTHETIC DYE FROM BATIK WASTEWATER USING TRICHODERMA VIRIDE IMMOBILIZED ON CA-ALGINATE 42
G Rahmaniah, C Mahdi, A Safitri

DETECTION AND ANALYSIS OF DELAYED-RIPENING GENES IN THE GROWTH OF STRAWBERRIES (FRAGARIA × ANANASSA) TREATED 1-NAPHTALENEACETIC ACID 48
G R Aristya, N T A Maulina, M F Arif, R S Kasiamdari, A Widiastuti

HEPATOPROTECTIVE EFFECT OF EUGENOL ON ACETAMINOPHEN-INDUCED HEPATOTOXICITY IN HEPG2 CELLS 58
I N E Lister, C N Ginting, E Girsang, A Amansyah, L Chiuman, N L W E Yanti, R Rizal, W Widowati

MARTAPURA RIVER WATER LEADS TO TESTES ALTERATION IN RATS 65
I Yuliana, H Khatimah, L Rosida, N S Skripsiana, E Suhartono

DNA BARCODING AS A METHOD FOR SPECIES IDENTIFICATION: CASE STUDY IN AHAETULLA SNAKE 71
L I N Indahsari, F Fatchiyah, N Kurniawan

SUSTAINABILITY STRATEGY OF AGARWOOD-PRODUCTION TREES IN LOMBOK ISLAND USING HARVESTING MATRIX 79
M U Romdhini, N Fitriyani, A Kurnia, I K A Baskara, M Ito, I Yamada, S Hadi, T Mulyaningsih

SNAKE FRUIT SEEDS FLOUR-FORTIFIED JELLY FOR SUPPLEMENTARY FEEDING IN ADOLESCENT GIRLS WITH ANEMIA 88
M A Nugraheni, D Indarto, E P Pamungkasari

KELAKAI EXTRACT PROTECTS SKIN FROM UV-INDUCED OXIDATIVE DAMAGE 94
M Mashuri, L D M Sihombing, S Alfaqihah, E Edyson, E Suhartono

THE EFFECT OF KAFFIR LIME (CITRUS HYSTRIX DC) ESSENTIAL OIL ON BIOPLASTIC DERIVED FROM CASSAVA PEEL WASTE 100
M Masruri, A Z Azhar, I Rosyada, A Febrianto

ANTIOXIDANT ACTIVITY OF VACUUM COLUMN CHROMATOGRAPHY FRACTIONS OF ETHANOL EXTRACT OF SARANG BANUA (CLERODENRUM FRAGRANS VENT WILLD) LEAVES 106
M Simorangkir, B Nainggolan, S Silaban

THE DIFFERENCE OF AMINO ACID PROFILING FROM TWO MORPHOLOGICAL PURPLE SWEET POTATOES FROM KAWI MOUNTAIN CULTIVARS, EAST JAVA, INDONESIA 111
N Kurnianingsih, R Ratnawati, F Fatchiyah, W Barlianto, M M Ali, A Safitri, E Suyanto

PHYTOCONSTITUENTS INVESTIGATION ON THE ETHANOLIC EXTRACT OF AZADIRACHTA INDICA VAR. INDONESIAN AND PHILIPPINES 116
R Raissa, A Safitri, M A G Beltran, A Aulanni'am

GROWTH OF SUNU GROUPER (PLECTROPOMUS LEOPARDUS) LARVAE THAT GIVEN ROTIVERA (BACHIONUS ROTUNDIFORMUS) ENRICHED WITH TAURINE AND GLUTAMINE 122
R Rostika, Iskandar, T Herawati, I Bangkit, G Banthani, L P Dewanti

ANTIOXIDANT POTENTIAL ETHANOLIC EXTRACT OF GLYCINE MAX (L.) MERR. VAR. DETAM AND DAIDZEIN 129
S Prahastuti, M Hidayat, S T Hasianna, W Widowati, A Amalia, D T Yusepany, R Rizal, H S W Kusuma

THE ANTIBACTERIAL ACTIVITIES OF BACTERIOCIN PEDIOCOCCUS ACIDILACTICI OF BREAST MILK ISOLATE TO AGAINST METHICILLIN-RESISTANT STAPHYLOCOCCUS AUREUS 141
S D Lestari, A L O Sadiq, W A Safitri, S S Dewi, M E Prastiyanto

BIOCELLULOSE ISOLATED FROM THE WASTE OF PINECONE FLOWER (PINUS MERKUSII JUNGH ET DE VRIESE) 147
S E Lusiana, M M Aisiyah, U Z 'Uyunin, Z D Nasihin, A Srihardyastutie, M F Rahman, M Masruri

CELLULOSE NANOCRYSTAL (CNC) PRODUCED FROM THE SULPHURIC ACID HYDROLYSIS OF THE PINE CONE FLOWER WASTE (PINUS MERKUSII JUNGH ET DE VRIESE) 154
S E Lusiana, A Srihardyastutie, M Masruri

COMPARATIVE ANALYSIS OF WHARTON'S JELLY MESENCHYMAL STEM CELL (WJ-MSCS) ISOLATED USING EXPLANT AND ENZYMATIC METHODS 159
W Widowati, R F Gunanegara, R Rizal, W S Widodo, A Amalia, S H B Wibowo, K Handono, M Marlina, I N E Lister, L Chiuman

GENOTYPE AND SEROTYPE IDENTIFICATION OF HEPATITIS B VIRUS IN CHRONIC HEPATITIS B PATIENTS TREATED WITH TELBIVUDINE 170
Y Purnamasari, Tien, Arimaswati, A N K Syarifin, A Ali, N Massi

MEDICAL PHYSICS

MOLECULAR DOCKING STUDY FROM LUNACRIDINE, SCOPOLETIN AND SKIMMIANINE AS ANTIDIABETES THROUGH A-GLUCOSIDASE INHIBITOR 175
Adriani, Noorhamdani, S Winarsih, T Ardyati

PREPARATION OF NANOPARTICLES FROM CURCUMA LONGA L. AND COSMOS CAUDATUS EXTRACTS 181
A S Putri, T D Octavianty, N T Wahyudi, A Safitri

THE EFFECT OF TOPICAL TOMAN FISH (CHANNA MICROPELTES) EXTRACT ON MACROPHAGES AND LYMPHOCYTES IN DIABETES MELLITUS WOUND HEALING 187
A N Carabelly, I W A K Firdaus, P C Nurmardina, D A Putri, M L Apriasari

THE PREVENTIVE EFFECT OF ACTIVATED CHARCOAL ON HDL LEVELS AND AORTA HISTOPATHOLOGICAL PROFILES IN HYPERCHOLESTEROL RAT MODELS 196
A Roosdiana, D Vidiastuti, H Herenda

ANTI-DIABETIC PROPERTIES OF ROOT EXTRACTS OF RUELLIA TUBEROSA L: EFFECTS ON SERUM ENZYME ACTIVITY 201
A Safitri, A Roosdiana, N Arrochmah, S S Nur'Adya

TOTAL AND DIFFERENTIAL CELLS OF LEUKOCYTE MICE (MUS MUSCULUS) ON EVALUATION IN VIVO ANTICANCER EXTRACTS ETHANOL MARINE SPONGES AAPTOS SUBERITOIDES 207
A P D Nurhayati, R Pratiwi, Soekardiman

THE EFFECTS OF ISOLATE OF GOAT CASEIN YOGHURT MILK TO MDA LEVEL AND GASTRIC HISTOPATHOLOGY OF RATS EXPOSED TO 2,3,7,8-TETRACHLORINEDIBENZO-P-DIOXIN (TCDD) 213
C Mahdi, A E P Haskito, M H Robbie

ANTI-INFLAMMATORY ACTIVITY OF QUERCITRIN ON HYPOXIA-INDUCED EA.HY926 220
C N Ginting, I N E Lister, E Girsang, M S Mutia, Y E P Lubis, A Amalia, R Rizal, W Widowati

INTERACTION OF CADMIUM-CYSTEINE BINDING AND OXIDATION OF PROTEIN CAUSES OF BLOOD THROMBOSIS 227
D Marisa, I Aflanie, N Muthmainah, E Suhartono

CHANGES OF BIOLOGIC MARKER IN NEONATAL SEPSIS: IS IT SIGNIFICANCE? 232
D I N Pratiwi, P W Nurikhwan, W R Devi

EFFECT OF DHA SUPPLEMENTATION ON THE MDA AND SOD LEVELS IN PROTEIN MALNOURISHED RATS 238
D D Sanyoto, A Asnawati, T Triawanti

THE ROLE OF PROPHYLACTIC ANTIBIOTICS ON ANC VALUE AND FEVER DURATION IN CANCER PATIENTS WITH POST-CHEMOTHERAPY NEUTROPENIA FEVER 243
E P Yunita, G J R Subagiyo, S O Wardhani

IMMUNOLOGICAL RESPONSES IN FEMALE RATS MODEL WITH ANEMIA TREATED WITH SNAKE FRUIT SEED FLOUR EXTRACT 250
I K Ristanti, D G Tamtomo, D Indarto

EFFECT OF FAT SUPPLEMENTATION CONTAINING HIGH PALMITIC ACID ENRICHED WITH LECITHIN ON QUAIL PERFORMANCES 256
I Syarif, W Hermana, Y Retnani

DIHYDROEPONEMYCIN INHIBITS UBIQUITIN PROTEASOME SYSTEM OF PLASMODIUM FALCIPARUM IN SILICO 265
K Mardhiyyah, M H Arif, L E Fitri, S Winarsih, D Chandradikusuma, T Nurseta

IN SILICO: γ-ORYZANOL AS ANTI-INFLAMMATORY DURING FOLLICULOGENESIS IN RATTUS NOVERGICUS EXPOSED TO PYRETHROID AEROSOL 269
L Lisnawati, M I E Santoso, S Poeranto, A T Endharti

THE EFFECT OF HIGH FAT HIGH FRUCTOSE DIET (MODIFICATION OF AIN-93M) ON NUCLEAR FACTOR KAPPA BETA EXPRESSION IN THE LIVER TISSUE OF MALE SPRAGUE DAWLEY RATS 275
M M S Engel, I Kusumastuty, K W Anita, D Handayani

THE IMPROVEMENT OF INS GENE SEQUENCES ON RAT MODEL OF TYPE 1 DIABETES IN GOAT MILK CASEIN PROTEIN TREATMENT, BUT NOT IN GLIBENCLAMIDE 288
M Mulyati, E Y Mu'jizah, F Fatchiyah

SYSTEMIC LUPUS ERYTHEMATOSUS: PKCA IS AN INHIBITION PATHWAY FOR MTOR BY THE ACTIVE INGREDIENT OF GREEN TEA 292
H Mawarti, J Nugraha, D A Purwanto, J Soeroso

FSH RECEPTOR GENE POLYMORPHISM IN INDONESIAN WOMEN WITH POLYCYSTIC OVARIAN SYNDROME (PCOS) 300
R Panghiyangani, M Kurniati, P Soeharso, A Andrijono, D A Suryandari, B Wiweko

GENOMIC PROFILE OF RIR AND RIRR IN TYPE 2 DIABETES MELLITUS RAT MODEL TOWARD EFFECT OF GOAT MILK CSN1S2 PROTEIN 311
S H Tambunan, H N Meidinna, R N Rohmah, F Fatchiyah

CONSTRUCTION AND OPTIMIZATION OF PRETHROMBIN-2 HUMAN GENES IN E. COLI FOR THE PRODUCTION OF ACTIVE THROMBIN 316
S Silaban, S Gaffar, M Simorangkir, I P Maksum, T Subroto

PROTEIN-BASED VIRTUAL SCREENING: SIRT-1 AS MODULATOR MATURATION OF CIRCULATING ENDOTHELIAL PROGENITOR CELLS 322
T A Wihastuti, W Nurwidyaningtyas, C T Tjahjono, T Heriansyah

IDENTIFICATION AND MOLECULAR DOCKING ANALYSIS ALKALOIDS POLYALTHIA LONGIFOLIA LEAVES FROM INDONESIA AND THE PHILIPPINES AS ANTI-INFLAMMATORY 327
T Z Anita, C Mahdi, M A G Beltran, A Aulanni'am

SWALLOW NEST EXTRACT CAN BE ABLE TO PREVENT NEPHROPATHY DIABETIC OF THE WHITE RATS STREPTOZOTOCIN-INDUCED 334
T Triawanti, F D Alexandra, A Frethernity, S Mahmud

EFFECTS OF PROBIOTIC TREATMENT ON HISTOPATHOLOGY OF KOI CARP (CYPRINUS CARPIO) INFECTED BY MYXOBOLUS SP. 343
U Yanuhar, N S Junirahma, K Susilowati, N R Caesar, M Musa

ANTI-INFLAMMATION ASSAY OF BLACK SOYBEAN EXTRACT AND ITS COMPOUNDS ON LIPOPOLYSACCHARIDE-INDUCED RAW 264.7 CELL 349
W Widowati, S Prahastuti, N L W Ekayanti, U Z Munshy, H S W Kusuma, S H B Wibowo, A Amalia, W S Widodo, R Rizal

THE DECREASE IN BACTERIALCOUNT IN ADULT CHRONIC RHINOSINUSITIS PATIENTS AFTER NASAL IRRIGATION THERAPY WITH NACL 0,9% SOLUTION 359
N Qamariah, N Muthmainah, R B B Tanjoto

MOLECULAR DOCKING OF THE ACTIVE COMPOUND GARCINIA MANGOSTANA ON THE RANKL/RANK/OPG SYSTEM 364
B Setiawan, E Suhartono, S Kaidah, I Z Akbar, Z Noor

UROLOGICAL CANCER: MOLECULAR DOCKING OF THE ACTIVE COMPOUND SCURRULA ATROPURPUREA AGAINST NUCLEAR FACTOR ERYTHROID2-RELATED FACTOR2 (NRF2) 369
H Prasetya, F R P Dewi, B Setiawan

EFFECT OF COAL DUST EXPOSURE TO SERUM INTERLEUKIN 6 (IL6) AND VASCULAR ENDOTHELIAL GROWTH FACTOR (VEGF) LEVEL AS PREECLAMPSIA INDUCTION IN PREGNANT RATTUS NOVERGICUS 375
S Wasilah, F Rachman, D Rosadi, A O Puteri, N Laily, F Yulidasari, L Anggraini, B Setiawan, M S Noor
ESTIMATION OF LEAF ANTIOXIDANT ACTIVITY USING IMAGE PROCESSING 380
E Suhartono, B Setiawan, P B Santosa, R Idroes, M S Indrawan
ABDOMINAL CIRCUMFERENCE, BODY FAT PERCENT, AND VO_2 MAX IN PILGRIMS OF HULU SUNGAI TENGAH REGENCY 384
Huldani, Asnawati, D Auliadina, F R Amilia, N Nuarti, R Jayanti
Author Index

Journal of Physics: Conference Series **1374** (2019) 011001 doi:10.1088/1742-6596/1374/1/011001

PREFACE
1ST INTERNATIONAL SEMINAR SMART MOLECULE OF NATURAL RESOURCES (ISSMART) 2019

The ISSMART 2019 is an annual scientific meeting organized by Research Centre of SMONAGENES (Smart Molecules of Natural Genetic Resource), Brawijaya University, Indonesia. This meeting aimed to promote mutual exchange between researchers and also experts, to discuss innovative ideas in scientific research.

The ISSMART 2019 runs for 2 days, 11-12 July 2019, and consists of 6 mini symposia, they are:

- Molecular genetics
- Bioinformatics
- Biotechnology
- Bionatural
- Biodiversity
- Biomedical

The selected papers of the conference will be published in Journal of Physics: IOP Conference Series. Around 100 people are attending this conference. They consist of 64 presenting participants, 25 non-presenting participants, 4 keynote speakers, and invited guests.

The theme of the ISSMART 2019 is "Smart Molecules of Natural Resources for Biotechnology Perspective in Industrial Revolution 4.0" which covers broad range of research in the topics of natural genetic resources and biotechnology. This particular theme is intended to promote recent advances in the field of natural genetic resources and biotechnology as a transformative trend to address the revolution industry 4.0. This international forum also provides a platform where national and international academia or researchers, policy makers, and other stakeholders to translate technology, exchange ideas, and help stimulate multidisciplinary international collaborations in a convergent-manner for shaping a worldwide sustainable development.

The ISSMART 2019 could not become a reality without the help and assistance of many parties. Thus, in this occasion I would like to sincerely thank the Rector of Universitas Brawijaya, Research Centre SMONAGENES, all members of the Organizing Committee, and all sponsors, who have provided meaningful help and assistance for the implementation of this conference. And last but not least, I would like to thank all of the conference participants who will contribute to making this the most memorable event of ISSMART 2019.

Last but not least, we have tried to do our best to prepare ISSMART 2019. Nevertheless, there is nothing completely perfect in the world, including this conference. Therefore, please accept our deep apologizes for any inconveniences found in this conference.

Journal of Physics: Conference Series **1374** (2019) 011001 doi:10.1088/1742-6596/1374/1/011001

LIST OF REVIEWER

No	Name	Affiliation	Email
1	Prof. James Ketudat Cairns,	Suranaree University of Technology, School of Chemistry, Nakhon Ratchasima, Thailand	jrkcairns@yahoo.com
2	Assoc. Prof. Mariena Ketudat-Cairns	Suranaree University of Technology, School of Biotechnology, Nakhon Ratchasima, Thailand	ketudat@sut.ac.th
3	Dr. Sakesit Chumnarnsilpa	School of Biochemistry, Suranaree University of Technology, Thailand	dr.sakesit@gmail.com
4	Takeshi Ohta, Ph.D	Japan Tobacco, Biological/Pharmacological Research Laboratories, Tokyo, Japan	takeshi.ota@jt.com
5	Tomohiko Sasase, Ph.D	Japan Tobacco, Biological/Pharmacological Research Laboratories, Tokyo, Japan	tomohiko.sasase@jt.com
6	Assoc. Prof. Dr. Roswanira Abdul Wahab	Universiti Teknologi Malaysia, Malaysia	roswanira@kimia.fs.utm.mychandra
7	Chandra Kirana, PhD	The University of Adelaide, Australia	chandra.kirana@adelaide.edu.au
8	Prof. Aulanni'am, DES	Brawijaya University, Faculty of Science, Department of Chemistry, Malang, Indonesia	aulani@ub.ac.id
9	Prof. Fatchiyah, M.Kes., Ph.D	Brawijaya University, Faculty of Science, Department of Biology, Malang, Indonesia	fatchiya@ub.ac.id
10	Dr. Ir. Gatot Ciptadi, DESS	Brawijaya University, Faculty of Animal Husbandry, Malang, Indonesia	ciptadi@ub.ac.id
11	Anna Safitri, Ph.D	Brawijaya University, Department of Chemistry, Malang, IndonesiaMalang, Indonesia	a.safitri@ub.ac.id

Journal of Physics: Conference Series **1374** (2019) 011001 doi:10.1088/1742-6596/1374/1/011001

12	Yoga Dwi Jatmiko, S.Si., M.App.Sc.Ph.D	Brawijaya University, Faculty of Science, Department of Biology, Malang, Indonesia	jatmiko_yd@ub.ac.id
13	Dian Siswanto, S.Si., M.Sc.M.Si.,Ph.D	Brawijaya University, Faculty of Science, Department of Biology, Malang, Indonesia	diansiswanto@ub.ac.id

EDITORIAL TEAM

1. **dr. Nia Kurnianingsih, S.Ked., M. Biomed (Coordinator)**
2. **Rista Nikmatu Rohmah, M.Si**
3. **Hazna Noor Meidinna, S.Si**
4. **Lidwina Faraline Triprisila, M.Si**

Journal of Physics: Conference Series **1374** (2019) 011001 doi:10.1088/1742-6596/1374/1/011001

CONFERENCE LOGO

ISSMART International Seminar on Smart Molecule of Natural Resources

SPONSHORSHIP

Supported by:

SCIENCEWERKE

Contact Info:
PT Sciencewerke
Jl. Palmerah Barat 25. Jakarta Barat 11480.
Telepon : +62 21 53667591. Fax: +62 21 5484232.

Journal of Physics: Conference Series **1374** (2019) 011001 doi:10.1088/1742-6596/1374/1/011001

COMMITTE
The 1st INTERNATIONAL SEMINAR SMART MOLECULE OF NATURAL RESOURCES (ISSMART)

July, 11-12th 2019, Swiss Bell Inn Hotel, Malang

Steering Committee	: Head of LPPM UB
	: Prof. Fatchiyah, M.Kes., Ph.D
Chairman Of Committee	: Eko Suyanto, S.Si., M.Sc
Secretary	: Lidwina Faraline Triprisila, S.Si, M.Si
Secretariat	: dr. Nia Kurnianingsih, M.Biomed
	drg. Veranda Susanti, M.Si.
	Gabriella Chandrakirana Krisnamurti, S.Si
Treasurer	: Rista Nikmatu Rohmah, S.Si, M.Si
Scientific Operational	: Anna Safitri, Msc, Phd
	Dian Handayani, Skm., Mkes Phd
	Hazna Noor Meidinna, S.Si
Accommodation And Transportation	: Bayu T. Rahayudi, Mt.Mm
Consumption	: Ema Pristi Yunita, S.Farm., M.Farm., Kiln., Apt
	: Dewi Ratih Tirto Sari, S.Si., M.Si
Documentation	: Setiawan Noerdajasakti, S.H., M.H.
Sponsorship	: Dr. Ir Ary Mustofa Ahmad, M.P
	Prof. Drs. Marjono, M.Phil., Ph.D

GALLERY

Journal of Physics: Conference Series **1374** (2019) 011001 doi:10.1088/1742-6596/1374/1/011001

MARTAPURA RIVER WATER LEADS TO TESTES ALTERATION IN RATS
ISSMART

Journal of Physics: Conference Series **1374** (2019) 011001 doi:10.1088/1742-6596/1374/1/011001

Molecule of National Resources for

Journal of Physics: Conference Series **1374** (2019) 011002 doi:10.1088/1742-6596/1374/1/011002

Peer review statement

All papers published in this volume of *Journal of Physics: Conference Series* have been peer reviewed through processes administered by the proceedings Editors. Reviews were conducted by expert referees to the professional and scientific standards expected of a proceedings journal published by IOP Publishing.

Separation of Flavonoids in The Extract *Polyalthia longifolia* (Sonn.) Thw. Leaves from Indonesia and The Philippines

D S Nahari[1], S Prasetyawan[1], M A G Beltran[2] and A Aulanni'am[3*]

[1]Chemistry Department, University of Brawijaya, Malang 65151, Indonesia
[2]College of Veterinary Medicine, Tarlac Agriculture University, Caliming, Tarlac, 2306 Philippines
[3]Faculty of Veterinary Medicine, University of Brawijaya, Malang 65151, Indonesia

*Corresponding author's email: aulani.fkhub@gmail.com

Abstract. *Polyalthia longifolia* (Sonn,) Thw. is a plant that has many benefits on health because every part of this plant contains secondary metabolites. Flavonoid class is one of the secondary metabolite that contained in the leaves. Flavonoids has pharmacological potencies or therapeutic effects. The objective of this research was to observe how many compound of flavonoid from the ethanolic extract *Polyalthia longifolia* (Sonn.) Thw. leaves, that obtained from Indonesia and The Philippines by using Thin Layer Chromatograpy and Liquid Chromatography-Mass Spectroscopy. Two leaves powder of *Polyalthia longifolia* (Sonn.) Thw. from Indonesia and The Philippines were extracted by using 70% of ethanol, respectively. The dry extract was obtained by rotary evaporation. The flavonoids of both dry extracts that obtained were analyzed by phytochemical assay. The separating of flavonoid class of both extract were analyzed by TLC and LC-MS. Based on the phytochemical screening, both of the ethanol extract of *Polyalthia longifolia* (Sonn.) Thw. from Indonesia and The Philippines showed that positive flavonoids. The result of TLC separation contained of flavonoids based on the yellowish-green, yellow, green, until blue spots under UV exposure. LC-MS separation showed that the both of extracts revealed the presence of flavonoids including to flavonols and cyanidine, with the similar pattern of LC-MS, respectively. The presence of flavonoids were quercetin with RT 5,39 and 5,05, respectively, and molecular weight m/z = 302,5 – 303,5. Rutin showed the single peak with RT 2,61 and 2,62, respectively, and molecular weight m/z = 610,5 – 611,5. Then, leucocyanidin with RT 5,16 and 5,15, and molecular weight m/z = 306,5 – 307,5. Beside that both of extracts also contained analogues compound. In conclusion, the both of ethanolic extract of *Polyalthia longifolia* (Sonn.) Thw. contain flavonoid class such as rutin, quercetin and analogues compounds.

1. Introduction

Glodokan tiang or *Polyalthia longifolia* (Sonn,) Thw. (Family Annonaceae) is a plant that is widely spreaded in tropical and sub-tropical countries in South Asia [1] and Southeast Asia, including Indonesia and the Philippines. This plant has potential as herbal medicines such as antiplasmodial [2], anti-inflammatory [3], antimicrobial [4], antifever, anti-diabetes and anti-hypertension. In addition, this plant is used as ornamental trees that effectively reduce noise pollution [4].

Previous studies reported that this plant contains alkaloids, sesquiterpenes, diterpenoids, flavonoids, and saponins [5]. Based on [6], stem bark also contains proantocyanidin. The content of this secondary metabolite which provides benefits to this plant. Several types of flavonoid compounds are detected in the leaves of this plant, such as quercetin and rutin, and some analog compounds [7]. Flavonoids are a large group of antioxidant compounds named as polyphenols consisting of anthocyanidins, biflavones, catechins, flavanones, flavones, and flavonoids. Quercetin is included in the flavonol compound group, which is distributed with quercetin and glycoside content of approximately 60-75% of the total flavonoids. Flavonoids consist of aromatic compound with antioxidant activity. This activity will prevent body tissue damage on the degenerative disease patient.

Based on the benefit of flavonoid compound, this study will separate the flavonoid compounds from ethanol extract *Polyalthia longifolia* (Sonn.) Thw. leaves from Indonesia and the Philippines. Then, both of extract obtained will analyzed its flavonoid by using Thin layer Chromatography (TLC) and would be confirmed by Liquid Chromatography Mass Spectroscopy (LC-MS).

2. Materials and Methods

2.1. Plant material

The leaves of the plants, especially Indonesia variant, were collected from area Malang, Indonesia at August until September. The dried leaves from The Philippines were deposited at Institute Biosains, University of Brawijaya, Malang, Indonesia. Based on the taxonomical identification, the leaves that obtained were Polyalthia longifolia (Sonn.) Thwaites.

2.2. Sample preparation and extraction

The leaves were air dried and powdered. The extracts of both of powder leaves were prepared by maceration using 70% ethanol for 3 days with occasional stirring, after that the filtrates were obtained by filtration using Whattman filter No. 1, respectively. The ethanol was evaporated by rotary vacuum evaporator at 50 °C to obtain the dry extract. The dry extract were stored at 4°C until use.

2.3. Identification of flavonoids compound

One milliliter of extract 5 % was diluted in the 50% hot methanol 2 mL. Then, it added some magnesium powder and 4-5 drops of HCl. Identification of the flavonoid in the both of extracts were showed red-orange solution.

2.4. Identification of flavonoids by Thin Layer Chromatography (TLC)

The dry extracts were dissolved in 70% ethanol (10 mg/mL), three mobile phases were prepared in the chamber that consist of (I) Toluene : Ethyl Acetate : Formic Acid (5 : 4 : 0,2); (II) ethyl acetate : methanol : water (5 : 1 : 5); (III) n-Butanol : Acetic acid glacial : water (5 : 1: 4), and those were saturated in 30 min. Thin layer chromatography was performed according [5], with modifications. On 3 x 10 cm plates pre-coated with silica gel G, the samples of extract were spotted on the lower left of the TLC plat. Then, the plats were run one dimensionally in the mobile phases at room temperature. Identification of the flavonoids in the extracts was done under UV light after the application of ammonia.

2.5. Identification of Flavonoids by Liquid Chromatography-Mass Spectroscopy (LC-MS)

PL ethanol extract samples were analyzed using LC-MS to obtain qualitative data. The results of LC-MS obtained by chromatogram and peak molecular weight of compounds contained in both extracts. The operational conditions of LC-MS equipment were adjusted for running. The column used was Hypersil Gold (50 mm x 2.1 mm x 1.9 µm). UHPLC brand ACELLA Thermo Scientific type 1250. Elution was carried out with mobile phase A consisting of 0.1% formic acid in aquabidest and mobile phase B consisting of 0.1% formic acid in acetonitrile, with linear gradient and flow rate of 300 µL/minute. The column was controlled at 30 °C and the autosampler compartment was set for C16. Mass Spectroscopy was used MS/MS Quadrupole Mass Spectrometer TSQ QUANTUM ACCESS MAX from Thermo Finnigan with ESI (Electrospray Ionization). The ionization source controlled by

TSQ Tune software which was operated by Negative ionization mode. The determination of targeted compounds was carried out by the SRM (Selected Reaction Monitoring) method.

3. Results and Discussion

3.1. Sample preparation

The leaves were air dried and powdered to get high surface area, that it can be increasing the powder contact with solvent. Extraction was done through maceration or soaking methods using 70% ethanol. The maceration method was chosen because of the easy and simple tools used. This method can maximize solvent contact and materials and can be used to extract substances that are not heat resistant. The extracts of Indonesian and Filipino PL extract obtained were 26,80 % and 23 %.

3.2. Identification of flavonoids

Identification of flavonoids was done by addition of Hydrochloric acid (HCl) solution and magnesium (Mg) powder to produce a red-orange solution. Based on this analysis, the two ethanol extracts of the leaves of *Polyalthia longifolia* (Sonn.) Thw. contain flavonoids by producing a yellowish-orange solution after adding Mg powder and HCl. The ethanol extract of *Polyalthia longifolia* leaves from two countries contain flavonoids which were have biological activities such as antimicrobial, antiradical and have cytotoxic activity against cancer cells [7].

3.3. Thin Layer Chromatography analysis

Thin-layer chromatography separation was a separation process based on the distribution of active compounds in two phases, the mobile phase (consist of a mixture of solutions that have a similar polarity with the compound to be separated) and the stationary phase was a silica gel coated plate (usually gives luminescence under UV light). The best separation product was showed by spots without tailing and no overlapping spots. Separation of flavonoid compounds using the TLC method was carried out with 3 mobile phases or eluent, and produced spots which can be luminescence under UV (ultraviolet) lamps at wavelengths of 254 nm and 366 nm. Table 1 showed the TLC results of the flavonoid compounds from the ethanol extract of leaves of *Polyalthia longifolia* (Sonn.) Thw from Indonesia and the Philippines.

Table 1. TLC Separation of Flavonoids Compound.

Source Plant	Mobile Phases	Spots	Assumption of Flavonoid spots	Rf Number	Spot Appearance under UV
Indonesia	Ethyl Acetate : Methanol : Water (5 : 1 : 5)	11	7	0.05; 0.07; 0.37; 0.49; 0.58; 0.67; 0.80	Green; Yellow; Yellowish orange; Yellowish green
The Philippines		7	4	0.04; 0.56; 0.73; 0.81	Green; Orange; Yellow; Dark Green under UV_{254}
Indonesia	Toluene : Ethyl Acetate : Formic Acid (5 : 4 : 1)	12	7	0.05; 0.08; 0.56; 0.81; 0.82; 0.86; 0.88	Yellowish green; Green; Yellow; Yellowish orange; Dark Green under UV_{254}
The Philippines		12	7	0.06; 0.26; 0.58; 0.66; 0.75; 0.85; 0.89	Yellowish green; Yellow; Yellowish orange;

					Dark green under UV_{254}
Indonesia	n-Butanol : Acetic Acid Glacial : Water (4 :1 : 5)	6	2	0.33; 0.76	Greenish; Yellowish green
The Philippines		7	4	0.10; 0.51; 0.77; 0.82	Dark green; Violet; Yellow

Based on previous study, flavonoids on the TLC system would show yellowish green [5], dark chocolate under UV_{254} [10], yellow and blue [8]. Besides that, positive flavonoid showed the fluorescence spot yellow, green, and blue [9]. The first eluent (EA: MeOH: W-5: 1: 5) was a polar solvent and it was capable of producing yellowish-green stains under UV_{366}. Retention factor (Rf) from *P. longifolia* leaves ethanol extract from Indonesia and the Philippines were 0.67 and 0.73, respectively, which were assumed to be rutin and analogs. According to Sampath and Vasanthin [5] flavonoid rutin, as standard compound, showed Rf 0.676 with a yellowish-green color under UV lamps. Secondary eluent (Toluene: EA: Acid Format-5: 4: 1) was a polar mobile phase and produces 12 spots from both of *P. longifolia* leaves ethanol extract, Indonesia and the Philippines. The result of this study was flavonoids spots showed on *P. longifolia* leaves from the Philippines, that showed spot at Rf 0.66 with a dark color under UV_{254} light. According to [10], flavonoid of quercetin from *P. longifolia* plants showed at Rf 0.62 with dark brown spot under UV_{254} light. The third eluent (nButOH: AAG: W-4: 1: 5) was a polar solvent that was able to provide good separation of polar compounds in the TLC. According to [8] separation by TLC using this mobile phase would show a clear separation between O-glycosides and C-glycoside flavones that were not hydrolyzed (mid low Rf) and aglycone (high Rf), while the expected color was light brown, bright yellow, and yellow-green. The result of separating using the third eluent showed yellow and green spots that suspected as flavonoids. Figure 1 shows the best separating of the second mobile phase.

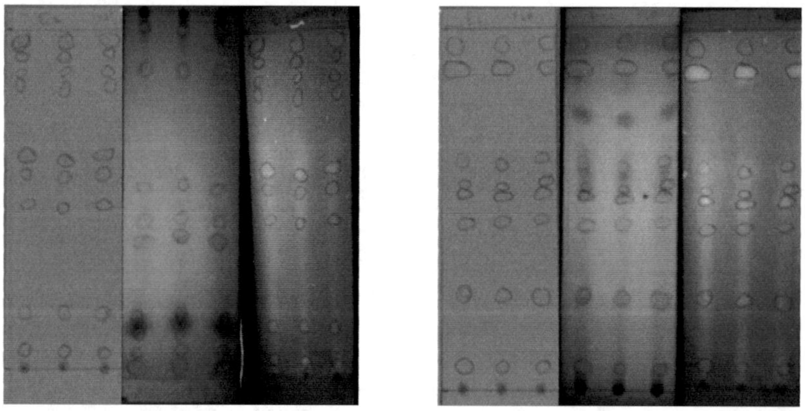

(A) (B)

Figure 1. Separating pattern of flavonoid by TLC using Toluene: Ethyl Acetate : Formic Acid (5:4:1). (A) *P.longifolia* from Indonesia, (B) *P. longifolia* from the Philippines.

3.4. Liquid Chromatography-Mass Spectroscopy analysis

Liquid Chromatography-Mass Spectroscopy (LC-MS) was advanced procedure to analysis a compound that confirmed from TLC result. In the LC-MS system, the separated compound, from LC, would go into MS system, then identified based on molecular mass. The molecular mass of flavonoids compound would be identified by ESI/MS, then showed as fragments with m/z (M^+). There were six target compounds that detected by LC-MS analysis. **Table 2** showed the flavonoid compounds and also analog compounds that was contained in the *Polyalthia longifolia* leaves ethanol extract from Indonesia and The Philippines based on LC-MS separation.

Table 2. Flavonoid compounds from *Polyalthia longifolia* leaves ethanol extract from Indonesia and The Philippines based on LC-MS.

Prediction compound	Source plant	M^+	Retention time (RT)
Quercetin ($C_{15}H_{10}O_7$)	Indonesia	302,5 – 303,5	5,39
	The Philippines		5,05
Rutin ($C_{27}H_{30}O_{15}$)	Indonesia	610,5 – 611,5	2,61
	The Philippines		2,62
Vicenin-2 ($C_{27}H_{30}O_{15}$)	Indonesia	595,5 – 596,5	2,80
	The Philippines		2,80
Quercetin-3-O-glucoside ($C_{12}H_{20}O_{12}$)	Indonesia	464,5 – 465,5	2,69
	The Philippines		4,88
Quercetin-O-O-galloyl hexoside ($C_{30}H_{30}O_{16}$)	Indonesia	614,15 – 615,5	5,57
	The Philippines		5,57
Leucocyanidin-(+)2,3-3,4-cis-3,4,5,7,3',4'-hexahydroxyl flavan ($C_{15}H_{14}O_7$)	Indonesia	306,5 – 307,5	5,16
	The Philippines		5,15

Based on LC-MS identification, the extract of *P.longifolia* from Indonesia and the Philippines founded quercetin. Based on the table, quercetin was found at m/z 302,5-303,5 as $[M+H]^+$ and rutin at m/z 610,5 – 611,5 as $[M+H]^+$. The similar compounds were characterized too from both of extracts. This study showed Vicenin 2, one of 4 analogue compounds of flavonoids at m/z 595,5 – 596,5 $[M+H]^+$. Quercetin derivative vicenin 2 was showed at m/z 594,8 (M) which usually following molecular peaks at m/z 611,7 $[M+O^+]$, the addition of oxygen ion, and m/z 649,2 with addition three water molecules, $[M+3H_2O]$. Quercetin-3-O-glucoside with m/z 464,5 – 465,5 as [M+H] and Quercetin-O-O-galloyl hexoside with m/z 614,5 – 615,5 as molecular ion, were detected. The molecular ion of Quercetin-O-O-galloyl hexoside was identified as M-31. That was mean MS spectra of those compounds was obtained another fragments which referred as additional of formic acid, elimination of CO_2, elimination of hexoside, and additional of acetate ion [10]. Leucocyanidin-(+)2,3-3,4-cis-3,4,5,7,3',4'-hexahydroxyl flavan was another compound that inculinding to the flavonoid class. Based on the LC-MS study, it was observed at m/z 306,5 – 307,5 as molecular ion [M+H]. **Figure 2** showed the structure of flavonoid compounds.

LC-MS analysis usually revealed the identity of compounds based on fragmentation behavior with process of protonation which throughout dehydration, loss or additional of some functional groups, and C-ring fission. Pattern of fragmentation was associated with the applied collision energy. If the collision energy is less than main fragment in the MS, the spectra produced was $(M + H)^+$. However, by enhancing collision energy, a complete fragmentation of the protonated molecule can be obtained [10].

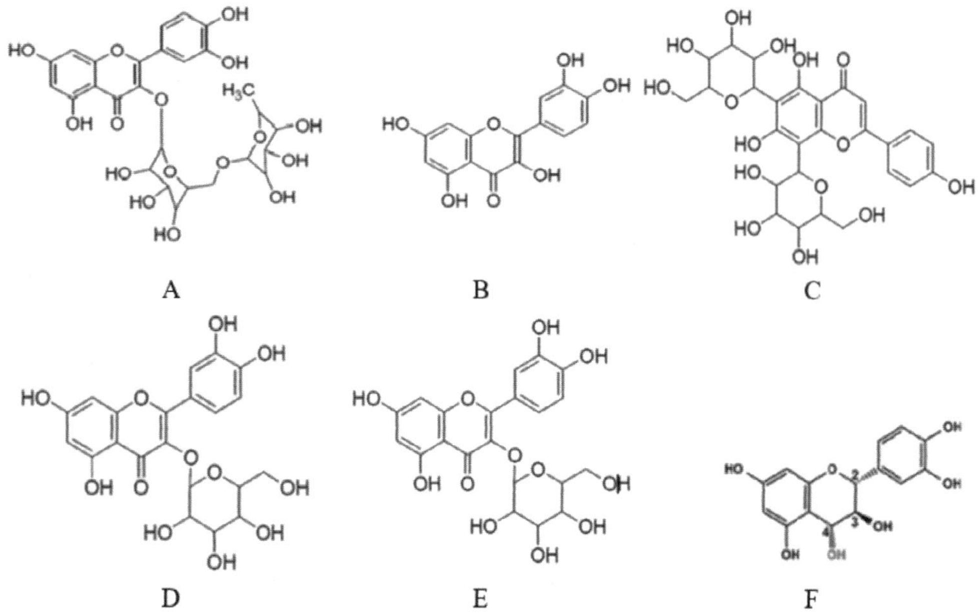

Figure 2. Structure of Flavonoid Compound. (A) Rutin, (B) Quercetin, (C) Vicenin-2, (D) Quercetin-3-O-glucoside, (E) Quercetin-O-O-galloyl hexoside, (F) Leucocyanidin-(+)2,3-3,4-cis-3,4,5,7,3',4'-hexahydroxyl flavan.

4. Conclusion

In conclusion, the ethanol extracts of *Polyalthia longifolia* (Sonn.) Thwaites. from Indonesia and the Philippines revealed the present of flavonoid compounds. Based on LC-MS analysis, the both of extracts contain of quercetin, rutin, and the analogue compounds such as Vicenin-2, Quercetin-3-O-glucoside, Quercetin-O-O-galloyl hexoside, and Leucocyanidin-(+)2,3-3,4-cis-3,4,5,7,3',4'-hexahydroxyl flavan.

References

[1] Katkar K V, Suthar A C and Chauhan V S 2010 The chemistry, pharmacologic, and therapeutic applications of P. longifolia *Pharmacognosy Reviews* **4** (7) 62-68

[2] Annan K, Ekuadzi E, Asare C, Sarpong K, Pistorius D, Oberer L, Gyan B A and Ofori M 2015 *Phytochemistry Letters* **11** 28-31

[3] Vezza T, Rodriguez-Nogales A, Algieri F, Utrilla M P, Cabezaz M E R and Galvez J 2016 Flavonoid in Inflammatory Bowel Disease : A Review *Nutrients* **8** 1-22

[4] Ghosh A, Dash B K, Chatterjee S K and Chandra G 2008 Antibacterials Potentiality and Phytochemichal Analysis of Mature Leaves of Polyalthia longifolia (Magnoliales : Annonaceae) *South Pac. J. Nat. Sci.* **26** 68-72

[5] Sampath M and Vasanthi M 2013 Isolation, Structural Elucidation Of Flavonoids From Polyalthia Longifolia (Sonn.) Thawaites And Evaluation Of Antibacterial, Antioxidant And Anticancer Potential *Int J Pharm Pharm Sci* **5** (1) 336-41

[6] Agrawal S and Misra K 1979 Proantocyanidin From Polyalthia longifolia Stem Bark *Current Science* **48** 141-143

[7] Doshi G M, Nalawade V V, Mukadam A S, Chaskar P K, Zine S P and Somani R R 2016 Elucidation of Flavonoids from Carissa congesta, Polyalthia longifolia, and Benincasa hispida Plant Extracts by Hyphenated Technique of Liquid Chromatography-mass Spectroscopy *Phcog. Res.* **8** 281-286

[8] Harborne J B 1987 *Metode Fitokimia Cara Modern Menganalisis Tumbuhan* Terjemahan Kosasih Padmawinata dan Iwang Soediro (Bandung: Penerbit ITB)

[9] Wagner H and Bladt S 1996 *Plant Drug Analysis:A Thin Layer Chromatography Atlas, Second Edition* pp 359, 362, 364 (New York: Springer)
[10] Gaurav, Doshi M and Une H D 2015 High Performance Thin Layer Chromatography and High Performance Liquid Chromatography determination of Quercetin from Polyalthia longifolia leaves *Free Radicals and Antioxidants* **2**

The Effect of Snake Fruit *(Salacca edulis Reinw.)* Seed Extract On Body Weight In Female Rats Model with Iron Deficiency Anemia

D Melati[1,2], D G Tamtomo[2,3] and D Indarto[2,4,5*]

[1]Human Nutrition, Postgraduate Program of Nutrition Sciences, Universitas Sebelas Maret, Surakarta
[2]Postgraduate Program of Nutrition Sciences, Universitas Sebelas Maret, Surakarta
[3]Department of Anatomy, Faculty of Medicine, Universitas Sebelas Maret, Surakarta
[4]Department of Physiology, Faculty of Medicine, Universitas Sebelas Maret, Surakarta
[5]Biomedical Laboratory, Faculty of Medicine, Universitas Sebelas Maret, Surakarta

*Corresponding author's email: dono@staff.edu.ac.id

Abstract. Anemia is a global nutritional problem in developing countries. Oral iron supplementation is recommended for anemia treatment in pregnant woman, and adolescent girl. However, this supplementation frequently has some side effects. In a previous study, snake fruit seeds naturally contain high level of iron, which is potentially used for alternative anemia supplementation. This study aimed to investigate the effect of snake fruit seed extract (SSE) on body weight (BW) in female rat model with anemia. The study was a pre-posttests control group design, 14 female rats strain Wistar were selected based on inclusion criteria. After 10 days modelling of anemia, all rats were randomly divided to 2 groups: control and treatment groups. For the remaining days, all rats received low iron feed, but added 1.75 g/kg BW/day SSE in the T group 14 days treatment. Body weight was measured before and after treatment, and all data were analyzed using independent and paired T-tests with $p<0.05$. Before treatment, the average of BW in the C group is lower than T group ($p=0.005$). After treatment, the average of BW in T and C groups were statistically differed from BW before treatment ($p<0.001$). In addition, a greater number of BW was found in the T group, compared with the C group ($p<0.001$). The administration of 0.175 g/100 g BW/day SSE can increase BW in female rats model with anemia.

Keywords : Anemia, Iron Deficiency Anemia, Snake Fruit Seed Extract, Body Weight

1. Introduction

Anemia, which is characterized by reduction of hemoglobin levels below normal, becomes a global health problem from 1990 to 2013 [5]. In 2013, 1.93 billion people around the world are estimated to suffer anemia and 59.5% among them have iron deficiency anemia (IDA) [6,1]. In addition, the prevalence of anemia in Indonesian population who aged over 1 year is 21.7% [7]. Pregnant women is one of the most vulnerable age groups and the prevalence of anemia increases from 37.1% in 2013 to 48.9% in 2018 [8].

Iron deficiency anemia is mainly found in children and childbearing women including adolescent girls and pregnant women [9,10]. The cause of IDA in those age groups can be low nutritional intake, decreased iron absorption or increased iron loss throughout the human body [11], which result in some IDA symptoms such as fatigue, weakness, low concentration and low productivity. Therefore, the IDA symptoms will cause several serious health problems like premature birth, low birth weight, miscarriage and pregnancy mortality [12].

Furthermore, recent studies have exibited that decreased appetite, delayed growth and low weight gain were reported in rats model with IDA is known can affect growth. Rat with IDA will shown limitations in growth as expressed by low weight gain [13,14]. Based on Moreno-Fernandez et al (2019) study, rats with IDA can increase some metabolic hormones such as Thyroid-stimulating hormone (TSH), Glucose-dependent Insulinotropic Polypeptide (GIP), glucagon, insulin, cortisol and adrenocorticotropic hormone (ACTH) and decrease ghrelin, which plays an important role in regulation of metabolism and appetite. The changes of those hormones lead to decrease of muscle mass, body fat and energy expenditure.

Iron supplementation is recommended by WHO for anemia treatment in public setting because of simple, cheap and easy [2,16]. In rats with IDA, iron supplementation improves BW for 5 weeks of treatment [13]. However, several unwanted effects such as nausea, vomiting, constipation, diarrhea and abdominal pain are reported in patients who received iron supplementation [2,17]. Therefore, herbal medicine becomes an alternative treatment for IDA. Some natural compounds have intensively been studied around the world. Snake fruit seeds are industrial waste of home industries that process snake fruits to produce chips, pie and snack bar. From data of laboratory analysis, 100 g of snake fruit seed flour (SSF) contains 26.88 mg iron, 4.56 mg zinc, 3.71 % w/w protein and 0.88 mg vitamin C [18]. Rats model with IDA which was given 37.2 g/kg BW/day SSF for 2 weeks shows a significant higher hemoglobin levels (14.0±0.4 g) and BW (213.0±3.3 g), compared to the control group (8.6±0.1g) ($p < 0.001$); (203.5±4.2 g) ($p < 0.001$) [3]. However, the amount of SSF is too much if given to patients with IDA. Extraction of SSF is the best way to increase iron concentration to overcome that problem [19]. We have successfuly generated seed extract (SSE) that contains 211 mg iron/100 g extract, 7.87 times higher than iron content in SSF [20]. Therefore, the aim of this study was to investigate the effect of SSE on BW in rats with IDA.

2. Methods

2.1. Extraction of Snake Fruit Seeds

Snake fruit seeds were collected from 3 home industries at Turi district, Sleman Regency, Yogyakarta. Snake seed extract (SSE) used a modification of extraction process, was done by Nursucihta et al (2014). Briefly, simplicia of snake fruit seeds was extracted using the maceration method with ethanol solvent [21]. Collected extracts of snake fruit seeds were dried using rotary vaccum evaporator for 60 minutes in Food Enginering Laboratory, Faculty of Agriculture, Universitas Sebelas Maret, Surakarta. For the next drying process, SSE was placed into oven for 12 hours in Physiology Laboratory, Faculty of Medicine, Universitas Sebelas Maret, Surakarta.

2.2. Animals Model of Anemia Care

This research study used female rats, which were obtained from Nutrition Laboratory Center for Food and Nutrition Studies, Universitas Gadjah Mada, Yogyakarta. These rats were placed in to a plastic cage with a wire cover and adaptation was done by controlling the temperature (12-hour light and 12-hour dark), providing unlimited access of water and controlling standard food [22]. The standard food was an AIN-93M and provided by the Nutrition Laboratory Center. The protocol of this study has been approved by the Ethical Committe of Medical Research, Faculty of Medicine, Universitas Sebelas Maret.

Rat model of anemia was adopted from Susanti et al (2018), which iron in AIN-93M standard food was removed. The amount of the standard food was given 10-15 g/day or 10% of BW rat/day for 10 days [23].

2.3. Experiment Design
This research study was an experimental laboratory with pre-posttests control group design. We used 14 female rats, which were Wistar species, aged 8-10 weeks old, had ± 200 g body weight, and were healthy condition. Selected female rats were randomly divided into 2 : control (C) and treatment (T) groups. The C group was given low iron feed, while the T group was given low-iron feed and 1.75 g /kg BW/day SSE for 14 days. The SSE dose was based on the calculation of daily iron requirement, performed by Susanti (2018) [3].The SSE was disolved in mineral water and given in the morning via intragastric administration using a 3 ml gastric bulb. The body weight of rats in both groups was periodically monitored every week using a digital Camry weight scale, to ensure that all rats were healthy.

2.4. Data Analysis
Data of body weight before and after treatment were presented as mean ± standard deviation (SD). We used independent and paired T-tests to analyze body weight since those data were normal and homogenous. Because the average of body weight in C and T groups were significantly difference, multiple regression test was used and the p value was set up <0.05.

3. Results and Discussion
After 10 days treatment with low iron feeding, all rats showed IDA signs such as pale in the skin, claws, ears, tails and eyes, rough and sparse fur, and inactive or lethargic. The average of hemoglobin (Hb) levels in the C group was 9.4±0.6 g/dL lower than that of T group (9.9±0.7 g/dL) with p=0.206. These results indicate that low iron feed is able to reduce Hb levels to become iron deficiency anemia. This finding indicates that there is no difference between C and T groups. Table 1 shows the results of multiple linear regression among treatment group and BW before treatment to delta BW.

Table 1. The Result of Linear Multiple Regression Analysis Among Treatment Group and BW Before Treatment to Δ Body weight.

Variable	B Coefficients	*p*	R	R square	Sig
Treatment group	3.067	0.01			
BW before treatment	-0.011	0.892			
(constant)	1.270		0.884	0.781	0.000

The results of multiple regression analysis show that BW before treatment was not significantly effected to weight gain, with p value was 0.892. The weight gain of female rats were effected with administration of 1.75 g /kg BW/day SSE. Based on the value of R square indicated that BW before treatment and treatment group influence amount 78% of the weight gain. Whereas other variables were not examined in this study.

From Figure 1, BW of all rats was evaluated before and after treatment with SSE. The C group had 188.28±3.72 g the average of BW while the average of rats BW in the T group was higher 194.42±2.82 g. A significant difference was observed in the average of rats BW with p = 0.005.

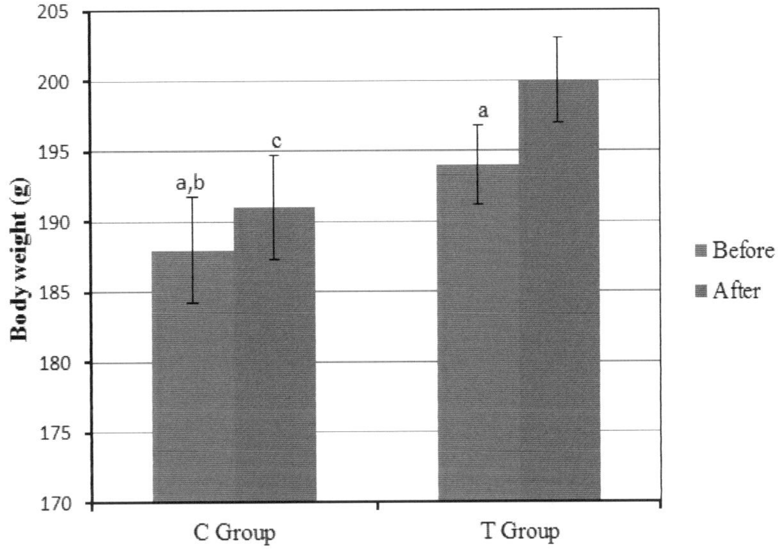

Figure 1. Body Weight Changes in Female Rats with IDA Treated with 1.75 g/kg BW/day SSE for 2 weeks. The C group (negative control) was only given low iron feed, while the T group (treatment) was given low-iron feed and 1.75 g /kg BW/day SSE. Each bar represented mean ± SD in both group. Independent and paired T-tests were used to analyze differences between C and T groups. a indicated significant differences between before and after treatments within groups ($p<0.001$). b and c indicated significant differences of before treatment and after treatment between C and T groups respectively ($p=0.005$ and $p<0.001$).

After treatment, the average rats BW increased by 2.28±0.48 g in the C group and 5.28±1.11 g in the T group. The increased BW in both group reached significantly with $p<0.001$, compared to the average of BW before treatment. Rats in the C group had 190.57 ± 3.69 g the average of BW, significantly lower than the average of BW in the T group (199.71 ± 3.03 g; $p <0.001$).

After the administration of SSE no abnormal clinical signs or macroscopic findings were observed. Female rats that receive SSE survived during the study period, the behaviour was normal, and the condition of their skin and fur, eyes, nose, oral cavity and abdomen were normal.

IDA rats model without iron supplementation, seen in the C group had low metabolism, which results in growth disorders like BW. Iron deficiency in the IDA rats model can cause reduction of enzyme activities and oxygen supply to all cells in the rats tissues. As a result, cellular energy will decrease and distrub all metabolisms in the rats body [16,15]. The results of this study show that 1.75 g /kg BW/day administration of SSE increases rats BW for 14 days. It is not surprised that iron administration derived from fruit and plant extracts can increase BW. Cui et al (2018) use polysaccharide-iron III extracted from *Enteromorpha prolifera* for treatment of anemic rats. Higher average of BW was observed in rats treated with polysaccharide-iron III, compared to the negative control group [14]. The same results of increased BW are also found in anemic rats given cocoa and mango extracts [24].

The decrease of body growth and weight loss on IDA can be considered with several mechanism. Tang et al (2014) study found that rats with IDA would be loss of appetite and shown apathy, this was due to decrease on ghrelin hormone. Ghrelin hormone will decrease if low iron stores in the body such as on IDA patient [25,26,27]. Ghrelin is a hormone secreted in the stomach and duodenum, and stimulates the central and peripheral nervous system pathways. Ghrelin functions is regulate appetite by stimulates hunger. The ghrelin plasma will increase before eating and decrease after eating [28]. Decreasing ghrelin levels from normal limits can cause cachexia, increase corticosterone secretion and induce catabolism [29,30].

The lack of iron intake causes changes in body composition on rats. Rats with IDA, shown descrease in cortisol secretion, which impact in decreased accumulation of body fat and increase energy expenditure. A decrease in the cortisol hormone causes an increase in lipolysis, plasma triglyceride and cholesterol, thus affecting the body fat composition in rats with anemia [31,32].

Other studies found IDA impaired the decrease in thyroid hormone. Thyroid hormones cause central nervous system changes, which play a role in the thermoregulatory response. Low thermoregulation responses cause a decrease in basal metabolic rate (BMR) and BW [33]. Increased TSH levels in IDA patient cause a compensatory mechanism by increasing T3 and T4 which relates in decreased bone formation and increased resorption [34]. Delayed skeletal development causes low weight gain [15].

After administration of Fe therapy in IDA patients, Hb levels returned to the normal level [35]. The normal Hb levels are associated with O2 consumption, CO2 production and energy expenditure are higher in IDA patients [15]. Increased appetite also was found in IDA patients [36] so that there was an increase in energy, protein and carbohydrate intake [26]. Administration of Fe therapy for IDA has been shown to improve physiological function and metabolism of IDA patients so that normal growth and weight gain can be achieved.

4. Conclusion

Administration of 1.75 g /kg BW/day SSE for 2 weeks increases body weight in IDA female rats model, higher than the control group. Further investigation is needed to find out the mechanism of SSE increasing body weight.

References

[1] Kassebaum N J 2016 The global burden of anemia *Hematol Oncol Clin North Am.* **30** 247-308
[2] Jimenez K, Kulnigg-Dabsch S and Gasche C 2015 Management of iron deficiency anemia *Gastroenterol Hepatol (NY).* **11** 241-250
[3] Susanti T 2018 *Pengaruh dosis dan durasi pemberian tepung biji salak terhadap kadar hemoglobin dan indeks eritrosit pada tikus model anemia* Thesis of Postgraduate Program of Nutrition Sciences, Universitas Sebelas Maret
[4] Susanti T, Paramasari D and Indarto D 2017 Development of rat model with iron deficiency anemia by modification of its standard food *2nd PHICo 2017.* **9**
[5] WHO 2017 *Nutritional anaemias: Tools for effective prevention and control* (Geneva: WHO Document Production Services - World Health Organization)
[6] Kassebaum N J, Jasrasaria R, Naghavi M, Wulf S K, Johns N, Lozano R, Regan M, Weatherall D, Chou D P, Eisele T P et al 2014 A systematic analysis of global anemia burden from 1990 to 2010 *Blood* **123** 615–624
[7] Kemenkes RI 2013 *Riset Kesehatan Dasar 2013* (Jakarta : Balai Penelitian dan Pengembangan Kesehatan Kementrian Kesehatan RI)
[8] Kemenkes RI 2018 *Hasil Utama Riskesdas 2018* (Jakarta: Badan Penelitian dan Pengembangan Kesehatan Kementrian Kesehatan RI)
[9] Gozzelino R and Arosio P 2016 Iron homeostasis in health and disease *Int. J. Mo.l Sci.* **17** 130
[10] Lopez A, Cacoub P, Macdougall I C and Peyrin-Biroulet L 2016 Iron deficiency anaemia *The Lancet* **387** 907-916
[11] Clarke L and Dodds A J 2014 Iron deficiency anaemia: Causes, symptoms and treatment. *Medicine Today* **15** 36-42.
[12] Camaschella C 2015 Iron-deficiency anemia *N Engl J Med* **372** 1842-1843
[13] Tang N, Chen L Q, Zhuang H, 2014 Effects of heme iron enriched peptide on iron deficiency anemia in rats, *Food Funct.* **5** 390–399

[14] Cui J, Li Y, Yu P, Zhan Q, Wang, J, Chi Y and Wang P 2018 A novel low molecular weight *Enteromorpha* polysaccharide-iron (III) complex and its effect on rats with iron deficiency anemia (IDA) *International Journal of Biological Macromolecules* **108** 412–418.

[15] Moreno-Fernandez J, Díaz-Castro J, Alférez M J M and López-Aliaga I 2019 Iron deficiency and neuroendocrine regulators of basal metabolism, body composition and energy expenditure in rats *Nutrients* **11** 631

[16] Polin V, Coriata R, Perkinsa G, Dhoogea M, Abitbola V, Leblanca S, Prat F and Chaussade S 2013 Iron deficiency: From diagnosis to treatment. *Dig Liver Dis* **45** 803-809

[17] Tolkien Z, Stecher L, Mander A P, Pereira D I A and Powell J J 2015 Ferrous sulfate supplementation causes significant gastrointestinal side-effects in adults: A systematic review and meta-analysis *PLoS ONE* **10** e0117383

[18] LPPT UGM 2018 *Laporan Hasil Uji-Tepung Biji Salak No Sertifikat 01748.01/X/UN1/LPPT/2018 No Pengujian 18080101748* (Yogyakarta: Laboratorium Penelitian dan Pengujian Terpadu Universitas Gadjah Mada)

[19] Azwanida N N 2015 A review on the extraction methods use in medicinal plants, principle, strength and limitation *Med Aromat Plants* **4** 196

[20] Lab. SEFa UMS 2019 *Hasil Pembacaan Sampel Ekstrak Biji Salak* (Surakarta: Laboratorium Standardisasi Ekstrak Fakultas Farmasi Universitas Muhammadiyah Surakarta)

[21] Nursucihta S, Thai'in H A, Putri D M, Utami N U and Ghani A P 2014 Antianemia activity of *Parkia speciosa* Hassk seed ethanolic extract *Traditional Medicine Journal* **19** 49-54

[22] Smith J B and Mangkoewidjojo S 1988 Pemeliharaan, Pembiakan dan Penggunaan Hewan Percobaan di Daerah Tropis. Tikus Laboratorium (Rattus norvegicus) (Jakarta: Penerbit Universitas Indonesia) pp 35-37

[23] Widiartini W, Siswati E, Setiyawati A, Rohmah I M and Prastyo E 2013 Pengembangan usaha produksi tikus putih *(Rattus norvegicus)* tersertifikasi dalam upaya memenuhi kebutuhan hewan laboratorium *Prosiding Elektronik PIMNAS Dikti*, PKM-K **2013**

[24] Modupe O, Olupo A O and Oladiji T A 2018 Dose-dependent effects of *Theobroma cacao* in iron deficient anemia treatment in rats *Journal of Nutrition & Intermediary Metabolism* **12** 1-7

[25] Isguven P, Arslanoglu I, Erol M, Yildiz M, Adal E and Erguven M 2007 Serum levels of ghrelin, leptin, IGF-I, IGFbp-3, insulin, thyroid hormones and cortisol in prepubertal children with iron deficiency *Endocr J* **54** 985–990

[26] Dogan A, Alioglu B, Dindar N, and Dallar Y 2013 Increased serum hepcidin and ghrelin levels in children treated for iron deficiency anemia *J. Clin. Lab. Anal.* **27** 81–85

[27] Ugur K, Aydin S, Donder E, Sahin I, Yardim M, Kalayci M, Gozel N, Ulu R, Dag M S and Sarikaya M 2018 Saliva and serum ghrelin and obestatin in iron deficiency anemia patients *Preprints* **2018** 2018070422

[28] Mahan L K, Raymond J L, Stopler T and Weiner S 2017 *Krause's 14 edition - Food & The nutrition care process : Medical Nutrition Therapy for Anemia* (Missouri : Elsevier) pp 635

[29] Otto B, Tschop M, Heldwein W, Pfeiffer A F and Diederich S 2004 Endogenous and exogenous glucocorticoids decrease plasma ghrelin in humans *Eur. J. Endocrinol.* **151** 113–117.

[30] Akarsu S, Ustundag B, Gurgoze M K, Sen Y and Aygun A D 2007 Plasma ghrelin levels in various stages of development of iron deficiency anemia *J. Pediatr. Hematol. Oncol.* **29** 384–387.

[31] Campos M S, Barrionuevo M, Alferez M J, Gomez-Ayala A E, Rodriguez-Matas M C, Lopez-Aliaga I and Lisbona F 1998 Interactions among iron, calcium, phosphorus and magnesium in the nutritionally iron-deficient rat *Exp. Physiol.* **83** 771–781.

[32] Ikeuchi M, Koyama T, Takahashi J and Yazawa K 2007 Effects of astaxanthin in obese mice fed a high-fat diet *Biosci. Biotechnol. Biochem.* **71** 893–899

[33] Beard J L, Brigham D E, Kelley S K. and Green M H 1998 Plasma thyroid hormone kinetics are altered in iron-deficient rats *J. Nutr.* **128** 1401–1408.

[34] Diaz-Castro J, Lopez-Frias M R, Campos M S, Lopez-Frias M, Alferez M J, Nestares T, Ojeda M L and Lopez-Aliaga I 2012 Severe nutritional iron-deficiency anaemia has a negative effect on some bone turnover biomarkers in rats *Eur. J. Nutr.* **51** 241–247.
[35] da Silva Lopes K, Takemoto Y, Garcia-Casal M N, Ota E 2018 Nutrition-specific interventions for preventing and controlling anaemia throughout the life cycle: an overview of systematic reviews *Cochrane Database of Systematic Reviews* **8**
[36] Andrews N C 2015 Hungry irony *J Clin Invest.* **125** 3422–3423

Effects Of 25%, 37.5% and 50% *Musa Acuminata* Extract as a Denture Cleanser on the Flexural Strength and Surface Roughness of Acrylic Resin

D Puspitasari[1]*, A Setiawan[1], D F Annisa[1], S R Pramitha[2] and M L Apriasari[2]

[1]Department of Dental Material, University of Lambung Mangkurat, Banjarmasin, Indonesia
[2]Department of Oral Medicine, University of Lambung Mangkurat, Banjarmasin, Indonesia

*Corresponding author's email: dewident@gmail.com

Abstract. Mauli banana (*Musa acuminata*) stem extract contains potential bioactive compounds against Streptococcus mutans and Candida albicans in which can be used as an alternative for synthetic denture cleanser. The use of denture cleanser may weaken flexural strength and surface roughness of heat-cured acrylic resin. This study was subjected to analyse the effect of 25%, 37.5% and 50% Mauli bananas stem extract immersion as a denture cleanser to surface roughness and flexural strength of heat-cured acrylic resin. Seventy specimens of heat-cured acrylic resin were fabricated in rectangular shape with 65 mm x 10mm x 3.3 mm in size according to ISO standard 1567. Treatment groups comprised of Mauli banana stem extract at the concentration of 25%, 37.5%, 50% while 0.2% Chlorhexidine gluconate (CHX) and aquadest were used as control groups. Heat-cured acrylic resin immersion was performed for 5 days. Thirty-five specimens were tested for flexural strength using universal testing machine and other 35 specimens were tested for surface roughness using surface roughness tester SJ-30. Analyzing results with one-way ANOVA test, no significant differences (p=0.986) of surface roughness was obtained from Mauli banana extract of 25% (0.158±0.020 μm), 37.5% (0.160±0.030 μm) and 50% (0.162±0.037 μm), 0.2% CHX (0.155±0.020 μm), aquadest (0.157±0.022 μm). The result flexural strength was significantly differed between 0.2% CHX (79.05±3.10MPa) and aquadest (94.36±6.31MPa) (p=0.003) with no significant differences (p>0.05) between Mauli banana stem extract concentration group 25% (85.30±8.65MPa), 37.5% (84.91±6.60MPa) and 50% (83.60±8.36MPa). No effects of 25%, 37.5% and 50% mauli banana stem extract immersion was presented upon flexural strength and surface roughness of heat-cured acrylic resin therefore it can be used as an alternative for denture cleanser material.
Keywords: *Musa acuminata*, denture cleanser, flexural strength, surface roughness, heat-cured acrylic resin

1. Introduction
Oral and denture hygiene illustrate direct effect upon the health of oral tissues, teeth and the associate denture [1]. Based on Basic Health Research (RISKESDAS) 2007, the percentage of fixed denture and removable denture users in Indonesia is 4.5% [2]. South Borneo is reported with a number of 3.3% population with removable denture ranging from 0.6% - 10.8% [3]. Denture hygiene should be maintained so that plaque on denture base may not accumulate and may not promote the growth of

Candida albicans as the cause of oral disease such as candidiasis or denture stomatitis [1,3]. Chemically-formulated denture cleansers are generally employed by denture wearers to clean the denture [4]. One of the most common denture cleansers used is 0.2% CHX. Notwithstanding fungicidal and broad-spectrum bactericidal effect of CHX, numerous side effects may arise due to chemical ingredients in 0.2% CHX. Discomfort, hypersensitivity reaction and discoloration of the teeth may be experienced by CHX customers. Prolonged usage may also indicate a loss of taste [4]. Denture cleanser solution contact with denture base may cause changes in physical and mechanical properties of acrylic resins such as changes in color stability, an increase in surface roughness and a reduction in flexural strength of acrylic resin [5].

Adverse effects of chemically-formulated drugs are the main reason in continuing traditional medicines development for alternative substitute [6]. Mauli banana plant is one of medicinal plants originated from South Kalimantan which may advance as an alternative for chemical medicine. Apriasari et al (2015) research mentioned that the compounds in Mauli banana stem extract acquire antioxidant, antibacterial, antifungal and non-toxic properties to the cell [7]. Mauli Banana stem extract with dosages between 125-1000 mg / kg BW are proved to have non-toxic effect to the liver of mice [6]. Other studies have also displayed no toxicity of 25% extract concentration to BHK 21 fibroblasts [8]. Mauli banana stem extract at concentration of 37.5 % and 50% moreover exhibit immunomodulatory effect [9]. Research by Septianoor et al (2013) revealed that Mauli banana extract with concentrations of 6.25%, 12.5% and 25% has antifungal capability against Candida albicans. The ability of Mauli banana stem extract to inhibit and kills microorganism especially Candida albicans on (60-100%) denture users assist the reliance of Mauli banana stem extract utilization as a denture cleanser [6].

Despite of Mauli banana stem extract predilection as an alternative for denture cleanser, no research has been conducted to elaborate its constitution. Several types of classes of phenol compounds might be discovered in Mauli banana stem extract such as saponins, tannins and flavonoids. These compounds hold antiseptic property which enable the inhibition of bacterial growth and further eliminate respective microorganism [7]. Based on a research by Wulandari (2012), surface roughness on acrylic resin might be exerted by the contact with phenolic compounds [10]. Extensive elaboration in the use of Mauli banana stem extract as denture cleanser is unable to attain by merely analyzing its effect on surface roughness and flexural strength of heat-cured acrylic resin.

Ideally, no side effect should be emerged toward the properties of denture base material following disinfection methods [11]. Various advantages of heat-cured acrylic resin forged this resin as one of high demand material for artificial denture bases manufacture [12]. It is able to absorb water through a diffusion reaction [13]. Phenols contact on acrylic resin surface will promote diffusion and chemical destruction through hydrolysis reaction. This condition leads into the degradation of polymer matrix creating roughness on the surface of acrylic resin [14]. Surface roughness in denture is a crucial aspect owing to its potential in resisting the growth of plaque which determines the quality and duration of denture use [15]. Flexural strength describes the resistance of denture base over chewing load as one of several aspect encompassing heat-cured acrylic resin mechanical properties. Many factors may affect flexural strength of acrylic resins, one of those is water absorption [10]. Immersion in denture cleansing solution may cause changes in flexural strength of the acrylic resin due to its absorbing ability as denture base material [11]. An in-vitro study of five days immersion of acrylic resin denture shows comparable result with one-year actual condition simulation that denture user experienced [16].

The use of Mauli banana as a denture cleanser is expected not to affect the roughness of heat-cured acrylic resin surface. Based on above backgrounds, it is necessary to identify the effect of heat-cured acrylic resin immersion in Mauli banana stem extract 25%, 37.5% and 50% as denture cleanser solution on the surface roughness and flexural strength of acrylic resin. The result of this study is anticipated to be a reference for the use of Mauli banana stem extract as an alternative natural denture cleanser solution.

2. Materials and Methods

2.1. Specimens preparations

The research had passed ethical clearance issued by the Ethics Committee of Faculty of Dentistry, University Lambung Mangkurat no. 017/KEPKG/FKGULM/EC/VIII/2017. Seventy specimens of heat cured acrylic resin were fabricated according to the manufacturer's instructions using conventional flasking and pressure-pack technique. Heat-cured acrylic resin was packed in the flask in a rectangular shaped with length of 65mm, width of 10mm and thickness of 3.3mm according to International Organization for Standardization (ISO): 1567. Polymerization of the resin was conducted in boiling water for 45 minutes, subsequently the curing flask were cooled to room temperature and the specimens were deflasked. The surfaces were finished and polished using 1000-, 2000- grit sandpaper. The dimension of specimens was checked with a digital caliper, subsequently specimens were stored and incubated in saline solution at 37°C for 24 h before immersion process.

2.2. Immersion process

Seventy specimens of acrylic resin (n=7/group) were prepared and distributed into three treatment groups and two control group, Mauli banana stem extract solution at the concentration 25%; Mauli banana stem extract solution at the concentration 37,5%; 0,2% CHX solution as positive control and aquadest solution as negative control group for 5 days to represent a 20 minute simulation of the use of denture cleanser solution for 1 year.

2.3. Measurement of flexural strength value

The flexural strengths were measured using Universal Testing Machine with 3-point-bending technique, load 50 kgf was applied vertically with cross head speed of 5mm/min, until fracture. Thirty-five specimens were measured and maximum load value was recorded to calculate the flexural strength value. The calculation of flexural strength is determined using the formula of:

$$S=\frac{3PL}{2bd^2}$$

Abbreviation:
S: flexural strength (MPa = kg / mm2)
P: fracture force or load (N) or (kgf)
L: length / width specimens spacing (cm)
b: specimen width (cm)
d: thickness of test plate (cm)

2.4. The measurement of surface roughness value

Surface roughness were measured using profilometer Surface Roughness Tester SJ-301 on another thirty-five specimens (n=7/group). The specimen was placed on a flat plane and the stylus tip was set on the surface of the specimen. The tool was activated and the stylus tip will move along a straight line (horizontally) along the surface and back again under constant pressure. Three measurements were conducted for each specimen, the average of three reading was considered as the final Ra value for the specimen. Data result was processed by using a computer software.

3. Results and Discussion

3.1. Flexural strength test

The mean value of the flexural strength of acrylic resin in Mauli banana extract immersion group was higher than 0.2% CHX (79.05±3.10MPa) and lower than Aquadest group (94.36±6.31 MPa). The flexural strength test of acrylic resin in Mauli banana extract immersion group showed the highest value at 25% concentration (85.30±8.65MPa) and the lowest value at 50% concentration (83.60±8.36MPa). There were no significant differences ($p>0.05$) among Mauli banana stem extract concentration group 25% (85.30±8.65); 37,5% (84.91±6.60 MPa) and 50% (83.60±8.36 MPa), but significantly differed between 0.2% CHX (79.05 ± 3.10 MPa) with Aquadest (94.36±6.31 MPa) ($p=0.003$) (Table 1).

Table 1. Mean value and standart deviation of all groups.

Group	Flexural Strength (MPa)	P
0.2% CHX	79.05 ± 3.10^{A}	0.005
Aquadest	94.36 ± 6.31^{B}	
25% Mauli Banana Extract	85.30 ± 8.65^{AB}	
37.5% Mauli Banana Extract	84.91 ± 6.60^{AB}	
50% Mauli Banana Extract	83.60 ± 8.36^{AB}	

One-way ANOVA test and post Hoc Bonferroni
*Value with different superscript letters shows significant difference at $p < 0.05$

The results of the study showed there was significant difference between the flexural strength value of acrylic resin immersed in aquadest compared with 0.2% CHX group. The research from Carvalho (2013), Oyar (2014) and Bhat (2015) mention that immersion of acrylic resin in CHX solution showed not altered in the flexural strength value of acrylic resin [17,18,19]. This result difference occured in the study depends on the type and length of contact, in our study the flexural strength of acrylic resin was evaluated after long-term immersion in denture cleanser, longer immersion in denture cleanser may cause structural changes in the inner matrix of the polymer [4,20]. Use of denture cleansers such as 0.2% CHX is known may affect the properties of acrylic resins if it used for a long time [21]. Meanwhile, in our study, aquadest does not altered flexural strength which is in line with the study of Wulandari et al, 2012 stated immersion of acrylic resin in Aquadest solution for 4 days, 12 days and 19 days not altered in the flexural strength [10]. Aquadest does not contain active ingredients that may destroy acrylic structure. Therefore, no significant changes on the properties of the acrylic resin shall be observed [14].

Chlorhexidine is a bisbiguanide compound with symmetrical molecules of 2 chlorophenyl rings and 2 biguanides chains linked to hexamethylene chain [22]. This compound is a soluble strong base [12]. Solution of CHX as a denture cleanser has a high ionic concentration which may dissolve the residual monomer of acrylic resin in a denture cleanser solution. This reaction will promote the acrylic resin to compensate and shrinkage [18]. Compounds of CHX that acts as an antimicrobial agent may have a contact with the acrylic resin and able to diffuse into the polymer bond through the gap on the surface of the acrylic resin. The compounds may disrupt the bonding of the polymethyl-methacrylate chain and subsequently degradation occurred in the polymer bonds resulting in reduction of the acrylic resin flexural strength [21].

In our study, Mauli banana stem extract showed not altered flexural strength compared to aquadest. Mauli banana stem extract contains of tannin, saponin and flavonoid compounds that potentially fight against Candida albicans [7]. Saponin is an active compound which occupies 14.49% of total compounds in Mauli banana stem extract [23]. It has glycosyl groups as polar groups and triterpenoids as non-polar groups that are active on its surface. This polar-non polar groups will form a foam such as soap if contact with water and also act not only as a detergent but also as a denture cleanser by releasing CO2 [23,24,25]. The release of CO2 result in oxidation reaction that may disrupt electron conjugation of polymethyl-methacrylate bond chain and subsequently change the chemical structure of acrylic resin [16]. This small amount of saponin content allegedly not provide significant effect on the flexural strength. This is in line with the research of Khalaf et al, 2013 which shows that immersion of acrylic resin in Siwak solution containing saponin content does not show significant effect on the flexural strength [26,27].

Flavonoids are acidic polyphenol compounds which may react with ester compounds of polymethyl-methacrylate in acrylic resin. Flavonoids contains of 0.25% in a total of mauli banana stem extract. Acid contacted with acrylic resin may cause the surface of acrylic resin to become weak and easily soluble, this may create a porous on the surface of denture and facilitates phenol compounds to penetrate on the acrylic resin, subsequently hydrolysis reactions occurred and interfere with acrylic resin polymer chain bonds. this reaction may result in reduction of the acrylic resin flexural strength. Mauli banana stem extract has only a small number of flavonoids, therefore not significantly affect on the flexural strength of acrylic resins [23,27]. Tannin is the highest compound (67.59%) identified in Mauli banana stem extract. It contains of acidic polyphenols, but classified as a weak acid therefore does not have a significant effect on the acrylic resin polymer bonds refer to flexural strength and surface roughness [23,28].

3.2. Surface roughness test

We found that the highest mean value for Surface roughness of heat-cured acrylic resin specimens were immersed in Mauli banana stem extract with concentration of 50 % (0.162 ± 0.027μm) while the lowest mean value was immersed in aquadest solution (0.155 ± 0.0207μm). *One way ANOVA* test result with significance value of 0.986 (p> 0,05) which mean there was no significant difference of surface roughness value of Mauli banana extract of 25% (0.158 ± 0.020 μm), 37.5% (0.160± 0.030 μm) and 50% (0.162 ± 0.0027 μm), 0.2% CHX (0.157 ± 0.022 μm) also Aquadest (0.155±0.020 μm) (Table 2).

Table 2. Mean value and standart deviation of all groups.

Group	Surface Roughness (μm)	P
0.2% CHX	0.155±0.020	
Aquadest	0.157±0.022	0.986
25% Mauli Banana Extract	0.158 ± 0.020	
37.5% Mauli Banana Extract	0.160± 0.030	
50% Mauli Banana Extract	0.162 ± 0.037	

One-way ANOVA test

The results of the study showed the value of surface roughness after immersion in 0.2% CHX (0.15±0.02μm) for 5 days was no significant difference to aquadest immersion. Our result was in line with previous studies that show there is no significant effect on the surface roughness of heat-cured acrylic resin by immersion in 0.2% CHX solution even with longer immersion times and even with a slight increase in concentration [29]. Azevedo, 2006 proved an increase in the surface roughness of acrylic resin, but not significant on heat-cured acrylic resins with a 7-day immersion in 4% CHX solution [30].

Surface roughness value after immersion in the aquadest solution for 5 days has the lowest surface roughness value compared to other immersion groups even statistically no difference. Notwithsatnding aquadest show the lowest, it has a role in the degradation which is initiated by penetration and absorption of water molecules into intermolecular chains of acrylic resin and result in decreased polar interactions [31]. Heat-cured acrylic resin material has a long polyester repeating bond form with a methyl methacrylate unit. The reaction leads to create the distance between polymer that causes the expansion of the matrix and separate the polymer chain bonds on the surface. A study by Sundari et al, 2016 proves that there is a role of water in the hydrolytic degradation reaction of heat-cured acrylic resin material. The penetration depth level of molecules into micro porosity is influenced by the duration of immersion. This molecule may fill the place between the polymer chains therefore the chemical bond structure could be disrupted and microporosity increase in surface roughness [12,32].

Measurements of surface roughness dental material require appropriate parameters as reference. According to Hilgenberg (2008) clinically the ideal dental material has a surface roughness level of ≤ 0.2 μm after polishing. This parameter is commonly used in previous research therefore the comparison of research results can be more suitable [33]. Based on the parameters, our study shows the value of surface roughness under ideal condition.

The ideal denture cleanser requires no negative effect or damage on the denture material. The compounds such as tannins, saponins, flavonoids contained in the mauli banana stem extract solution react with acrylic resin for five days of immersion did not have a significant effect on the surface roughness and flexural strength, therefore mauli banana stem extract with concentration of 25%, 37.5% and 50% has fulfilled the requirements as a denture cleanser solution in terms of surface roughness and flexural strength.

4. Conclusion

There were no effects of 25%, 37.5% and 50% Mauli banana stem extract immersion to the flexural strength and surface roughness of heat-cured acrylic resin therefore it can be used as an alternative to denture cleanser material.

References

[1] Suresan V, Mantri S, Deogade S, Sumathi K, Panday P, Galav A and Mishra K 2016 Denture hygiene knowledge, attitudes, and practices toward patient education in denture care among dental practitioners of Jabalpur city, Madhya Pradesh, India *J. Indian Prosthodont. Soc.* **16** (1) 30-5

[2] Agtini M G 2010 Persentase Pengguna Protesa Di Indonesia *Artikel Media Litbang Kesehatan* **20** 54-55

[3] Yunanto M Y G 2016 Frekuensi Terjadinya Gingivitis pada Pemakai Gigi Tiruan Sebagian Lepasan *Dentino Jurnal Kedokteran Gigi* **1** 210

[4] Jeyapalan K, Kumar J K and Azhagarasan N S 2015 Comparative Evaluation of the Effect of Denture Cleansers On the Surface Topography of Denture Base Materials: An in-vitro study *Journal of Pharmacy & BioAllied Sciences* **7** 1

[5] Indah Y F, Marsono and Yusuf M 2015 Efektifitas Ekstrak Lengkuas Putih (Alpina galangal L stuntz var. alba) Dan Kunyit (Curcuma Domestica L) Terhadap Pertumbuhan Candida albicans Pada Plat Akrilik *Medali Jurnal* **2** 38

[6] Septianoor M H, Carabelly A N and Apriasari M L 2013 Uji Efektivitas Antifungi Ekstrak Metanol Batang Pisang Mauli (Musa sp) terhadap Candida albicans *Jurnal PDGI* **62** 7-10

[7] Apriasari M L 2015 *Potensi Batang Pisang Mauli (Musa acuminata) Sebagai Obat Topikal Pada Penyembuhan Luka Mulut* (Banjarmasin: Grafika Wangi Kalimantan) p 34, 43

[8] Apriasari M L, Adhani R and Savitri D 2014 Uji Toksisitas Ektrak Metanol Batang Pisang Mauli (Musa sp) terhadap Sel Fibroblas BHK (Baby Hamster Kidney) 21 *Dentino Junal Kedokteran Gigi* **2** 211-213

[9] Apriasari M L, Dachlan Y P and Ernawati D S 2016 Effect of Musa acuminata Stem by Immunohistochemistry Test in Ulcer *Asian Journal of Biochemistry* **1** 140

[10] Wulandari F, Rostiny and Soekobagiono 2012 Pengaruh Lama Perendaman Resin Akrilik Heat Cured dalam Eugenol Minyak Kayu Manis Terhadap Kekuatan Transversa *Journal of Prosthodontics* **3** 3

[11] Porwal A, Khandelwal M, Punia V and Sharma V 2017 Effect of Denture Cleansers on Color Stability, Surface Roughness and Hardness of Different Denture Base Resins *The Journal of Indian Prosthodontic Society* **17** 62-65

[12] Sitorus Z and Dahar E 2012 Perbaikan Sifat Fisis dan Mekanis Resin Akrilik Polimerisasi Panas dengan Penambahan Serat Kaca *Dental Journal* **17** 25-26

[13] McCabe J F and Walls A W G 2008 *Applied Dental Materials, Ninth Edition* (Blackwell Publishing Ltd) p 110-120

[14] Puspitasari D, Saputera D and Anisyah R N 2016 Perbandingan Kekerasan Resin Akrilik Tipe Heat-Cured pada Perendaman Larutan Desinfektan Alkalin Peroksida dengan Ekstrak Seledri (Apium Graveolens L.) 75 % *Odonto Dental Journal* **3** (1) 34-41

[15] Durkan R, Ayaz E A, Bagis B, Gubruz A, Ozturk N and Korkmaz F M 2013 Comparative Effects of Denture Cleansers on Physical Properties of Polyamide and Polymethyl Methacrylate Base Polymers *Dental Materials Journal* **32** 367-373

[16] Pisani M X, Lovato C D and Paranhos H 2012 The Effect of Experimental Denture Cleanser Solution Ricinus communis on Acrylic Resin Properties *Material Research* **13** 370

[17] Carvalho C F, Vanderlei A D, Marocho S M, Pereira S M B, Nogueira L and Paes-Junior T J A 2012 Effect of disinfectant solutions on a denture base acrylic resin *Acta Odontologica Latinoamericana* **25** 255-260

[18] Oyar P and Can G 2014 The Effects of Disinfectant Solutions on The Viscoelastic Properties of Acrylic Resins *Cumhuriyet Dental Journal* **17** 210

[19] Bhat V, Suhaim K S and Shenoy K K 2015 Comparative study on effect, denture cleanser and disinfectant have on flexural strength of PMMA *International Journal of Applied Dental Sciences* **1** (3) 24-26

[20] Savabi O, Attar K, Nejatidanesh F, Goroohi H and Badrian H 2013 Effect of different chemical disinfectants on the flexural strength of heatpolymerized acrylic resins *Eur J Prosthodont Restor Dent* **21** 105- 108

[21] Morweng R P N, Essop A R M and Motloba D 2016 Effect of Denture Cleansers on Flexural Strength of Heat-Polimerized and Auto-Polimerized Acrylic Resins *Journal of South African Dental Assosiation* **71** 518

[22] Bettencourt A F, Neves C B, Almeida M S, Pingheiro L M, Oliveira S A, Lopes L P and Castro M F 2010 Biodegradation of Acrylic Based Resins: A Review *Elsevier Journal* **26** 172, 176

[23] Apriasari M L, Dachlan Y P and Ernawati D S 2017 *Potensi Bahan Alam Terhadap Penyembuhan Ulser Mukosa Mulut* (Jakarta: Salemba Medika)

[24] Sangi M, Runtuwene M R J, Simbala H E I and Makang V M A 2008 Analisis Fitokimia Tumbuhan Obat di Kabupaten Minahasa Utara *Progress in the Chemistry of Organic Natural Products journal* **1** 52

[25] Chairunnisa R and Chailess S 2015 Pengaruh Waktu Perendaman Basis Gigi tiruan Resin Akrilik Polimerisasi Panas Dalam Ekstrak Buah Lerak 0,01% Terhadap Kekuatan Impak *dentika Dental Journal* **18** 275

[26] Khalaf H A R 2013 Effect of Siwak on Certain Mechanical Properties of Acrylic Resin *Journal of Oral Research* **1** 46-47

[27] Indiani S R 2008 The Transversal Strength of Acrylic Resin Plate After Being Immersed Soaking in Noni Fruit (Morinda Citrifolia Linn.) *Juice. Dent. Journal* **4** 87

[28] Rajaee N, Vojdani M and Adibi S 2014 Effect of Food Simulating Agents on the Flexural Strength and Surface Hardness of Denture Base Acrylic Resins *Journal of Oral Health and Dental Management* **13** 1044

[29] Vincken J P, Heng L, Groot A and Gruppen H 2007 Saponins, classification and occurrence in the plant kingdom *Phytochemistry* **68** 276

[30] Azevedo A, Machado A L, Vergani C E, Giampaolo E T, Pavarina A C and Magnani R 2006 Effect of disinfectants on the hardness and roughness of reline acrylic resins *J. Prost.* **15** 236

[31] Raj N and Souja M D 2011 Comparison of Transverse Strength of Denture Base Resin on Immersion for Varying Time Period in Water and Denture Cleansers – An In Vitro Study. *Asian Journal of Oral Health* **1** 97-99

[32] Sundari I, Sofya P A and Hanifa M 2016 Studi Kekuatan Fleksular antara Resin Akrilik Heat Cured dan Termoplastik Nilon setelah direndam dalam Minuman Kopi Uleekareng (Coffea robusta). *Journal of Syiah Kuala Dentistry Society* **1** (1) 51-58

[33] Hilgenberg S P, Jimenez E E O, Navarro W F S, Correa B E A, Alves D C T and Campanha N H 2008 Evaluation of surface physical properties of acrylic resins for provisional prosthesis *Mat. Res.* **11** 259

Effect of Active Compound of Pasak Bumi Root (*Euricoma longifolia*, Jack) as an inhibitor of CDK2 methylation*: In Silico* Study

E Y Rahman[1,2]*, M Ali[3], B B Purnomo[4], N Kania[5], D H Utomo[6] and P W Nurikhwan[7]

[1]Doctoral Program of Medical Science, Faculty of Medicine, University of Brawijaya Malang, Indonesia
[2]Department of Surgery, Urology Division, Faculty of Medicine, University of Lambung Mangkurat Banjarmasin, Indonesia
[3]Department of Pharmacology, Faculty of Medicine, University of Brawijaya Malang, Indonesia
[4]Department of Urology, Faculty of Medicine, University of Brawijaya Malang, Indonesia
[5]Department of Anatomical Pathology, Urology Division, Faculty of Medicine, University of Lambung Mangkurat Banjarmasin, Indonesia
[6]Department of Biology, Faculty of Mathematics & Natural Science, University of Brawijaya Malang, Indonesia
[7]Department of Medical Education, Faculty of Medicine, University of Lambung Mangkurat Banjarmasin, Indonesia

*Corresponding author's email: eyurologi05@gmail.com

Abstract. Active compound of Pasak Bumi's root, a native plant of Indonesia, is used as antitumor by triggering cell apoptosis and reactivating the silence tumor suppressor gene caused by hypermethylation, inhibiting cancer cell proliferation. The aim of this study is to know the role of quassinoid from Pasak Bumi's root (*Eurycoma longifolia* Jack) as an inhibitor of CDK2 methylation *in silico*. This is a descriptive study. CDK2 samples are obtained from Protein Data Bank (RSCB.org) with ID3SWR, samples of natural quassinoid are obtained from PUBCHEM NCBI and controlled by Sunitinib (®Sutent). Autodock Vina program PyRx 0.8 is used to analyze Molecular Docking. The process of analyzing molecular interactions is carried out using the LigandScout V.2.0 program. Visualization process are carried out using LigandScout V.2.0 program. The affinity of quassinoid and sutinab to CDK are -6.1 and -9.4, respectively. The more negative the binding affinity value, the better the ability of the compound (ligand) to bind to the receptor (macromolecules). From this case, Sutinab has better value compared to quassinoid. Target protein analysis using HITPICK shows quassinoid's target predictor is JUN protein. Protein interaction analysis are obtained, and the compound is using stitch. JUN protein and Sunitib could bind with CDK2. The conclusion of this study is Sutinib has greater affinity compared to quassinoid.
Keywords : herbal compound, quassinoid, Pasak bumi root, antiproliferation, CDK2, prostate cancer cell line

1. Introduction

In United States 2014, around 233,000 new cases and around 29,480 deaths from prostate cancer [1]. The etiology of prostate cancer is still controversial. Genetic and epigenetic play a role in the process of prostate cancer carcinogenesis. Genes that experience mutations in prostate cancer are the p53, p16 (CDKN2A), PTEN, bcl-2, race, p27 (CDKN1B), and caspase genes [2,3]. The p27 tumor suppressor gene (p27 gene) is a gene that has hypermethylation in prostate cancer. The p27 gene is proven to be activated by hypermethylation promoter in prostate cancer [2], whereas PTEN inhibits AKT phosphorylation, which is required for activation and targets many effectors [4]. PTEN may loss (events that are common in the treatment of poorly differentiated and resistant prostate cancer) and cause activation of the P13K / AKT pathway and subsequent apoptotic resistance [5]. Restoration of PTEN activity in prostate cancer cells PTEN deficiency has been shown to increase sensitivity and encouraging apoptosis [6].

Based on above, prostate cancer epigenetic changes occur due to hypermethylation of tumor suppressor genes p27 and PTEN. The reversible nature of epigenetic changes in cancer results in the possibility of epigenetic therapy as an alternative treatment. The aim of epigenetic therapy is to reverse the causes of epigenetic changes that occur in cancer so that the normal epigenome condition returns. DNA methylation inhibitor compounds are the first epigenetic drugs proposed as drugs for cancer therapy.

Herbal therapy can be an alternative. Several types of bioactive components derived from herbs that have the ability to inhibit the work of the DNA methyltransferase (DNMT) enzyme can affect DNA methylation processes and have barriers to cancer activation through reactivation of tumor suppressor genes that experience silencing [6-14]. Flavonoids can reactivate suppressor tumor genes that have undergone silencing due to hypermethylation [15,16].

Eurycoma longifolia Jack or the Pasak Bumi plant can be an alternative herbal therapy choice. Pasak bumi is a tropical plant belonging to the Simaroubuceae family that is distributed in Southeast Asian countries, native plants in South Kalimantan [17]. *Eurycoma longifolia* has biological active ingredients found in the roots. Pasak Bumi (*Eurycoma longifolia* Jack) is one of the plants that has been used as a drug for detoxification, free radical antioxidants, and anticancer [18,19]. The compounds contained in E. longifolia are quassinoid [18,20,21] and 9-methoxycantin-6-on alkaloids [22,23] and canthinone alkaloids [24]. Quassinoid is a triterfenoid compound that has a flavonoid-like structure so that the Pasak Earth root quassinoid compound has the potential as a demethylating agent.

Cheah and Azihmahtol's [25] study showed that methanol extract of E. longifolia can induce apoptosis in MCF-7 breast cancer cell line through a decrease in Bcl-2 expression [25]. 9-methoxycantine-6-on compounds from methanol extract and chloroform E. longifolia have a cytostatic effect on ovarian cancer cells (CaOv3), epidermoid carcinoma (KB), breast cancer (MCF-7) and rhabdosarcoma (RB) [22]. Conscience research in vitro and in vivo using ethanol extract on breast cancer cells produced inhibitory activity of COX-2 expression, decreased BCl-2 expression, increased Caspase 3, increased expression of p53, increased expression of p21, increased expression of GADD45 and decreased Ras [26]. This shows that the *E. longifolia* Jack has the potential as a new active drug for cancer suppression.

Based on the thinking above the active compound of the Pasak Bumi root (*E. Longifolia* Jack) has cytotoxic effects on various cancers such as colon cancer, breast cancer, lung cancer, skin cancer (melanoma), ovarian cancer, and others, it is necessary to know whether it is also cytotoxic in prostate cancer that is independent androgen (hormonal resistant therapy). According to author's knowledge, study on the potential of the Pasak Bumi's roots is still few especially study for hypermethylation, and inhibiting cancer cell proliferation in Prostate Cancer (PC-3 CDK2). This later became the background of the authors to conduct research on the potential of Pasak Bumi's roots in inhibiting proliferation and increasing apoptosis in PC-3 Prostate Adenocarcinoma cells. The aim of this study is to know the role of quassinoid from Pasak Bumi's root (*Eurycoma longifolia* Jack) as an inhibitor of CDK2 methylation in Prostate Cancer (PC-3) *in silico*.

2. Materials and Methods

2.1. Design

This study using descriptive method to identify the Quassinoid compound found in the roots of Pasak Bumi (*E. longifolia* Jack) and identify its potential in antiproliferation *in silico study,*. For positive control ligand, this study used Oral, small-molecule Sunitinib (marketed as Sutent by Pfizer), is a multi-targeted receptor tyrosine kinase (RTK) inhibitor that is approved for treatment of renal cell carcinoma (RCC) and imatinib-resistant stromal gastrointestinal tumor (GIST) on January 26, 2006. Sunitinib was approved as the first line for cancer drug in two different indications.

2.2. Procedure

2.2.1. The Interaction between Pasak Bumi's Quassinoid and PTEN(CDK2). CDK2 samples are obtained from Protein Data Bank (RSCB.org) with ID3SWR. This model is human CDK2 protein which has been linked to an inhibitor, while a sample of quassinoid natural ingredients is obtained from PUBCHEM NCBI and controlled by Sunitinib (®Sutent). The potential of each compound in inhibiting CDK2 was analyzed by looking at the affinity of the compound in binding to the active side of CDK2. The Molecular Docking analysis process is done using Autodock Vina on the PyRx 0.8 program. The Docking process is carried out on the active side of the CDK2.

3. Result and Discussion

The aim of this study is to know the role of quassinoid from Pasak Bumi's root (*Eurycoma longifolia* Jack) as an inhibitor of CDK2 methylation in silico. This study found that the quassinoid has an affinity as inhibitor of CDK2. In this study, the composition of silico study are ie Quassinoid Analog as ligand, CDK2 (3TI1) as macromolecule, and **SUNITINIB** *(*C_{22} H_{27} F N_4 O_2) as Positive control ligand. The affinity of quassinoid and Sunitinib to CDK are -6.1 and -9.4, respectively (Table 1). Ligand's ability to bind with receptor (macromolecule) is shown by its negativity towards binding affinity, the more negative the binding affinity value, the better the ability of the compound (ligand) to bind to the receptor (macromolecules). In this case, Sunitinib has better value compared to quassinoid means Sunitinib has better ability to bind than quassinoid

Table 1. Result.

	Quassinoid (Binding affinity)	Sunitinib (Binding affinity)
CDK2	-6.1	-9.4

To analyse target protein of quassinoid after binding, HITPICK was used. As seen in figure 1, Protein target analysis using HITPICK shows that the target prediction of the quassinoid is JUN protein. Then the interaction analysis between protein and compounds is done using stitch. JUN proteins and Sunitinib could bind with CDK2. CDK2 Serine / threonine-protein kinase is involved in cell cycle control, meiosis, and triggers duplication of centrosomes and DNA. CDK2 has an important role in regulating a good balance between cellular proliferation, cell death, and DNA repairing in human embryonic stem cells (hESCs). By binding the quassinoid to CDK, the cellular profileration and cell death may be balanced. Figure 2 explained this mechanism, If P21 / P27 is not phosphorylated, the CDK2 activity increases, increasing tumorigenesis and abnormal cell cycle regulation. P27 is a CDK inhibitor. Drugs that work in inhibiting CDK2 have a function to arrest the cell cycle.

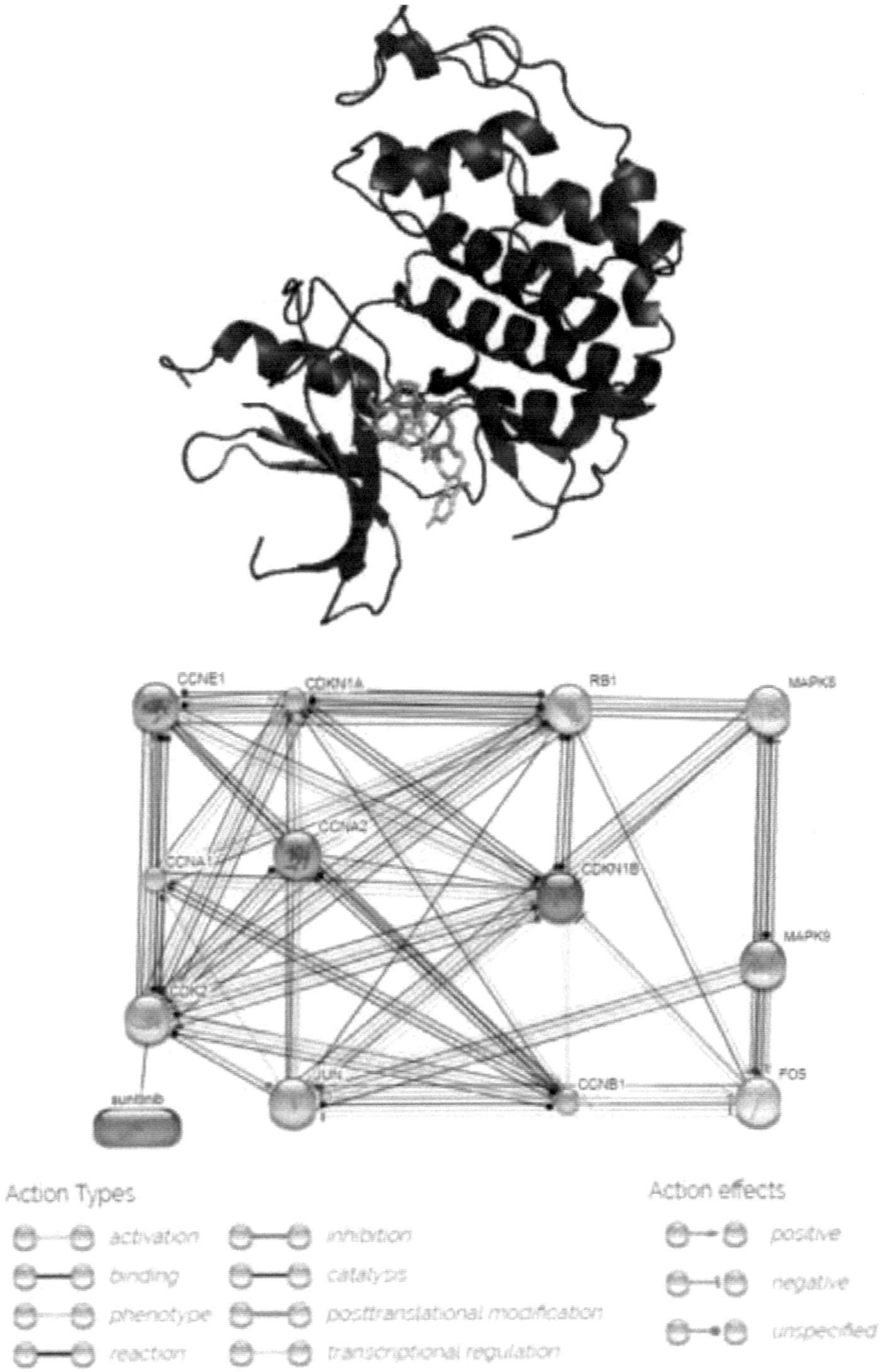

Figure 1. Action type and effect toward CDK2. Green line: Sunitinib. Pink line: Quassinoid.

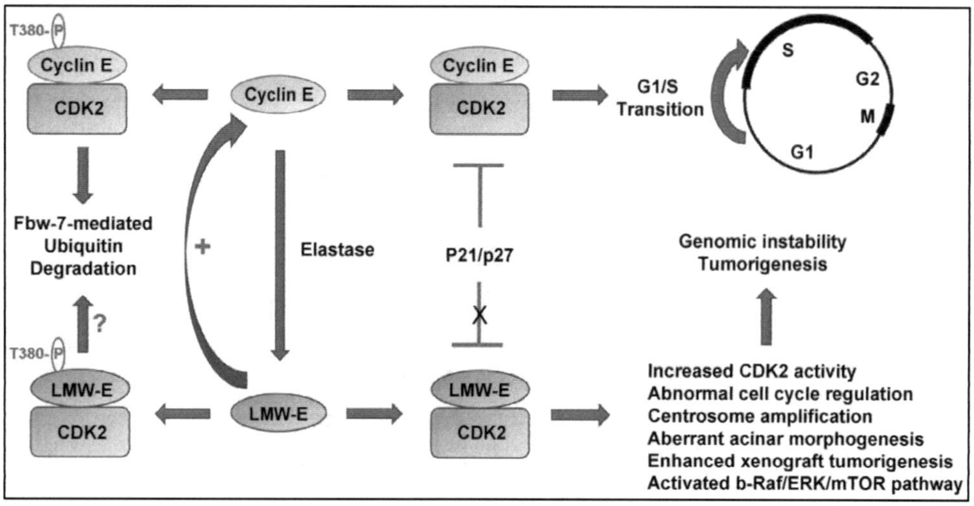

Figure 2. CDK2 Regulation in tumorigeneses.

The limitation of this study is there is no other marker or protein whom assessed. Future study with more protein involvement is needed.

4. Conclusion

The role of quassinoid of Pasak Bumi's root (*Eurycoma longifolia* Jack) is an inhibitor of CDK2 methylation as in silico study with a lesser affinity compared to Sunitinib.

References

[1] Isaacs W and Kainu T 2001 Oncogenes and tumor suppressor genes in prostate cancer *Epidemiol. Rev.* **1** 36-41

[2] Dharmayanti N L P 2003 Kajian Biologi Molekuler : gen suppressor tumor (p53) sebagai target gen dalam pengobatan kanker *J. Wartazoa* **13** (30) 54-60

[3] Wang S, Gao J, Lei Q, Rozengurt N, Pritchard C, Jiao J, Thomas G V, Li G, Roy-Burman P, Nelson P S, Liu X and Wu H 2003 Prostate-specific deletion of the murine Pten tumor suppressor gene leads to metastatic prostate cancer *Cancer Cell* **4** (3) 209–221

[4] Davies M A, Koul D, Dhesi H, Berman R, McDonnell T J, McConkey D, Yung W K and Steck P A 1999 Regulation of Akt/PKB activity, cellular growth, and apoptosis in prostate carcinoma cells by MMAC/PTEN *Cancer Re.s* **59** (11) 2551–2556

[5] Yuan X J and Whang Y E 2002 PTEN sensitizes prostate cancer cells to death receptor-mediated and drug-induced apoptosis through a FADD-dependent pathway *Oncogene* **21** (2) 319–327

[6] Yoo C B and Jones P A 2006 Epigenetic therapy of cancer: past, present and future *Nature review* **5** 37-50

[7] Esteller M 2008 Molecular origins of cancer: Epigenetic in cancer *The New England Journal of Medicine* **358** (11) 1148-1159

[8] Li Y and Tollefsbol T O 2010 Impact on DNA methylation in cancer prevention and therapy by bioactive dietary component *Curr. Med. Chem.* **17** (20) 2141-2151

[9] Meeran S M, Ahmed A and Tollefsbol T O 2010 Epigenetic targets of bioactive dietary component for cancer prevention and therapy *Clin. Epigenetic* **1** 101-116

[10] Sharma S, Kelly T K and Jones P A 2010 Epigenetic in cancer *Carcinogenesis* **31** (1) 27-36

[11] Azad N, Zahnow C A, Rudin C M and Baylin S B 2013 The future of epigenetic therapy in solid tumours-lesson from the past *Nat. Rev. Clin. Oncol.* **10** 1-11

[12] Lee W J, Shim J Y and Zhu B T 2005 Mechanism for the inhibition of DNA methyltransferases by tea catechins and bioflavonoids *Mol. Pharrmacol.* **68** 108-1030

[13] Fang M, Chen D and Yang C S 2007 Dietary pholiphenol may affect DNA methylation *The Journal of Nutrition* **137** 223-228
[14] Rege Y D and Rangnekar V M 2004 Molecular therapy intervention prospects in prostate cancer *Curr. Pharm. Des.* **10** 523-30
[15] Zhang C, Gao C, Xu Y and Zhang Z 2004 CtBP2 could promote prostate cancer cell proliferation through c-Myc signaling *Gene* **546** 73-79
[16] Rahman A S, Sim Yap M M, Md. Shakaff A Y, Ahmad M N, Dahari Z, Ismail Z and Hitam M S 2004 A microcontrollerbased taste sensing system for the verification of *Eurycoma longifolia Sens. Actuators B* **101** 191-198
[17] Ang H H, Ngai T H and Tan T H 2003 Effects of *Eurycoma longifolia* Jack on sexual qualities in middle aged male rats *Phytomed.* **10** 590-593
[18] Schmitt B, Bennett C L, Seidenfeld J, Samson D and Wilt T 2000 Role of Maximal androgen blockade for advanced prostate cancer *Cochrane Database Syst. Rev.* **2** D001526
[19] Bouillet P and Strasser A 2002 BH3-only proteins – evolutionarily conserved proapoptotic Bcl-2 family members essential for initiating programmed cell death *Journal of Cell Science* **115** 1567-1574
[20] Nuryamin A 2000 *Studi Potensi Tumbuhan Obat Akar Kuning (Arcangelisia flava L. Merr), Pasak Bumi (Eurycoma longifolia Jack), Seluang belum (Luvunga eleutheranda Dalz) dan Gin Kalimantan (Psychotria valetonii I locrhr)* Skripsi Jurusan Konservasi Sumberdaya Hutan Fakultas Kehutanan IPB
[21] Nurkhasanah M and Azimahtol H L P 2008 Eurycomanone Induces Apoptosis through the Up-Regulation of p53 in Human Cervical Carcinoma Cells *Journal of cancer molecules* **4** 109-115
[22] Tannock I F, de Wit R, Berry W R, Horti J, Pluzanska A, Chi K N, Oudard S, Théodore C, James N D, Turesson I, Rosenthal M A and Eisenberger M A 2004 Docetaxel plus prednisone or mitoxantrone plus prednisone for advanced prostate cancer *New Engl. J. Med.* **351** (15) 1502-12
[23] Chan K L, Choo C Y, Abdullah N R and Ismail Z 2004 Antiplasmodial studies of *Eurycoma longifolia* Jack using the lactate dehydrogenase assay of Plasmodium falciparum *J. Ethnopharmacol,* **92** 223-227
[24] Chuen C S and Pihie A H L 2004 Eurycomanone exert antiproliferative activity via apoptosis upon MCF-7 cells *Proc. Sym. Biol. Kebangsaan Malaysia ke-7* 73-77
[25] Nurani L H and Mubarika S 2011 *Mekanisme Molekuler Kemopreventif dan Antikanker Senyawa Aktif Akar Pasak Bumi (Eurycoma longifolia Jack) Kajian In vitro pada sel T47D dan In vivo pada Kanker Payudara pada Tikus SD yang diinduksi DMBA* Universitas Gajah Mada
[26] Chan K L, Choo C Y and Abdullah N R 2005 Semisynthetic 12-O-Acyl- and 1,15-Di-O-acyleurycomanones from *Eurycoma longifolia* as potential antimalarial *Planta Med.* **71** 967-69

Journal of Physics: Conference Series **1374** (2019) 012005 doi:10.1088/1742-6596/1374/1/012005

Production and Optimization of Lipase by *Aspergillus niger* using Coconut Pulp Waste in Solid State Fermentation

E Suyanto[1,2*], E S Soetarto[3] and M N Cahyanto[4]

[1]Biology Department, Faculty of Mathematics and Natural Sciences Universitas Brawijaya, Malang, Indonesia
[2]Research Centre for Smart Molecules of Natural Genetic Resources, Universitas Brawijaya, Malang, Indonesia
[3]Faculty of Biology, Universitas Gadjah Mada, Yogyakarta, Indonesia
[4]Faculty of Agricultural Technology, Universitas Gadjah Mada, Yogyakarta, Indonesia

*Corresponding author's email: e.suyanto@ub.ac.id

Abstract. Lipase has high economical value and an attribute that treasures wide range applications. Among the microbes, fungi to be the best source for lipase production in agro-industrial residue using solid state fermentation (SSF). One of agro-industrial residue is coconut pulp waste. The aims of this study is to explore the possibility of coconut pulp waste for producing of lipase using fungi isolate, *Aspergillus niger* and find the partial optimum condition of lipase production based on 2 parameters only. The crude enzyme was extracted by mixing fermented substrate with 50 mL sterile aquades pH 7 containing 0.01% Tween-80. The protein concentrations are determined by Bradford method using spectrophotometric. Lipase activity was followed the titrimetric method. The SSF was carried out with variation of initial spore concentration of $1x10^5$, $1x10^6$, $1x10^7$, and $1x10^8$ (spores.mL^{-1}) and crude enzyme was harvested by adding 25, 50, 75, 100, and 125 (mL) of extract solution. The result showed that initial spore concentration and extract solution affected of lipase production. The SSF in coconut pulp waste using *A. niger* has successfully produced lipase with optimum condition were achieved by adding spore concentration inoculum $1x10^6$ spore.mL^{-1} in acidic condition at 30°C for 7 days and extracted by 75 mL of extract solution that was result 72.5 mL of crude enzyme with lipase activity 10.83 U.mL^{-1} and protein content 0.074 mg/mL^{-1}.
Keywords: lipase, fungi, SSF, coconut pulp waste

1. Introduction

Lipase (*triacylglycerol acyl hydrolases,* E.C 3.1.1.3) has high economical value and also catalyse the hydrolysis of triglycerides to glycerol and free fatty acids [1]. The application of lipase including food additives, detergents, cosmetics, pharmaceuticals, leather processing, biodiesel, organic synthesis, medical area, treatment of domestic and industrial wastes [2],[3],[4]. Due to their distinctive catalytic potential, lipase is one of the most crucial industrial enzyme [5].

Lipase produced by plant, animal and microbe such as bacteria, yeast and fungi. One of the lipolytic fungi is *Aspergillus* producing lipase for industrial application [6]. Lipase originated from microbes are the better sources to provide high stability and offer ease of cultivation. Among the microbes, fungal lipases to be the best source for lipase production because thermally stable with high turnover number and easy recovery of extracellular enzymes [7].

Generally, lipase have been produced using submerged fermentation (SmF) method although it is high cost of equipment, media and higher probability of contamination. The alternative technology for producing lipase using solid state fermentation (SSF) to improve yield of enzyme and its activity which require less energy than SmF [8]. SSF also offers advantages including ease of extraction and resistance to contamination. SSF technology is currently the best method to obtain fungi spore on the non-soluble matrix that acts as a natural habitat for filamentous fungi [9]. Substrate of SSF achieved by using low cost accessible agro-industrial residues and provide nutrient source to microbial cells. These are being prospect in developing countries and problems associated with their disposal also important environmental concern. One of agro-industrial residue is coconut pulp waste prospective for developing as substrate of SSF.

Therefore, in the present study we explored the possibility of coconut pulp waste for producing of lipase using fungi isolate as *Aspergillus niger* and find the optimum condition of lipase production based on parameters such as initial spore concentration and volume of extract solution.

2. Materials and Methods

2.1. Microorganism and growth media

The filamentous fungi strains were isolated from leaf litter located in Arboretum (7°45'58.6"S 110°22'31.3"E), Yogyakarta, Indonesia. The fungi isolates were cultured in Potato Dextrose Agar (PDA) media with composition (g.L^{-1}) 4-potato extract, 20-dextrose, 15-agar then incubated at 30°C for 7 days. We found KLC-33 isolate, *Aspergillus niger* so high for lipolytic enzyme on based on result of screening (qualitative method and quantitative method) that we used it for advanced study.

2.2. Solid substrates and chemicals

Coconut pulp waste was obtained from UD Barepan, Moyudan, Yogyakarta, Indonesia. Preparation of coconut pulp as substrate for SSF carried out by drying coconut pulp under the sun's heat and then drying in incubator at 60°C for 3 h until the moisture content around 5-10%, grounded and sieved to provide particle size 250 μm prior to use. It was stored at room temperature. All the media components and chemicals were of analytical reagent grade.

2.3. SSF and enzyme extraction

The spores of *A. niger* on PDA slant were harvested using 0.01% Tween-80 solution then suspended in 3 mL of sterile distilled water with 0.01% Tween-80 then added 7 mL again for release spores from agar surface [10],[11]. The basal fermentation media of coconut pulp adjusted to pH 7 with moisture content to 60% then it sterilized. SSF was carried out by adding 1 mL of spores biomass into 5 g of coconut pulp then incubated at 30°C for 7 days. Crude enzyme was extracted by mixing fermented substrate with 50 mL sterile aquades pH 7 containing 0.01% Tween-80 on rotary shaker (150 rpm) at 30°C for 30 min. The suspension was filtered by sterile filter paper then filtrat was centrifuged (4000 rpm) at 4°C for 30 min. The supernatant (crude enzyme) was used for lipase assay and then stored at -20°C [12].

2.4. Enzyme assays

2.4.1. Determination of protein. Determination of protein of crude enzyme determined by Bradford method using Bovine Serum Albumin standard based on the reaction of the enzyme with 0.5 mL Bradford reagent at 30°C for 2 min. The absorbance read at λ 595 nm. The soluble protein was determined by converting the absorbance value into standard curve.

2.4.2. Lipase activity. Determination of lipase activity in crude enzyme followed the titrimetric method [13]. Lipase activity determined in substrate solution with ratio of arabic gum: 0.2 M phosphate buffer pH 7: coconut oil was 2:1:1. All components of substrate solution mixed until homogenous. It taken 0.1 mL and put into flask then added 1 mL of 1 M $CaCl_2$. The solution mixed with 0.01 mL of crude enzyme, homogenized by vortex and incubated on rotary shaker (150 rpm) at 30°C for 1 h. The reaction was stopped by adding 5 mL of acetone:ethanol (1:1) mixture after the incubation time was reached then homogenized by shaking. Titration was performed with 0.01 N NaOH until the color of solution become pink. One unit (U) lipase activity was defined as the amount of enzymes which were used to produce one μmol of fatty acids equivalent per minute per mL under specified assay conditions.

$$\text{Lipase activity} = \frac{\Delta V \ x \ N \ x \ 1000}{V \ sample \ x \ T}$$

where $\Delta V = V_2 - V_1$. V_1 the volume of NaOH as control, V_2 the volume of NaOH as experimental, N the normality of NaOH, V_{sample} the volume of enzyme and T incubation time.

2.5. Effect of initial spore concentration. The effect of spore concentrations for inoculum studied by using variation of spore concentration such as $1x10^5$, $1x10^6$, $1x10^7$, and $1x10^8$ (spores.mL^{-1}). It was inoculated into 5 g coconut pulp pH 7 with moisture content of 60% then the fermentation was incubated at 30°C for 7 days. Crude enzyme harvested by adding 50 mL of sterile distilled water pH 7 with 0.01% Tween-80. The optimum condition was used for further treatment.

2.6. Effect of extract solution. The effect of extract solution based on result of effect of spore concentration. The optimum of spore concentration was inoculated into 5 g coconut pulp pH 7 with moisture content of 60% then fermentation was incubated at 30°C for 7 days. Crude enzyme was harvested by adding 25, 50, 75, 100, and 125 (mL) of extract solution.

3. Results and Discussion

3.1. Effect of variations of spore concentration

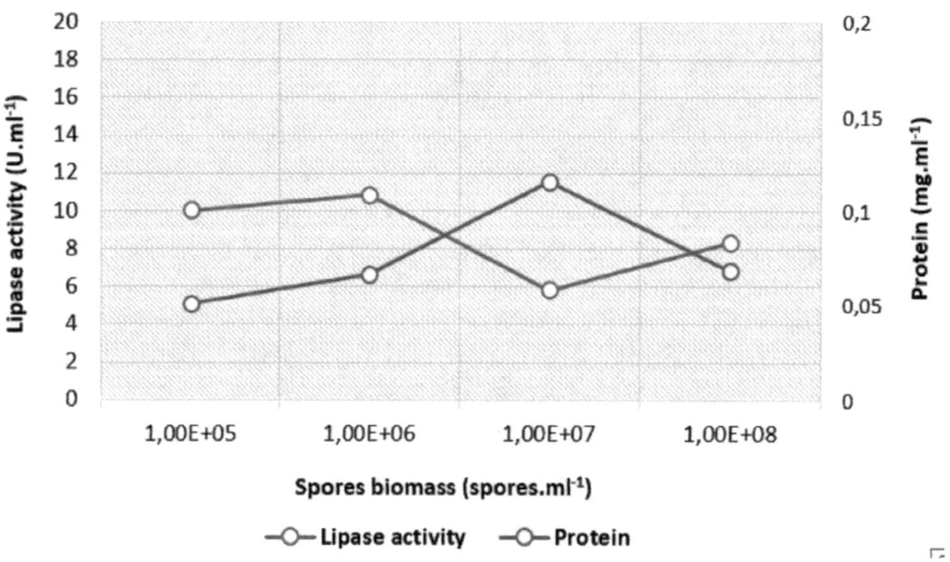

Figure 1. The correlation of lipase activity and protein in crude enzyme using different variation of spore concentration inoculum.

The effect of different concentration of spores is shown in Figure 1. It is shown that optimum of spore concentration for this fermentation was achieved at spore concentration of $1x10^6$ spore.mL^{-1} with lipase activity 10.83 U.mL^{-1} and protein content 0.067 mg.mL^{-1}. In contrast, at spore concentration of $1x10^8$ was resulted lipase activity 8.33 U.mL^{-1} and protein content 0.068 mg.mL^{-1}. Lipase activity was produced by the addition of spore concentration inoculum $1x10^6$ spore.mL^{-1} was higher than lipase activity of *Rhizopus pusilus*, 10.8 U.mL^{-1} on live-bagasse pulp at same of spore concentration [14]. Lipase production and lipase activity was influenced by microorganism, growth environment, substrate

[15], kind of carbon and nitrogen sources [4]. The result also showed that lipase activity was influenced by adding spore concentration inoculum at the beginning of SSF.

The effect of spore concentration to lipase activity and protein content in crude enzyme was not linear. Increasing of spore concentration inoculum tends to reduce lipase activity and protein content. In this case, there was an optimum condition of spore concentration inoculum to produce a high amount lipase activity. The additional spore concentration inoculum not only increase spore biomass but also decrease enzyme production along with limited nutrient in the substrate [16]. The result also showed that spore biomass production by *A. niger* was not linearly correlation with lipase activity (table 1) although the pH of substrate was gradually decreased during SSF using *A. niger* for all treatments of spore concentration inoculum (Figure 2)

Table 1. Spore biomass of *A. niger* after treated by adding inoculum of spore concentration into substrate at 30°C for 7 days.

Isolate	Conc. (spores.mL^{-1})	Spores biomass (spores.mL^{-1})		Crude enzyme (mL)
		Ho	H+7	
A. niger	1x10^5	1x10^5	0,44x10^8	43.5
	1x10^6	1x10^6	0,51x10^8	45
	1x10^7	1x10^7	0,85x10^8	43
	1x10^8	1x10^8	1,19x10^8	46

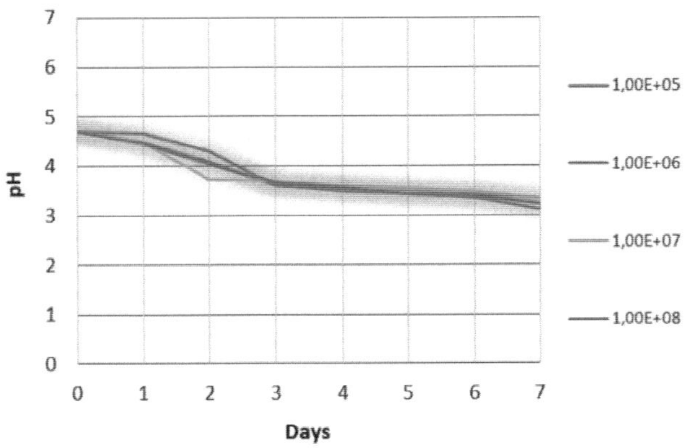

Figure 2. Change of pH in substrate of SSF was incubated at 30°C for 7 days with additional spore concentration inoculum of *A. niger* (spore.mL^{-1}).

3.2. Effect of extract solution

The effect of extract solution shown in Figure. 3. It shown that optimum of extract solution for this fermentation achieved at 75 mL with lipase activity 10.83 U.mL^{-1} and protein content 0.074 mg.mL^{-1}. The result also showed that increasing of extract solution more than optimum limit of crude enzyme extraction of *A. niger* decreased protein concentration and lipase activity due to dilution factor.

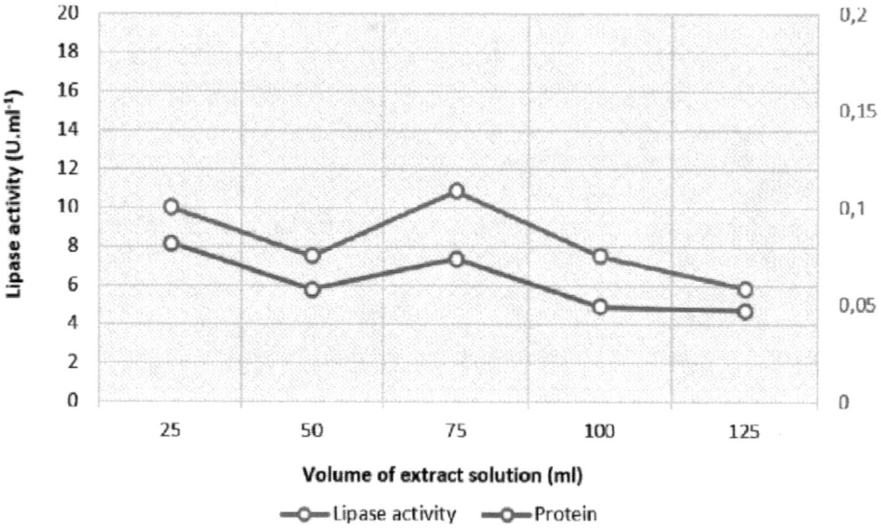

Figure 3. The correlation of lipase activity and protein in crude enzyme using different of volume extract solution.

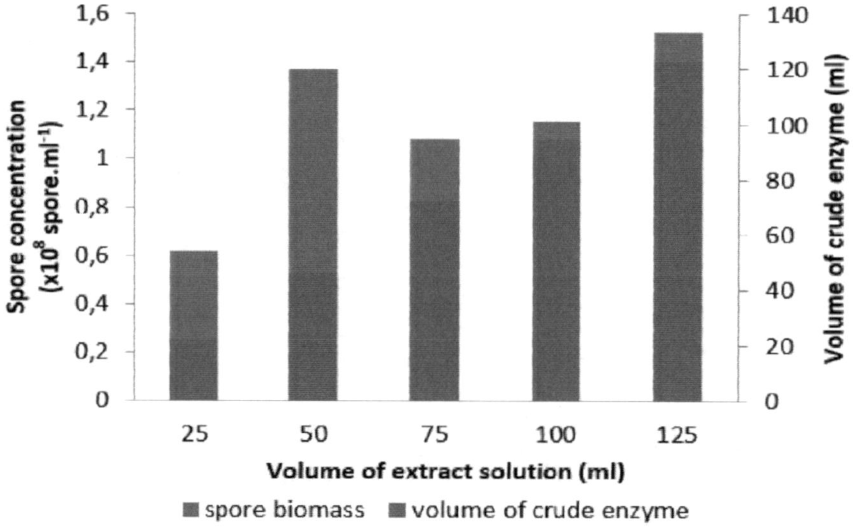

Figure 4. Spore biomass and volume of crude enzyme of *A. niger* after treated by adding gradually extraction solution into substrate fermentation.

The correlation of volume of extracting solution with spore biomass showed that the highest spore biomass was produced by treatment of 125 mL, 1.51×10^8 spore.mL^{-1} and crude enzyme volume 123 mL. The additional extract solution more highly able increase of spore biomass and crude enzyme production (Figure. 4) but decreased protein concentration and lipase activity due to dilution.

4. Conclusion

This paper demonstrated *A. niger* has successfully produced lipase with optimum condition were achieved by adding spore concentration inoculum 1×10^6 spore.mL^{-1} in acidic condition at 30°C for 7 days and extracted by 75 mL of extract solution that it resulted 72.5 mL of crude enzyme with lipase activity 10.83 $U.mL^{-1}$ and protein content 0.074 $mg.mL^{-1}$.

References

[1] Kapoor M and Gupta M 2012 Lipase promiscuity and its biochemical applications *Process Biochem.* **47** 555–69
[2] Selvakumar P and Sivashanmugam P 2017 Optimization of lipase production from organic solid waste by anaerobic digestion and its application in biodiesel production *Fuel Process Technol.* **165** 1–8
[3] Zhang K, Lai J, Huang Z and Yang Z 2011 Bioresource Technology Penicillium expansum lipase-catalyzed production of biodiesel in ionic liquids *Bioresour. Technol.* **102** (3) 2767–72
[4] Treichel H, Oliveira D De, Mazutti M A, Luccio M Di and Oliveira J V 2010 A Review on Microbial Lipases Production *Food Bioprocess Technol.* **3** 182–96
[5] Andualema B G A 2012 Microbial lipases and their industrial applications: review *Biotechnology* **11** 100–118
[6] Nema A, Patnala S H, Mandari V, Kota S and Devarai S K 2019 Production and optimization of lipase using Aspergillus niger MTCC 872 by solid-state fermentation **43** (82)
[7] Salihu A, Bala M and Bala S M 2013 Application of Plackett-Burman experimental design for lipase production by Aspergillus niger using shea butter cake *ISRN Biotechnol.* **2013** 1–5
[8] Ashok A, Doriya K, Rao D R M and Kumar D S 2017 Design of solid state bioreactor for industrial applications: an overview to conventional bioreactors *Biocatal Agric Biotechnol.* **9** 11–18
[9] Colla L M, Ficanha A M, Rizzardi J, Bertolin T E, Reinehr C O and Costa J A V 2015 Production and characterization of lipases by two new isolates of Aspergillus through solid-state and submerged fermentation *Biomed. Res. Int.* **2015** 1–9
[10] Hernández-rodríguez B, Córdova J, Bárzana E and Favela-torres E 2009 Effects of organic solvents on activity and stability of lipases produced by thermotolerant fungi in solid-state fermentation *Journal of Molecular Catalysis B Enzymatic* **61** 136–42
[11] Durbar S and Rai B 2012 *Screening, optimization and pilot scale production of enzyme Lipase in a bioreactor from Aspergillus species* A project report submitted to: Ministry of Science and Technology
[12] Gutarra M L E, Godoy M G, Maugeri F, Isabel M, Freire D M G and Castilho L R 2009 Bioresource Technology Production of an acidic and thermostable lipase of the mesophilic fungus Penicillium simplicissimum by solid-state fermentation *Bioresour. Technol.* **100** (21) 5249–54
[13] Kundu A K and Pal N 1970 Isolation of lipolytic fungi from soil *J Pharm Ind.* **32** 96–7
[14] Singh A K and Mukhopadhyay M 2012 Overview of Fungal Lipase : A Review *Appl Biochem Biotechnol* **166** 486–520
[15] Salihu A, Alam Z, Abdulkarim M I and Salleh H M 2012 Lipase production : An insight in the utilization of renewable agricultural residues *Resources, Conserv Recycl.* **58** 36–44
[16] Iftikhar T, Niaz M, Zia M A and ul-Haq I 2010 Production of extracellular lipases by Rhizopus oligosporus in a stirred fermentor *Braz. J. Microbiol.* **41** 1124–1132

Journal of Physics: Conference Series **1374** (2019) 012006 doi:10.1088/1742-6596/1374/1/012006

Antioxidant and Anti-Inflammatory activity of Chlorogenic Acid on Lead-Induced Fibroblast Cells

E Girsang[1]*, I N E Lister[1], C N Ginting[1], S L Nasution[1], S Suhartina [1], U Z Munshy[2], R Rizal[2] and W Widowati[3]

[1]Universitas Prima Indonesia, Jl. Belanga No. 1, Medan 20118, North Sumatera, Indonesia
[2]Biomolecular and Biomedical Research Center, Aretha Medika Utama, Jalan Babakan Jeruk II No. 9, Bandung 40163, West Java, Indonesia
[3]Faculty of Medicine, Maranatha Christian University, Jl Surya Sumantri No. 65, Bandung 40164, West Java, Indonesia

*Corresponding author's email: ermigirsang@unprimdn.ac.id

Abstract. Lead (Pb) can be a cause of oxidative stress, because it is a toxin found in the environment and can have a harmful effect on the body. increased reactive oxygen species (ROS) and nitrogen species (RONS) levels can cause cellular aging and inflammation. natural compound found in plant is one of the sources of antioxidant and anti-inflammatory agent that has ability to prevent aging including chlorogenic acid (CA). Cytotoxicity assay of CA against fibroblast cells (BJ) was conducted using MTS (3-4,5-dimethylthiazol-2-yl)-5-(3-carboxymethoxyphenyl)-2-(4-sulfophenyl)-2H-tetrazolium). The intracellular ROS levels were detected by flow cytometry using a DCF-DA fluorescent probe. BJ cells were incubated at 37^0C, 5% CO_2, treated by 25 and 6.25 μg/ml CA for 4 h and followed by 400 μg Pb for 3 days. The anti-inflammatory potential was determined using ELISA to measure IL10 and TNFα. CA at 3.13-25 μg/ml were nontoxic to the BJ cells. CA treatment at 6.25 and 25 μg/ml was capable to reduce the accumulation of ROS in lead-induced BJ cells. CA at 6.25 and 25 μg/ml increased IL10 and reduced TNFα compared to positive control (lead-induced cells). This study shows that chlorogenic acid has the potential as protective effect through suppression of ROS levels and have anti-inflammatory properties related to Pb poisoning.
Key words : lead, reactive oxygen species, chlorogenic acid, IL10, TNFα

1. Introduction

Lead (Pb) can be a cause of oxidative stress, because it is a toxin found in the environment and can have a harmful effect on the body. Oxidative stress is a condition where there is an imbalance between the level of ROS and antioxidant defenses, and causes oxidative damage [1]. Oxidative stress caused by lead can be induced through reactive oxygen species (ROS) and nitrogen species (RONS) generation, which has become important mechanism underlying lead toxicity [2], moreover a decrease in the activity of the antioxidant system can also lead to lead-induced oxidative stress. ROS is a term used to indicate the presence of free radicals originating from O_2 or for non-radical species [3].

In cells, the generation of ROS exists in equilibrium with a variety of antioxidant defenses. Low ROS levels are often found in organism with long life span [4]. Inversely, excessive ROS generation can primed senescent cells proliferate in aging, as a stress response [5]. ROS plays an important role in the induction of apoptosis, oxidative stress can cause cellular apoptosis through extrinsic cell death receptor pathways and also intrinsic cell death pathways [6]. The accumulation of ROS which induces apoptosis is then an important contributor to several diseases and aging [7]. In aging process, the accumulation of ROS can leads to loss of skin elasticity and cause formation of wrinkle, brown spots, uneven pigmentation, and even skin cancer [8]. Moreover, production of reactive oxygen species (ROS) have an important role to the advancement of many inflammatory disease [9].

Natural compound especially phytochemicals are eminent for its biological activities including antioxidant, antiaging, and anti-inflammatory [8,10,11]. Chlorogenic acid (CA), one of the most abundant polyphenol compounds in the human diet, CA is a group of phenolic secondary metabolites produced by certain plant species [12]. Most natural products containing CA displayed anti-inflammatory effects, indicating that CA may be a potential antioxidant, antiaging, and anti-inflammatory agent [13]. Here, we studied the antioxidant and anti-inflammatory effects of CA on lead-induced fibroblast cells (BJ) as aging role model.

2. Materials and Methods

2.1. Fibroblast Cells Cytotoxicity assay

Human fibroblast cell line (BJ) (ATCC® CRL-2522™) was obtained from Aretha Medika Utama, Biomolecular and Biomedical Research Center, Bandung, Indonesia. We observed to determine the maximum tolerance concentration of CA on BJ cells and to determine the optimal oxidative damage concentration of lead(Pb) for the following experiments. BJ Cells were cultured in MEM (Biowest, L0416-500) supplemented with 10% fetal bovine serum (FBS) (Biowest, S1810-500), 1% Antibiotic/antimycotic (ABAM, Biowest, L0010100), 1% Nanomycopulitine (Biowest, L-X16-100), 1% Amphotericin B (Gibco, 1%), 0.1% Gentamicin (Gibco, 15750045). Cells were incubated at 37^{0}C in a humidified atmosphere with 5% CO_2. After that, 80% of cells confluency, 5000 cells were seeded in each well of 96-well plate. After 24 h incubation, the cells were treated with CA at various concentrations (3.13, 6.25, 12.5, 25, 50, and 100 µg/mL) for 24 h. To determine cell viability, 3-(4,5-dimethylthiazol-2-yl)-5-(3-carboxymethoxyphenyl)-2-(4-sulfophenyl)-2H-tetrazolium (MTS) assay (Promega, Madison, WI, USA) was used. MTS was added to each well at a ratio of 1:10 [8,14] The plate was incubated in 5% CO_2 at 37^{0}C for 4 h. Absorbance was measured at 490 nm on a microplate reader. The data are presented as the percentage of viable cells (%) and data were analyzed using ANOVA and continued by Tukey post hoc test.

2.2. Measurement of intracellular reactive oxygen species

The ROS levels were detected by flow cytometry using a DCF-DA fluorescent probe (invitrogen) according to the method of Jie *et al* [10,15,16] with minor modification. After 7 days of culture, BJ cells were digested with trypsin-EDTA and 10^5 cells were incubated with 10 µM DCF-DA at 37^{0}C for 30 min, after that incubated with CA (25 and 6.25 µg/ml) for 4 h and followed by 400 µg Pb for 3 days. The intracellular ROS levels were measured using Miltenyi Flow Cytometer (MAQS quant). BJ Cells treated with Pb without CA treatment showed as controls. The measured fluorescence values were expressed as a percentage of control.

2.3. Quantification of IL10 and TNFα level

Quantification of Interleukin 10 (IL10) and Tumor necrosis factor-α (TNFα) were assessed using ELISA Kit, IL10 (Elabsci, E-EL-H0103) and TNFα (Elabsci, E-EL-H0109). The procedure was in accordance with manufacturer protocol. Sample absorbances were read at 450 nm using spectrophotometer (Multiskan GO, ThermoScientific). Color changes of samples are observed then read immediately at 450 nm wavelength and the IL10, TNFα level can be calculated based on a protein standard curve [17–19].

3. Results and Discussion

Skin aging is a natural process due to environmental factors such as lead exposure. The repetitive exposure to lead toxicity cause accelerated physical changes in the skin and connective tissue through the cell contents and ROS [8], increased generation of ROS can reduce antioxidant system activity or changes in both can lead to lead-induced oxidative stress. different antioxidant enzymes and molecules have been evaluated to analyze lead-induced oxidative stress in both clinical and experimental studies [20].

3.1. Fibroblast Cells Cytotoxicity Assay

Fibroblasts have been used as a standard cell line for many cell biological studies. A member of the connective-tissue cell family, fibroblasts are responsible for the synthesis and metabolism of most connective-tissue components and also play an active role in the body's general inflammatory/immune responses. During inflammation, fibroblasts are key cells in granulation tissue and scar formation. In this study we describe the cytotoxicity of CA toward fibroblast cells, based on the results CA suppressed cell growth only at the highest concentration (100 μg/ml), the results obtained are incordance with the result of previous study [21].

The solution used in this assay contains a tetrazolium compound 3-(4,5-dimethylthiazol-2-yl)-5-(3-carboxymethoxyphenyl)-2-(4-sulfophenyl)-2H-tetrazolium (MTS). The MTS compound is bio-reduced by cells into a coloured formazan product due to conversion by dehydrogenase enzymes in metabolically active cells [14].

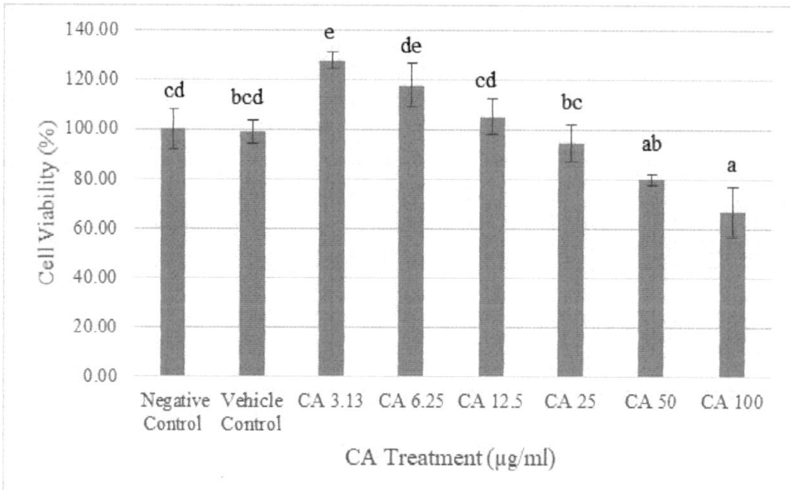

Figure 1. Effect chlorogenic acid (CA) on cytotoxicity of fibroblast cells.

* The histograms are presented as mean ± standard deviation. The data were analyzed with ANOVA and continued with Tukey post hoc test. Different letters (a,ab,bc,bcd,cd,de,e) on figure A and (a,ab,b,c) on figure B indicate significant differences among treatment. Negative Control: cells without any treatments; Vehicle Control: cells with DMSO 10% treatment.

* Negative Control: cells without any treatments; Vehicle Control: cells with DMSO 10% treatment.

Number of viable fibroblast cells (BJ) were decreased in CA concentration dependent manner, (Figure 1). The lower number of cell viability was obtained from the highest concentration of CA. The data of viable cell number were served in percentage of viable cell. Based on the results of cytotoxicity assay and Tukey post hoc test results, the safe concentration that can be used for further experiments and not damaging cells are found in CA concentrations 6.25 and 25 μg/ml.

3.2. Intracellular ROS level

The accumulation of ROS which induces apoptosis is then an important contributor to several diseases and aging [7]. In aging process, the accumulation of ROS can leads to loss of skin elasticity and cause formation of wrinkle, brown spots, uneven pigmentation, and even skin cancer [8]. Our results in accordance with previous study which state that lead exposure can increase the ROS levels [22]. Addition of CA can reduce ROS levels, previous study state that CA can reduce ROS levels approximately 20.3% after exposure with hydrogen peroxide [23].

ROS production can be seen based on level of fluorescence intensity as an indicator. Based on figure 2, ROS levels increased significantly (approximately 25%) in cells induced lead (400 uM) for 3 days compared to negative controls (Cells stained using DCF-DA). The results of treatment using chlorogenic acid with concentrations of 25 and 6.25 μg/ml can significantly reduce ROS levels (approximately 13% and 8%). The most optimal CA concentration in reducing ROS content was 6.25 μg/ml, but the concentration did not differ significantly from CA concentrations of 25 μg/ml based on Tukey post hoc test.

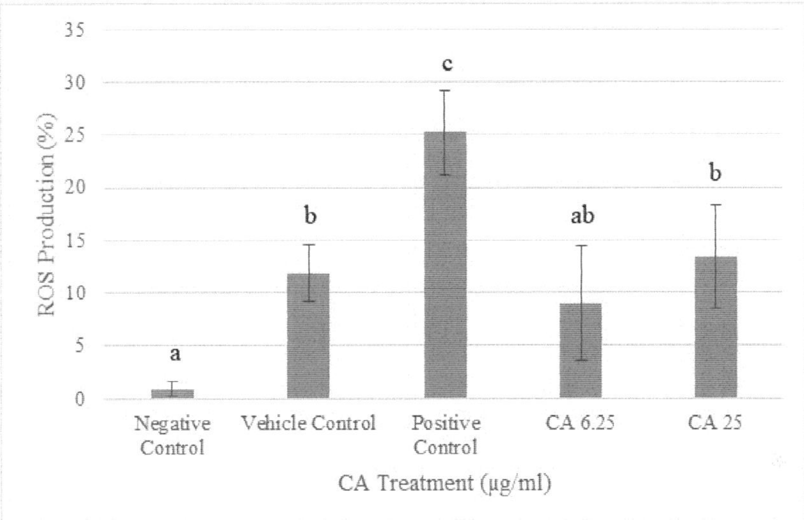

Figure 2. Effect CA toward ROS production on lead-poisoned fibroblast cell.

* The histograms are presented as mean ± standard deviation. The data were analyzed with ANOVA and continued with Tukey post hoc test ($p<0.05$). Different letters (a,ab,b,c) indicate significant differences among treatment.

* Negative Control: cells without any treatments; Vehicle Control: cells with DMSO 10% treatment; Positive Control : cells with Pb; CA 6.25: cells treated Pb + CA 6.25 μg/mL treatment; CA 25: cells treated Pb + CA 25 μg/mL treatment

Negative control (Cell only+DCFDA) (0.63%)	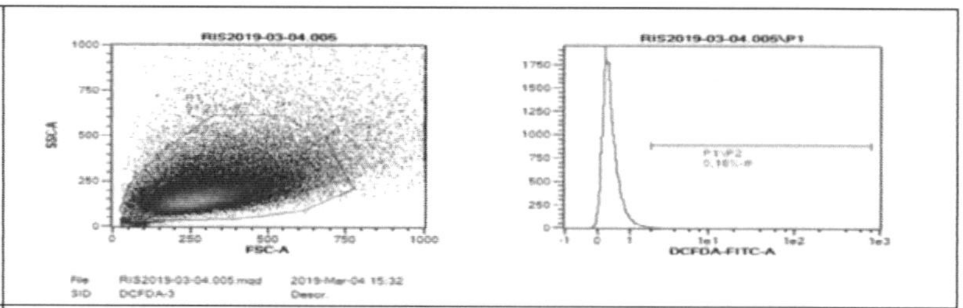
Positive control (Cell only+Pb) (35.63%)	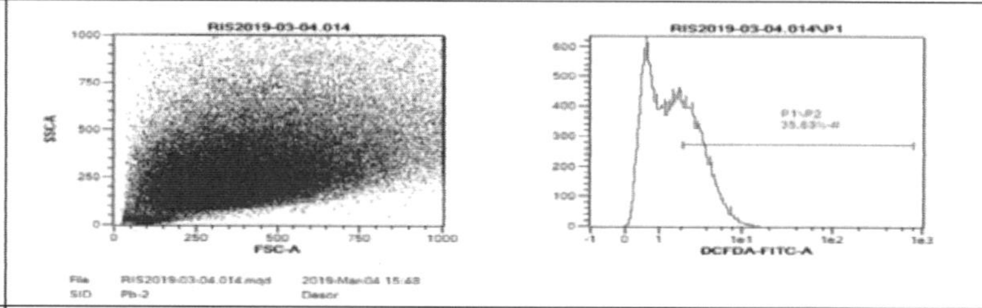
Chlorogenic acid induced 6.25 µg/ml (8.29%)	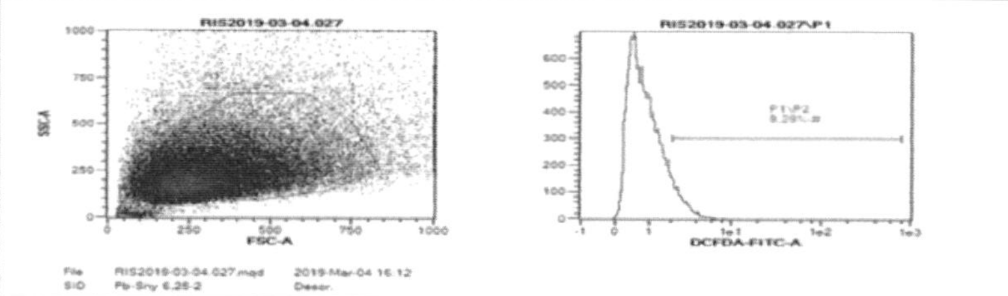
Chlorogenic acid induced 25 µg/ml (12.79%)	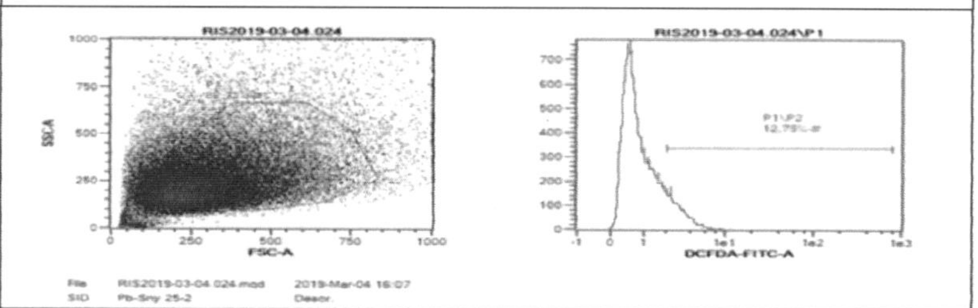

Figure 3. The representative of dot blot of various concentrations of CA treatment on lead-poisoned fibroblast cell toward ROS level.
TBHP = tert-butyl hydroperoxide, DCFDA = 2',7'-dichlorofluorescin diacetate

3.3. Level of IL10 and TNFα

Increases in cellular ROS production can result in an increase in the expression of NF-kB, leading to the upregulation of factors involved in inflammation. During aging process, TNFα are potent releasers of IL6 which is one of the first cytokines to be linked to the aging process [24]. The effect of TNFα can turn triggers effects that increase inflammation that can indicate with increased ROS levels [25]. In contrast, IL10 is a cytokine that acts as an anti-inflammatory and works by stimulating antagonist proteins against TNFα as pro-inflammatory cytokine [26].

IL10 and TNFα are cytokines that acts during inflammation, TNFα is a member of the pro-inflammatory cytokine group [27], while IL10 is a member of anti-inflammatory cytokine [24]. Figure 3 showed the results of TNFα and IL10 levels using the ELISA method. It can be seen that fibroblast cells induced with lead has the highest levels of TNFα among others, while addition CA shows a significant decrease in TNFα. In the other hand, fibroblast cells induced with lead has the lowest levels of IL10 and addition of CA can elevate the IL10 levels. CA at 25 and 6.25 µg/ml can reduce the TNFα levels but no significant, same as in IL10 levels, both CA concentrations can increase IL10 levels but the results are not significant based on post hoc test.

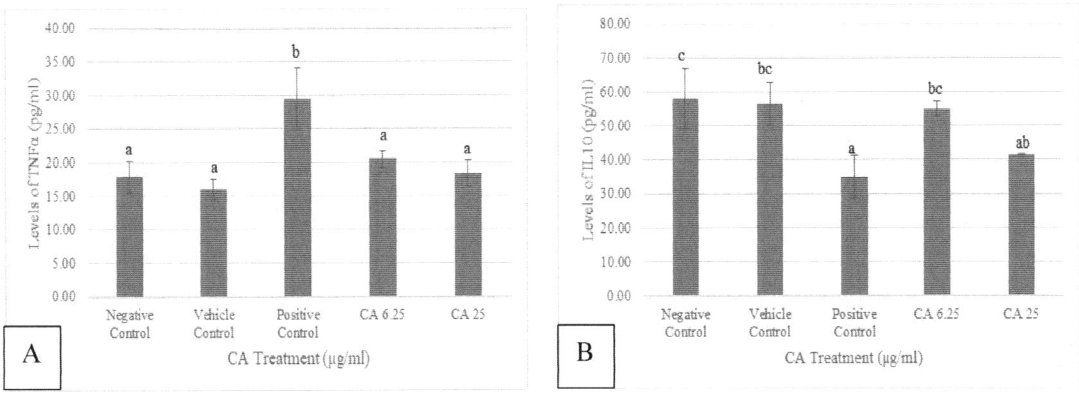

Figure 4. Effect of CA toward TNFα level (A) and IL10 level (B) on lead-poisoned cells model.
* The histograms are presented as mean ± standard deviation. The data were analyzed with ANOVA and continued with Tukey post hoc test ($P<0.05$). Different letters (a,b) on figure A and (a,ab,bc,c) on figure B indicate significant differences among treatment.
* Negative Control : cells without any treatments; Vehicle Control : cells with DMSO 10% treatment; Positive control : cells with Pb; CA 6.25: cells treated Pb + CA 6.25 µg/mL treatment; CA 25: cells treated Pb + CA 25 µg/mL treatment

4. Conclusion

This study shows that chlorogenic acid as natural compound has the potential as protective effect through suppression of ROS levels and have anti-inflammatory properties to decrease TNF-α level and increase IL-10 on lead-poisoned human fibroblast cells.

5. Acknowledgement

We are gratefully acknowledge the financial support of the Research Center and Service Community, Universitas Prima Indonesia, Medan, North Sumatera, Indonesia for research grant 2018. This study also was funded, facilitated and supported by Biomolecular and Biomedical Research Center, Aretha Medika Utama, Bandung, West Java, Indonesia. We are thankful to Wahyu Setia Widodo, Satrio Haryo Benowo Wibowo, Hanna Sari Widya Kusuma, Annisa Amalia, Ni Luh Wisma Ekayanti, Ika Adhani Sholihah, Dewani Tediana Yusepany, Dwi Surya Artie from Biomolecular and Biomedical Research Center, Aretha Medika Utama, Bandung, West Java, Indonesia for their valuable assistance.

References

[1] Redza-Dutordoir M and Averill-Bates DA 2016 Activation of apoptosis signalling pathways by reactive oxygen species *Biochim. Biophys. Acta. Mol. Cell. Res.* **12** 2977–92
[2] De Voogt P 2015 Reviews of Environmental Contamination and Toxicology *Rev. of Env. Cont. Toxicol.* 1–297
[3] Halliwell B and Gutteridge J M C 2015 Free Radicals in Biology and Medicine (USA: Oxford University Press)
[4] Davalli P, Mitic T, Caporali A, Lauriola A and D'Arca D 2016 ROS, cell senescence, and novel molecular mechanisms in aging and age-related diseases *Oxid. Med. Cell Longev.* **2016** 1–18
[5] Kuilman T, Michaloglou C, Mooi W J and Peeper D S 2010 The essence of senescence *Genes. Dev.* **22** 2463–79
[6] Sinha K, Das J, Pal P B and Sil P C 2013 Oxidative stress: the mitochondria-dependent and mitochondria-independent pathways of apoptosis *Arch. Toxicol.* **7** 1157–80
[7] Orr W and Sohal R 1994 Extension of life-span by overexpression of superoxide dismutase and catalase in Drosophila melanogaster *Science.* **5150** 1128–30
[8] Widowati W, Fauziah N, Herdiman H, Afni M, Afifah E, Kusuma H S W, Nufus H, Arumwardana

S and Rihibiha D D 2016 Antioxidant and anti aging assays of oryza sativa extracts, vanillin and coumaric acid *J. Nat. Remedies.* **3** 88

[9] Mittal M, Siddiqui M R, Tran K, Reddy S P and Malik A B 2014 Reactive oxygen species in inflammation and tissue injury *Antioxid. Redox Signal.* **7** 1126–67

[10] Widowati W, Darsono L, Suherman J, Yelliantty Y and Maesaroh M 2014 High performance liquid chromatography (HPLC) analysis, antioxidant, antiaggregation of mangosteen peel extract (Garcinia mangostana L.) *Int. J. Biosci. Biochem. Bioinforma.* 458–66

[11] Widowati W, Widyanto R M, Husin W, Ratnawati H, Laksmitawati D R, Setiawan B, Nugrahenny D and Bachtiar I 2014 Green tea extract protects endothelial progenitor cells from oxidative insult through reduction of intracellular reactive oxygen species activity *Iran J. Basic Med. Sci.* **9** 702–9

[12] Olthof M R, Hollman P C H and Katan M B 2001 Chlorogenic acid and caffeic acid are absorbed in humans *J. Nutr.* **1** 66–71

[13] Francisco V, Costa G, Figueirinha A, Marques C, Pereira P, Miguel Neves B, Celeste Lopes M, Garcia Rodriguez C, Teresa Cruz M and Teresa Batista M 2013 Anti-inflammatory activity of Cymbopogon citratus leaves infusion via proteasome and nuclear factor-κB pathway inhibition: Contribution of chlorogenic acid *J. Ethnopharmacol.* **1** 126–34

[14] Novilla A, Djamhuri D S, Nurhayati B, Rihibiha D D, Afifah E and Widowati W 2017 Anti-inflammatory properties of oolong tea (Camellia sinensis) ethanol extract and epigallocatechin gallate in LPS-induced RAW 264.7 cells *Asian Pac. J. Trop. Biomed.* **11** 1005–9

[15] Jie G, Lin Z, Zhang L, Lv H, He P and Zhao B 2006 Free radical scavenging effect of pu-erh tea extracts and their protective effect on oxidative damage in human fibroblast cells *J. Agric. Food Chem.* **21** 8058–64

[16] Prahastuti S, Hidayat M, Hasiana S T, Widowati W, Amalia A, Qodariah R L, Rizal R, Kusuma H S W and Khoiriyah Z 2019 Ethanol extract of jati belanda (Guazuma ulmifolia L.) as therapy for chronic kidney disease in in vitro model *J. Reports. Pharm. Sci.* XX XX

[17] Widowati W, Widyastuti H, Murti H, Laksmitawati D, Sari H, Kusuma H S W, Rizal R, Afifah E, Sumitro S, Widodo M A and Bachtiar I 2017 Interleukins and VEGF secretome of human wharton's Jelly mesenchymal stem cells-conditioned medium (hWJMSCs-CM) in different passages and oxygen tensions *Biosci. Res.* **14**

[18] Widowati W, Afifah E, Mozef T, Sandra F, Rizal R, Amalia A, Arinta Y, Bachtiar I and Murti H 2018 Effects of insulin-like growth factor-induced wharton jelly mesenchymal stem cells toward chondrogenesis in an osteoarthritis model *Iran J. Basic Med. Sci.* **7** 745–52

[19] Noverina R, Widowati W, Ayuningtyas W, Kurniawan D, Afifah E, Laksmitawati D R, Rinendya P, Rilianawati R, Faried A, Bachtiar I and Wirakusumah F F 2019 Growth factors profile in conditioned medium human adipose tissue-derived mesenchymal stem cells (CM-hATMSCs) *Clin. Nutr. Exp.* **24** 34–44

[20] Wang C, Liang J, Zhang C, Bi Y, Shi X and Shi Q 2007 Effect of ascorbic acid and thiamine supplementation at different concentrations on lead toxicity in liver *Ann. Occup. Hyg.* **6** 563–9

[21] Koganov M M, Dueva O V and Tsorin B L 1999 Activities of plant-derived phenols in a fibroblast cell culture model *J. Nat. Prod.* **3** 481–3

[22] Lopes A C B A, Peixe T S, Mesas A E and Paoliello M M B 2016 Lead exposure and oxidative stress: a systematic review 193–238

[23] Hoelzl C, Knasmüller S, Wagner K-H, Elbling L, Huber W, Kager N, Ferk F, Ehrlich V, Nersesyan A, Neubauer O, Desmarchelier A, Marin-Kuan M, Delatour T, Verguet C, Bezençon C, Besson A, Grathwohl D, Simic T, Kundi M, Schilter B and Cavin C 2010 Instant coffee with high chlorogenic acid levels protects humans against oxidative damage of macromolecules *Mol. Nutr. Food Res.* **12** 1722–33

[24] Morley J E and Baumgartner R N 2011 Cytokine-related aging process *J. Gerontol. Ser. A .Biol.*

Sci. Med. Sci. **9** M924–9

[25] Marcu K B, Otero M, Olivotto E, Borzi R M and Goldring M B 2010 NF-kB signaling: multiple angles to target OA *Curr. Drug Targets.* **5** 599–613

[26] Wojdasiewicz P, Poniatowski Ł A and Szukiewicz D 2014 The role of inflammatory and anti-inflammatory cytokines in the pathogenesis of osteoarthritis *Mediators Inflamm.* **2014** 1–19

[27] Laksmitawati D R, Prasanti A P, Larasinta N, Syauta G A, Hilda R, Ramadaniati H U, Widyastuti A, Karami N, Afni M, Rihibiha D D, Kusuma H S W and Widowati W 2016 Anti-inflammatory potential of gandarusa (Gendarussa vulgaris Nees) and soursoup (Annona muricata L) extracts in LPS stimulated-macrophage cell (RAW264.7) *J. Nat. Remedies* **2** 73

Biosorption of Synthetic Dye from Batik Wastewater Using *Trichoderma viride* Immobilized on Ca-Alginate

G Rahmaniah[1], C Mahdi[1] and A Safitri[1,2*]

[1]Chemistry Department, Brawijaya University, Indonesia
[2]Research Centre for Smart Molecules of Natural Genetic Resources, Brawijaya University, Indonesia

*Corresponding author's email: a.safitri@ub.ac.id

Abstract. Batik and textile industry is one of the sources of liquid waste derived from the staining process. One of the color substances commonly used is remazol brilliant violet which is an azo-type substance that is difficult to be degraded in the environment and is toxic pollutants. A method that can be used to decolorize of synthetic dye is biosorption. *Trichoderma viride* can be used as biosorbent, since it is resistant to toxic compound. The immobilization of *T. viride* is applied to increase their stability and to minimize cellular damage. Calcium alginate can be used for immobilization matrix because it is affordable, non-toxic and biocompatible. This study focused on determining the optimum conditions on biosorption process of remazol brilliant violet using immobilized *T. viride*. Remazol brilliant violet solution were absorbed at pH 4, 5, 6, 7, 8, and 9, contact time for 10, 20, 30, 50, 60, 70, and 90 min, and the amount of adsorbent used were varied at 0.3; 0.4; 0.6; 0.8; and 1 g. The optimum condition was obtained at 60 min biosorption time, at pH 4 with the amount of adsorbent was 1 g, with percent of adsorption was 8.76%. The adsorption capacity of remazol brilliant violet using *T. viride* immobilized with Ca-alginate was 0.089 mg/g. FTIR spectra results indicated functional groups that involved in immobilization of *T. viride* were O-H and C=O carboxylic, while in biosorption process violet were C-H sp3, C=O carboxylic, C=O amide and C=C benzene.
Keywords: biosorption, remazol brilliant violet, *Trichoderma viride*, immobilization, Ca-alginate

1. Introduction

At the present time, batik industry has become popular and has many contributions to the economic growth for Indonesia. Batik industrial is mostly manufactured in a small scale industry. Nonetheless, awareness on the importance of clean practices in the production of batik is degrading. This is because, some of the batik manufacturers release batik waste into water directly without proper pre-treatment. It is known that batik industries use large amount of synthetic dyes during their coloring process [1]. As a result, the presence of synthetic dyes in the wastewaters leads to harsh environmental pollution and also cause serious health risks, since synthetic chemical dyes have a complex molecular structure which makes them more stable, and can stay in the environment for a prolonged time period [2].

Because of this serious water pollution, there is persistent need from the researchers to find an alternative and friendly method to preserve the environment. At this time, there are many technologies available for the solution to the problems caused by the textile industry, including chemical precipitation, ultrafiltration, carbon adsorption, coagulation, flocculation and ozone oxidation [3]. Nevertheless, these treatments were either non-economical to apply, or create a huge chemical sludge production [3]. Thus, it is important to protect the water resources by seeking more environmental friendly and affordable solution to solve this problem.

Remazol brilliant violet (RBZ) is frequently used in textile industry, including batik home industry. This synthetic dye has aromatic azo groups (-N=N-) [4]. These azo groups make this molecule non-degradable. Decolorization of synthetic dyes, therefore, is important. The biosorption process of synthetic dyes can be performed with the use of microorganisms, fungi, or yeast, *i. e. Trichoderma viride* [5]. However, direct use of these microorganisms have some drawbacks, such as they are unstable and easily damage due to extreme environment. Thus, the immobilization is conducted to increase the surface of the fungi, to improve fungi resistant from the environmental stress live in toxic environment [6]. Cell entrapment in Ca-alginate is one the immobilization methods. The entrapment method with Ca-alginate offers many benefits, including this is easy to apply, economical, and bio-degradable [7].

Therefore, this study investigates the efficiency of immobilized *T. viride* on decolorizing of batik wastewater model, using RBZ dye. In addition, in the biological adsorption processes, numerous physicochemical parameters including pH, temperature, oxygen, time period, dye structure and dye concentration, directly influence the capability of bacterial decolorization of dye wastewater and also effects on microbial growth [8]. Therefore, in particular, the aims of the current study are to investigate the effects of pH, contact time, and the amount of biosorbent on the biosorption process of RBZ on immobilized *T. viride*.

2. Materials and Methods

2.1. Chemicals and Instrumentation

Materials used in this study were culture of *Trichoderma viride* obtained from microbial stock collection of Microbiology Laboratory, Universitas Negeri Malang. All reagents of analytical or higher purity grade were purchased from Merck or Sigma-Aldrich and were used as received: HCl (37% aqueous solution), HNO_3 (trace pure, 65% w/w in H_2O), NaOH (99.9%), H_2SO_4 (98%, d = 1.84 g/mL), ethanol (96%), $CaCl_2{\cdot}2H_2O$, KH_2PO_4, remazol brilliant violet, Na-alginate. Water was purified using distillation technique. The growth media for was potatoes dextrose agar (PDA). The determination of remazol brilliant violet concentration before and after biosorption process was conducted using UV-Vis spectrophotometer (1601-Shimadzu). The functional groups change in the biosorbent before and after biosorption process was determined using FTIR spectrophotometer (8400S/Shimadzu).

2.2. Immobilization Process of T. viride with Ca-alginate

A 1.25 g of Na-alginate was added to 80 mL of liquid media, heated and stirred with magnetic stirrer. Culture of *T. viride* growth in solid media was harvested with sterile water. These were mixed with a mixture of Na-alginate and liquid media. The mixture was shaken for 36 h in room temperature. The solution mixture was added to 20 mL of 0.15 M $CaCl_2$ solution drop wise, until the biosorbent beads were formed.

2.3. Biosorption Process of Remazol Brilliant Violet with Immobilized T. viride

A 25 mL of RBZ solution (40 mg/L) was transferred into Erlenmeyer flask. A 0.4 g of immobilized *T. viride* was added as biosorbent. Biosorption process was carried out at room temperature, pH 4, in variations of contact time for 10, 20, 30, 50, 60, 70, and 90 min. For the pH influence, the biosorption process was conducted in similar ways, the pH was varied at pH 4, 5, 6, 7, 8, and 9, in 60 min of time contact. All experiment was conducted in triplicates. Finally, the amount of biosorbent used were determined in the variations of 0.3, 0.4, 0.6, 0.8, and 1 g. The concentration of RBZ was calculated after biosorption process. Determination of the amount of RBZ after biosorption process was calculated using this formula:

Biosorption (%) = ((Co-Ce))/Co x 100%; where Co = initial concentration (mg/L), Ce=final concentration (mg/L). Adsorption capacity of immobilized *T. viride* was determined as follow:

Qt = (Co-Ce) V/W; where Qt = adsorption capacity (mg/g), Co = final concentration (mg/L), Ce = final concentration (mg/L), V = sample volume (L), and W = adsorben mass (g). The precipitates from biosorption process were collected and analyzed using FTIR spectrophotometer at a range of 4000-400 cm^{-1} (8400S/Shimadzu).

3. Results and Discussion

Figure 1 shows the effect of time on the biosorption process of RBZ on immobilized *T. viride*. Biosorption of the dye increased time by time, since the beginning of the process (10 min), until at the end of the assay (90 min). However, statistical analysis revealed that from 60 min onward, the percentage of adsorption was not significantly different ($p<0.05$). Therefore, 60 min was the optimum contact time for the biosorption process, with the amount of RBZ adsorbed was 9.4%. The biosorption process of RBZ was found to be in a two-stages. In the beginning, biosorption process was fast, since the active sites of the adsorbent were still available to adsorp the dye, from 10 to 60 min. After that, 60 to 90 min, the equilibrium of adsorption and desorption was achieved. Therefore, the biosorption process started to slowing down, due the active sites were already occupied with the dye. The biosorption kinetics observed in biosorption process is proposed by involving no energy-mediated reactions and dye removal from solution is due to purely physico-chemical interactions between microorganism and dye solution [9].

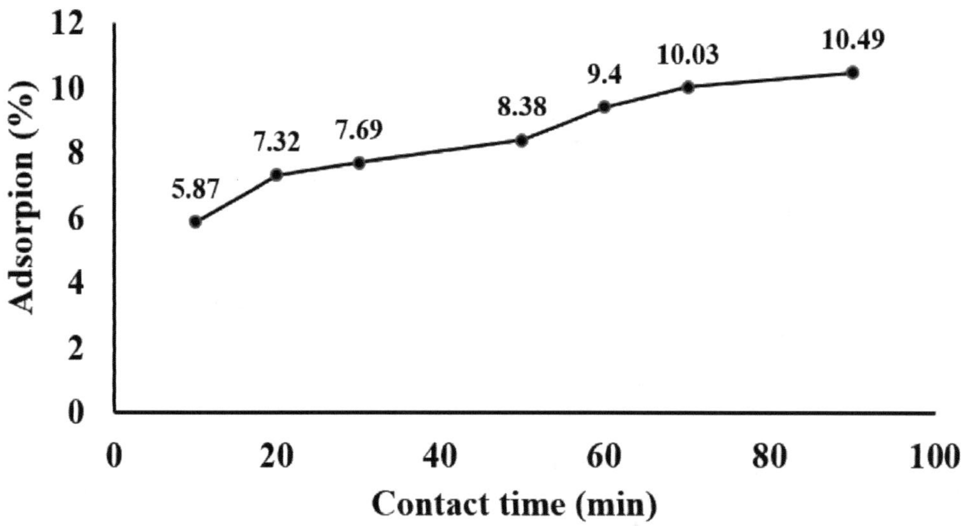

Figure 1. Result of biosorption of remazol brilliant violet using immobilized T. viride, under the influence of time (10, 20, 30, 50, 60, 70, and 90 min).

The influence of pH (pH 4, 5, 6, 7, 8, and 9) on the biosorption process of RBZ is shown in Figure 2. Interestingly, from the lowest pH used, pH 4, to the highest pH, pH 9, the adsorption of RBZ showed a decrease. This is because at low pH, surface of the biosorbent has been protonated, due to the abundance of H^+ ion in acidic condition. The pH of the medium affects both the solubility of RBZ and ionization of the functional groups such as $-COO^-$, $-OPO_3^{4-}$, and $-NH^-$, of the cell wall of *T. viride*; which are acidic and carry negative charges rendering cell walls [10]. At acidic pH, the protonated adsorbent resulted in electrostatic interaction between the dye and adsorbent, since RBZ is anionic dye that has negative –S group. This result is in agreement with the previous study by Bello. *et. al.* [4], that investigated adsorption of synthetic dye using activated carbon, the adsorption of the dye decreased as the pH increased. In another study [11], RBZ adsorption using rice hulls resulted in the highest adsorption at acidic pH, pH 1.

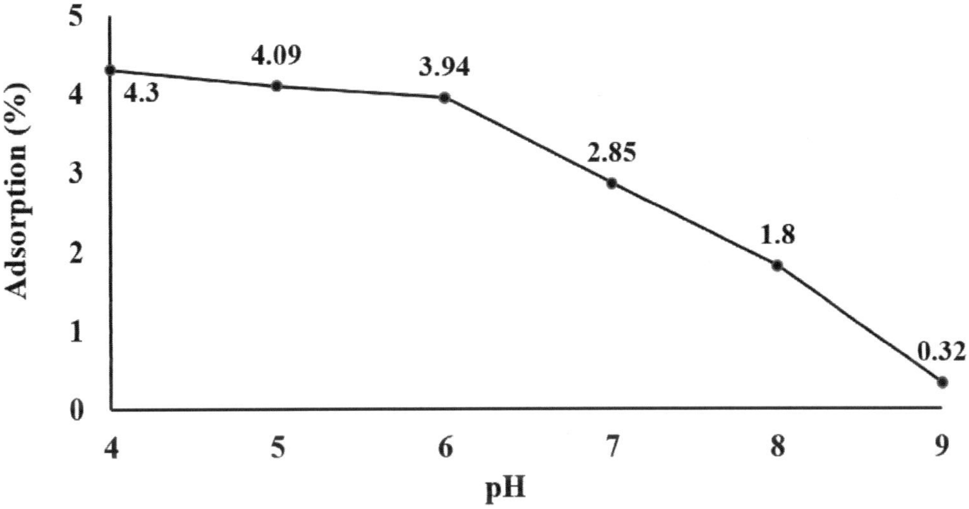

Figure 2. Result of biosorption of remazol brilliant violet using immobilized *T. viride*, under the influence of pH.

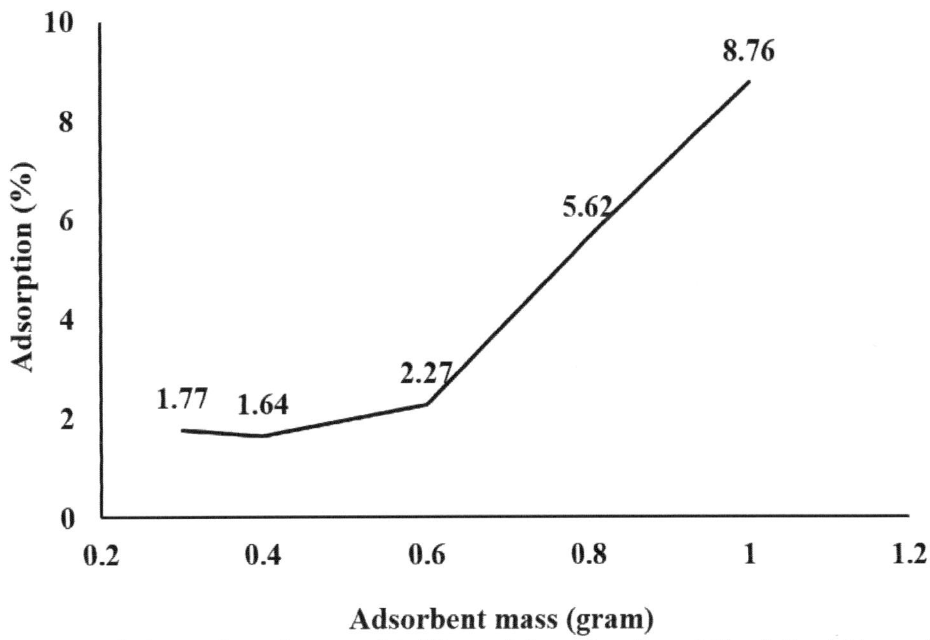

Figure 3. Result of biosorption of remazol brilliant violet using immobilized *T. viride*, under the influence of mass of adsorbent.

Effect of the amount of biosorbent (immobilized *T. viride*) was investigated by contacting RBZ and *T. viride* with previous optimum conditions that obtained before (pH 4, contact time 60 min), with the amount of biosorbent ranging from 0.3 to 1 g. It was found that the amount of RBZ taken up by the immobilized *T. viride* increased slowly with an increase of the amount of biosorbent from 0.3 to 0.6 g.

A rapid increase was observed when the amount of biosorbent greater than 0.6 g. The highest amount of biosorbent used, 1 g, resulted in the highest amount of RBZ adsorbed, at 8.76%. In this study, the percent of adsorption was not optimal (less than 50%). There are several reasons for this. Firstly, the pores size of Ca-alginate used to immobilize *T. viride* are relatively small, 5.2 nm ± 0,9 nm [12]. Moreover, the molecular weight of RBZ is *remazol brilliant violet* is a bulky molecule. Based on National Center for Biotechnology Information [13], RBZ has relatively high molecular weight of 735.563 g/mol.

Adsorption capacity of the immobilized *T. viride* was calculated at optimum conditions, at pH 4, time contact of 60 min, and the amount of adsorbent was 1 g. The adsorption capacity obtained indicates the capacity of immobilized *T. viride* can absorb RBZ in those conditions. In this study, the adsorption capacity was 0.089 mg/g. In contrast, a study conducted by Bishnoi *et. al.* [14] reported biosorption of Cr(VI) metal using *T. viride* which was immobilized with Ca-alginate, that resulted in higher adsorption capacity, 16.075 mg/g. The differences may be caused by the size of Cr(VI) metal ions which are smaller than RBZ.

The analysis of functional group involved in the biosorption process was conducted using FTIR spectrophotometry. Figure 4 shows the FTIR spectra of immobilized *T. viride* before and after biosorption process, while assignment of the bands of interest is listed in Table 1.

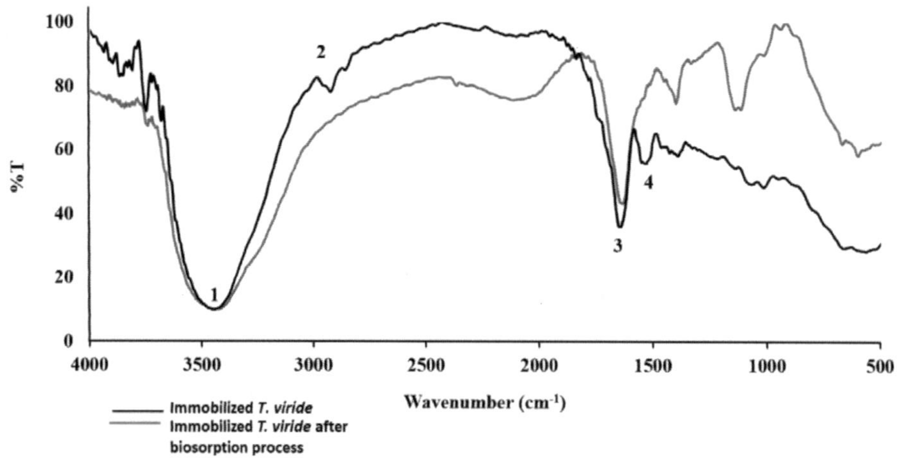

Figure 4. The FTIR spectra from immobilized *T. viride* before and after biosorption process of remazol brilliant violet, at pH 4 and contact time of 60 min.

Table 1. FTIR spectra assignment of bands of interests from Figure 4.

No	Wavenumber (cm^{-1})		Assignment [15, 16]
	Before biosorption	After biosorption	
1	3423.41	3452.34	-O-H carboxylate
2		2929.67	-C-H sp3
3	1631.67	1643.24	-C=O amide
4		1533.30	-C=C aromatics rings

FTIR spectra of *T. viride* (non-immobilized) have been reported in our previous study [rensani paper]. In the FTIR spectra of immobilized *T. viride,* there were only appeared two bands of interest, at around 3430 cm^{-1}, assigned as -O-H stretching from carboxylate groups, and ~1630 cm^{-1}, assigned as C=O amide, that may come from functional groups in Ca-alginate. The difference between FTIR soectra of Ca-alginate and immobilized *T. viride* is that in the -O-H alcohol band of *T. viride* there is an overlapping band of N-H amide. This can be seen from the FTIR spectra of immobilized *T. viride* which has a shoulder in the wavenumber at around 3500 cm^{-1}.

After biosorption process, there was band at ~2930 cm^{-1}, indication of -C-H sp3. In addition, there were bands at wavenumber around 1643 cm^{-1} and1530 cm^{-1} from C=C benzene or aromatic rings from the RBZ. The functional groups of C=C aromatics have specific two bands at around 1600 cm^{-1} and 1500-1450 cm^{-1}. The C=C aromatics bands indicated that there was bond formation between RBZ and immobilized *T. viride*. The bond formation is from positively charged amide group from *T. viride* with sulphonate group from RBZ. This bond may also cause a shift in wavenumber and decrease in percent transmittance (%T).

4. Conclusion

This study has demonstrated that biosorption process of immobilized *T. viride* on Ca-alginate is affected by time, pH, and the amount of biosorbent. The optimum conditions achieved in the study were at pH 4, at 60 min, and 1 g of biosorbent mass, with the percent of adsorption at 8.76%. The adsorption capacity was 0.089 mg/g. The functional groups involved in the biosorption process were -C-H sp3, C=O amide, and -C=C benzene. In order to increase adsorption capacity, the use of natural polymers or matrices that have larger pore size is needed, since the synthetic dye is generally has large molecular weight. The biosorption process of RBZ using immobilized *T. viride* can be applied as one of alternative solutions for synthetic dye removal in the environment.

References

[1] Handayani W, Kristijanto A I and Hunga A I R 2018 *Sustainable Water Resources Management* **4** 1011-1021
[2] Susanty A, Hartini S, Puspitasari D and Arsiwi P 2015 *Mediterr. J. Soc. Sci.* **6** 318–327
[3] Yagub M T, Sen T K, Afroze S and Ang H M 2014 *Adv. Colloid and Interface Sci.* **209** 172–184
[4] Bello O S and Ahmad M A 2011 *Chem. Ecol.* **27** 481–492
[5] Safitri A, Mahardini P, Prasetyawan S and Roosdiana A 2018 *IOP Conf. Series: Mat Sci Eng* **299** 012023 8 pages
[6] Przystas W, Zabłocka-Godlewska E and Grabinska-Sota E 2017 *Brazilian J. Microbiol* **49** 285-295
[7] Anwar A, Ul Qader S A, Raiz A, Iqbal S and Azhar A 2009 *World Appl. Sci. J.* **7** 1281-1286
[8] Saratale R G, Saratale G D, Chang J S and Govindwar S P 2011 *J. Taiwan Institute of Chem. Eng.* **42** 138-157
[9] Gadd G M 2009 *J. Chem. Technol. Biotechnol.* **84** 13–28
[10] Prasad A A, Varatharaju G, Anushri C and Dhivyasree S 2013 *Br. Biotechnol. J.* **3** 66–70
[11] Ribeiro G A C, Silva D S A, dos Santos C C, Vieira A P, Bezerra C W B, Tanaka A A and Santana S A A 2017 *Polímeros* **27** 16-26
[12] Simpliciano C, Clark L, Asi B, Chu N, Mercado M, Diaz S, Goedert M and Mobed-Miremadi M 2013 *J. Surface Engineered Mat. Adv. Technol.* **3** 1-12
[13] National Center for Biotechnology Information, PubChem Database *Remazol Brilliant Violet 5R,* CID=135445762,https://pubchem.ncbi.nlm.nih.gov/compound/Reactive-Violet-5 (accessed on May 29, 2019)
[14] Bishnoi N R, Rajender K and Bishnoi K 2007 *Indian J. Experimen. Biol.* **45** 657-664
[15] Talari A C S, Martinez M A G, Movasaghi Z, Rehman S and Rehman I U 2017 *J. Appl. Spectros. Rev.* **52** Pages 456-506
[16] Ovchinnikov O V, Evtukhova A V, Kondratenko T S, Smirnov M S, Yu V, Oksana K and Erina V 2016 *Vib. Spectrosc.* **86** 181-189
[17] Taloin R, Safitri A, Prasetyawan S, Kamulyan B and Andayani U 2018 *J. Pure App. Chem. Res.* **7** 70-78

Detection and Analysis of Delayed-Ripening Genes in the Growth of Strawberries (*Fragaria* x *ananassa*) Treated 1-Naphtaleneacetic Acid

G R Aristya[1*], N T A Maulina[1], M F Arif[1], R S Kasiamdari[2] and A Widiastuti[3]

[1]Laboratory of Genetics and Breeding, Faculty of Biology, Gadjah Mada University, Yogyakarta, Indonesia
[2]Laboratory of Plant Systematics, Faculty of Biology, Gadjah Mada University, Yogyakarta, Indonesia
[3]Laboratory of Integrated Pest Control, Faculty of Agriculture, Gadjah Mada University, Yogyakarta, Indonesia

*Corresponding author's email: ganies_riza@ugm.ac.id

Abstract. Strawberries (*Fragaria* spp.) are non-climatic fruits that are popular due to its good taste, fragrance and attractive colors. It contains anthocyanins and other phenolic compounds capable of preventing such diseases as cancers and heart diseases. Rapid decay of fruits is among farmers' post-harvest problems. 1-Naphtaleneacetic acid (NAA) is a hormone known to be capable of delaying fruit ripening. Molecular studies of growth and development control genes have been carried out, including color pigment control in strawberry fruits. *FaCHS* and *FaPYR1* are genes that play a role in fruit development, especially in ripening strawberry fruits. The purpose of the present study was to detect fruit ripeness-delaying genes in *Fragaria* x *ananassa* treated with 1-Naphtaleneacetic acid (NAA). It was conducted by inducing strawberry plants from Banyuroto, Magelang of Central Java, with 10 ppm, 30 ppm and 60 ppm of NAA. DNA was isolated from fruit using the Nucleon Phytopure Kit and then tested qualitatively and quantitatively using electrophoresis and spectrophotometry. Genes were detected using the MyTaqTM-Bioline PCR method. The phenotypic characteristics showed that there was an increase in the leaves area and stem diameter in the cultivar 'KP Brite' and the fruit size in the cultivar 'Crystal' after 10-ppm NAA treatment. Results of amplification of *FaCHS* DNA fragments showed *FaCHS* was not detected in the cultivar 'KP Brite' after 60-ppm NAA treatment, while *FaPYR1* could only be detected in the red stage of the cultivar 'Crystal' after 10-ppm NAA treatment and the pink and red stages of the cultivar 'KP Brite' after 30-ppm NAA treatment.
Keywords: delayed-ripening genes, *FaCHS*, *FaPYR1*, *Fragaria* x *ananassa*, 1-Naphtaleneacetic acid

1. Introduction

Strawberries (*Fragaria* spp.) belong to herbaceous plants firstly discovered in Chile, America. Strawberry fruits are non-climatic fruits that are popular due to their good taste, fragrance and attractive colors, making them potential for development of fruit-picking agro-tourism in various places. Strawberry fruits contain anthocyanins and other phenolics capable of preventing such diseases as cancers and heart diseases. Ellagic acid in their seeds is capable of beautifying the skin, whitening teeth, eliminating bad breath, and increasing strength [1]. Among the widely consumed strawberry species is *Fragaria* x *ananassa*. This species is quite large and easily found during the summer. Its unique flavor make this fruit much sought after and liked [2].

Relatively fast decay of strawberry fruits is one of the obstacles that should be overcome by farmers in the post-harvest fruit handling. 1-Naphtaleneacetic acid, or NAA, is one of auxin synthetic hormones. In addition to its known function in the induction of cell expansion and root elongation, this hormone is known to be capable of delaying fruit ripening [3, 4], making it effective in delaying decay. This hormone can prevent fruit ripening since it works to inhibit ABA [5].

Molecular studies of growth and development control genes have been carried out, including color pigment control in strawberry fruits. *FaCHS* and *FaPYR1* are genes that play a role in fruit development, especially in ripening strawberry fruits.

The purpose of the present study was to screen and detect fruit-ripeness delaying genes in F. x ananassa of cultivars "Crystal' and 'KP Brite' induced with NAA at concentrations of 10 ppm, 30 ppm and 60 ppm.

2. Materials and Methods

2.1. Time and place

Strawberry planting and germination were carried out in the Agro-tourism areas of Desa Banyoroto, Magelang, Central Java. Strawberries (*Fragaria* x *ananassa*) of cultivars 'Crystal' and 'KP Brite' were induced with 10 ppm, 30 ppm and 60 ppm of NAA after they were ±3 months old. The present study was conducted from January to May 2019 at Genetic and Breeding Laboratory, Faculty of Biology, Gadjah Mada University.

2.2. Observation of NAA induction results

Such physiological characters as leaf length, leaf width, leaf number, plant height, stem diameter, fruit diameter, and fruit weight were measured and observed.

2.3. DNA isolation

DNA was isolated from fruits using plant genome isolation methods from Illustra™ Nucleon Phytopure Kit. Those fruits were taken from the four stages of fruit ripeness, namely the green, white, pink and red stages.

2.4. Detection of FaCHS and FaPYR1

FaCHS and *FaPYR1* were amplified using the MyTaqTM-Bioline PCR method. PCR used the target *FaCHS*, 26S-18S RNA housekeeping gene, *FaPYR1* gene, and Actin primers (Table 1).

Table 1. Primers used in PCR [6, 7].

Name of Primer		Oligonucleotide Sequence
FaCHS	F	5'-GCCTTTGTTTGAGCTGGTCT-3'
	R	*5'-CCCAGGAACATCTTTGAGGA-3'*
26S-18S RNA *housekeeping gene*	F	5'-ACCGTTGATTCGCACAATTGGTCATCG-3'
	R	5'-TACTGCGGGTCGGCAATCGGACG-3'
FaPYR1	F	5'-ATGGAGAAACCATCATCGGC-3'
	R	5'-TCAGACCTGGGGAGTTAGCG-3'
Actin	F	5'-TGGGTTTGCTGGAGATGAT-3'
	R	5'-CAGTAGGAGAACTGGGTGC-3'

2.5. ANOVA analysis
Physiological characters observed were analyzed using the analysis of variance (ANOVA) at a significance level of 0.05 or a confidence level of 95% in order to determine the significance of the effect of various concentrations on physiological characters of plant.

2.6. Qualitative analysis
DNA isolates were qualitatively tested by electrophoresis using 0.8% agarose gel. PCR amplification products were qualitatively tested by electrophoresis using 2% agarose gel. Electrophoresis used a 100-bp marker or ladder to determine the successful amplification of *FaCHS* (118 bp), *26S-18S RNA housekeeping gene* (146 bp), *FaPYR1* (627 bp), and *Actin* (262 bp).

2.7. Quantitative analysis
Isolates were analyzed using an Ultraviolet-Visible (UV-Vis) spectrophotometer by observing the ratio of purity of A260/A280 nm absorbance values in the range 1.8 to 2.0 [17, 18]. The DNA concentration calculation formula is:

$$\text{DNA } (\mu l/ml) = A_{260} \text{ x fp x } 40\ (\mu l/ml)$$

Notes:
A_{260} = absorbance value at λ 260 nm fp = dilution factor

3. Results and Discussion

3.1. Physiological Characteristics of NAA-Induced Strawberries
Results of ANOVA of the effects of NAA induction on the growth of cultivars 'Crystal' and 'KP Brite' showed that NAA concentration had effects on leaf length, leaf width, leaf number, plant height, stem diameter, fruit diameter, and fruit wet weight (Table 2). Table 2 shows that NAA treatment at concentrations of 10 ppm, 30 ppm, and 60 ppm did not indicate any interaction with vegetative growth of plants, such as leaf length, leaf width, leaf number, plant height, and stem diameter of cultivar 'Crystal'. The role of NAA in the cultivar "Crystal" was not shown in the vegetative phase, but it did in the generative phase of plants. NAA showed a significant effect on fruit development in terms of fruit diameter and wet weight of the cultivar 'Crystal'.

The 10-ppm NAA-induced cultivar 'Crystal' showed a significant difference from the control plants. The 10-ppm NAA treatment had an effect on increased fruit diameter of 2.365 ± 0.092 cm relative to the control fruits of 1.952 ± 0.037 cm. Similarly, the 10-ppm NAA treatment significantly affected the fruit weight of cultivar 'Crystal'. The concentration can increase the wet weight of the fruits to 6.006 ± 0.964 grams, or nearly twice the weight of the control fruits of 3.748 ± 0.251 grams. This shows that the addition of exogenous NAA can increase fruit diameter and weight. This is in consistent with Satriowibowo *et al.* (2014) [8] saying that NAA can increase fruit weight. However, at concentrations of 30 and 60 ppm, NAA did not show a significant difference in fruit diameter or wet fruit weight of the control. Wilkins (1992) [9] argued that healthy plants have optimal levels of substances that regulate growth, including auxin (NAA). NAA treatment at a higher concentration will show a different reaction from plants treated with low concentrations of NAA. Responses of plants to NAA may depend on the concentration given. In the present study, high concentrations of NAA (30 and 60 ppm) led to different results from low concentrations of NAA (10 ppm), where the concentration of 10 ppm was the optimum concentration to increase the size and weight of cultivar 'Crystal'.

Different results were shown by the cultivar 'KP Brite' (Table 3), in which leaf length and leaf width of the controls were not significantly different at NAA concentrations of 10 ppm and 30 ppm, but it was significantly different at NAA concentration of 60 ppm. In addition, observation of leaf length and leaf width did not show a significant difference between NAA concentrations of 30 ppm and 60 ppm. The 30-ppm and 60-ppm NAA treatments had an effect on decreasing the leaf length and leaf width relative to those of controls. The leaf length and leaf width of cultivar 'KP Brite' induced with 60 ppm of NAA were 4.875 ± 1.217 cm, smaller than those of controls with a leaf length and leaf

Journal of Physics: Conference Series **1374** (2019) 012008 doi:10.1088/1742-6596/1374/1/012008

width of 6.500 ± 1.134 cm. NAA induction at various concentration also affected the leaf number of cultivar 'KP Brite'. The leaf number of controls did not show a significant difference at NAA treatments of 10 ppm and 30 ppm, but it was significantly different from those at the NAA treatment of 60 ppm. The 10-ppm, 30-ppm and-60 ppm NAA treatments were not significantly different. The three concentrations showed a decrease in leaf number relative to that of controls, but the effect of 60-ppm NAA treatment showed a significance, which was 21.625 ± 13.016, far lower than those of controls of 41.625 ± 20.170. The plant height of controls was not significantly different at the 10-ppm NAA treatment, but it was significantly different at 30-ppm and 60-ppm NAA treatments. The 10-ppm NAA treatment was not significantly different from the NAA 30 ppm treatment, while the NAA 30 ppm treatment was not significantly different from the 60-ppm NAA treatment. The stem diameter of cultivar 'KP Brite' of controls at 10-ppm and 30-ppm NAA treatments was not significantly different but it was significantly different from that of the 60-ppm NAA treatment (Table 3).

Table 2. Physiological characteristics of the NAA-induced cultivar 'Crystal'.

NAA Treatment (ppm)	Parameters						
	Leaf Length (cm)	Leaf Width (cm)	Number of Leaf	Plant Height (cm)	Stem Diameter (cm)	Fruit Diameter (cm)	Fruit Weight (g)
Control	**6.143 ± 1.376^{a}**	**5.357 ± 1.281^{a}**	36.143 ± 19.608^{a}	17.714 ± 4.261^{a}	2.643 ± 0.627^{a}	1.952 ± 0.037^{a}	3.748 ± 0.251^{a}
10 ppm	6.143 ± 1.069^{a}	5.286 ± 1.185^{a}	50.000 ± 31.091^{a}	17.857 ± 2.410^{a}	**2.714 ± 0.567^{a}**	**2.365 ± 0.092^{b}**	**6.006 ± 0.964^{b}**
30 ppm	5.286 ± 1.524^{a}	4.929 ± 1.427^{a}	**62.429 ± 37.398^{a}**	16.929 ± 3.220^{a}	2.643 ± 0.690^{a}	1.962 ± 0.049^{a}	2.626 ± 0.594^{a}
60 ppm	5.786 ± 0.756^{a}	4.929 ± 0.535^{a}	42.714 ± 19.164^{a}	**17.857 ± 2.545^{a}**	2.500 ± 0.408^{a}	1.941 ± 0.115^{a}	3.188 ± 1.626^{a}

Means followed by similar letters show no significant difference based on the Duncan's multiple range test at a significance level of 5%

Table 3. Physiological characteristics of the NAA-induced cultivar 'KP Brite'.

NAA Treatment (ppm)	Parameters						
	Leaf Length (cm)	Leaf Width (cm)	Number of Leaf	Plant Height (cm)	Stem Diameter (cm)	Fruit Diameter (cm)	Fruit Weight (g)
Control	6.500 ± 1.134^{a}	6.500 ± 1.134^{a}	**41.625 ± 20.170^{a}**	**19.313 ± 1.907^{a}**	3.938 ± 0.623^{a}	2.132 ± 0.064^{a}	5.350 ± 0,644^{a}
10 ppm	**6.625 ± 1.061^{a}**	**6.625 ± 1.061^{a}**	37.875 ± 16.120^{ab}	16.938 ± 2.243^{ab}	**4.000 ± 0.802^{a}**	2.227 ± 0.032^{a}	5.350 ± 0.646^{a}
30 ppm	5.313 ± 1.193^{ab}	5.313 ± 1.193^{ab}	35.250 ± 15.755^{ab}	15.375 ± 3.335^{bc}	3.250 ± 1.069^{a}	2.323 ± 0.032^{a}	6.460 ± 0.942^{a}
60 ppm	4.875 ± 1.217^{b}	4.875 ± 1.217^{b}	21.625 ± 13.016^{b}	12.438 ± 2.821^{c}	2.125 ± 1.157^{b}	**2.524 ± 0.192^{b}**	**8.007 ± 2.637^{a}**

Means followed by similar letters show no significant difference based on the Duncan's multiple range test at a significance level of 5%

Induction with 10 ppm of NAA can increase the leaf length, leaf width, and stem diameter of the plants. This is in line with the fact that NAA is among the auxin hormones serving to stimulate cell elongation and expansion, especially in vegetative cells [10]. NAA is capable of increasing the plasticity of cell walls. NAA can stimulate certain proteins in the plasma membrane of plant cells to pump H+ ions onto the cell walls. The H+ ions located on the cell walls lead to the activation of certain enzymes that are able to break some of the hydrogen cross-links of cellulose molecular chains that make up the cell walls. This leads to the plasticity of cell walls to increase and the cells absorb more water through osmosis, causing the plant cells to increasingly elongate [11]. Table 3 shows that an increase in NAA concentrations can lead to a decrease in the leaf number. This is because an increase in auxin (NAA) concentration can trigger ethylene production and stimulation. This excess ethylene may result in leaf falls and decrease in leaf number [12, 13].

Observation of the physiological character of the cultivar 'KP Brite' showed that NAA induction of 60 ppm had a significant effect. The cultivar 'KP Brite' treated with NAA of 60 ppm experienced a decrease in plant growth in terms of leaves, stems and height. An increase in auxin concentration did not necessarily lead to an increase in plant growth. An increase in auxin at certain concentrations could reduce plant growth. NAA at high concentrations can inhibit plant growth due to an excessive high NAA. At a high dose of auxin, growth inhibition can also be caused by excessive ethylene production. In addition to leaf falls, this excess ethylene may cause the plants to die through leaf damage and aging [14].

3.2. Detection of FaCHS and FaPYR1

In gene analysis and detection, strawberry fruit samples were collected to obtain DNA isolates. Samples were collected from the 4 stages of fruit development, including the Green (8 days), White (18 days), Pink (22 days), and Red (25 days) stages. The collected fruits were subsequently extracted using Nucleon Phytopure Plant DNA Extraction Kit. The purpose of the DNA extraction was to separate genomic DNA from such other components as protein, carbohydrates, phenols, etc. This kit is designed with the ability to prevent contamination of polysaccharides commonly contaminating the plant DNA. Those polysaccharides need to be removed to prevent further inhibition of enzymatic analysis of DNA [15]. Genomic DNA was quantitatively tested by spectrophotometry using a spectrophotometer to determine the concentration and purity of genomic DNA isolated from cultivars 'Crystal' and 'KP Brite' (Table 4).

Nucleic acids, both in the form of DNA and RNA, can absorb maximum UV lights at a wavelength (λ) of 260 nm. Contaminants in the form of proteins and phenols can absorb maximum lights at a wavelength (λ) of 280 nm, while contaminants in the form of polysaccharides will absorb maximum lights at a wavelength (λ) of 230 nm. Thus, the absorbance ratio of $A_{260/280}$ indicates the degree of genomic DNA purity relative to protein/phenolic contaminants, and the absorbance ratio of $A_{260/230}$ indicates the degree of genomic DNA purity relative to polysaccharide contaminants [16]. Genomic DNA purity can be considered good if the value of the $A_{260/280}$ ratio is 1.8 to 2.0 [17; 18]. The value of the $A_{260/280}$ ratio above 1.8 indicates the presence of contamination with RNA, while the value of the $A_{260/280}$ ratio of less than 1.8 indicates the presence of contamination with proteins [19].

Table 4 shows that the value of the $A_{260/280}$ ratio for cultivar 'Crystal' induced with NAA of 60 ppm at the green and white stage and that of cultivar 'KP Brite' induced with NAA of 30 ppm at the red stage were 1.8 to 2.0. According to Singh and Singh (2015) [18], the DNA purity can be considered quite good and can be used for PCR analysis. A large amount of DNA isolates showed values below 1.8, indicating contamination with proteins or phenols or polysaccharides [20].

The isolated genomic DNA was subsequently amplified by the polymerase chain reaction (PCR) method using primers specific for the target genes *FaCHS* and *FaPYR1* (Table 1). PCR amplification of DNA is a method to amplify a sequence of nucleotides exponentially by means of *in vitro* enzymatic processes [21].

FaCHS plays a role in fruit ripening [22]. Fruit ripening occurs simultaneously with discoloration due to the activation of anthocyanins that provide fruit colors. Meanwhile, *FaPYR1* serves as an ABA receptor coding gene. NAA has the role of inhibiting ABA, in which ABA is capable of initiating the synthesis of ethylene that plays a role in decay [23]. The present study used *28S-16S RNA housekeeping gene* as the control gene for *FaCHS* and *Actin* as the control gene for *FaPYR1*.

The annealing temperature in the present study for *FaCHS* and *28S-16S RNA housekeeping gene* was 54°C. The annealing temperature for *FaPYR1* and *Actin* was 60°C. Annealing temperature is an important part of the amplification process. An excessively high and low annealing temperature may cause the primers cannot attach to a specific site and, therefore, target gene amplification products cannot be obtained [24].

The PCR products were then electrophoresed using 2% agarose gel. Electrophoretic visualization showed that *FaCHS* was amplified at 118 bp, *FaPYR1* at 627 bp, and *28S-16S RNA housekeeping gene* at 146 bp, while *Actin* was amplified at 262 bp. Therefore, the present study used 100-bp markers to visualize PCR products.

Amplification of *FaCHS* DNA fragments showed that *FaCHS* was detected in both cultivars at each stage of development except at a concentration of 10 ppm in the green stage and strawberries induced with 30 ppm of NAA at the white stage. Amplification of *FaPYR1* DNA fragments in the

Table 4. Spectrometric data of DNA isolates.

Crystal					KP Brite				
Sample		ng/µl	260/230 Ratio	260/280 Ratio	Sample		ng/µl	260/230 Ratio	260/280 Ratio
Treatment	Fruit				Treatment	Fruit			
Control	Green	210.28	0.34	1.44	Control	Green	112.28	0.40	1.69
	White	144.65	0.25	1.37		White	216.19	0.42	1.77
	Pink	105.84	0.21	1.53		Pink	84.93	0.33	1.37
	Red	89.10	0.25	1.35		Red	64.62	0.23	1.5
10 ppm	Green	261.85	0.29	1.31	10 ppm	Green	786.65	0.38	1.24
	White	72.10	0.33	1.65		White	178.07	0.25	1.45
	Pink	238.34	0.27	1.38		Pink	81.55	0.26	1.69
	Red	198.35	0.23	1.20		Red	138.27	0.22	1.30
30 ppm	Green	156.43	0.29	1.67	30 ppm	Green	276.21	0.33	1.34
	White	18.16	1.64	1.30		White	168.02	0.24	1.32
	Pink	65.23	0.30	1.45		Pink	139.29	0.36	1.70
	Red	122.38	0.14	1.17		Red	113.48	0.28	1.85
60 ppm	Green	323.75	0.41	1.80	60 ppm	Green	1007.31	OVER	1.18
	White	280..41	0.40	1.82		White	214.82	0.34	1.59
	Pink	201.65	0.23	1.08		Pink	108.40	0.33	1.37
	Red	1.82	-0.15	1.77		Red	180.12	0.33	1.61

Table 5. PCR Amplification of *FaCHS* and *FaPYR1*.

Treatment		Crystal				KP Brite			
		FaCHS	*Houskeeping*	*FaPYR1*	*Actin*	*FaCHS*	*Houskeeping*	*FaPYR*	*Actin*
Control	Green	v*	v	v	v	v	v	-	v
	White	v	v	-	v	v	v	-	v
	Pink	v	v	-	v	v	v	-	v
	Red	v	v	-	v	v	v	-	v
10 ppm	Green	-	v	-	-	-	v	-	v
	White	v	v	-	-	v	v	v	v
	Pink	v	v	v	v	v	v	-	-
	Red	v	v	v	v	v	v	-	v
30 ppm	Green	v	v	-	v	v	v	v	v
	White	-	-	-	-	v	v	v	v
	Pink	v	v	-	v	v	v	v	v
	Red	v	v	-	v	v	v	v	v
60 ppm	Green	v	v	-	v	-	-	-	-
	White	v	v	-	v	-	-	-	-
	Pink	v	v	-	v	-	-	-	-
	Red	v	-	-	-	v	-	-	-

* v = amplified

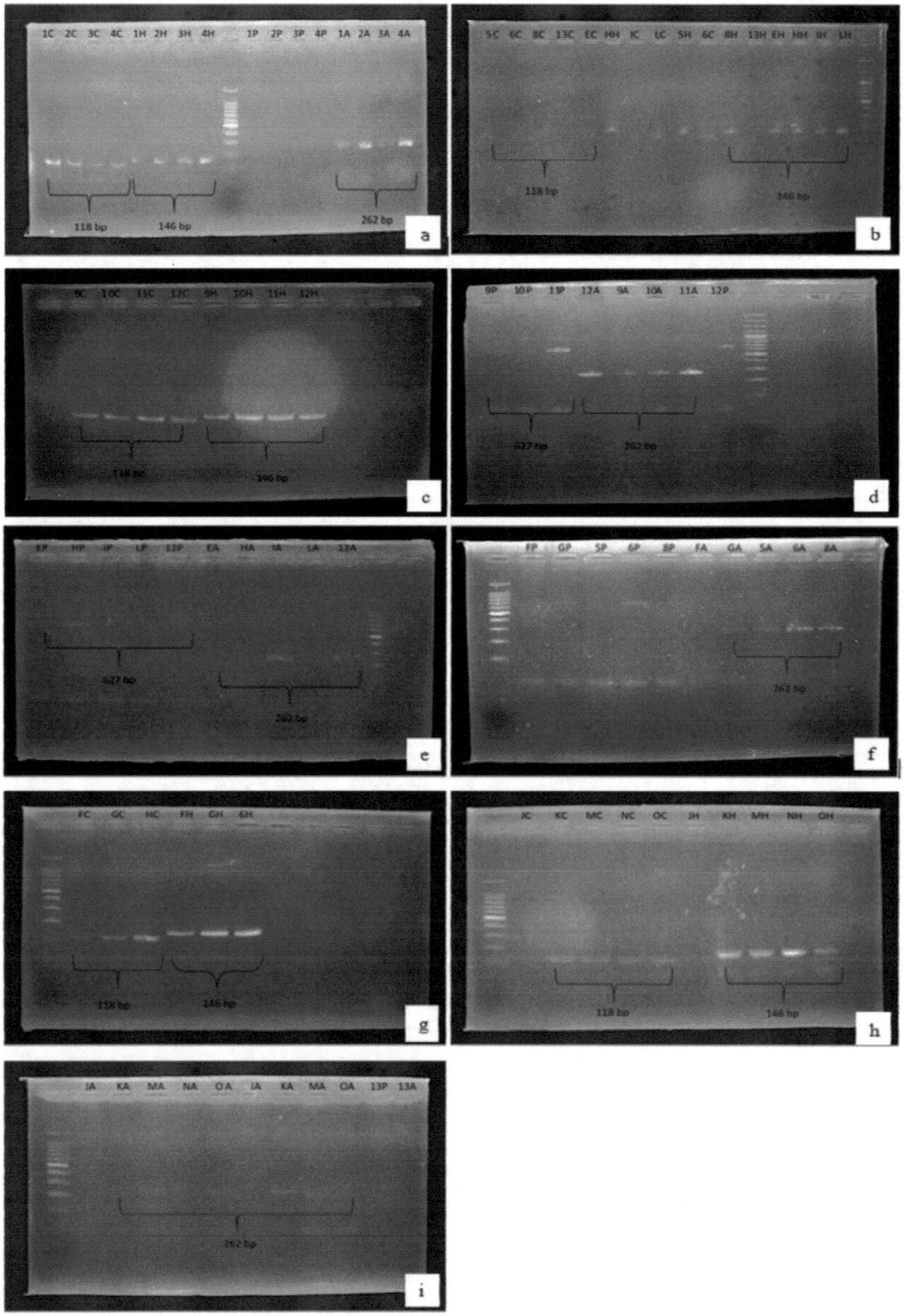

Figure 1. Detection results of *FaCHS*, *26S-18S housekeeping gene*, *FaPYR1*, and *Actin*. 100-bp Marker

Notes:
a. Control cultivar 'KP Brite' (1C-4C = target *FaCHS*; 1H-4H = *26S-18S housekeeping gene*; 1P-4P = target *FaPYR1*; 1A-4A = *Actin*) 1 = green, 2 = white, 3 = pink, 4 = red.
b. Cultivar 'KP Brite' (5C-8C = target *FaCHS*, 10 ppm; 13C = target *FaCHS*; 60 ppm; 5H-8H = *26S-18S housekeeping gene*, 10 ppm; 13H = *26S-18S housekeeping gene*, 60 ppm) 5 = green, 6 = white, 8 = red, 13 = green.
c. Cultivar 'Crystal' (EC = target *FaCHS*, 10 ppm; EH-HH = *26S-18S housekeeping gene*, 10 ppm; IC-LC = target *FaCHS*, 30 ppm; IH-LH = *26S-18S housekeeping gene*, 30 ppm) E, I = green, H, L = red.
d. Cultivar 'KP Brite' (9C-12C = target *FaCHS*; 9H-12H = *26S-18S housekeeping gene*) 9 = green, 10 = white, 11 = pink, 12 = red.
e. Cultivar 'Crystal' (EP-HP = target *FaPYR1*, 10 ppm; EA-HA = *Actin*, 10 ppm; IP-LP = target *FaYR1*, 30 ppm; IA-PA = *Actin*, 30 ppm) E, I = green, H, L = red
f. Cultivar 'KP Brite' (5P-8P = target FaPYR1, 10 ppm; 5A-8A = *Actin*, 10 ppm) 5 = green, 6 = white, 8 = red, 13 = green.
Cultivar 'Crystal' (FP-GP = target *FaPYR1*, 10 ppm; FA-GA = *Actin*, 10 ppm) F = white, G = pink.
g. Cultivar 'Crystal' (FC-HC = target *FaCHS*, 10 ppm; FH-GH = *26S-18S housekeeping gene*, 10 ppm) F = white, G = pink, H = red.
h. Cultivar 'Crystal' (JC-KC = target *FaCHS*, 30 ppm; JH-KH = *26S-18S housekeeping gene*, 30 ppm; MC-OC = target *FaCHS*, 60 ppm; MH-OH = *26S-18S housekeeping gene*, 60 ppm) M = green, J, N = white, O, K = pink.
i. Cultivar 'Crystal' (JA-KA = *Actin*, 30 ppm; MA-OA = *Actin*, 60 ppm) M = green, J, N = white, O, K = pink.
Cultivar 'KP Brite' (13P = target *FaPYR1*, 60 ppm; 13A = *Actin*, 60 ppm) 13 = green.

controls showed that *FaPYR1* could only be detected in the green and red stages. Treatment results showed that *FaPYR1* gene was only amplified in strawberries treated with 10 ppm of NAA (Table 5). This indicates that NAA is able to inhibit ABA by inhibiting *FaPYR1* formation in strawberries induced with 30 or 60 ppm of NAA. Inhibition of ABA will prevent the synthesis of ethylene which plays a role in strawberry fruit decay. NAA concentrations of 10, 30 and 60 ppm cannot inhibit the formation of *FaCHS* which plays an active role to provide colors via activation of anthocyanins. At the NAA concentration of 30 ppm at the white stage *FaCHS* could not be amplified since the DNA isolates obtained had a fairly low concentration of below 20 ng/μl.

FaCHS could not be amplified in cultivar 'KP Brite' induced with 30 ppm of NAA at the white stage. In addition, *FaCHS* was not amplified in the isolates of strawberry fruits treated with 60 ppm of NAA at the green, white, and pink stages.

FaPYR1 fragments of cultivar 'KP Brite' could only be detected in DNA isolates of strawberry fruits treated with 30 ppm of NAA and 10 ppm of NAA at white stage (Table 5). These results indicate that the 60-ppm NAA induction treatment can inhibit the formation of *FaCHS* in the accelerated decay of fruits. *FaCHS* could be detected in the strawberry fruits of cultivar "KP Brite" treated with 60 ppm of NAA at the red stage due to the formation of anthocyanins that provide strawberry fruits with the red color.

4. Conclusion

Treatment of NAA induction in Crystal cultivar of strawberry plants has no effect on physiological characters (such as leaf length and width, number of leaves, plant height, and stem diameter) of plants in the vegetative stage. The treatment of 10 ppm has increasing the fruit diameter and weight for the Crystal cultivar strawberry plants.The treatment of NAA 30 and 60 ppm induction affected the *FaPYR1* gene profile in Crystal cultivars. While the 60 ppm NAA induction treatment was able to influence the *FaCHS* gene profile in Kp-Brite cultivars.

5. Acknowledgement
The authors would like to thank the Applied Research Program No.: 2838/UN1.DITLIT/DIT-LIT/LT/2019 of 2019 and the Final Assignment Recognition program No.: 2129/UN1/DITLIT/DIT-LIT/LT/2019 of 2019 for the funding assistance provided at the time of data collection and publication. The authors also express their gratitude to the Faculty of Biology, Gadjah Mada University and the Banyuroto Agro-tourism, Desa Banyuroto, Kecamatan Sawangan, Magelang City, Central Java for discussion and planting and care of strawberry plants.

References
[1] Sutopo 2016 *Stroberi Tanpa Tanah Ramah Lingkungan, Balitbangtan, Kementrian Pertanian*, 2016 Pp. 1-43

[2] Forney C F, Kalt W and Jordan M A 2000 The Composition of Strawberry Aroma Is Influenced by Cultivar, Maturity, and Storage *Hortscience* **35** (6) 1022-1026

[3] Manning K 1994 Change in Gene Expression During Strawberry Fruit Ripening and Their Regulation by Auxin *Planta* **194** 62-68

[4] Given N K, Venis M A and Gierson D 1988 Hormonal regulation of ripening in the strawberry, a non-climacteric fruit *Planta* **174** 402–406

[5] Ziliotto F, Corso M, Rizzini F M, Rasori A, Botton A and Bonghi C 2012 Grape berry ripening delay induce by a preveraison NAA treatment is paralleled by a shift in the expression pattern of auxin- and ethylene- related gene *BMC Plant Biology* **12** (18) 1-15

[6] Munoz C, Sanchez-Sevilla J F, Botella M A, Homann T, Schwab W and Valpuesta V 2010 Polyphenol Compotition in the Ripe Fruits of Fragaria Species and Transcriptional Analyses of Key Gene in Pathway *Journal of Agricultureal and Food Chemist*

[7] Chai Y, Jia H, Li C, Dong Q and Shen Y 2011 FaPYR1 is Inovolved in Strawberry Friut Ripening *Journal of Experimental Botany* **62** (14)

[8] Satriowibowo E A, Nawawi M and Koesrihar 2013 Pengaruh waktu aplikasi dan konsentrasi NAA (Napthalene Acetic Acid) pada pertumbuhan dan hasil tanaman cabai besar (Capsicum annuum l.) varietas Jet Set *Jurnal Produksi Tanaman* **2** (4) 282-291

[9] Wilkins M B 1992 *Fisiologi Tanaman. Penerjemah Sutedjo M.M dan Kartasapoetra A.G.* (Jakarta: Bumi Aksara)

[10] Gardner F P, Pearce R B and Mitchell R L 1991 *Fisiologi Tanaman Budidaya. Alih Bahasa: Herawati Susilo* (Jakarta: Penerbit UI)

[11] Wattimena G A 1987 *Diktat Zat Pengatur Tumbuh Tanaman, Laboratorium Kultur Jaringan Tanaman PAU Bioteknologi IPB* Bogor

[12] Goldsworthy P R and Fisher N M 1992 *Fisiologi Tanaman Budidaya Tropik* (Yogyakarta: Universitas Gadjah Mada Press)

[13] Lakitan B 1996 *Fisiologi Pertumbuhan dan Perkembangan Tanaman* (Jakarta: PT Raja Grafindo Persada)

[14] Gergoff G, Chaves A and Bartoli C G 2010 Ethylene regulate ascorbic acid content during dark-induced leaf senescence *Plant Science* **178** 207-212

[15] Nucleon 2010 *Data File: Nucleic Acid Purification. Nucleon Phytopure. DNA from Plant Tissue, Amersham Biosciences, UK*

[16] Kaeppler-Hanno K 2010 *Troubleshooting Guide for the measurement of nucleic acids with the BioPhotometer plus, Eppendorf,* Hamburg

[17] Keer J T and Birch L 2008 *Essentials of Nucleic Acid Analysis: A Robust Approach* (Cambridge: RSC Publishing)

[18] Singh B D and Singh A K 2015 *Marker-Assisted Plant Breeding: Principles and Practices, Springer* New Delhi p 60

[19] Corbridge D E C 2013 *Phosphorus: Chemistry, Biochemistry and Technology* (Boca Raton: CRC Press) p 1339

[20] Teare J M, Islam R, Flanagan R, Gallagher S and Davies M G 1997 Measurement of Nucleic Acid Concentrations Using the DyNA Quant and the GeneQuant *BioTechniques* **22** 1170-1174

[21] Handoyo D and Rudiretna A 2001 Prinsip umum dan pelaksanaan polymerase chain reaction (PCR) *Unitas* **9** (1) 17-29

[22] Carvalho R F, Carvalho S D, Kevin O and Kevin M F 2016 *Agroinfiltration of Strawberry Fruit: a Powerful Transient Expression System for Gene Validation, Elsivier*

[23] McAtee P, Karim S, Schaffer R and David K 2013 A Dynamic Interplay between Phytohormones is Require for Fruit Development, Maturation and Ripening *Plant Science* **4** (79) 1-7

[24] Hikmah R U, Retnoningsih A and Habibah N A 2016 Keragaman durian berdasarkan fragmen Internal Transcribed Spacers (ITS) DNA ribosomal melalui analisis PCR-RFLP *Jurnal MIPA* **39** (1) 11-18

Hepatoprotective effect of Eugenol on Acetaminophen-Induced Hepatotoxicity in HepG2 cells

I N E Lister[1*], C N Ginting[1], E Girsang[1], A Amansyah[1], L Chiuman[1], N L W E Yanti[2], R Rizal[2] and W Widowati[3]

[1]Universitas Prima Indonesia, Jl. Belanga No. 1, Medan 20118, North Sumatera, Indonesia
[2]Biomolecular and Biomedical Research Center, Aretha Medika Utama, Jalan Babakan Jeruk II No. 9, Bandung 40163, West Java, Indonesia
[3]Faculty of Medicine, Maranatha Christian University, Jl Surya Sumantri No. 65, Bandung 40164, West Java, Indonesia

*Corresponding author's email: yysunpri@gmail.com

Abstract. Imbalance in liver metabolism lead to oxidative stress mainly caused by free radicals or termed as reactive oxidative oxygen (ROS). Prolonged ROS exposure without proper treatment induce severe liver damage and serious hepatic diseases including cirrhosis. Eugenol (4-allyl 2-methoxyphenol) is phenolic derivative compound that showed antioxidant, anti-inflammatory, analgesic, antibacterial, antifungal, and antitumor activities. This study aims to evaluate the hepatoprotective effect of eugenol through biochemical markers analysis. Cytotoxic assay was performed in various concentrations of eugenol (3,125; 6,25; 12,5; 25; 50; 100 μg/mL) using (3-(4,5-dimethylthiazol-2yl)-5-(3-carboxymethoxyphenkyl)-2-(4-sulfophenyl)-2H-tetrazolium) to determine the safe concentrations for next assays. Aspartate aminotransferase (AST), alanin aminotransferase (ALT), and lactate dehydrogenase (LDH) assay were performed using colorimetric method to evaluate the levels and activity of liver-related enzymes which are elevated in damaged liver as they were used as hepatotoxicity markers. The viability of HepG2 cells increased in eugenol concentration 3.125 μg/mL and then decreased along with the rise of eugenol concentrations. From this cytotoxic assay, two concentrations of eugenol were choosen (6.25 and 25 ug/ml) to be evaluated in the next assays. The level of LDH, ALT, and AST decreased after eugenol treatment compared to negative control. The most effective concentration of eugenol to seemed different in certain hepatotoxicity markers. This study suggests that eugenol was safe to use for cells culture environment in large ranges of concentrations and shows hepatoprotective effect in APAP-induced hepatotoxicity model by the decrease of LDH level and AST and ALT activities.
Keywords: Hepatoprotective, Hepatotoxicity, Eugenol, HepG2, LDH, ALT, AST

1. Introduction

Liver is one of crucial organ in human body. It plays important role in maintaining metabolism, especially in toxic substances elimination and biotransformation. Imbalance in liver metabolism lead to oxidative stress which mainly caused by free radicals or termed as reactive oxidative oxygen (ROS). ROS generation within hepatocytes induces oxidative damage, cell metabolism, cell necrosis, moreover cell death [1]. Prolonged ROS exposure without proper treatment can cause severe liver damage and serious hepatic diseases including cirrhosis [2]. According to World Health Organization (WHO), around 800 thousand of 2.4 million deaths yearly that related to liver disease are caused by cirrhosis [3]. The incidence of hepatic diseases has not decreased although there were many modern medications available nowadays [4].

Acetaminophen (APAP), or widely known as paracetamol, is one of most common drugs that is prescribed for mild pain treatment. It has analgesic and antipyretic properties which is safe and effective at specific doses. APAP overdose produces the toxic effects through mitochondrial damage and increased oxidative stress [5][6]. Several studies have revealed that the mechanism of APAP hepatotoxicity is initiated by the formation of reactive and toxic metabolite called Nacetyl-P-benzoquinoneimine (NAPQI) that is mainly occurred in hepatic cytochrome P-450. Normal amount of NAPQI could be detoxified by conjugating with glutathione (GSH) – natural antioxidant – to form mercapturic acid and excreted in bile [7][8]. The excess generation of NAPQI is occurred when the availability of GSH is insufficient. They increased the reaction of NAPQI with sulfhydryl groups of proteins to form protein adducts. Next, protein adducts formation is the critical process in hepatotoxicity because it induces mitochondrial oxidative especially in superoxide production [8][9]. Other studies revealed that the initial event of APAP-induced liver injury is the changes of macrophages microenvironment and secondary innate immune response activation by upregulating proinflammatory cytokines production [10][11].

Eugenol (4-allyl 2-methoxyphenol) is phenolic derivative compound that showed antioxidant, anti-inflammatory, analgesic, antibacterial, antifungal, and antitumor activities [12]. It is derived from *Eugenia caryophyllata*, especially from the leaves and buds, and is formed clove essential oil. Eugenol is considered safe to use in the food industry, cosmetics industry and in dentistry by the US Food and Drug Administration (FDA). Eugenol is known to have an inhibitory effect of proinflammatory cytokines production and potential to decrease the activity of dehydrogenase enzymes at dose dependent manner [13]. Other studies showed that eugenol could elevate the level of GSH as the important protein to maintain during the occurrence of stress oxidative in hepatotoxicity process [12]. Therefore, this study aims to evaluate the hepatoprotective effect of eugenol through analysis of specific enzymes production which elevated during hepatotoxicity induced by APAP.

2. Materials and Methods

2.1. Cytotoxic Assay

Human hepatocellular carcinoma (HepG2) cells line (ATCC, HB-8065™) was obtained from Biomolecular and Biomedical Research Center, Aretha Medika Utama, Bandung, Indonesia. The cells were thawed and grown in Dulbecco's Modified Eagle Medium (DMEM) (Biowest, L0102-500) supplemented with 10% (v/v) fetal bovine serum (FBS) (Biowest, S1810), addition of 1% (v/v) antibiotic-antimycotic (Gibco, 15240062) and 1% (v/v) nanomycopulitine (Biowest, LX16). The cells were incubated in 37°C and 5% CO_2. The cells were maintained by changing its medium once per three days. Cytotoxic assay was performed in various concentrations of eugenol (Chengdu BioPurify, BP0569) (3,125; 6,25; 12,5; 25; 50; 100 μg/mL) using MTS (3-(4,5-dimethylthiazol-2yl)-5-(3-carboxymethoxyphenyl)-2-(4-sulfophenyl)-2H-tetrazolium) (Abcam, ab197010) to determine the safe concentrations for next assays. The cells were harvested and 5 x 10^4 of cells were seeded in 96-wells plate. The cells were treated with various concentration of eugenol and incubated for 24 hours in 37°C and 5% CO_2. Next, 20 μl of MTS solution was added into each well and the culture was incubated for 3 h in 37°C and 5% CO_2. This assay was conducted with spectrophotometric measurement and the absorbance was read at 490 nm ((Thermo MultiSkan GO Microplate Reader) [14][15][16].

2.2. LDH, ALT, and AST Activity

Lactate dehydrogenase (LDH), Aspartate Aminotransferase (AST), and Alanin Aminotransferase (ALT) assay were performed using colorimetric method to evaluate the levels and activity of metabolic enzymes which are elevated in damaged tissues especially in liver tissue. HepG2 cells were plated in 6-wells plate at cells density of 1 x 10^6 cells/well and incubated in 37°C and 5% CO_2 to attain confluency.

After 24 hours, cells were treated with 1,8 mL 2% FBS supplemented DMEM containing 40 mM acetaminophen (Sigma Aldrich, A7085), 1% (v/v) DMSO (Merck, 1029521000), and medium only for 24 hours in 37°C and 5% CO_2. The cells were added with eugenol solution at concentrations 6.25 μg/mL and 25 μg/mL for 24 h. LDH levels, ALT and AST activity were determined using LDH assay kit (Elabscience, E-BC-K045), ALT assay kit (Elabscience, E-BC-K235), and AST assay kit (Elabscience, E-BC-K236). All assays were conducted as described in manufacturer protocol [2][7].

3. Results and Discussion

3.1. Cytotoxicity of Eugenol on HepG2 Cells

In this study, APAP-induced HepG2 cells line is used as hepatotoxic model. It is widely known that APAP can induce liver injury through oxidative stress mechanism which is initiated by toxic metabolite, NAPQI [8]. APAP induced hepatotoxicity in overdose condition and there was a variability result of induction. Therefore, this study used acetaminophen at concentration 40 mM because it resulted in best inducer for cells injury as reported by Gonzalez et al. (2017) [7].

The cytotoxic assay was measured based on the MTS assay and colorimetric principle for assesing cell metabolic activity. In the presence of the viable cells, yellow-colored MTS will be reduced to purple-colored formazan product by NAD(P)H-dependent dehydrogenase enzymes that is produced by metabolically active cells [15][17]. From the Figure 1, the viability of HepG2 cells increased significantly around 120% in eugenol at concentration 3.125 μg/mL compared to the normal cells (without eugenol treatment). But, it decreased along with the rise of eugenol concentrations, and result in lowest cell viability after treating with eugenol at concentration 100 μg/mL among other groups of concentrations.

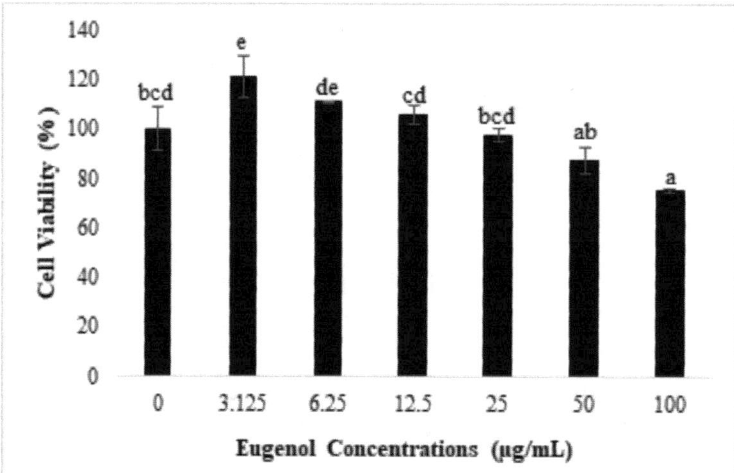

Figure 1. Cell viability of HepG2 after treatment by various concentrations of eugenol *The data was presented as a histogram of mean±standard deviation. Different letters (a, ab, bcd, cd,de e) symbolized a significant different among various treatments of eugenol concentrations, based on One-Way Anova analysis and it was followed by Tukey HSD post hoc test ($p<0.05$).

Based on the result of viability assay (Figure 1), eugenol is considered safe in broad range of concentrations for HepG2 cells line. FDA also approved the use of eugenol in several types of industry [13]. It has been reported that eugenol has toxic effect at dose dependent level and less cytotoxic effect compared to its derivative, isoeugenol [18]. from this assay, two concentrations of eugenol were chosen (6.25 and 25 μg/mL) to be evaluated in the next assays. Those two concentrations were chosen because they had similar percentage of viability with the normal one.

3.2. LDH, AST, and ALT activity

The toxic effect of APAP are shown in Figure 2 based on the level of LDH, as it is known as marker for the damage tissues. The induction of APAP significantly increased LDH level compared to normal cells and vehicle control. The addition of both concentrations of eugenol reduced significant level of LDH. It is evident from the result that eugenol at concentration 6,25 µg/mL reduced LDH level. Furthermore, at concentration 25 µg/mL had better to approaching the normal level of LDH in control groups.

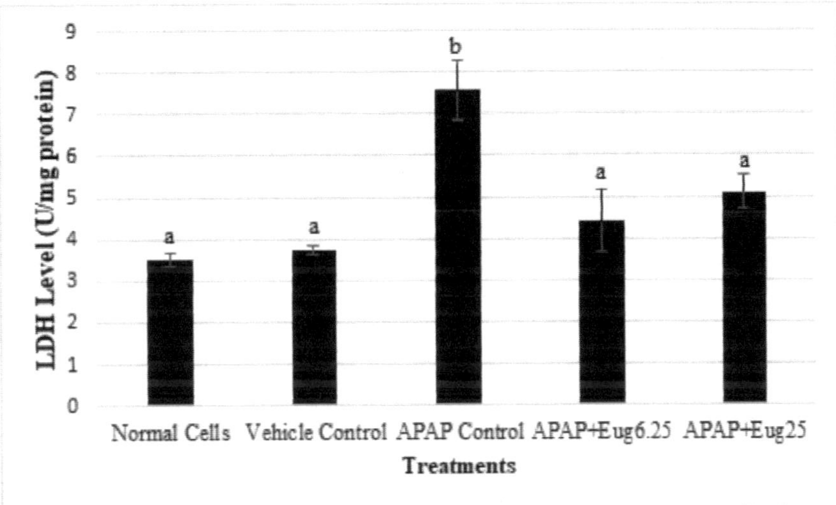

Figure 2. The effect of eugenol concentrations on APAP-induced HepG2 cells towards LDH production. The data was presented as a histogram of mean±standard deviation. Different group of letters (a, b) symbolized a significant different among various group treatments, based on One-Way Anova analysis and it was followed by Tukey HSD post hoc test ($P<0.05$). Normal cells: cells without any treatments; Vehicle control: cells with DMSO 1% treatment; APAP control: cells with APAP treatment; APAP+Eug6.25: cells with APAP + Eugenol 6.25 µg/mL treatment; APAP+Eug25: cells with APAP + Eugenol 25 µg/mL treatment.

LDH is a non-organ specific enzyme distributed throughout the body and elevated in different types of damage tissues including skeletal, cardiac muscle, kidney, and liver. LDH levels indicates the increased level of cells necrosis and was regarded as biomarker of cell cytotoxicity [19]. In this study, the levels of LDH significantly decreased in the addition of 6.25 µg/mL eugenol. The antioxidant properties of eugenol such as elevating GSH level may cause the reduction of enzymes leakage production by the damage cells [2].

ALT is a specific metabolic enzyme that predominantly found in liver organ specifically located in the cytoplasm. As it is found mainly in liver, ALT was considered as valuable biomarker of liver damage. AST is another enzyme that is used to monitor various liver diseases including alcoholic hepatitis and cirrhosis in combination with ALT. However, AST is not only found in liver tissue which is similar to LDH, but also in several damage tissues such as heart, skeletal muscles, kidney, pancreas, lungs and brain tissues [19]. The ratio of AST and ALT can be an indication of liver-related disease, higher level of ALT than AST indicates acute hepatitis caused by virus. On the other hand, alcoholic hepatitis or cirrhosis caused by stress damage result in higher AST:ALT ratio. The result of this study revealed that AST produced higher range of activities in U/mg protein than that of ALT [20][19].

Journal of Physics: Conference Series **1374** (2019) 012009 doi:10.1088/1742-6596/1374/1/012009

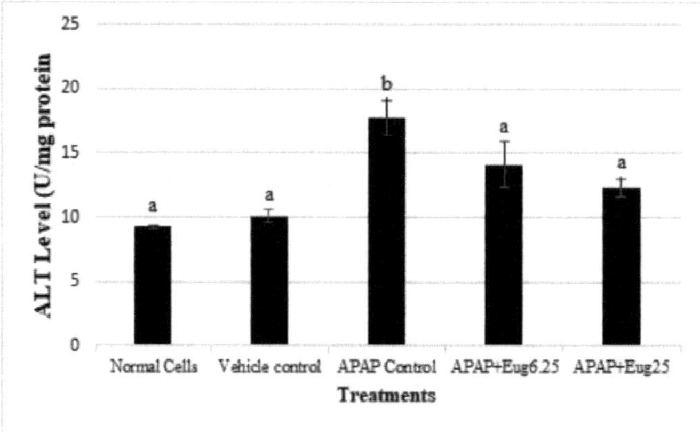

Figure 3. The effect eugenol concentrations on APAP-induced HepG2cells toward ALT activity. *The data was presented as a histogram of mean±standard deviation. Different group of letters (a, b) symbolized a significant different among various group treatments, based on One-Way Anova analysis and it was followed by Tukey HSD post hoc test (p<0.05). Normal cells: cells without any treatments; Vehicle control: cells with DMSO 1% treatment; APAP control: cells with APAP treatment; APAP+Eug6.25: cells with APAP + Eugenol 6.25 µg/mL treatment; APAP+Eug25: cells with APAP + Eugenol 25 µg/mL treatment.

The ALT and AST activities can be seen at Figure 3 and 4 after APAP induction for 24 hours and it was followed by the treatment with two different concentrations of eugenol. Eugenol at concentration 25 µg/mL could reduce the activity of ALT compared to the negative control (APAP control), and it was significant different with the ALT activity that slightly reduce in a group treatment of eugenol at concentration 6,25 µg/mL (Figure 3).

It suggested that eugenol at concentration 25 µg/mL had better capacity of reducing ALT activity. The AST activity show the same result, eugenol at concentration 25 µg/mL showed drop of AST activity among the other group treatments and APAP control (Figure 4). This finding clearly revealed that eugenol at concentration 25 µg/mL had better capacity to lower the AST activity.

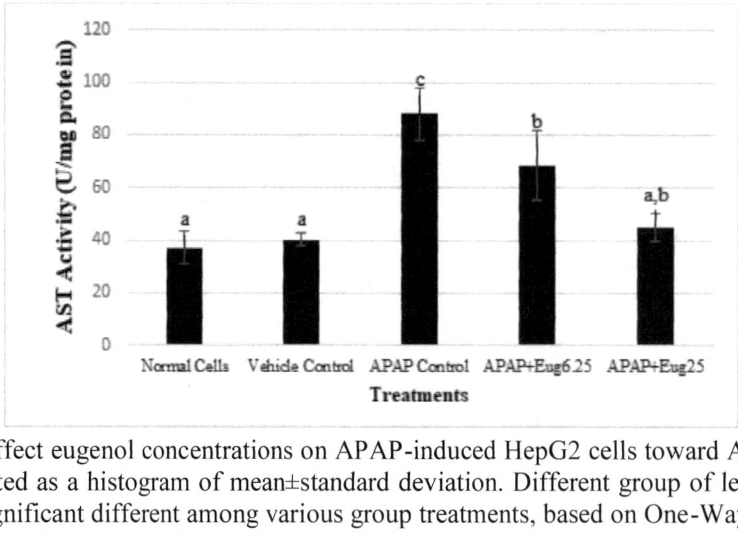

Figure 4. The effect eugenol concentrations on APAP-induced HepG2 cells toward AST activity. The data was presented as a histogram of mean±standard deviation. Different group of letters (a, ab, b, c) symbolized a significant different among various group treatments, based on One-Way Anova analysis and it was followed by Tukey HSD post hoc test (P<0.05). Normal cells: cells without any treatments; Vehicle control: cells with DMSO 1% treatment; APAP control: cells with APAP treatment; APAP+Eug6.25: cells with APAP + Eugenol 6.25 µg/mL treatment; APAP+Eug25: cells with APAP + Eugenol 25 µg/mL treatment.

The addition of eugenol in APAP-induced HepG2 showed reductions in ALT and AST activities (Figure 3 and 4). This finding has similar result with the study demonstrated by Binu et al. (2018) that eugenol exhibited an hepatoprotective effects through the rise of specific antioxidant markers including GSH, catalase (CAT), and glutathione peroxidase (GPX) as well as reduction in AST and ALT activities, decreased NO and Thiobarbituric acid reactive substances (TBARS) in arsenic trioxide-induced hepatotoxicity [12]. The elevation of AST and ALT activities may cause by the leakage of plasma membrane as the consequence of defect transport function of hepatocytes [21]. The decrease of these enzyme levels in APAP-induced hepatotoxicity model after eugenol treatment has clearly proved its hepatoprotective effects at dose dependent. This study suggested that eugenol at concentration 25 µg/mL is way more effective in reducing the activity of ALT or AST. However, further studies of eugenol antioxidant and anti-inflammatory capacity will be necessary to support this finding.

4. Conclusion

This study suggests that eugenol was safe to use for cells culture environment in broad ranges of concentrations. The best hepatoprotective effects exhibited by eugenol at concentration 25 µg/mL by the decrease of LDH level and AST and ALT activities in APAP-induced hepatotoxicity model.

5. Acknowledgement

We are gratefully acknowledging the financial support from the Research Center and Service Community, Universitas Prima Indonesia, Medan, North Sumatera, Indonesia for research grant 2018. This study also was funded, facilitated and supported by Biomolecular and Biomedical Research Center, Aretha Medika Utama, Bandung, West Java, Indonesia. We are thankful to Wahyu Setia Widodo, Satrio Haryo Benowo Wibowo, Rr Anisa Siwianti, Hanna Sari Widya Kusuma, Annisa Amalia, Ika Adhani Sholihah, Dewani Tediana Yusepany, Dwi Surya Artie from Biomolecular and Biomedical Research Center, Aretha Medika Utama, Bandung, West Java, Indonesia for their valuable assistance.

References

[1] Chandrasekaran C, Dethe S, Mundkinajeddu D, Pandre M, Balachandran J, Agarwal A *et al.* 2012 Hepatoprotective and antioxidant activity of standardized herbal extracts *Pharmacogn. Mag.* **8** (30) 116-23

[2] Pareek A, Godavarthi A and Nagori B P 2013 In vitro hepatoprotective activity of Corchorus depressus L. against CCl4 induced toxicity in HepG2 cell line *Pharmacogn. J.* **5** (4) 191–5

[3] Delgado-Montemayor C, Cordero-Pérez P, Salazar-Aranda R and Waksman-Minsky N 2015 Models of hepatoprotective activity assessment *Med. Univ.* **17** (69) 222–8

[4] Abdallah H M, Ezzat S M, El Dine R S, Abdel-Sattar E and Abdel-Naim A B 2013 Protective effect of Echinops galalensis against CCl4-induced injury on the human hepatoma cell line (Huh7) *Phytochem. Lett.* **6** (1) 73–8

[5] Choi Y-H, Lee H S, Chung C-K, Kim E J and Kang I-J 2017 Protective effects of an ethanol extract of Angelica keiskei against acetaminophen-induced hepatotoxicity in HepG2 and HepaRG cells *Nutr. Res. Pract.* **11** (2) 97-104

[6] Yan M, Huo Y, Yin S and Hu H 2018 Mechanisms of acetaminophen-induced liver injury and its implications for therapeutic interventions *Redox Biol.* **17** 274–83

[7] González L T, Minsky N W, Espinosa L E M, Aranda R S, Meseguer J P and Pérez P C 2017 In vitro assessment of hepatoprotective agents against damage induced by acetaminophen and CCl4 *BMC. Complement. Altern. Med.* **17** (39) 1-10

[8] Ramachandran A and Jaeschke H 2017 Mechanisms of acetaminophen hepatotoxicity and their translation to the human pathophysiology *J. Clin. Transl. Res.* **3** 157-69

[9] McGill M R, Lebofsky M, Norris H-R K, Slawson M H, Bajt M L, Xie Y *et al* 2013 Plasma and liver acetaminophen-protein adduct levels in mice after acetaminophen treatment: Dose–response, mechanisms, and clinical implications *Toxicol. Appl. Pharmacol.* **269** (3) 240–9

[10] Raza H and John A 2015 Differential Cytotoxicity of Acetaminophen in Mouse Macrophage J774.2 and Human Hepatoma HepG2 Cells: Protection by Diallyl Sulfide. Ahmad A, editor *PLoS One* **10** (12) e0145965
[11] Imaeda A B, Watanabe A, Sohail M A, Mahmood S, Mohamadnejad M, Sutterwala F S *et al* 2009 Acetaminophen-induced hepatotoxicity in mice is dependent on Tlr9 and the Nalp3 inflammasome *J. Clin. Invest.* **119** (2) 305-14
[12] Binu P, Priya N, Abhilash S, Vineetha R C and Nair H 2018 Protective effects of eugenol against hepatotoxicity induced by arsenic trioxide: an antileukemic drug *Iran J. Med. Sci.* **43** (3) 305
[13] Nejad S M, Özgüneş H and Başaran N 2017 Pharmacological and toxicological properties of eugenol *Turkish J. Pharm Sci.* **14** (2) 201–6
[14] Endrini S, Suherman and Widowati W 2014 Annona muricata leaves have strongest cytotoxic activity against breast cancer cells *Universa Med.* **33** (3) 179–84
[15] Widowati W, Darsono L, Suherman J, Fauziah N, Maesaroh M and Erawijantari P P 2016 Anti-inflammatory effect of mangosteen (*Garcinia mangostana* l.) peel extract and its compounds in lps-induced raw264.7 cells 2016 *Nat. Prod. Sci.* **22** (3) 147–53
[16] Novilla A, Djamhuri D S, Nurhayati B, Rihibiha D D, Afifah E and Widowati W 2017 Anti-inflammatory properties of oolong tea (*Camellia sinensis)* ethanol extract and epigallocatechin gallate in LPS-induced RAW 264.7 cells *Asian Pac. J. Trop. Biomed.* **7** (11) 1005–9
[17] Widowati W, Wijaya L, Laksmitawati D R, Widyanto R M, Erawijantari P P, Fauziah N, Bachtiar I, Sandra F Tea flavonoids protect endothelial progenitor cells (EPCs) from oxidative damage by reducing reactive oxygen species (ROS) level *Nat. Prod. Sci.* **22** (2) 87-92
[18] Atsumi T, Fujisawa S and Tonosaki K 2005 A comparative study of the antioxidant/prooxidant activities of eugenol and isoeugenol with various concentrations and oxidation conditions *Toxicol. Vitr.* **19** (8) 1025–33
[19] Arika W M A, Nyamai D W, Osano K O, Ngugi M P and Njagi E N M 2016 Biochemical markers of in vivo hepatotoxicity *J. Clin .Toxicol.* **6** (2) 1-8
[20] Shimizu Y 2008 Liver in systemic disease *World J. Gastroenterol.* **14** (26) 4111
[21] Mohammadyari A, Razavipour S, Mohammadbeigi M, Negahdary M and Ajdary M 2014 Exploring vivo toxicity assessment of copper oxide nanoparticle in Wistar rats *J. Biol. Today's World* **3** (6) 124-128

Journal of Physics: Conference Series 1374 (2019) 012010 doi:10.1088/1742-6596/1374/1/012010

Martapura river water leads to testes alteration in rats

I Yuliana[1*], H Khatimah[1], L Rosida[1], N S Skripsiana[1] and E Suhartono[2]

[1]Biomedical Department,Faculty of Medicine, Lambung Mangkurat University, South Kalimantan, Indonesia
[2]Biochemistry and Biomolecular Department, Faculty of Medicine, Lambung Mangkurat University, South Kalimantan, Indonesia

*Corresponding author's email: dr.idayuliana@gmail.com

Abstract.The Martapura River is the main river in the Banjar district. Local residents use the water for life, even though the river has been polluted. This study aims to analyze changes in testicular morphometry and histology due to drinking water from the Martapura River.A total of 32 male Sprague Dawley rats were randomly divided into two groups ($n = 16$ per group). The control group was given aquades as drinking water. The treatment group is a group that has been drinking from the water of the Martapura River for 30 days. After euthanasia, the testes were processed and stained with hematoxylin-eosin. Testicular size, number of seminiferous tubules, spermatogonia cells, primary spermatocyte cells, and Leydig cells were analyzed microscopically.Martapura river water significantly induces testicular enlargement ($p = 0.0036$). In addition, there was a decrease in the number of seminiferous tubules ($p < 0.001$), spermatogonia cells ($p = 0.002$), primary spermatocytes cells ($p < 0.001$), and Leydig cells ($p < 0.001$) significantly compared to the control group.We conclude that drinking the water from the Martapura River triggers changes in morphometry and histological structure of the testis so that it can disrupt male reproductive function.

1. Introduction

Martapura and Banjarmasin are areas in South Kalimantan that is passed by the Martapura River. This river has been polluted by various pollutants such as heavy metals Pb, Cd, Hg and Fe[1], [2]. The heavy metal has been known to have a bad influence on health, for example Pb affects tissue structure of the testes [3-7]. Besides Pb, mercury also causes the reduction of purkinje cerebellum diameter [8]. Cadmium causes interference in various organs including brain[9],[10],kidney[6],[11], liver[12],[13], ovary[11],[14-16] bone and muscle [17].

The mechanism of organ damage due to heavy metals dissolved in the Martapura River water is still unclear. However, some previous studies have stated that pathomechanism of organ damage to cellular level through changes in cell signal cascade pathways, cell to cell adhesion, damage to protein structure, and oxidative stress [10],[18]. Cellular damage by heavy metals begins with mild injury stage occurs due to activation of the autofagi process and non-proliferative repair. If exposure to heavy metal concentrations increases, there will be a mechanism for releasing cell-to-cell bonds, and disruption of cell signaling pathways that have an impact on cell necrosis and apoptosis [19],[2].

Testes are reproductive organs in men who are susceptible to damage due to heavy metals. The testes function to produce male germ cells in the seminiferous tubules and the hormone testosterone by Leydig cells [21],[22]. These organs can experience various disorders that can cause a decrease in function. There are many factors that influence decreased testicular function caused by cell damage. One of the factors is testicular poisoning by heavy metals. Based on research reports it is known that heavy metals can cause testicular germ cell damage in the form of apoptosis and cell necrosis in rodents, banded knife fish, Pseudosciaenacrocea (large yellow croaker), and Oreochromisniloticus (Nile tilapia) [22-24].

Research on the impact of heavy metals on the testes has been widely carried out, but the impact of the Martapura river water on the testes has not been widely studied (because there is no primary and secondary data on the incidence of male infertility in the Martapura region). Therefore, this research needs to be conducted as primary research.

2. Subject and methods

Total number of study subjects were 32 rats (Rattus novergicus), males, aged 2-6 months, and weight 250 - 300 g; obtained from BVET Banjarbaru. The study group was divided into 2 groups (16 groups each) namely the control group (given aquades) and treatment (given the Martapura river) for 30 days. Before the treatment period, the rats were acclimatized for 7 days. Rats are maintained according to maintenance standards in standard laboratories. Given a special cage made of plastic boxes with a size of 60 cm x 50 cm x 22 cm given a cage under the form of dry husk and given a lid from the wire each cage is only filled with 4 tails. For cage beds are replaced once a week continuously. Feeded specifically for animals, given an ad libitum drink. Evaluated physical and psychological conditions of rats every day.

Rats were given Martapura river water as a substitute for drinking water ad libitum. Rats were sacrificed on the 31st day and continued the process of making testicular histopathology with HE staining.

Rats were sacrificed by anesthesia using pentobarbital sodium drug. After the rat was declared dead, an autopsy of the abdomen was performed to isolate the testes organ. The size of testes was measured using a ruler (units of cm). Testes organ pieces were immediately fixed by soaking the organ in fixative solution (formol saline) for 24 hours or more.

Starting with the deparafinization process by means of preparations that are ready to be soaked first in a solution of xylol I and xylol II for 2 minutes. Continued the hydration process in an alcohol solution with a decreasing concentration of concentration, namely from 100% -95% -90% -80% -70% for 2 minutes each. Then the preparation is incubated in Hematoxylin Harris solution for 15 minutes, then rinsed in running water for 5-7 minutes. Preparations are dipped in ammonium or 0.4% carbonate lithium quickly as much as 3-10 x, then checking color differentiation under a light microscope. The preparation is rinsed again in running water for 1 minute. The preparation is dipped in lithium carbonate solution so that the preparation is bright blue. Preparations are rinsed in running water for 10-20 minutes. The preparation is incubated in 1% eosin for 2 minutes. The process of dehydration is continued by dipping the preparation into alcohol with a concentration that increases slowly, each for 2 minutes. The preparations were soaked in xylol I and xylol II solution for 2 minutes each. Continued sealing and covered with a cover glass. The preparation is ready to be observed under a light microscope.

The number of seminiferous tubules, Leydig cell, spermatogonium cells , and primary spermatocytes cells was analyzed microscopically using the binocular camera^{Nikon} and software$^{Immage\ Raster}$, with a 10 μm and 100 μm scale and magnification of 10 x and 40 x.

The results of the research data were analyzed by independent t-test to evaluate alteration in size of testes, the number of seminiferous tubules, Leydig cell, spermatogonium cells , and primary spermatocytes cells in the control and treatment groups. In this study using a 95% confidence interval and significant data if p value <0.05.

3. Result and discussion

Morphological alterations from both control and treatment groups can be seen in and Table 1 and Figure 1.

Table 1. Morphological alterations from both control and treatment groups.

Alteration on testes	Control	Treatment	P
Testicular size	2.49±0.58	2.893±0.62	0.03
Number of seminiferous tubules	8.44±1,22	6.19±1.03	0.00
Number of Leydig Cells	11.3±1.92	8.12±1.03	0.00
Number of spermatogonium cells	28.05±1.05	18.30± 1.78	0.00
Number of primary spermatocytes cells	42.08±1.00	32.05± 0.58	0.00

The testes function to produce male germ cells in the seminiferous tubules and testosterone by Leydig cells [21],[22]. These organs can experience various disorders that can lead to a decrease in function because it is an organ that is susceptible to damage due to heavy metals contaminating river water.Testicular damage in the study results was marked by enlargement of testicular size (Table 1). These conditions indicate the process of cell damage that begins with cell degeneration conditions include hydropic degeneration, fatty degeneration, swollen degeneration and vacuole degeneration[26],[27]. This is consistent with the study of Vergilio CS *et al* (2015)which shows that heavy metal exposure can cause changes in size testis [22].The phenomenon of enlargement of the macroscopic size of the testes in the treatment group compared to the testis size of the treatment group due to the odem process in the interstitial tissue in the testicular organs. This is confirmed by the existence of microscopic evidence of hydrophic degeneration, swollen degeneration and distance between seminiferous tubules which causes enlargement of testicular size in the treatment group (Figure 1).

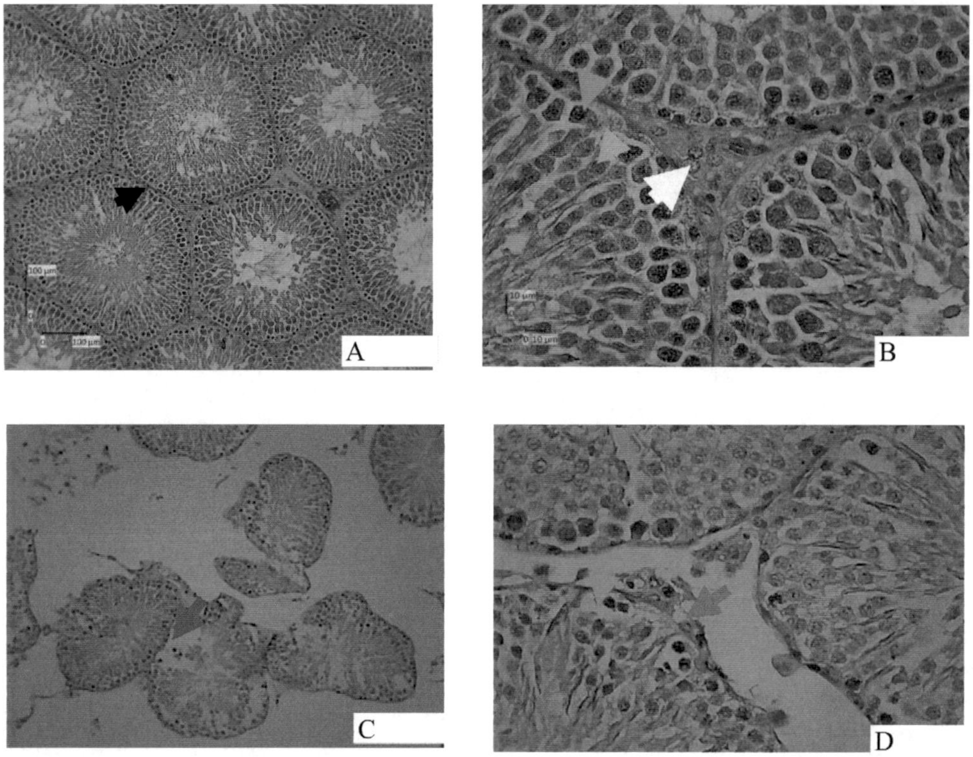

Figure 1. Microscopic analysis of testes control group (A, B) and treatment (C, D); black arrow (seminiferous tubule); yellow arrow (Leydig cell); brown arrow (spermatosit primer); green arrow (spermatogonium); blue arrow (odem); green arrow (hypertopy cell); red arrow head (vacuole cell); (HE stain, binocular camera with 10 μm and 100 μm with magnification of 10 x and 40 x).

Martapura River water administration causes cell and tissue damage in the form of a significant decrease in the number of seminiferous tubules, Leydig cells, spermatogonia cells and primary spermatocytes cells in the treatment groups (Table 1 and Figure 1). In Figure 1 found fatty degeneration in the form of vacuole cells and hypertrophy cells as a marker of hypertrophic degeneration. The decrease in the number of seminiferous tubules and Leydig cells indicates the occurrence of germ cell death and Leydig cells due to the process of necrosis and apoptosis. This is in accordance with the research of Teiichiro Aoyagi *et al.* (2002); N Shiraishi*et al* (2001), FW Santos *et al* (2004), Wu X *et al* (2017) which statement that heavy metals cause a decrease in the number of germ cells in the seminiferous tubules and interstitial tissue proliferation [20],[28-30]. Wu X *et al* research (2017) shows that heavy metals can cause a decrease in the number of Leydig cells [30].

There are several pathomechanism theories mentioning cell damage in organs due to heavy metals. Exposure to heavy metals with low concentrations causes the process of oxidative stress, changes in the cascade pathway of cell signals, changes in cell adhesion to cells and damage to protein structure. The mechanism of celldamage through the pathway of oxidative stress because heavy metals such as iron, mercury, copper, nickel, cadmium and arsenic can be explained because of the ability of these molecules to form Reactive Oxygen Species (ROS). Reactive Oxygen Species causes cell damage through a mechanism of decreasing lipid bilayer and changes in DNA[31]. The accumulation of ROS in tissues characterized by the formation of active species such as hydroxyl radicals ($^{\cdot}OH$), hydrogen peroxidase (H_2O_2) and single oxygen (o=o) will cause disruption of balance between oxidants - antioxidants.

The active species will act with biomolecules such as proteins, nucleic acids, and lipids which cause lipid peroxidation, protein denaturation, and DNA mutations which result in changes in the normal activity of cells, tissues, organs and even the physiological system of the body [10],[18],[32]. At the beginning exposure to heavy metals in cells will cause mild injury stage injury that occurs due to activation of the autofagi process and non-proliferative repair. If exposure to heavy metal concentrations increases, there will be a mechanism for releasing cell-to-cell bonds, and disruption of cell signaling pathways that have an impact on cell necrosis and apoptosis [19],[20].

4. Conclusion

Conclusion from research is administration of Martapura river water can cause alteration morphometry and histological structure of the testis in rats. Major conclusion as follows :

1. Significantly increased testicular size in the treatment groups (p = 0.03)
2. Significantly decreased number of seminiferous tubules, Leydig cells, spermatogonia cells and primary spermatocyte cells in the treatment groups (p = 0.00)
3. Microscopic features in the testis treatment groups obtained hydrophic degeneration, swollen degeneration and distance between seminiferous tubules, vacuole cells and hypertrophy cells as a marker of hypertrophic degeneration.

References

[1] Festiani S 2015 Martapura river high pollution. Republika.Co.Id. 19 April 2015
[2] Fokus Batulicin 2011 River water quality in South Kalimantan decreases. 11 April 2011. Available www.Fokusbatulicin.Com
[3] Triadayani A L, RirisAryawati and GustiDiansyah 2010 The Effect of Lead Metal (Pb) on Testis Tissue of Duck Grouper (Cromileptes Altivelis) *Maspari Journal* 0142-47
[4] Wiguna A, Hadi and Amarwati S 2017 Effect Of Administration Of Mercury Per Oral On Histopathological Features Of Wistar Rat Liver. *Jurnal Kedokteran Diponegoro*. **16** 673-681
[5] Wang L, Xu T, Lei W *et al* 2011 Cadmium-Induced Oxidative Stress And Apoptotic Changes In The Testis Of Freshwater Crab, Sinopotamon Heannense *Plos One* **6** E27853
[6] Karantika Ea, Supianur, Edyson and Suhartono E 2016 Kinetic paramaters analysis of liver and kidney catalase under the influence of cadmium (Cd) and mercury (Hg) in vitro *The Journal Of Tropical Life Science* **6** (2) 84
[7] Hirako A, Takeoka Y, Furukawa S and Sugiyama A 2017 Effects Of Cadmium Exposure OnMedaka (*Oryzias Latipes*) Testes *J. Toxicol. Pathol.* **30** 255–260
[8] Widyasari, Sunarti and Mufidah 2015 Purkinje cell diameter of rat (Rattus Wistar L) after induced water polluted with gold ore processing waste *Jurnal Biotik* **3** (1) 15-20
[9] Somade P M, Adnaik R S, Mohite Sk *et al* 2014 Cadmium toxicity exposure-induced oxidative stress in postnatal development of wistar rats *Journal Of Toxicology And Environmental Health Sciences* **3** 269
[10] Kamel M and Padhy P K 2013 Oxidative stress and heavy metals: an appraisalwith reference to environmental Biology *Int. Res. J. Biological Sci.* **2** (10) 91-101
[11] Suhartono E, Faturahman and Iskandar T 2015 *Reproductive system disorders due to cadmium exposure* (Yogyakarta: Gosyen Publishing)
[12] Koyu A, Gokcimen A, Ozguner F *et al* 2006 Evaluation of the effect of cadmium on rat liver *Moll. Cell Biochem.* **284** (1-2) 81-5
[13] Kar R, Seema G, Halder S *et al* 2015 Cadmium exposure induces oxidative stress by decreasing expression of antioxidant enzyms in mice liver *International Journal Of Clinical Biochemistry And Research* **2** (2) 89-96
[14] Ramadhani R, Eko S and Meitria S 2014 Perbedaan kadar advanced oxidation protein product (Aopp) ovarium tikus putih (Rattus novergicus) yang terpajan cadmium dan tidak terpajan cadmium *Jurnal Berkala Kedokteran* **10** (1) 111-120
[15] Kara, Karata F and Canatan H 2005 Effet of single dose cadmium chloride administration on oxidative stress in male and female rats *Turk. J. Vet. Anim. Sci.* **29** 37-42

[16] Arizal H M, Eko S and Bakhriansyah 2014 Effects of cadmium exposure to MDA levels on white rat ovaries *Berkala Kedokteran* **2** 1-8

[17] Morsy A A, Salama K H A, Kamel H A and Mansour M M F 2012 Effect of heavy metals on plasma membrane lipids and antioxidant enzymes of Zygophyllum species *Eurasia J.Biosci.* **6** 1-10

[18] Djohan C R T 2015 Accumulation of black lead in meat and bone of domestic and local chicken (lead accumulation in meat and bones of domestic and broiler chicken) *Jurnal Veteriner* **16** (4) 542-552

[19] Xu C, Johnson J E, Singh P K *et al* 1996 In Vivo studies of cadmium-induced apoptosis in testicular tissue of the rat and its modulation by a chelating agent *Toxicology* **107** (1) 1–8

[20] Aoyagi T, Ishikawa H, Miyaji K, Hayakawa K and Hata M 2002 cadmium-induced testicular damage in a rat model of subchronic intoxication *Reproductive Medicine And Biology* **1** (2) 59–63

[21] Ross M H and Pawlina W 2006 *Histology A Text And Atlas With Collerated Cell And Molecular Biology* (Philadelphia :Lippincott Williams & Wilkins)

[22] Junqueira L C 2000 *Basic Histology* (Jakarta: EGC)

[23] Vergilio C S, Moreira R V, Carvalho C E and Melo E J 2015 Evolution of cadmium effects in the testis and sperm of the tropical fish Gymnotus Carapo *Tissue Cell* **47** 132–139

[24] Luo Y, Shan D, Zhong H, Zhou Y, Chen W, Cao J *et al* 2015 Subchronic effects of cadmium on the gonads,expressions of steroid hormones and sex-related genes in Tilapia Oreochromis Niloticus *Ecotoxicology* **24** 2213–2223

[25] Sheng Z, Yang W X and Zhu J Q 2015 Metallothionein from pseudosciaenacrocea: expression and response to cadmium induced injury in the testes *Ecotoxicology* **24** 779–794

[26] Price S A and Lorraine M W 2006 *Pathophysiology of the Clinical Concept of Disease Processes (*Jakarta: EGC)

[27] Spector 2006 *Introduction to General Pathology* (Yogyakarta: Gadjah Mada University)

[28] Shiraishi N and Waalkes M P 1996 Acquired tolerance to cadmium-induced toxicity in the rodent testes *Toxic Subst. Mech.* **15** 27-42

[29] Santos F W, Oro T, Zeni G, Rocha J B, DoNascimento P C and Nogueira C W 2004 Cadmium induced testicular damage and its response to administration of Succimer and diphenyldiselenide in mice *Toxicol. Lett.* **152** 255-263

[30] Wu X, Guo X, Wang H, Zhou S, Li L, Chen X *et al* 2017 A brief exposure to cadmium impairs Leydig cell regeneration in the adult rat testis *Sci. Rep.* **7** (1) 6337

[31] Valko M, Morris H and Cronin M T D 2005 Metals,Toxicity And Oxidative Damage *Current Medicinal Chemistry* **12** (10) 1161-208

[32] Faix S, Faixova Z, Boldizarova K and Javorsky P 2005 The effect of long-term high heavy metal intake on lipid peroxidation of gastrointestinal tissue in sheep *Vet. Med.–Czech.* **50** (9) 401–405

DNA barcoding as a method for species identification: case study in *Ahaetulla* snake

L I N Indahsari[1,2*], F Fatchiyah[1,3] and N Kurniawan[1,2]

[1]Department of Biology, Faculty of Mathematics and Natural Science, Brawijaya University, Malang, Indonesia
[2]NK Research, Department of Biology, Faculty of Mathematics and Natural Science, Brawijaya University, Malang, Indonesia
[3]Research Center of Smart Molecule of Natural Genetics Resource, Brawijaya University, Malang, Indonesia

*Corresponding author's email: lilin.annovasho@gmail.com

Abstract. Species identification using molecular analyses has recently been developed, one of which is DNA barcoding method. The purpose of this study was to apply the DNA barcoding method to identify *Ahaetulla* snake which previously had not been identified molecularly. The specimens used came from Sumatra, Java, Bali, Lombok and additional samples obtained from Genbank. Cytochrome B gene was extracted and amplified to obtain DNA barcode. Phylogenetic analysis using ML method was used to determine the kinship relationship of the *Ahaetulla* snake. Morphological analyses were added to confirm the results of molecular identification. The results of molecular analysis using DNA barcoding method showed the presence of two *Ahaetulla* species in this study, there are *A. prasina* and *A. mycterizans*. This difference is indicated by the formation of two separate main clades, clade I contains *A. prasina* and clade II contains *A. mycterizans*. The results of the morphological analyses also showed differences in the character of the two different clade species. Thus, molecular analyses result agrees with morphological analyses. It can be concluded that the DNA barcoding method can be used as a valid method to identify *Ahaetulla* snake species.

1. Introduction

Traditional method for species determination is by identifying morphological characters compare to the other species. However, this traditional method fails to discriminate some species due to several conditions, for example, species that having various external body colorations especially when specimens are not fresh [1], museum preserved species [2] and samples showing phenotypic plasticity [3].

Nowadays, DNA barcode attracts the taxonomies attention as the latest system in the identification of almost all fauna species, both interspecific and intraspecific, quickly and accurately. DNA barcoding uses small regions of mitochondrial DNA (mtDNA) that used as a barcode to amplify a gene. Various primers set of mitochondrial gene for barcoding are available in published studies [4]. With DNA barcode, identification by taxonomies will be easier. This identification technique fully supports the improvement of animal classification as well as helps to sort out any ambiguity at the species level. However, DNA barcoding with the support of traditional taxonomy has the capability to identify species complexes within populations [5].

The COI gene is a part of mtDNA which is commonly used as a barcode to identify species, however, the cyt b gene is also part of mtDNA which has a high mutation rate, therefore it also can be used as a barcode. The variation in the Cyt B sequence causes this gene to be widely used to compare species in the same genus or family. The uniqueness of the Cyt B gene sequence is that there are parts that are conserved in species level, so it can be used for grouping based on animal species or for determining kinship relationships between animal species [6]. The results of the study by Laopichienpong et al. [7] showed that the use of the Cyt B gene was more effective for the identification of *Ahaetulla* snakes than the COI gene.

In this study, Cyt B gene was isolated from *Ahaetulla* snake that widely distributed in Indonesia. There are at least three species distributed in this country, *A. prasina, A. mycterizans,* and *A. fasciolata* [8]. Although it is widely distributed with vulnerable population, this snake is also threatened by human killed and traded. Faith and Williams [9] argued that the most significant contribution of DNA barcoding to conservation efforts is its role in improving and speeding up phylogenetic diversity assessments. Previous study by Leo et al.[10] showed that there are no significant morphological differences among *A. prasina* of Java and Sulawesi. Therefore, it is necessary to do research about identification of *Ahaetulla* snake using molecular analyses.

This study aimed to hold molecular analyses by using Cyt B gene as barcode in *Ahaetulla* snake of Indonesia. Morphological analyses also added to confirm the validity of molecular analyses.

2. Materials and methods

2.1. Samples collection

All samples were collected from 10 localities along Sumatra, Java, Bali, and Lombok Island (Figure 1). The total of 10 samples including of 3 samples of Sumatra, 5 samples of Java, a sample of Bali and a sample of Lombok. Four samples were taken from Genbank used as outgroup (www.ncbi.nlm.nih.gov). There are *A. prasina*, *A. mycterizans*, *Chrysopelea ornata*, and *Dendrelapis pictus*.

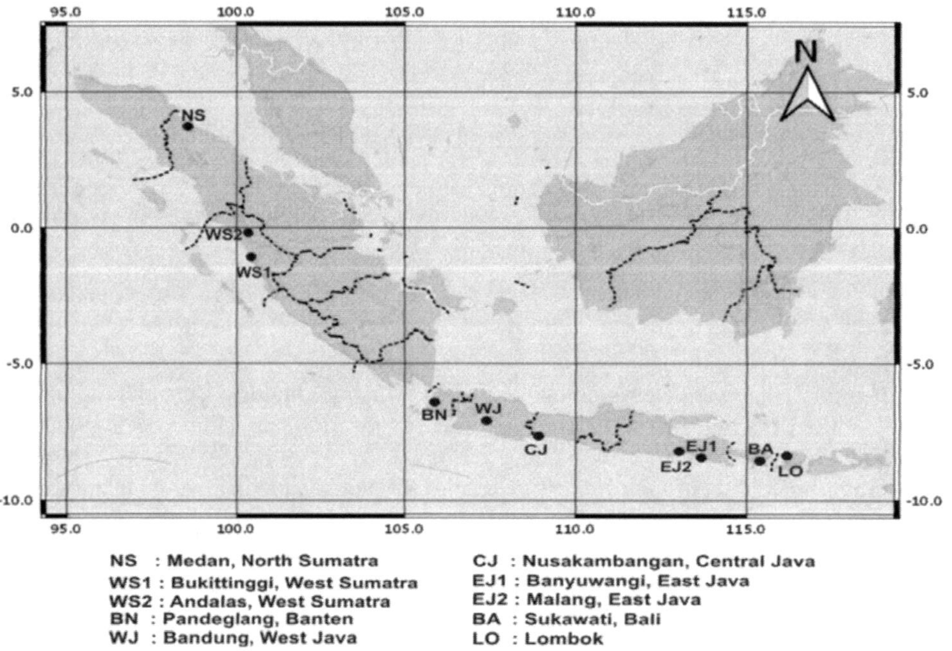

Figure 1. Distribution area of *Ahaetulla* Genus in Indonesia used in this study edited by QGIS software. Source of map: www.naturalearthdata.com.

Table 1. List of specimens used as in group and outgroup in the phylogenetic analysis.

Specimens	Locality	Genbank Accession no.	Source
Ahaetulla sp.	Lombok, NTB		This study
Ahaetulla sp.	Sukawati, Bali		This study
Ahaetulla sp.	Banyuwangi, East Java		This study
Ahaetulla sp.	Malang, East Java		This study
Ahaetulla sp.	Nusakambangan, Central Java		This study
Ahaetulla sp.	Bandung, West Java		This study
Ahaetulla sp.	Pandeglang, Banten		This study
Ahaetulla sp.	Padang, West Sumatra		This study
Ahaetulla sp.	Bukittinggi, West Sumatra		This study
Ahaetulla sp.	Medan, North Sumatra		This study
A. prasina	Thailand	LC105637	[7]
A. mycterizans	Thailand	KX660437	[11]
Chrysopelea ornate	-	KX694558	[12]
Dendrelaphis pictus	Malang, East Java	KY700863	[13]

2.2. DNA extraction

Cyt B fragments was amplified by PCR using Forward – L14910 (5' -GAC CTG TGA TMT GAA AAC CAY CGT TGT -3') and Reverse – H16064 (5'- CTT TGG TTT ACA AGA ACA ATG CTT TA -3') primer [14].The cycle of PCR was set for pre denaturation 940 for 7 mins then followed by 40 cycles of denaturation 940 C 30 seconds, annealing 460 C 30 seconds, extension 720 C 60 seconds, and post extension 720 7 mins. [14]. The amplification product was sequenced using the same primer as PCR amplification.

2.3. Phylogenetic analyses

Cyt B sequences were contiged by Sequencher (Gene codes, Ann Arbor, Michigan, USA). The sequences were aligned using MEGA7 by Clustal W method and genetic distance was calculated as uncorrected pairwise distance [15]. Phylogenetic tree was inferred using maximum likelihood (ML) method. For ML, tree was constructed by RAxML using automatically bootstrap option by RAxML. Bootstrap value equal or more than 70% was defined as significant value [16].

3. Result and Discussion

3.1. DNA barcoding analyses

The results of the genetic distance analysis showed that there were two identified species based on the genetic comparisons with *A. prasina* and *A. mycterizans* sequences from Genbank. The genetic distance between *A. prasina* and *A. mycterizans* is about >6%. According to Jeong [17], Cyt B sequences in reptiles that have a genetic distance >5% are considered as different species. The results of this genetic distance have identified that *A. prasina* and *A. mycterizans* in this study have >6% genetic distance. The next analyses is the construction of phylogenetic trees to determine the relationship among Genus *Ahaetulla* in Indonesia.

Journal of Physics: Conference Series **1374** (2019) 012011 doi:10.1088/1742-6596/1374/1/012011

Table 2. Genetic distances among Genus *Ahaetulla* based on Cyt B sequences.

No	Samples	1	2	3	4	5	6	7	8	9	10	11	12	13
1	*A. prasina*_NS													
2	*A. prasina*_WS2	1.02												
3	*A. prasina*_WS1	1.21	0.18											
4	*A. prasina*_WJ	1.98	1.69	1.88										
5	*A. prasina*_EJ1	2.28	2.27	2.46	2.08									
6	*A. prasina*_EJ2	2.37	2.37	2.56	2.18	0.65								
7	*A. prasina*_CJ	2.39	2.39	2.59	2.19	0.67	0.02							
8	*A. prasina*_BA	2.86	3.15	3.34	2.47	1.50	1.78	1.77						
9	*A. prasina*_Thailand	4.56	4.45	4.45	4.86	4.98	6.28	6.26	5.28					
10	*A. mycterizans*_Genbank	6.48	6.15	6.15	6.95	6.92	6.39	6.37	6.59	6.44				
11	*A. mycterizans*_BN	7.56	7.54	7.54	7.12	7.24	7.24	6.26	7.13	7.42	5.7			
12	*A. mycterizans*_LO	7.78	7.76	7.77	7.34	7.47	7.47	7.45	7.25	7.64	5.71	0.74		
13	*C. ornata*	13.75	13.75	13.87	13.76	13.90	13.90	13.92	13.76	13.3	13.97	13.92	13.67	
14	*D. pictus*	12.94	12.94	13.82	12.80	12.96	13.96	12.98	13.08	12.8	12.04	12.5	12.37	14.99

The results of phylogenetic analysis show that there are two separated main clades with high reliability values (ML= 100). The separation of those two clades was based on the two *Ahaetulla* species found in this study. Based on clustering with *Ahaetulla* sequences from Genbank, two *Ahaetulla* species in this study are *A. prasina* and *A. mycterizans*. Clade I is consist of *A. prasina* of Sumatra, Java, Bali, and Thailand, while clade II is consist of *A. mycterizans* of Banten and Lombok. In the clade of *A. prasina*, there two subclades was formed (ML=95). Subclade I in *A. prasina* (ML=98) contains *Ahaetulla* of Indonesia (Sumatra and West Java, Central Java, East Java, and Bali), while subclade II only contains *A. prasina* of Thailand. This result indicates that Cyt B gene of *A. prasina* of Indonesia significantly different with *A. prasina* of Thailand. The presence of geographical barrier like Hindia Ocean made the snake of Indonesia can't easily migrate to Thailand or vice versa.

In *A. prasina* of Indonesia clade, there are 2 subclades formed based on their localities (ML=99). Subclade I consist of *A. prasina* of Sumatra and West Java, while subclade II consist of *A. prasina* of Central Java, East Java, and Bali. The clustering joining of *A. prasina* of Sumatra and West Java is predicted because of the integration of the mainland of Sumatra and Java during glacial period. At that time, it was estimated that there were many species migrations which made it easier for species from Java and Sumatra to move each other [18]. This makes a pattern of adaptation which then affects the genetic structure of the species. Likewise, the Bali island which once joined with the mainland of Java allowed species migration from Java to Bali or vice versa. The island is part of the mainland of Sundaland that was once connected to become a major mainland in the past [19].

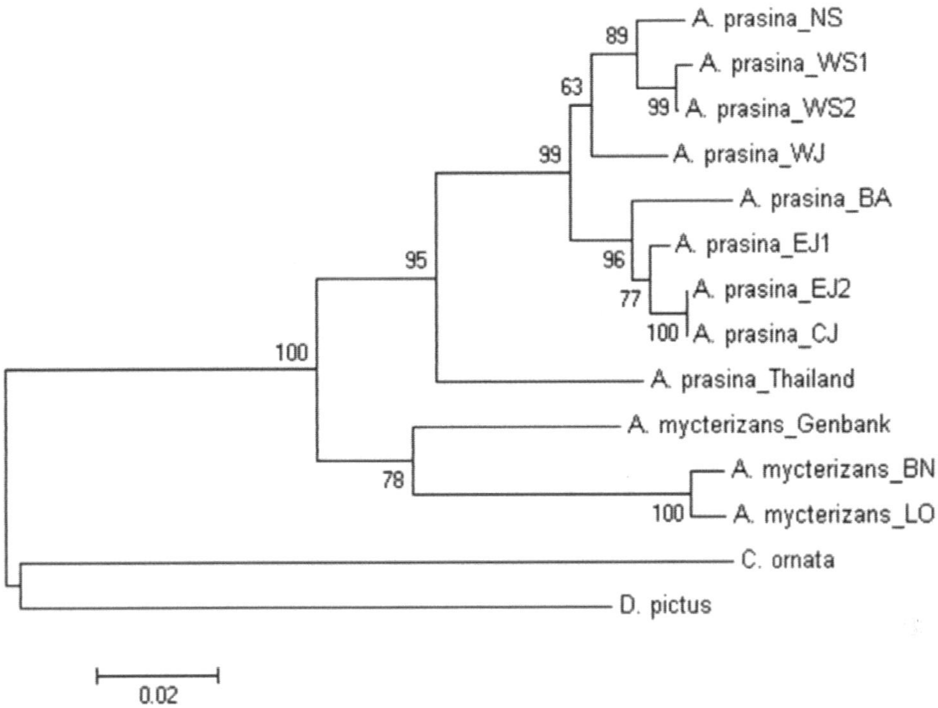

Figure 2. Phylogenetic tree of *Ahaetulla* Genus in Indonesia based on Cyt-B gene. Tree was constructed by ML analysis. Nodal supports represent ML bootstrap value.

Within *A. mycterizans* clade, 2 subclades were formed (ML=78) consisting of *A. mycterizans* sequence from Genbank and *A. mycterizans* of Lombok and Banten. *A. mycterizans* of Lombok and Banten formed a clade with significant bootstrap (ML=100) although Lombok is located in Lesser Sunda and Banten in Sundaland. This analyse indicates that *A. mycterizans* has high similarity Cyt B sequence with *A. mycterizans* of Banten. It can be predicted that A. *mycterizans* of Lombok could be originating from Sundaland. We refer to Inger [20] which proposed that fauna in Wallace's line, including Lesser Sunda region, originated from Sundaland region.

This result of molecular identification adds evidence that molecular analyses with Cyt-B gene as barcode can be used as alternative method for identifying species

3.2. Morphological analyses

To confirm the identification result, we also add morphological identification. The specimens used are representative of *A. prasina* and *A. mycterizans* used in this study. The results of morphological identification are shown in Figure 3 and Table 2. As illustrated in Figure 3, the differences between *A. mycterizans* and *A. prasina* are: 1) *A. mycterizans* eyes is bigger than *A. prasina*, 2) upper surface of snout is convex in *A. mycterizans* but flat or even depressed in *A. prasina*. The differences between *A. mycterizans* and *A. prasina* not illustrated in Figure 3 and not mentioned in Table 4 are: 1) *A. mycterizans* has anal entire, while *A. prasina* has anal divided. 2) *A. mycterizans* has paler grey longitudinal lines in venter than *A. prasina*.

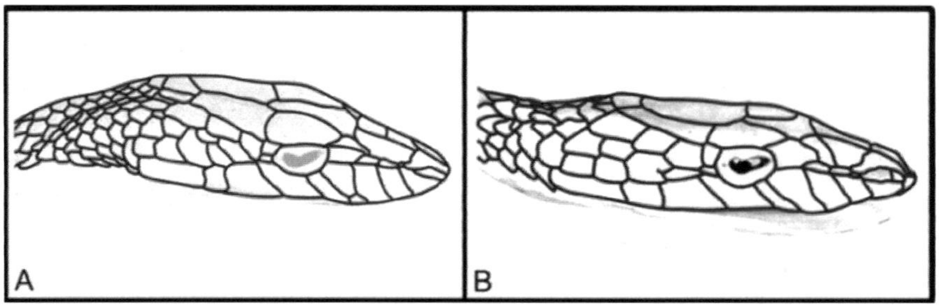

Figure 3. Morphological comparison between *Ahaetulla* Genus of Indonesia: A) *A. prasina* of Bali (BA), B) *A. mycterizans* of Banten (BN).

Table 3. The morphological comparison between *A. prasina* and *A. mycterizans* identified in this study based on morphometry and meristic measurement. Morphometry measurement in cm.

Specimen	Locality	Sex	SVL	TaL	TL	TaL/TL	VEN	DSR	SC
A. prasina	Bali	F	43,2	22,5	65,7	0,34	200	16:16:13	162
A. mycterizans	Banten	F	75,0	40,3	115,3	0,25	184	15:15:13	134

SVL:Snouth-Vent Length; TaL: Tail Length; TL: Total Body Length (SVL + TL); DSR Dorsal Scale Rows; VEN:Ventral Scale; SC: Sub caudal Scale

Based on the results of morphological identification, our specimen of Lombok agrees with five diagnostic morphological characters given for *A. mycterizans* by Miralles and David [21]: 1) ventral scale less than 200, 2) Anal entire, 3) Snout convex above, 5) venter with grey longitudinal lines. Moreover, our identification of *A. mycterizans* also agrees with recent studies that *A. mycterizans* eyes is wider than *A. prasina* and the number of ventral scales has a big influence in distinguishing *A. prasina* and *A. mycterizans* [22].

The morphological characters of *Ahaetulla* of East Java agree with diagnostics characters of *A. prasina* given by De Rooij [23] (who mentioned as *Dryophis prasinus*): 1. number of ventral scales 194-235 (206), 2. anal divided, 3. Snout more than twice the diameter of eye, 4. Sub caudal scales 151-207 (162).Therefore, we agree that there are 2 species found in this study based on morphological characters, *A. prasina* and *A. mycterizans*.

4. Conclusion

From those both analyses, it can be concluded that molecular analysis using Cyt B gene as a barcode can be used to identify species accurately because it shows appropriate result with morphological identification. Among animal taxonomists, the COI gene is the most often used for species identification. Based the results of this study, it can also be added that the Cyt B gene is also an effective gene used as a DNA barcode for species identification. Besides to be used for species identification, DNA barcoding method may also be used to analyse the biogeography of species in the distribution area.

5. Acknowledgments

Field sampling was funded by the NSF to Eric N. Smith (NSF Award Number: DEB-1146324) and PEER Science USAID grant to Nia Kurniawan (Subgrant number: 2000005071). We thanks to A.M. Kadafi, B. Priambodo, and M. Fathoni for collecting all specimens used in this study, D. R. Tirtosari and Y. Bare for molecular work, M. Fahmi for phylogenetic analysis, F. A. D. Nugraha and A. Maulidi for sharing discussions.

References

[1] Sriwattanarothai N, Steinke D, Ruenwongsa P, Hanner R and Panijpan B 2010 Molecular and morphological evidence supports the species status of the Mahachai fighter Betta sp. Mahachai and reveals new species of Betta from Thailand *J Fish Biol* **77** 414-424

[2] Hebert P D, Zakharov E V, Prosser SW, Sones J E, McKeown J T, Mantle B and La Salle J 2013 A DNA 'Barcode Blitz': Rapid digitization and sequencing of a natural history collection *PLoS One* **8** e68535

[3] Weigand A M, Jochum A, Pfenninger M, Steinke D and Klussmann-Kolb A 2011 A new approach to an old conundrum—DNA barcoding sheds new light on phenotypic plasticity and morphological stasis in microsnails (Gastropoda, Pulmonata, Carychiidae) *Mol. Ecol. Res.* **11** 255-265

[4] Imtaz A, Nor SAM and Naim AM 2017 Review: Progress and potential of DNA barcoding for species identification of fish species *Biodiversitas* **10** 1394-1405

[5] Iwatsuki Y, Tanaka F and Allen GR 2015 Lutjanus xanthopinnis, a new species of snapper (Pisces: Lutjanidae) from the Indo-west Pacific, with a redescription of Lutjanus madras (Valenciennes 1831) *J Ocean Sci Found* **17** 22-42

[6] Stuart J A 2009 *Mitochondrial DNA Methods and Protocol* 2nd Edition (Canada: Humana Press)

[7] Laopichienpong N, Muangmai N, Supikamolseni A, Twilprawat P, Chanhome L et al. 2016 Assessment of snake DNA barcodes based on mitochondrial COI and Cytb genes revealed multiple putative cryptic species in Thailand *Genes* **2016**

[8] Das I 2015 *A field Guide to The Reptiles of South-East Asia* (London: Bloomsbury Publishing)

[9] Faith D P and Williams K J 2005 *How large-scale DNA barcoding programs can boost biodiversity conservation planning: linking phylogenetic diversity (PD) analyses to the Barcode of Life Database (BOLD)* In Australian Entomological Society's 36th AGM and Scientific Conference/7th Invertebrate Biodiversity and Conservation Conference/Australian Systematics Society. Canberra, Australia. December 2005

[10] Leo S, Amarasinghe T and Supriatna J 2015 *Morphological variation of Ahaetulla prasina (Boie, 1827) (Squamata: Colubridae) in Java.* The 1st Symposium on South East Asia Herpetology and Envenomation (SEASHE) "Spesies Diversity and Animal Bites in Emergency Medicine" in conjunction with The 4th Congress of Herpetological Society of Indonesia, University of Indonesia. August 2015

[11] Figueroa A, Mckelvy A D, Grismer L L, Bell C D and Lailvaux SP 2016 A Species Level Phylogeny of Extant Snakes with Description of a New Colubrid Subfamily and Genus *PLos ONE* **11** 9

[12] Alencar LRV, Quental TB, Grazziotin FG, Alfaro ML, Martins M et al. 2016 Diversification in vipers: phylogenetic relationships, time of divergence and shifts in speciation rates *Molecular Phylogenetics and Evolution* **105** 50-62

[13] Nugraha F A D, Fatchiyah F, Smith E N, Kurniawan N 2018 Phylogenetic analysis of colubrid snakes based on 12S-rDNA reveals distinct lineages of Dendrelaphis pictus (Gmelin, 1789) populations in Sumatra and Java *Biodiversitas* **19** 303 – 310

[14] Burbrink F T, Lawson R and Slowinski J B 2000 Mitochondrial DNA phylogeography of the polytypic north American rat snake (Elaphe obsoleta): a critique of the subspecies concept *Evolution* **54** *2107–2118*

[15] Kumar S, Stecher G and Tamura K 2016 MEGA7: Molecular Evolutionary Genetics Analysis version 7.0 for bigger datasets *Mol. Biol. Evol.* **33** 1870-1874

[16] Drummon A J, Ho S Y, Phillips M J and Rambaut A 2006 Relaxed phylogenetics and dating with confidence *PLoS Biol.* **4**: e88

[17] Jeong T A, Jun J, Han S, Kim H, Oh K and Kwak M 2013 DNA barcode reference data for korean herpetofauna and their appications *Molecular Ecology Reseources* **3** 1019-1032

Journal of Physics: Conference Series **1374** (2019) 012011 doi:10.1088/1742-6596/1374/1/012011

[18] Hall R 2012 *Sundaland And Wallacea: Geology, Plate Tectonics and Palaeogeography Biotic Evolution and Environmental Change in Southeast Asia*, eds D. J. Gower et al. (London: Cambridge University Press)

[19] Lohman D J, Bruyn M, Page T, Rintelen K, Hall R et al. 2011 Biogeography of the Indo-Australian Archipelago *Annual Review of Ecology, Evolution, and Systematics* **42** 205-226

[20] Inger R F 2005 *The frog fauna of the Indo-Malayan region as it applies to Wallace's line. In: Wallace in Sarawak– 150 years later.* In: A. A. Tuen & I. Das, editors. Proceedings of an International Conference on Biogeography and Biodiversity. Institute of Biodiversity and Environmental Conservation, Universiti Malaysia Sarawak, Kota Samarahan pp:82-90

[21] Miralles A and David P 2010 First record of *Ahaetulla mycterizans* (Linnaeus, 1758) (Reptilia, Squamata, Colubridae) from Sumatra, Indonesia, with an expanded definition *Zoosystema* **32** 449-456

[22] Baker N 2019 Some snake records from Gunung Arong Forest Reserve, Johor, Peninsular Malaysia *Southeast Asia Vertebrate Records* 001-006

[23] De Rooij N 1917 *The Reptiles of the Indo-Australian Archipelago* (Leiden: E.J. Brill)

Sustainability Strategy of Agarwood-Production Trees in Lombok Island using Harvesting Matrix

M U Romdhini[1*], N Fitriyani[1], A Kurnia[1], I K A Baskara[1], M Ito[2], I Yamada[3], S Hadi[4] and T Mulyaningsih[5]

[1]Mathematics Department, Universitas Mataram, Mataram, Indonesia
[2]Graduate School of Pharmaceutical Sciences, Kyoto University, Japan
[3]Center for Southeast Asian Studies, Kyoto University, Japan
[4]Chemistry Department, Universitas Mataram, Mataram, Indonesia
[5]Biology Department, Universitas Mataram, Mataram, Indonesia

*Corresponding author's email: mamika_ur@yahoo.com

Abstract. Eaglewood (*Gyrinops verstiigee*) is one example of a commercially valuable aromatic plant in the form of agarwood. The high selling value of agarwood encourages people to use it. The distribution area of Agarwood-producing trees is found in several regions in Indonesia. One of agarwood producers has been widely exploited in West Nusa Tenggara Province. The problem is the number of population and the quality of agarwood production which is declining due to exploitation carried out continuously and excessively without calculations and improper harvesting techniques. Therefore, a forest management model that cares for sustainability, called a sustainable harvesting model, could be used. This research used harvesting matrix to build this model. This model considers the initial configuration of the forest to be the same as the final configuration of the forest, so that the harvesting could be done continuously without damaging the forest configuration. The location of data collection was conducted in several areas, such as Orong Puncak, Lembah Sari, Orong Selatan, Kekait, Sepakek, East Lombok, and Kerujuk. The data obtained were divided into 5 intervals based on the diameter of the eaglewood tree. The research found that the number of trees that could be harvested on one harvesting period at each interval of trees diameter, i.e. 0 tree in 0-10 cm interval; 388 trees in 10-20 cm interval; 270 trees in 20-30 cm interval; 17 trees in 30-40 cm interval; and 3 trees in 40-50 cm interval. Harvesting matrix could be used to help the government in order to build a strategy for the sustainability of eaglewood trees in Lombok Island.
Keywords: eaglewood, forest configuration, forest management, tree diameter

1. Introduction

Eaglewood is a commercially valuable aromatic plant in the form of agarwood (sapwood and kemedangan). Agarwood began to be known to the Indonesian people around 1200, which was only obtained from the collection of natural forest products by utilizing natural dead trees in the form of clumps, debris, and cinders, which were the waste of the cleaning process. As one of the commodities of forest products, agarwood is widely used for various purposes, such as perfume and traditional medicine. The form of trade in agarwood varies from chunks of wood, powder (ash), and agarwood oil. Agarwood is widely exported to Arab countries, Singapore, and China [1].

One of the producers of agarwood has been widely exploited in the area of West Nusa Tenggara and East Nusa Tenggara. Nowadays the population number and quality of agarwood-production decreases due to exploitation carried out continuously and excessively without calculations and improper harvesting techniques. Therefore, today the agarwood trees are increasingly scarce and the location for finding agarwood is getting farther into the forest, also the time taken for making taking is getting longer [2].

Therefore, a forest management model concerned with forest sustainability is needed. This model is called a sustainable harvesting model. This model is very suitable to be applied in forest management related to forest harvesting, because the sustainable harvesting model stipulates that the initial configuration of the forest must be the same as the final configuration of the forest, so that harvesting can be carried out continuously without changing the forest configuration [3].

2. Literature Review

According to [3], the compilation of sustainable harvesting models is based on several assumptions, i.e. not counting trees that die between harvest times and each plant planted survives and grows until it can be harvested. By this way, the number of trees in the forest will always be the same. Consider Figure 1 below.

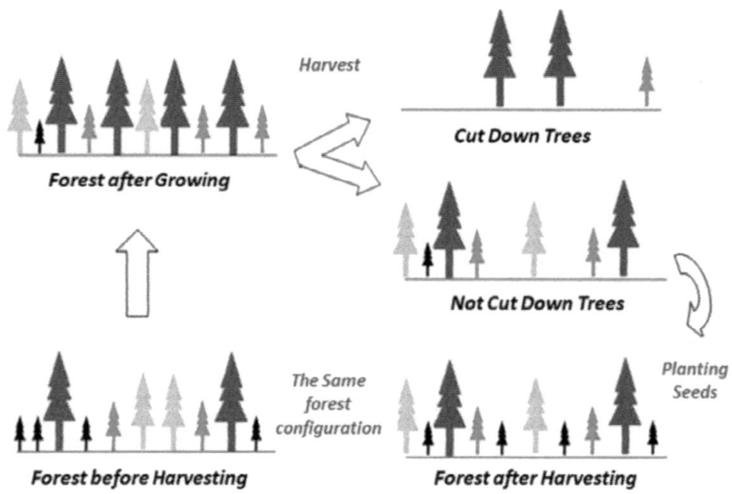

Figure 1. Sustainable Forest Harvest Cycle.

From Figure 1, it can be seen that the forest before harvest (the forest at the beginning of growth) has the same forest configuration as the forest after harvest (forest at the end of growth). The forest configuration is the number of trees in each diameter class that can be represented by vectors. Consider Table 1 below to see the configuration of forest before harvest (forest at the beginning of growth).

Table 1. Interval Class Description Based on Class and Diameter of the Eaglewood Tree.

Class	Interval Diameter
1	$[0, h_1)$
2	$[h_1, h_2)$
3	$[h_2, h_3)$

Class	Interval Diameter
$\vdots$	$\vdots$
$n-1$	$[h_{n-2}, h_{n-1})$
n	$[h_{n-1}, \infty)$

Let $x_1 (i = 1,2,\cdots,n)$ be the number of trees in the first – class before a period of growth, then a column vector of $\boldsymbol{x}$ can be formed, called a non-harvest vector.

$$\mathbf{x} = \begin{bmatrix} x_1 \\ x_2 \\ \vdots \\ x_n \end{bmatrix}$$

x_1 indicates the number of trees in the first – class at the beginning of the growth period, x_2 indicates the number of trees in the second – class at the beginning of the growth period, and x_n indicates the number of trees in the nth – class at the beginning of the growth period.

To construct a sustainable harvesting policy, after every harvest, the forest is returned to a fixed configuration, determined by a non-harvest vector $\boldsymbol{x}$. Since the number of trees in the forest is fixed, the following equation can be determined.

$$x_1 + x_2 + \cdots + x_n = s \tag{1}$$

When reviewed from the growth of the forest between harvest times, a tree in the first – class can grow and increase to a higher class or remain in the same class due to various natural factors. As a result, g_i and $1 - g_i$, can be defined as the growth parameters. For $i = 1,2,\cdots,n-1$, g_i is the fraction of the first – class trees that grow to $(i+1)th$ – class in one period of growth, and $1 - g_i$ is fraction of the first – class trees that remain in the first – class in one period of growth.

With *n – 1* growth parameters, a growth matrix of *n x n* can be formed as follows.

$$\mathbf{G} = \begin{bmatrix} 1-g_1 & 0 & 0 & \cdots & 0 & 0 \\ g_1 & 1-g_2 & 0 & \cdots & 0 & 0 \\ 0 & g_2 & 1-g_3 & \cdots & 0 & 0 \\ \vdots & \vdots & \vdots & \ddots & \vdots & \vdots \\ 0 & 0 & 0 & \cdots & 1-g_{n-1} & 0 \\ 0 & 0 & 0 & \cdots & g_{n-1} & 1 \end{bmatrix} \tag{2}$$

Since the elements of vector x are the number of trees in n classes before a growth period, the input of the vector $\mathbf{Gx}$ can be written as follows.

$$\mathbf{Gx} = \begin{bmatrix} (1-g_1)x_1 \\ g_1x_1 + (1-g_2)x_2 \\ g_2x_2 + (1-g_3)x_3 \\ \vdots \\ g_{n-2}x_{n-2} + (1-g_{n-1})x_{n-1} \\ g_{n-1}x_{n-1} + x_n \end{bmatrix} \quad (3)$$

Information:
$\mathbf{Gx} =$ the number of trees in n classes after a period of growth;
$(1-g_1)x_1 =$ the number of trees in the first – class at the end of the growth period;
$g_1x_1 + (1-g_2)x_2 =$ the number of trees in the second – class at the end of the growth period;
$g_2x_2 + (1-g_3)x_3 =$ the number of trees in the third – class at the end of the growth period; and
$g_{n-1}x_{n-1} + x_n =$ the number of trees in the nth – class at the end of the growth period.

During the harvesting period, for example, $y_i (i = 1,2,\cdots,n)$ number of trees taken from the ith – class. So that a vector ***y*** can be formed as a harvest vector.

$$\mathbf{y} = \begin{bmatrix} y_1 \\ y_2 \\ \vdots \\ y_n \end{bmatrix}$$

Information:
$y_1 =$ the number of trees harvested in the first – class after the end of the growth period;
$y_2 =$ the number of trees harvested in the second – class after the end of the growth period; and
$y_n =$ the number of trees harvested in the nth – class after the end of the growth period.

So, there is a total number of $y_1 + y_2 + \cdots + y_n$ trees cut down each harvest period. This amount of number is also the total number of trees added to the first – class (new seedlings) after each harvest period of time. For example,

$$\mathbf{R} = \begin{bmatrix} 1 & 1 & \cdots & 1 \\ 0 & 0 & \cdots & 0 \\ \vdots & \vdots & \cdots & \vdots \\ 0 & 0 & \cdots & 0 \end{bmatrix} \quad (4)$$

R is defined as a replacement matrix *n x n*. Then obtained a column vector,

$$\mathbf{Ry} = \begin{bmatrix} y_1 + y_2 + \cdots + y_n \\ 0 \\ \vdots \\ 0 \end{bmatrix} \quad (5)$$

with **Ry** determines the configuration of trees planted after each harvest period.

From the previous result on forest management policies, we can conclude the equation that characterizes a sustainable harvesting policy, as follows:

$$\begin{bmatrix} \textit{configuration} \\ \textit{at the end} \\ \textit{of the growth period} \end{bmatrix} - [\textit{harvest}] + \begin{bmatrix} \textit{subtitute} \\ \textit{for new seedlings} \end{bmatrix} = \begin{bmatrix} \textit{configuration} \\ \textit{at the beginning} \\ \textit{of the growth period} \end{bmatrix}$$

Mathematically written as:

$$\mathbf{Gx} - \mathbf{y} + \mathbf{Ry} = \mathbf{x} \tag{6}$$

Equation (6) can be described as the equation below:

$$(\mathbf{I} - \mathbf{R})\mathbf{y} = (\mathbf{G} - \mathbf{I})\mathbf{x} \tag{7}$$

More fully, equation (7) can be written as follows [3]:

$$\begin{bmatrix} 0 & -1 & -1 & \cdots & -1 & -1 \\ 0 & 1 & 0 & \cdots & 0 & 0 \\ 0 & 0 & 1 & \cdots & 0 & 0 \\ \vdots & \vdots & \vdots & \ddots & \vdots & \vdots \\ 0 & 0 & 0 & \cdots & 1 & 0 \\ 0 & 0 & 0 & \cdots & 0 & 1 \end{bmatrix} \begin{bmatrix} y_1 \\ y_2 \\ y_3 \\ \vdots \\ y_{n-1} \\ y_n \end{bmatrix} = \begin{bmatrix} -g_1 & 0 & 0 & \cdots & 0 & 0 \\ g_1 & -g_2 & 0 & \cdots & 0 & 0 \\ 0 & g_2 & -g_3 & \cdots & 0 & 0 \\ \vdots & \vdots & \vdots & \ddots & \vdots & \vdots \\ 0 & 0 & 0 & \cdots & -g_{n-1} & 0 \\ 0 & 0 & 0 & \cdots & g_{n-1} & 0 \end{bmatrix} \begin{bmatrix} x_1 \\ x_2 \\ x_3 \\ \vdots \\ x_{n-1} \\ x_n \end{bmatrix} \tag{8}$$

Since the trees in the first – class are seedlings that have no selling value or cannot be harvested, it is determined that:

$$y_1 = 0 \tag{9}$$

So, from equation (8), the following equation is obtained:

$$\begin{aligned} & y_2 + y_3 + \cdots + y_n = g_1 x_1 \\ & y_2 = g_1 x_1 - g_2 x_2 \\ & y_3 = g_2 x_2 - g_3 x_3 \\ & \vdots \\ & y_{n-1} = g_{n-2} x_{n-2} - g_{n-1} x_{n-1} \\ & y_n = g_{n-1} - x_{n-1} \end{aligned} \tag{10}$$

From equation (9) and equation (10), the following equation is obtained:

$$\begin{aligned} & y_1 = 0 \\ & y_2 = g_1 x_1 - g_2 x_2 \\ & y_3 = g_2 x_2 - g_3 x_3 \\ & \vdots \\ & y_{n-1} = g_{n-2} x_{n-2} - g_{n-1} x_{n-1} \\ & y_n = g_{n-1} - x_{n-1} \end{aligned} \tag{11}$$

Since $y_i \geq 0$, for $i = 1, 2, \ldots, n$, then equation (11) requires that [3]:

$$g_1 x_1 \geq g_2 x_2 \geq \cdots \geq g_{n-1} x_{n-1} \geq 0 \tag{12}$$

3. Research Methodology

This study aimed to find out: (a) sustainable harvesting models of eaglewood trees on Lombok Island, and (b) the number of trees that can be harvested at each interval of tree diameters in one harvest period from sustainable harvesting models obtained. This research was a study with literature studies, by looking for references on theories that were relevant to the case or problem raised, which were used as the basis for conducting research. The data used in this study were primary data and secondary data. Secondary data was used as preliminary data on the growth of agarwood-production trees on Lombok Island, obtained from inventory data on potential production forests in Mount Rinjani Forest Area in 2013 [4]. The primary data was conducted by the research team, which contained data on the growth of agarwood-production trees on Lombok Island. The steps used in this study were: (a) literature study; (b) data collection; (c) interval distribution based on tree diameter; (d) model application; and (e) conclusions.

4. Results and Discussion

Table 2 shows data on the number of trees in each of the 5 classes of tree diameter intervals based on the data obtained.

Table 2. Description of Agarwood-Production Tree Interval Classes on Lombok Island.

Class	Tree Diameter Interval (cm)	Number of Trees in the Beginning (x_i)	Number of Trees that are Remains in the *i*th-class (x_i^i)	Number of Trees that Move to *(i+1)*th-class (x_i^{i+1})
1	[0-10)	3843	3165	678
2	[10-20)	1099	809	290
3	[20-30)	53	33	20
4	[30-40)	12	9	3
5	[40-50)	6		

Based on Table 2, a column vector ***x*** can be formed as follows, by substituting each value of x_1, x_2, x_3, x_4, and x_5 from Table 2.

$$\mathbf{x} = \begin{bmatrix} x_1 \\ x_2 \\ x_3 \\ x_4 \\ x_5 \end{bmatrix} = \begin{bmatrix} 3843 \\ 1099 \\ 53 \\ 12 \\ 6 \end{bmatrix}$$

Information:

x_1 = number of agarwood-production trees in the 1st – class 1 at the beginning of the growth period;

x_2 = number of agarwood-production trees in the 2nd – class 1 at the beginning of the growth period;

x_3 = number of agarwood-production trees in the 3rd – class 1 at the beginning of the growth period;

x_4 = number of agarwood-production trees in the 4th – class 1 at the beginning of the growth period; and

x_5 = number of agarwood-production trees in the 5th – class 1 at the beginning of the growth period.

Furthermore, we can set up a *5 x 5* growth matrix as follows:

$$\mathbf{G} = \begin{bmatrix} 1-g_1 & 0 & 0 & 0 & 0 \\ g_1 & 1-g_2 & 0 & 0 & 0 \\ 0 & g_2 & 1-g_3 & 0 & 0 \\ 0 & 0 & g_3 & 1-g_4 & 0 \\ 0 & 0 & 0 & g_4 & 1 \end{bmatrix}$$

Then the following fraction values are obtained:

$$g_1 = \frac{678}{3843} = 0,1764\,; \qquad 1-g_1 = \frac{3165}{3843} = 0,8236$$

$$g_2 = \frac{290}{1099} = 0,2693\,; \qquad 1-g_2 = \frac{809}{1099} = 0,7361$$

$$g_3 = \frac{20}{53} = 0,3774\,; \qquad 1-g_3 = \frac{33}{53} = 0,6226$$

$$g_4 = \frac{3}{12} = 0,25\,; \qquad 1-g_4 = \frac{9}{12} = 0,75$$

By substituting the fraction values into the matrix ***G***, it is obtained:

$$\mathbf{G} = \begin{bmatrix} 0,8236 & 0 & 0 & 0 & 0 \\ 0,1764 & 0,7361 & 0 & 0 & 0 \\ 0 & 0,2639 & 0,6226 & 0 & 0 \\ 0 & 0 & 0,3774 & 0,75 & 0 \\ 0 & 0 & 0 & 0,25 & 1 \end{bmatrix}$$

Vector ***Gx*** is defined as the number of agarwood-production trees after one period of growth. So, we obtain:

$$\mathbf{Gx} = \begin{bmatrix} (1-g_1)x_1 \\ g_1x_1 + (1-g_2)x_2 \\ g_2x_2 + (1-g_3)x_3 \\ g_3x_3 + (1-g_4)x_4 \\ g_4x_4 + x_5 \end{bmatrix} = \begin{bmatrix} 3165 \\ 1487 \\ 323 \\ 29 \\ 9 \end{bmatrix}$$

Vector ***y*** as the harvest vector can be written as follows:

$$\mathbf{y} = \begin{bmatrix} y_1 \\ y_2 \\ y_3 \\ y_4 \\ y_5 \end{bmatrix}$$

R is defined as a *5 x 5* replacement matrix.

$$\boldsymbol{R}=\begin{bmatrix}1&1&1&1&1\\0&0&0&0&0\\0&0&0&0&0\\0&0&0&0&0\\0&0&0&0&0\end{bmatrix}$$

So, the $\boldsymbol{Ry}$ vector becomes,

$$\boldsymbol{Ry}=\begin{bmatrix}y_1+\quad y_2+\quad y_3+\quad y_4+\quad y_5\\0\\0\\0\\0\end{bmatrix}$$

$\boldsymbol{Ry}$ determines the configuration of trees planted after being ready for harvest. Based on equation (11), we obtain:

$y_1=0$;

$y_2=(0{,}1764\times 3843)-(0{,}2639\times 1099)=388$;

$y_3=(0{,}2639\times 1099)-(0{,}3774\times 53)=270$;

$y_4=(0{,}3774\times 53)-(0{,}25\times 12)=17$; and

$y_5=(0{,}25\times 12)=3$.

Suitability of the continuous harvesting model that sets the initial configuration must be the same as the final configuration, can be known through the application of equation (6), with the following details.

$$\mathbf{Gx}-\mathbf{y}+\mathbf{Ry}\;=\;\mathbf{x}$$

$$\begin{bmatrix}(1-g_1)x_1\\g_1-x_1+(1-g_2)x_2\\g_2x_2+(1-g_3)x_3\\g_3x_3+(1-g_4)x_4\\g_4x_4+x_5\end{bmatrix}-\begin{bmatrix}y_1\\y_2\\y_3\\y_4\\y_5\end{bmatrix}+\begin{bmatrix}y_1+\quad y_2+\quad y_3+\quad y_4\quad y_5\\0\\0\\0\\0\end{bmatrix}=\begin{bmatrix}x_1\\x_2\\x_3\\x_4\\x_5\end{bmatrix}$$

$$\begin{bmatrix}3165\\1487\\323\\29\\9\end{bmatrix}-\begin{bmatrix}0\\388\\270\\17\\3\end{bmatrix}+\begin{bmatrix}0+\quad 388+\quad 270+\quad 17\quad 3\\0\\0\\0\\0\end{bmatrix}=\begin{bmatrix}3843\\1099\\53\\12\\6\end{bmatrix}$$

$$\begin{bmatrix}3843\\1099\\53\\12\\6\end{bmatrix}=\begin{bmatrix}3843\\1099\\53\\12\\6\end{bmatrix}$$

The results prove that the initial configuration of the forest is the same as the final configuration of the forest.

Thus, the total agarwood-production trees on Lombok Island which can be harvested in one harvest period are 0 tree in 0-10 cm interval; 388 trees in 10-20 cm interval; 270 trees in 20-30 cm interval; 17 trees in 30-40 cm interval; and 3 trees in 40-50 cm interval.

5. Conclusion

Harvesting matrix obtained could be used to help the government in order to build a strategy for the sustainability of agarwood-production trees in Lombok Island.

References

[1] Suhartono T 2001 Aloes and Its Usability and Utilization Paper Presented at the Workshop on Agarwood Development, Directorate General of Land Rehabilitation and Social Forestry, Ministry of Forestry in Mataram

[2] Suarta I K and Soenarno 2011 Plantation of Eaglewood (Gyrinops versteegii (Gilg.) Domke) with Intercropping System at Rarung, West Nusa Tenggara Province *Journal of Forest Research and Nature Conservation* **8** 349-361

[3] Anton and Rorres 2005 Linear Algebra and Its Applications, translated by: Irham (Jakarta: Erlangga)

[4] West Rinjani Protection Forest Management Unit 2013 Inventory Data on the Potential of the Gunung Rinjani Production Forest Group, West Nusa Tenggara

Snake Fruit Seeds Flour-Fortified Jelly for Supplementary Feeding in Adolescent Girls with Anemia

M A Nugraheni[1*], D Indarto[2,3,4] and E P Pamungkasari[2,5]

[1] Human Nutrition, Post Graduate Program of Nutrition Sciences, Universitas Sebelas Maret, Surakarta, Indonesia
[2] Postgraduate Program of Nutrition Sciences, Universitas Sebelas Maret, Surakarta, Indonesia
[3] Department of Physiology, Faculty of Medicine Universitas Sebelas Maret, Surakarta, Indonesia
[4] Biomedical Laboratory, Faculty of Medicine Universitas Sebelas Maret, Surakarta, Indonesia
[5] Department of Public Health, Faculty of Medicine Universitas Sebelas Maret, Surakarta, Indonesia

*Corresponding author's email : martha8nugraheni@gmail.com

Abstract. Anemia still becomes a public health problem over the world. Adolescent girls are one of the most suspectible groups who suffer anemia and have significant impacts on their health and study achievement. Fortification with high Fe natural ingredients is commonly used to overcome anemia in public setting. Therefore, the aim of this study was to evaluate the effect of snake fruit seed flour (SSF) on body weight (BW) and BMI/age of adolescent girls with anemia. A total of 20 adolescent girls participitated in this study and randomly divided in control group and treatment group. They received 3 x 100 g jelly/week for 8 weeks. BW and BMI/age was measured in day 0, 30, and 60. Collected data were analyzed using paired t-test and ANOVA and p value < 0.05. Average of BW in control group was higher than in treatment group but it was not statiticaly different (p = 0,374). Average of BMI/age in control group was higher than in treatment group but it was not statiticaly different (p = 0,348). Administration of SSF does not influence BW and BMI/age in adolescent girls with anemia. Further investigation is required to verify wether or not SSF contains high carbohydrate and fibers.

1. Introduction

The global prevalence of anemia is 29.0% (496.3 million peoples) [1] so that anemia becomes a serious public health problem in developed and developing countries [2]. In Indonesia, adolescent girls are one of the highest risk age group with anemia and the prevalence is 26.5% [3]. Anemia is a condition that is charactarized by reduction of erythrocytes amount and hemoglobin (Hb) concentration as an oxygen transporter throughout the human body below normal [4].

Adolescence is the second fastest stage of growth and development after childhood period. For their physical and physiological changes, they need higher amount of nutrients increase nutritional needs [5, 6]. It is not surprising that adolescent girls have 10 times higher risk of anemia than adolescent boys because they undergo regular blood loss during menstruation [7]. However, other factors also contribute to anemia such as low nutrient intake, excessive physical activity, and anemia-limited knowledge [7,8].

Beside of increase of anemia incidence, low nutrition intake also affects nutritional status of adolescent girls. Body mass index (BMI) is the most important indicator for evaluation of nutritional status of adolescent girls [9]. Based on data of Riskesdas (2013), the prevalence of very thin (0.1%), thin (0.4%) and obese (5.9%) nutritional status has increased from 2010 to 2013 [4].

Anemia in adolescent girls can affect not only their human body but also school achievement [10]. In general, anemia can cause symptoms of lethargy, weakness, fatigue, tired and forgetful [11]. Anemia also decreases physical health and ability [5]. Even the incidence of anemia can also reduce economic productivity [12].

Oral iron supplementation becomes the standard therapy for anemia including in Indonesia. However, oral iron supplementation in adolescent girls does not distribute evenly in secondary schools in all areas in Indonesia. Iron-fortified foods are another alternative for anemia treatment for which snacks are often chosen like cookies, candys, chocolate and jelly [5, 13, 14, 15]. Adolescents are more likely to consume snacks than high densed foods. Jelly, for example, is more suitable for iron fortification in adolescent girls because of low energy and anytime consumption.

In Indonesia, snake fruits are cultivated in some regions but snake fruit pondoh is the most popular that is found in Pakem and Turi districts, Sleman Regency. Snake fruit seeds become waste and have not been used for medical purposes. From our research group, we have made SSF that has the content of Fe (26.8 mg / 100 g), Zn (4.65 mg / 100 g), v it C (0.82 mg / 100 g), and protein (3.7%). Therefore, the purpose of this study was to evaluate the effect of SSF jelly on BW and BMI/age in adolescent girls with anemia.

2. Methods

2.1. Experimental design

This study was a randomized control trial with pre-posttests group control design. It was performed in adolescent girls who came from senior high schools (SMA) in the Sukoharjo Regency from January to May 2019. A purposive sampling was used to select districts where representated the prevalence of anemia (high, moderate, and low) in adolescent girls and then selection of SMA used a simple random sampling. We got five SMAs: SMA N 1 Mojolaban, SMA N 1 Kartasura, SMA N 1 Nguter, SMA N 1 Weru, and SMA N 3 Sukoharjo. The protocol of this study was approved by the Ethical Committe of Medical Research, Faculty of Medicine, Unersitas Sebelas Maret, Surakarta (165/UN27.6/KEPK/2018). Selected adolescent girls agreed to follow this study if they signed the informed consent.

2.2. Selection of subjects

Sample size of our study was calculated using a formula from Jaykaran and Tamoghna [16] with 95% confidence level and 80% power test. Therefore, we needed at least 9 students for each group to avoid loss of follow up [17]. Adolescent girls were recruited as research subjects if they met inclusion criteria: 15-18 years old and having moderate anemia with Hb levels 80 - 109 g / dL. A total 38 adolescent girls met the inclusion criteria but only 22 adolescent girls participated in this study and were randomly divided into 2 groups. During 8 weeks intervention, there were two adolescent girls did not complete the study because of rejecting eating the jelly.

2.3. Intervention

SSF was the main ingredients of jelly and was added to agar-agar flour (PT. Dunia Bintang Walet – Jakarta, Indonesia), sugar, and mineral water, based on the procedure of Sani study [15]. Boiled jelly was put into a 100 ml plastic cup and was allowed to cool at room temperature for 1-2 hours. Every research subject in the C group was received 150 g jelly, which was equal to 129 mg Fe and were given three times/week for 8 weeks. Meanwhile T group received a cup of jelly, which contain 37 g of SSF and were equal to 30 mg Fe. Jelly was distributed to research subjects on Monday, Wednesday and Friday.

2.4. Data collection and statistical analysis

Data of subject characteristics including age, height, history of allergy and illness were collected using open questionnaries. Measurements of BW and height used a Gea weight scale and microtoise respectively before, during, and after intervention. Body Mass Index (BMI) was calculated by dividing BW/height^2 and nutrition status was based on z-score [9]. Characteristics data were presented as number, percentage or mean ± standard deviation. The differences of pre-posttest data were statistically analyzed using the paired t-test and one way anova followed by Tukey post hoc test was used to determine differences among four different groups with p value < 0.05.

3. Results and Discussion

In this study, adolescent girls with anemia were given snacks in the form of jselly fortified with Fe supplement in the control group and SSF for the treatment group. Fortification is the addition of a nutrient to food. Fortification can be used to prevent the occurrence of iron deficiency, so that it can increase the value of health. Jelly is made as it is in general with only added ingredients which refers to the manufacture of high Fe pudding with modifications [15]. Jelly with the selected formulation was then tested for safety by the Center for Industrial Pollution Prevention Technology, Semarang. The test results show that observing SFF jelly meets the quality requirements according to SNI 01-35532-1994.

In this RCT Study, 11 adolescent girls were in the C group whereas only 9 adolescent girls were in the T group due to not eating jelly completely (**Table 1**). All research subjects aged beetwen 15 and 17 years old. The mean age of research subject was 15.91 ± 0.83 years old in the C group and 15.89 ± 0.60 years old in the T group. From anthrophometric data, the mean height of research subjects in the C group is similar to the mean height of research subjects in the T group. However, higher mean BW was observed in the C group (49.73 ± 7.80 kg), compared to the T group (46.77 ± 6.44 kg). In Indonesia the average size of the weight and height of adolescent girls is 50 kg and 154 cm [5]. Average of BW and height subjects in this study showed a smaller than the national average. The average of WB in the C group was around 49 kg while in the T group it was in the range of 46 kg. The average of height in the C group and T group was in the same range around of 153 cm.

Table 1. Characteristics of Research Subjects who Participated in The RCT Study.

	Group		P value
	Control	Treatment	
Mean age (years old)	15.91 ± 0.83	15.89 ± 0.60	0.952
Sebaran umur :			
15 years old	4 (36%)	2 (22%)	0.414
16 years old	4 (36%)	6 (67%)	0.527
17 years old	3 (28%)	1 (11%)	0.317
Mean height (cm)	152.92 ± 5.66	153.01 ± 5.49	0.971
Mean BW (kg)	49.73 ± 7.80	46.77 ± 6.44	0.374
Nutritional status:			
Severely wasted	-	-	-
Wasted	-	-	-
Normal	9 (82%)	7 (78%)	0.617
Overweight	1 (9%)	2 (22%)	0.564
Obesity	1 (9%)	-	-
History of allergy			
Eggs	1 (9%)	1 (11%)	1.000
Seafoods	2 (18%)	2 (22%)	1.000
History of illness	1 (9%)	-	-

Based on the 24-hour food recall (**Table 2**), showed the T group higher than the C group in all macronutrient parameters (energy, carbohydrate, protein, fat, and fiber) but did not show a significant difference (p> 0.05). Macronutrient intake on day 60 showed an increase compared to day 0. One of the factors influencing the increase in intake on days 30 and 60 due to the administration of SSF jelly in the study subjects. The average energy intake in both groups showed lower results than the nutritional adequacy rate of Indonesia (AKG). Similarly, energy, carbohydrate, protein, fat, and fiber intake also shows results below the AKG for women in the age group of 15-18 years.

Table 2. Mean of Macronutrient Intake of Research Subjects.

Macronutrient	Day	Mean		P value
		C group	T group	
Energi (kcal)	0	1418.90 ± 679.27	1419.49 ± 530.30	0.998
	30	1328.26 ± 416.21	1639.07 ± 493.06	0.143
	60	1623.05 ± 729.93	1884.24 ± 542.48	0.385
Carbohydrate (g)	0	209.96 ± 105.82	181.50 ± 75.91	0.508
	30	208.77 ± 68.10	308.49 ± 69.58	0.005
	60	246.81 ± 117.40	327.14 ± 139.68	0.179
Protein (g)	0	46.78 ± 20.50	47.74 ± 14.44	0.907
	30	39.68 ± 13.62	42.80 ± 17.44	0.658
	60	49.44 ± 22.88	55.27 ± 18.71	0.547
Fat (g)	0	42.64 ± 22.80	55.01 ± 36.21	0.363
	30	39.76 ± 20.07	39.93 ± 26.03	0.987
	60	50.96 ± 39.55	55.48 ± 13.92	0.748
Fiber (g)	0	6.03 ± 3.23	6.10 ± 2.74	0.958
	30	10.58 ± 4.19	8.62 ± 3.86	0.296
	60	9.96 ± 5.22	11.16 ± 5.81	0.632

Nutritional status of research subjects showed different distribution. Distribution of normal, overweight and obesity nutritional status in the C group was 82%, 9% and 9% respectively. Whereas in T group distribution only in normal (78%) and overweight (22%) nutritional status. But the distribution of each nutritional status in the two groups did not show a significant difference with a P value of 0.617 (normal) and 0.564 (overweight). In this study there were no research subjects with nutritional status wasted and several wasted. Around 10% research subjects in both groups had egg allergy and 20% among research subjects had seafood allergy. From all research subjects, there was only 9% having history of illness.

Table 3. The Effect of SSF Jelly on BW and BMI/age in Research Subjects with Anemia.

	Day	Mean		P value
		C group	T group	
BW (kg)	0	49.73 ± 7.80	46.77 ± 6.44	0.374
	30	49.61 ± 8.36	46.99 ± 6.74	0.458
	60	49.15 ± 9.03	46.16 ± 6.38	0.421
	P value	0.985	0.963	
BMI (kg/m^2)	0	21.25 ± 3.15	20.00 ± 2.49	0.348
	30	21.08 ± 3.46	20.03 ± 2.61	0.407
	60	20.59 ± 3.63	19.53 ± 2.61	0.534
	P value	0.897	0.964	

Table 2 showed the BW average in the C group showed a decrease on the day 30 (49.61 ± 8.36) and returned to the day 60 (49.15 ± 9.03) but did not show a significant difference (p = 0.985), while the T group showed fluctuating results, increased in day 30 (46.99 ± 6.74) and decreased in day 60 (46.16 ±

Journal of Physics: Conference Series **1374** (2019) 012013 doi:10.1088/1742-6596/1374/1/012013

6.38). The increase in BW average in the treatment group was 0.12 ± 1.58 kg but on day 60 it decreased by 0.46 ± 1.19 kg from the day 30, but the difference in values did not show a significant difference (p = 0.963). The difference in mean BW in both groups on days 0, 30 and 60 did not show a significant difference, with P values 0.348, 0.407 and 0.534 respectively.

The nutritional status indicators used BMI/age which is based on anthropometric measurements of BW and height. The classification indicator of BMI/age according to RI Ministry of Health is divided into five: severely wasted (Z score < -3.0), wasted (Z score ≥ -3.0 until < -2.0), normal (Z score ≥ -2.0 until ≤ 1.0), overweight (Z score > 1.0 until ≤ 2.0) and obesity (Z score > 2.0) [9]. BMI/age showed the same pattern as BW, the average in the C group showed a decrease on the day 30 (21.08 ± 3.46) and day 60 (20.59 ± 3.63), while in the T group the BW increased on the day 30 (20.03 ± 2.61) but fell on the day 60 (19.53 ± 2.61). The difference in mean BMI/age on day 0, 30 and 60 days in the both groups did not showed significant differences, 0.897 and 0.964 respectively. Average of BMI/age on a daily basis did not show a significant difference in the C group and T group which was indicated by the P value on the day 0 (0.348), 30 (0.407) and 60 (0.534).

Basically, nutritional status (BMI/age) does not affect the incidence of anemia [18]. In this study around 80% of research subjects with anemia have a normal nutritional status and just around 20% who have poor nutritional status (overweight and obesity). This result not linear with Yunarsih and Sunny statement that poor nutritional status risks causing anemia [18]. Another statement also said adolescents with normal nutritional status have a small risk to anemia than adolescents with more nutritional status sufficient [19]. Adolescents with normal nutritional status or more who anemia are due to lack of consumption of macro nutrients (energy and protein) and micronutrients (such as nutrients), so that the ingredients used to form red blood cells are not sufficient [19].

SSF fortification in jelly is done to meet the needs of macronutrients and micronutrient adolescent girls with anemia. Macronutrients are obtained like carbohydrates and proteins, while the micronutrients obtained are Fe, Zn and vitamin C. In addition, jelly also contains fiber which is a macronutrient. Fortification is generally carried out by adding a nutrient to a food that is widely consumed [11]. In this study jelly was chosen as fortified food, considering the subjects in this study were adolescents who preferred consumtion a low density food (snacks).

This study shows that SSF jelly in anemic adolescent girls does not affect both BW and BMI/age on C group or T group because there was no significant increase or decrease. SSF is a snack that is a food with high energy density [20], because containing macronutrients like carbohidrat, protein, and fiber. This study is in line with the statement of Adhitya et al [21] which stated that giving snacks did not affect nutritional status but had a tendency towards increasing nutrient intake. In the study of Chaterine et al [22], the administration of snacks did not affect the nutritional status of girls and boys, and did not show significant differences in BMI/age. In a study conducted by Leon et al [23], was shown that iron supplementation in children aged 8-12 years with iron deficiency anemia did not show a significant difference in weight and height.

One of the factors that influence nutritional status is physical activity. Basically the nutritional status, physical activity and incidence of anemia are related. There is no significant difference in the administration of jelly, both in the C group and in the T group from days 0 to 60 can be caused by physical activity. According to Silvia et al [24], children with normal nutritional status have a tendency to be more active so that they engage in more physical activity. In this study the majority of research subjects were in normal nutritional status, so it is possible for them to have high physical activity. High physical activity helps the process of energy expenditure, so it is not possible to accumulate energy which causes an increase in BW [25].

In addition, the presence of fiber content in jelly can also be a factor that affects BW or BMI/age. The presence of fiber can reduce BW and reduce macronutrient absorption [26]. High-fiber foods require a longer digestive time, so that young women do not easily feel hungry and reduce excessive food consumption [27]. Therefore young women who were the subjects of this study did not experience an increase in BW, so that it did not affect nutritional status. The disadvantage of this study is that there has been no measurement of physical activity and daily food intake.

4. Conclusions

Jelly SSF as a snack does not show a significant difference of WB and BMI/age in day 0, day 30 and day 60, so observing SSF jelly can be used as a snack alternative for adolescent girl without worrying about increasing WB. However, further research is needed to determine the content in SSF jelly, especially the carbohydrate and fiber content.

References

[1] WHO 2015 The Global Prevalence of Anaemia in 2011 (Geneva : World Health Organization)
[2] Tapas C, Pinnamneni V Rao and Pradeep K Meshram 2012 *Public Health Nutrition* **16** 377-80
[3] Verarica S, Evawany A and Taufik A 2016 *Jurnal Kesehatan Masyarakat* **11** 96-102
[4] Ministry of Health 2013 Riset Kesehatan Dasar (Jakarta : Kemenenterian Kesehatan RI)
[5] Dodik B, Aris S, Elvira S and Dian H 2013 *MKB* **45** 206-12
[6] Melkam T, Tilahun Y, Wondinagegn A, Yaregal A and Lealem G 2015 *Adolescent Health, Medicine and Therapeutics* **6** 189-196
[7] Alfishar A and Sri S 2017 *Amerta Nutr* 105-16
[8] Desri S, Riska H and Rinseti J R 2015 *Jurnal Kesehatan Masyarakat Andalas* **10** 11-8
[9] Ministry of Health 2011 Keputusan Menteri Kesehatan Republik Indonesia Nomor: 1995/MENKES/SK/XII/2010 tentang Standar Antropometri Penilaian Status Gizi Anak (Jakarta : Kementerian Kesehatan RI)
[10] Pratiwi R A Dumilah and Sri S 2017 *Amerta Nutr* 331-40
[11] Masrizal 2007 *Jurnal Kesehatan Masyarakat* **2**
[12] UNHCR 2008 UNHCR Strategic Plan for Anaemia Prevention, Control and Reduction (United Nations High Commissionerfor Refugees)
[13] Nehad R El-Tahan and Nehal A A Alfky 2016 *International Journal of Advanced Research* **4** 6-11
[14] Dace P and Sandra M 2016 *Research for Rural Development* **1** 111-15
[15] Sani 2014 *Pengembangan Puding Instan Tinggi Fe sebagai Makanan Selingan untuk Remaja Putri* (Bogor : Institut Pertanian Bogor)
[16] Jaykaran C and Tamoghna B 2013 *Indian Journal of Psychological Medicine* **35** 121-6
[17] Sastroasmoro S and Ismael S 2014 *Dasar-dasar Metodologi Penelitian Klinis* (Jakarta: Sagung Seto)
[18] Yunarsih and Sumy D Antono 2014 *Jurnal Ilmu Kesehatan* **3** 25-33
[19] Basri A, Nihan W Siketang and Abidah N 2017 *Jurnal Penelitian Kesehatan* **4** 21-30
[20] Azizah A Pratiwi and Triska S Nindya 2017 *Amerta Nutr* 153-61
[21] Adhitya A Candra, B Setiawan and M R M Damanik 2013 *Jurnal Gizi dan Pangan* **8** 103-8
[22] Catherine M Nguu-Gutu, Jasper K Imungi and Edith M Ngatia 2014 *Food Science and Quality Management* **30** 48-58
[23] Leon A, Tiangsa , Ani A and Bidasari L 2009 *Paediatrica Indonesiana* **49** 160-4
[24] Silvia B S B, Lina E F P D L R, Gilberto P R and Ria D C LR 2006 *Rev Bras Med Esporte* **12** 80-4
[25] Weni K, Isti S, Nurul H S, Listiana P S, Mahardika M A, Diana M and Kurnia I S 2015 *Jurnal Gizi Klinik Indonesia* **11** 179-90
[26] Susilowati, Ayu L M, Astrina T and Tya N A 2017 *Jurnal Kesehatan Kartika* **12** 53-63
[27] Eiman A H 2017 *Arab Journal of Nutrition and Excercise* **1** 77-84

Journal of Physics: Conference Series **1374** (2019) 012014 doi:10.1088/1742-6596/1374/1/012014

Kelakai Extract Protects Skin From UV-Induced Oxidative Damage

M Mashuri[1], L D M Sihombing[2], S Alfaqihah[2], E Edyson[3] and E Suhartono[3*]

[1]Department of Radiology, Faculty of Medicine, Lambung Mangkurat University, Veteran Street No. 128, Banjarmasin 70232, South Kalimantan, Indonesia
[2]Faculty of Medicine, Lambung Mangkurat University, Veteran Street No. 128, Banjarmasin 70232, South Kalimantan, Indonesia
[3]Department of Medical Chemistry/Biochemistry, Faculty of Medicine, Lambung Mangkurat University, Ahmad Yani Street Km. 36, Banjarbaru 70712, South Kalimantan, Indonesia

*Coresponding author's email: esuhartono@ulm.ac.id

Abstract. Oxidative stress is an unbalance situation between oxidant and antioxidant. Oxidative stress can be caused by UV exposure until it can result skin damage. This damage caused by the production of reactive oxygen compound (ROC), such as excessive superoxide anion. Thus, natural exogen antioxidant is needed, that is Kelakai. This research is pure experimental research with rats *(Rattus norvegicus)* as its subject. There were 24 samples were divided into 4 treatment groups; samples were chosen used simple random sampling method and repetition value used Federe method. Kelakai leaves were extracted using ethanol with maseration method. Superoxide anion level and superoxide dismutase enzyme activities were measured with Misra and Fridovich method; carbonyl level was measured using modified DNPH (*dinitro phenilhidrazin*) method; and conjugated diena level was measured with Kwiat Kowska method. The data was analysed statistically, and resulted significant differences between the group that only exposed to UV and the group that was given Kelakai extract and exposed to UV. Based on that result, it can be concluded that Kelakai extract can influence superoxide dismutase activities, superoxide anion level, carbonyl level, and conjugated diena in rats skin exposed to UV.

1. Introduction

Oxidative stress is an unbalance situation between oxidant and antioxidant, which leads to oxidative damage (lipid peroxidation, DNA oxidation, protein oxidation, etc). Those oxidative damage can be caused by many factors, one of them is ultra violet (UV) radiation [1,2].

UV is an electromagnetic wave with a wavelength at 100-400 nm. UV that comes from sun rays can cause skin disorders, such as erythema, dry skin, wrinkles, hyperpigmentation, and also can leads to cancer [3]. Those disorders were started by the absorbance of UV by oxygen, which influence its electronical structure. The changed structure causes the forming of singlet oxygen and reactive superoxide anion [4]. Both oxygen types are reactive and can oxidate biomolecule compound, such as lipid and protein. One of indicators from protein damages is the forming of carbonyl compound [5], while lipid damage is marked by the forming of conjugated diena compound [6].

To prevent oxidative damage due to UV exposure, need natural ingredients that has function as antioxidant. The natural ingredients for example is Kelakai. Kelakai (*Stenochlaena palustris)* is one of typical plants for South Kalimantan that has been used beneficially as functional food, dismenore medication, reducing fever, etc. Adenan's research (2010) stated that Kelakai extract has been proven in reducing liver proxidative stress in rats induced fever [7]. Kelakai has flavonoid compound 14,5 QE/µg/mL which has role as peroxide [8]. Kelakai also capable to inhibit glication and fructation reactions; the reaction between glucose/fructose and protein [9].

Antioxidant in extract Kelakai allegedly capable to be a protector to skin that has been induced acute UV. However, no one has revealed this. Because of that, in this research Kelakai extract role will be analyzed in inhibiting oxidative stress for skin exposure with UV, by measuring superoxide anioncarbonyl, and conjugated diena level, also the activity of superoxide dismutase (SOD) enzyme [10].

2. Research methodology

2.1. Research subject

Research subject is male rat aged 8-10 weeks with weight 200-300 grms. Rats were acclimated for one week and given the same meals and drink. Next, rats were divided into four groups: control (P0); rats given UV exposure for 24 hours (P1), rats given Kelakai extract in dosage 4 mg/cm^2 (P2), rats rats given Kelakai extract in dosage 4 mg/cm^2 and exposured with UV for 24 hours (P3).

2.2. Treatment of Experimental Animals

Rats were shaved its back hair in width 3 x 3 cm^2 then rubbed with Kelakai extract in dosage 4 mg/cm^2. Next, they were put in box with 106x34x53 cm^3 in width where there was Philips UV lamp with wavelength of 253,7 nm. Radiation was given for 24 hours. After that, rats were dissected in order to extarct its skin.

2.3. Making of Rats Skin Homogenat

Rats skin was extracted by dissection, then it as cleaned up and smoothed using 1 ml of TCA (trichloroasetar acid) 20% and 3 ml of buffer phosphate pH 7,4. After that, the skin was put into tubes and centrifugated in 2000 rpm of rate for 10 minutes for its supernatant to be used. The supernatant was used to measured superoxide anion level, carbonyl compound, conjugated diena, and dismutase enzyme activity.

2.4. Measurement of Superoxide Anion Level

Superoxide anion level was measured by using Misra and Fridovich method. Supernatant was added into 500 µl of solution that contained 800 µl of buffer carbonate 0,1 M pH 10,2. Next, in that solution, was also added 100 µl of adrenaline (epinephrine 3 mM) and then was measured its initial uptake (A_0) using spectrophotometer at wavelength 480 nm [11,12] .

2.5. Measurement of Superoxide Dismutase Enzyme Activity

Superoxide dismutase enzyme activity in supernatant was measured by Misra and Fridovich method. The 500 µl of supernatant were added to 800 µl of buffer carbonate (100 nM pH 10,2) and 100 ml of epinephrine (3 mM). The absorbance changes in each sample were recorded at spectrophotometer with wavelength 480 nm in interval 15 seconds [11,12] .

2.6. Measurement of Carbonyl Level

Carbonyl level was measured using modified DNPH (dinitro phenilhidrazin) method. Each solutions were taken for 0,5 ml and divided into two tubes, tube A and tube B. tube A was added with DPNH fo 1 ml; and Tube B was added 1 ml of HCl 2,5 M. those solutions were incubated and shaken with vortex. The next step was adding 1 ml of TCA 20% then incubated in ice for 5 minutes and centrifugated for 5 minutes in rate 1400 rpm, and the supernatant was discarded. The further step was the cleaning by adding

1 ml of *metanol-etil asetat*. It was then added by 1 ml of urea to be incubated and centrifugated. The resulted color was measured it absorbance at wavelength 390 nm. Protein damage was calculated at equation C=A/(eb) (A=absorbance, e=extinction coefficient 22.000 mM cm^{-1}, and b=1 cm) [13].

2.7. Measurement of Conjugated Diena Level
Conjugated diena level was measured with *Kwiat Kowska* method. Supernatant for 0,5 ml was mixed with 7 ml of *choloroform/methanol* (1:1 u/u). Next, it was shaken for 2 minutes (1500 gr for 5 minutes). The 5 ml of the solution at the lower layer mixed with 2 ml of aquadest which acidified with HCl pH 2,5. Then the solution was shaken again for 2 minutes in the same way at the previous step. The recidue was dissolved with 2 ml of heptane (cyclohexane) and measured its absorbance 233 nm with heptane solution as its blank (cyclohexane). The result was stated as absorbance unit or resulted hydroperoxide [14].

3. Result and discussion
3.1. Superoxide Anion
Superoxide anion is a reactive oxygen derivative, resulted form UV absorbance by oxygen which caused the changes in its electronic strukture. Superoxide anion level in this research is shown in figure 1.

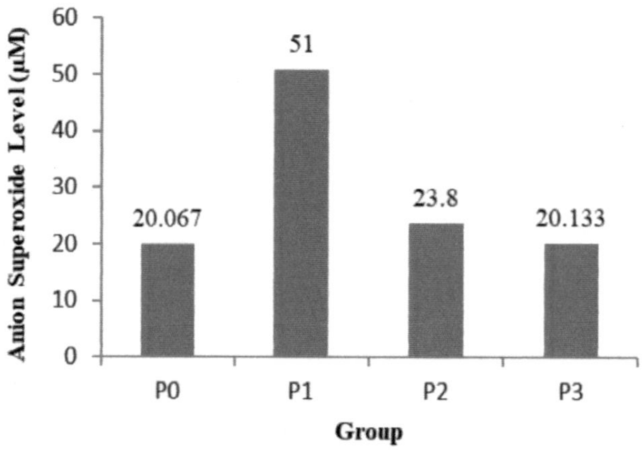

Figure 1. Superoxide anion level in rats skin homogenat (*= significantly different with P0).

One way ANOVA test result concluded that there was significant differences (p=0,000; $p< 0.05$). Skin contains of melanine that has chromophor group, which has ability to absorb UV rays. UV rays absorbance causes melanine changes into a more active compound. The melanine which more active will react to oxygen and forms superoxide anion.[15] The excess of UV exposure can stimulate the forming of ROC that leads to the increase of the formed superoxide anion level.[16] Kelakai extract contains flavonoid, which is able to capture superoxide anion resulted due to UV exposure.[10]

3.2. Activity of Superoxide Dismutase Enzyme
Superoxide anion can be neutralized using dismutase superoxide enzyme by forming H_2O_2. The activity of superoxide dismutase enzyme in this research can be seen in figure 2. One way ANOVA test result concluded that there was signifacant differences (p=0,000; $p< 0.05$).

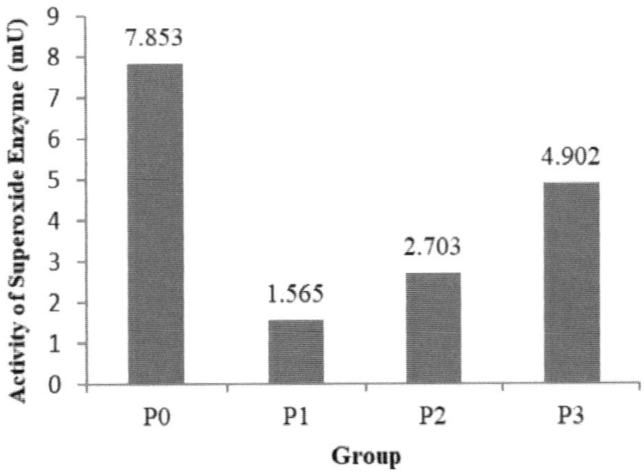

Figure 2. The activity of Superoxide Dismutase Enzyme in rats skin homogenate.
(*= significantly different to P0)

UV rays that reached skin will create the forming of ROC which is able to cause moleculer and celular damage. Skin has endogen antioxidant in order to againts that ROC that is superoxide dismutase enzym. UV exposure will cause the use of superperoxide dismutase enzyme increased, therefore the enzym activity is decreaed.[17] Flavonoid contents in Kelakai extract capable to capture free radicals (*free radical scavenging)*, so that the use of superoxide dismutase enzyme decreased.[10] Decreased using of superperoxide dismutase enzyme will increase superperoxide dismutase enzyme activitiy. But, flavonoid also can be prooxidant if aroxyl radical resulted from the work mechanism of flavonoid reacted to oxygen. That reaction will produce superoxide anion, so that will cause the decreased of superoxide dismutase enzyme activity.[18]

3.3. Conjugated Carbonyl and Diena
Carbonyl is compound formed due to protein damage. The level of carbonyl compound can be seen in figure 3.

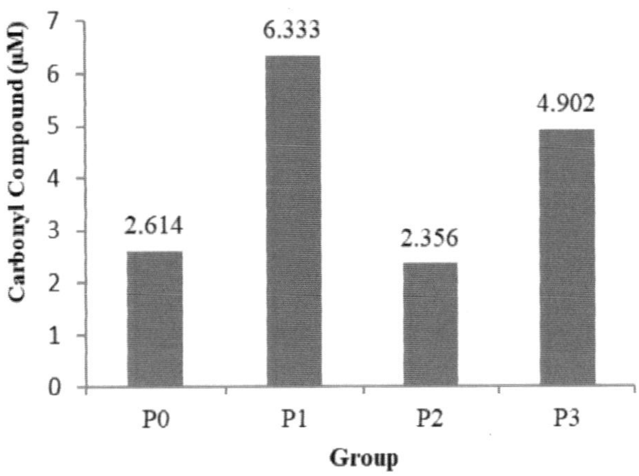

Figure 3. The average of carbonyl compound in rats skin homogenate.
(*= significantly different to P0)

Kruskal-Wallis result test concluded that there was singnificant differences (p=0,000; p< 0.05).

Conjugated diena is compound that formed due to lipid damage. The level of conjugated diena in this research can be seen in figure 4.

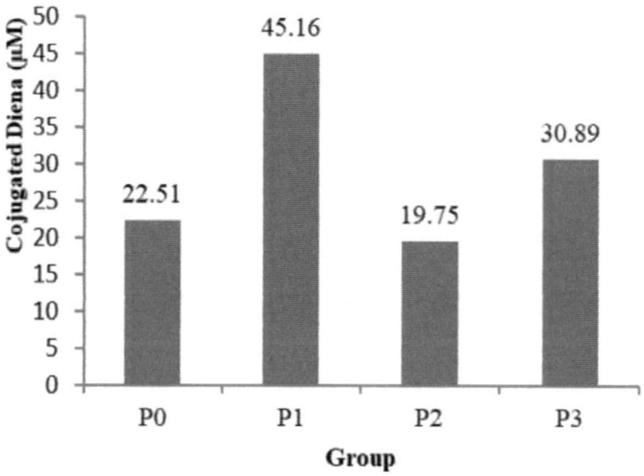

Figure 4. The average of conjugate diena compound in rats skin homogenate. (*= significantly different to P0)

Kruskal-Wallis result test concluded that there was singnificant differences (p=0,000; $p < 0.05$).

Oxidative stress happened in UV exposured rats skin valued form carbonyl compound level. UV compound can be absorbed by O_2 in biosphere, untill forms $^1O_2$, which for the next can transforms into $^\bullet O_2$. Both oxygen molecul will oxidize protein side chain, especially prolile, arginile, lysine and threonine, even carbonyl compound.[5] UV radiation can oxidize epithelial keratinocyte protein in normal human. That research also reported that protein GRP 78 radiated by UV caused the increasing of carbonyl level.[19]

Furthermore, oxidative stress happened in UV exposured rats skin cam be valued from conjugated diena compound level. UV compound absorbed by O_2 transforms into $^1O_2$, which for the next can transforms into $^\bullet O_2$. Both oxygen molecule are named reactive oxygen compound (ROC). The formed ROC does not only damage to protein, but also damage to lipid. Lipid damage will forms a compound which will be detected as lipid damage marker that is conjugated diena.[19] Bignon research said UV exposured on skin for 15-30 seconds, measured with wavelength 230 and 270 nm resulted the decrease of celulare methabolic activity and increase cell death in lipid and protein at maximal UV exposure for 230 nm. PARP marker formed as the result from cell apoptosis process.[20]

In order to decrease the damage due to oxidative stress, an additional antioxidant is needed. In this research, the additional antioxidant used was Kelakai extract. Kelakai extract containts a compound that is able to capture free radical, flavonoid. The mechanism of capturing free radical by flavonoid is started from electron release, forming reactive flavonoid radicals which at the end bind free radical, superoxide anion.[7] Superoxide anion binding by reactive flavonoid radicals will reduce damage happened in protein and lipid, which finally decrease carbonyl compound and cojugated diena levels.

4. Conclusion

Based on the research that has been done, it can be concluded that there is an influence from the administration of Kelakai extract to rats' skin exposed to UV rays, reviewed from superoxide dismutase enzym activity, superoxide anion level, carbonyl level, and conjugated diena level.

5. References

[1] Yoshikawa T and Naito Y 2002 What is oxidative stress *Journal of the Japan Medical Association* **45** (7) 271–276

[2] Edyson, Pardede A M E, Nugraha H G, Mashuri and Suhartono E 2019 In vivo antioxidant and UV Photoprotective of extract pasak bumi (Eurycoma Longifolia Jack.) *AIP Conference Proceedings* **2108** 020010
[3] D'Orazio J, Jarrett S, Amaro O A and Scott T 2013 UV radiation and the skin *International Journal of Molecular Science* **14** (6) 12–48
[4] Yokawa K, Kagenishi T and Baluška F 2016 UV-B induced generation of reactive oxygen species promotes formation of BFA-induced compartements in cell of arabidopsis root species *Front Plant Sci.* **6** 1-10
[5] Dalle D I, Rossi R, Giustarini D, Milzani A and Colombo R 2003 Protein carbonyl groups asbiomarkers of oxidative stress *Clinica Chimica Acta* **329** (1–2) 23–38
[6] Farhoosh R and Moosavi S M R 2009 Evaluating the performance of peroxide and conjugated diene values in monitoring quality of used frying oils *J. Agric. Sci. Technol.* **11** (2) 173-9
[7] Adenan and Suhartono E 2010 Stenochlaena palustris aqueous extraxt reduce hepaticperoxidative stress in Marmota caligata with induced fever *Universa Medicina* **29** (3) 123-128
[8] Suhartono E, Viani E, Rahmadhan M A, Gultom I S, Rakhman M F and Indrawardhana D 2012 Total flavonoid and antioxidant activity of some selected medicinal plants in South Kalimantan of Indonesian *APCBEE Procedia* **4** 235-239
[9] Suhartono E, Bahriansyah M and Triawanti 2016 The inhibition effect of kelakai *(Stenochlaena palustris)* extract on cadmium-induced glycation and fructation in vitro *International Journal of Pharmaceutical and Clinical Research* **8** (4) 248-253
[10] Margono D P N H, Suhartono E and Arwati H 2016 Potensi ekstrak kelakai (*Stenochlaena palustris (Burm.f) Bedd*) terhadap kadar tumor necrosis factor-alfa (TNF-α) pada mencit BALB/c yang diinfeksi Plasmodium berghei ANKA *Berkala Kedokteran* **12** (1) 77-85
[11] Misra H P and Fridovich I 1972 The role of superoxide anion in the autoxidation of epinephrine and a simple assay for superoxide dismutase *J. Biol. Chem.* **247** (10) 3170-3175
[12] Mashuri, Ruhullah M, Putera B D, Mega V P, Putra F A and Suhartono E 2018 The Role of Ceftazidime as a Photosensitizer in Human Erythrocytes through Oxidative Stress Mechanism *Indones. Biomed. J.* **10** (2) 128-132
[13] Noor W F, Aprianti N, Saputra S R, Apriasari M L and Suhartono E 2015 Oxidative Stress on Buccal Mucosa Wound in Rats and Rule of Topical Application of Ethanolic Extracts of Mauli Banana (Musa acuminata) Stem *Journal of Tropical Life Science* **5** (2) 84-87
[14] Kwiatkowska S, Piasecka G, Zieba M, Piotrowski W and Nowak D 1999 Increased serumconcentrations of conjugated diens and malondialdehyde in patients with pulmonary tuberculosis *Respir. Med.* **93** (4) 272–6
[15] Vilela F M P, Fonseca Y M, Jabor J R, Vicentini F T M C and Fonseca M J V 2012 Effect of ultraviolet filters on skin superoxide dismutase activity in hairless mice after a single dose of ultraviolet radiation *European Journal of Pharmaceutics and Biopharmaceutics* **80** 387-392
[16] Buettner G R 2011 Superoxide dismutase in redox biology the roles of superoxide and hydrogen peroxide *Anti cancer agents in medicinal chemistry* **11** (4) 341-346
[17] Pandel R, Poljšak B, Godic A and Dahmane R 2013 Skin photoaging and the role of antioxidants in its prevention *ISRN Dermatology* **2013** 1-11
[18] Biworo A, Atanta A W, Arianto I S, Hamidah S and Suhartono E 2019 Ameliorative effect of tuber extract from bawang dayak (*Eleutherine palmifolia* (L.) merr) against acute UV-induced skin oxidative damage in *Rattus norvegicus AIP Conference Proceedings* **2108** 020010
[19] Perluigi M, Domenico F D, Blarzino C *et al* 2010 Effect of UVB-induced oxidative stress on protein expression and specific protein oxidation in normal human epithelial keratinocytes: a proteomic approach *Proteome Science* **8** (13) 153-161
[20] Bignon E, Marazzi M, Besancenot V *et al* 2017 Ibuprofen and ketoprofen potentiate UVA induced cell death by a photosensitization process *Scientific Reports* **7** (8855) 1-10

Journal of Physics: Conference Series **1374** (2019) 012015 doi:10.1088/1742-6596/1374/1/012015

The effect of kaffir lime (*Citrus hystrix* DC) essential oil on bioplastic derived from cassava peel waste

M Masruri[1*], A Z Azhar[1], I Rosyada[1] and A Febrianto[1]

[1]Chemistry Department, Faculty of Mathematics and Natural Sciences, Brawijaya University
Jl. Veteran 65145 Malang

*Corresponding author's email: masruri@ub.ac.id

Abstract. Kaffir lime essential oil is reported contain a biologically active secondary metabolite from the class of terpenoid. It was reported active as antibacterial, antioxidant, and anticancer. Combination of this bioactive essential oil to afford bioplastic or biodegradable plastic will open their future applications in industry and medicine. This paper report the characteristic of the combining of kaffir lime essential with starch from cassava peel waste to produce bioplastic. The FTIR spectroscopy, SEM, XRD and elasticity test was reported to the bioplastic produced. In short, bioplastic prepared from both starting materials are able to be made. The appearance is clear and transparent. The FTIR spectra also provide a combination of both functional groups, and some fractures found to the bioplastic without addition of kaffir lime essential oil. Interestingly, elasticity improves for bioplastic composed of essential oil.

1. Introduction

Essential oil produces from kaffir lime (*Citrus hystrix* DC) by steam distillation shows many bioactivities, such as antibacterial [1, 2], antioxidant [3], and anticancer [4, 5]. This oil is composed of sesquiterpene (C15) and monoterpene (C10) class of the terpenoid secondary metabolite [6] group (Fig 2). The structure of both groups have hydroxyl- (OH), carbonyl- (C=O), and carboxyl- (O=CO-R) ester (Figure 1), that is able to be used interact to the other polar groups from the other materials. This interaction can undergo via hydrogen bonding, electronic induction and or nonpolar-nonpolar interaction, similar to that glycerin does act as plasticizer in plastic material[7, 8]. With the longer carbon chain from 10 in monoterpene to 15 in sesquiterpene structures increase their hydrophobicity. In other hand, East Java province also reported to be the producer of cassava (*Manihot esculenta*). The average production is about 3.2 kTon per year during the last decade, and is mostly used as raw material for cassava flour, traditional food-fermented product, and cracker-based snacks. However, this production lines provide the peel's waste in average 10-15% from the total of dry weight. The disposed waste still contains starch in about 10% as well (Fig 1). Previously, it has been reported to modification into the acetyloxy- and oleic acid ester of the starch peel's waste [9]. This paper is disclosed the application of starch prepared from cassava peel's waste for bioplastic production. The addition of kaffir lime essential oil (EO) toward the bioplastic increases not only the elasticity or strength and elongation stability of bioplastic, but also the smoothness and integrity of bioplastic surface.

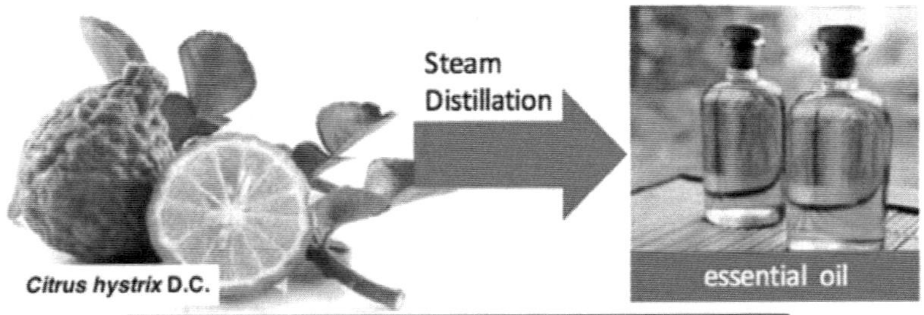

No	Name	Percentage
1	Nerolidol (C15)	58.3%
2	Citronellal (C10)	15.5%
3	Citronellol (C10)	2.78%
4	Linalool (C10)	2.28%
5	Isopulegol (C10)	1.61%
6	citronellyl acetate (C10)	1.28%
7	*trans*-Caryophyllene (C15)	1.17%
8	Geranyl acetate (C10)	0.69%
9	α-Copaene (C15)	0.34%
10	α-Farnesene (C15)	0.27%

Figure 1. Process production of kaffir lime essential oil from leave or fruit part of *Citrus hystrix* by steam distillation and its essential oil composition.

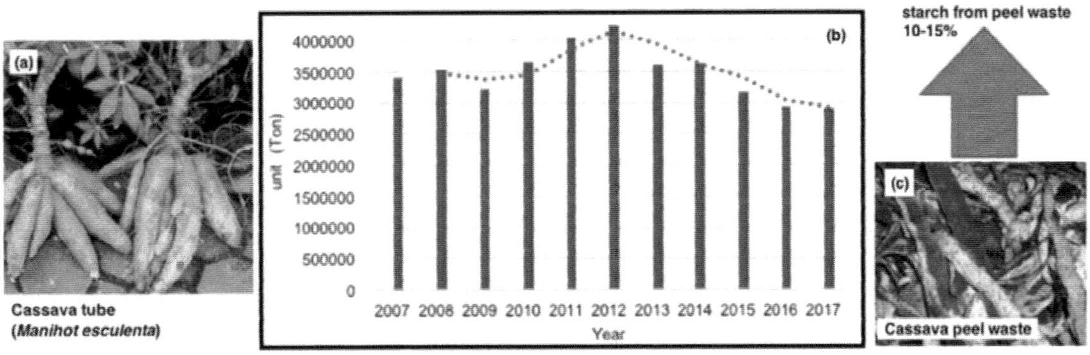

Figure 2. Cassava (a) production in East Java, Indonesia (b) and its potential peel's waste derived starch (c) in between 10 and 15% from the disposed waste.

2. Experiment section

2.1. *Material and chemicals*

Sample of cassava is bought from local farmer in Malang region, East Java. The taxonomy is known as Manihot esculenta as previously reported. Meanwhile the kaffir lime (Citrus hystrix DC) essential oil (EO) is bought from Institute Atsiri of Brawijaya University, Malang. Some chemicals used for research include citric acid (CA) (Merck), and glycerol (Merck).

2.2. *Instrument analysis*

Spectrometer used for analysis include infrared spectra measurement using FTIR spectrophotometer (Shimadzu), scanning electron microscope (SEM) brand of Hitachi TM-3000, x-ray diffraction (XRD) PanAnalytical type E'xpert Pro, elongation and tensil strength measurement using Imada Force ZP-200N.

2.3. Bioplastic preparation
A homogeneous solution of dried of peel's waste starch (5.0 g) is added glycerol (1.5 mL). Then, kaffir lime essential oil (EO) (0.5 mL) and citric acid (0.7 g) is added. This mixture is stirred for 45 min, and heated to 80 °C until a viscose mixture is formed. This mixture is poured into bioplastic template, made of a glass plate (10x10 cm) and dried in oven at 50 °C for 48 h. The bioplastic resulted is further analysis for characterization. This procedure is labelled as sample 1. Others separated procedures are undertaken with different addition. Sample 2, is prepared without addition of essential oil and using citric acid 0.7 g. Sample 3 uses essential oil 0.5 mL and without addition of citric acid. Sample 4 is prepared without the addition of essential oil and citric acid.

3. Result and discussion
The bioplastic prepared from cassava peel's waste are able to be produced with starch as the main composition, and additive plasticizing agent of glycerol (Figure 3). Four different mixtures of bioplastic are prepared with different addition of kaffir lime essential oil (EO) and citric acid. Addition of the EO on the bioplastic mixture made the surface smooth and denser. No wrinkle is observed (Figure 1a and b), however without addition EO made bioplastic feature crack and coarse (Figure 3c and d). SEM micrograph also give more detail features morphology of bioplastic in the surface (Figure 4). EO of kaffir lime is predicted to affect the smooth of bioplastic surface (Figure 4a and b). But a coarse and less dense surface with pore or hole observe is observed for bioplastic without addition of EO from kaffir lime (Figure 4c and d).

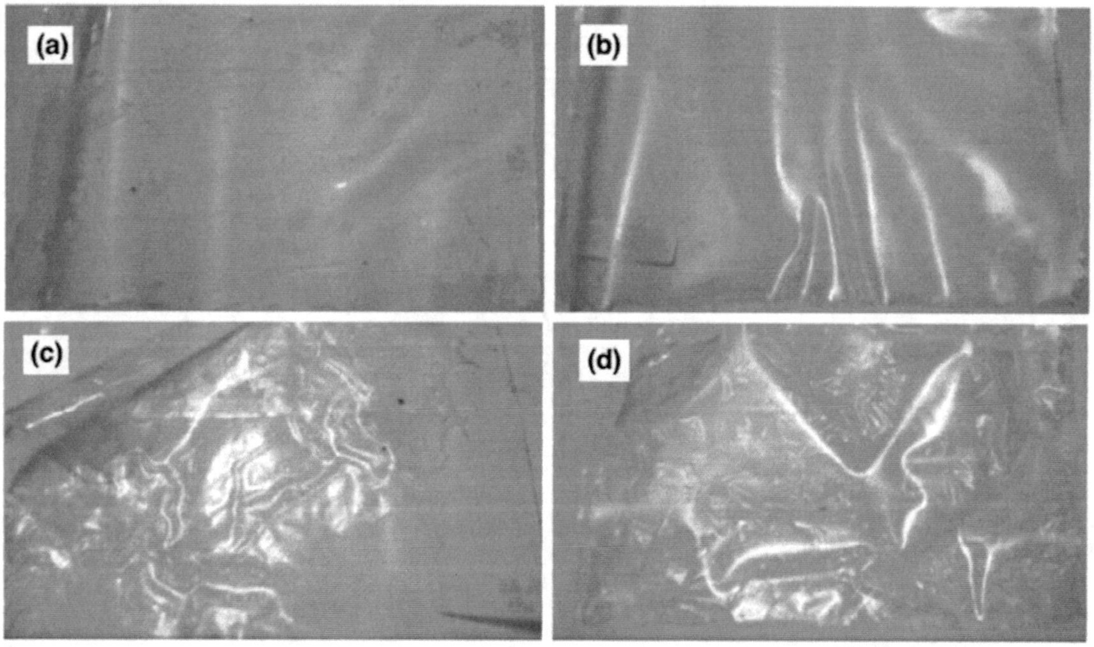

Figure 3. Bioplastic prepared from starch of cassava peel waste with and without addition of kaffir lime essential oil (EO) and citric acid. (a) sample 1, (b) sample 2, (c) sample 3, and (d) sample 4.

The physical analysis toward bioplastic is also undertaken using FTIR spectrophotometer. In some cases, blending process all of reactants such as starch, essential oil, plasticizer and citric acid can affect the integrity of bioplastic resulted. Homogeneous mixing can be represented by a complete solvation and is indicated from the combination of the FTIR spectra. The result (Figure 5), indicate FTIR spectra consists of the components of the starting materials. Hydroxyl-group is observed in between 3500-3200 cm^{-1}. It is a broad band that correlate to hydroxyl-group (O-H) of carboxylic acid (RCOOH) compounds as well. The carbonyl-group (C=O) is detected at 1726 cm^{-1} and the double bond of carbon-carbon (C=C) alkene group is recorded presence in 1647 cm^{-1}. Meanwhile the band for single bond of (C-O-C) ester found is between 1200-1001 cm^{-1}. All these bands correlate to the starch, EO, and citric acid that are mixed in bioplastic.

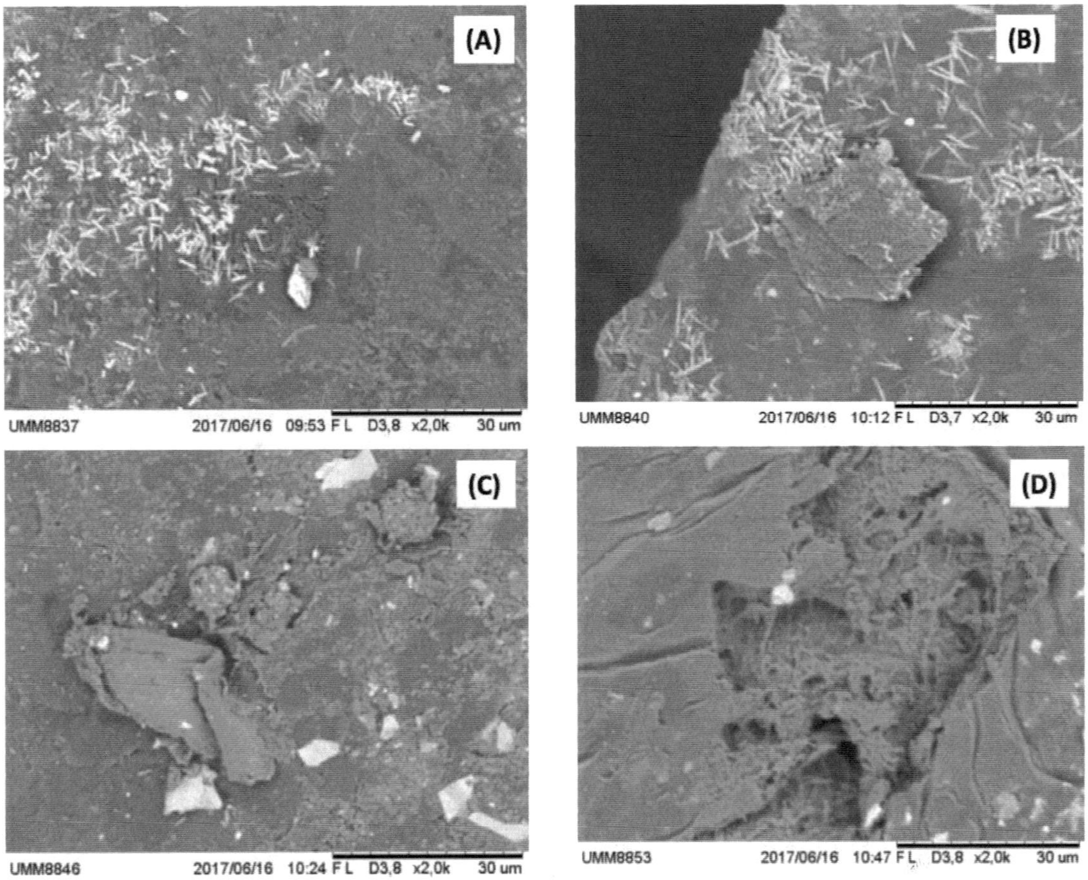

Figure 4. Photograph of scanning electron microscope.

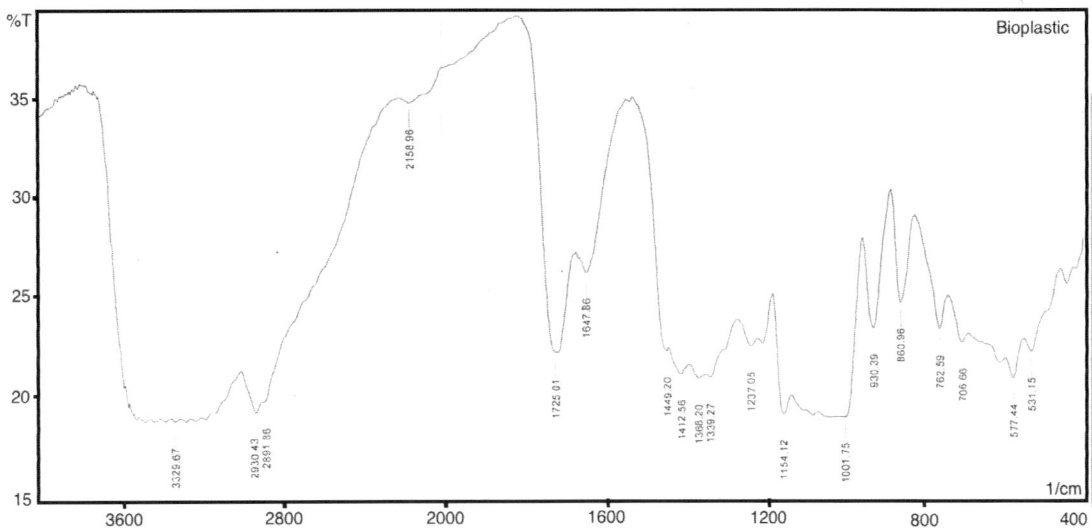

Figure 5. Infra red spectra of bioplastic prepared with kaffir lime essential oil and starch derived from cassava peel waste.

Even though the FTIR spectra indicate a blend and homogeneous mixtures, x-ray diffraction spectra slightly give a detail accuracy of homogeneous mixtures (Figure 6). The bioplastic has a low crystallinity. An amorphous mixture is generally observed, but combination peaks of all reagent composed of the bioplastic clearly represented (sample 1, sample 2 and sample 4)

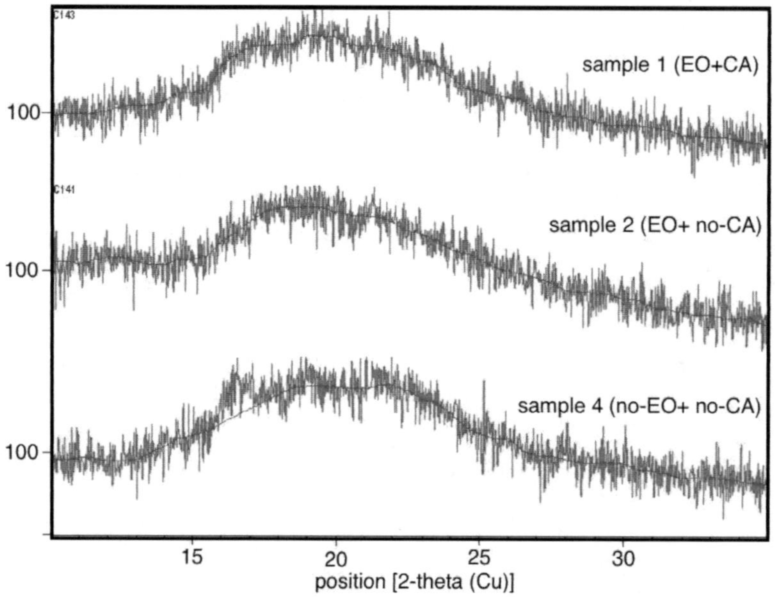

Figure 6. XRD diffractogram of bioplastic.

Moreover, the effect of addition of EO from kaffir lime increase the stability in tensile strength of bioplastic (Figure 7). The bioplastic made of EO resist the stretching energy until 1.8 N/cm and able to elongate 85% (Sample 1) and about 65% (Sample 2). But, a low stability is observed for bioplastic made of cassava peel's waste starch without addition of EO from kaffir lime nor citric acid.

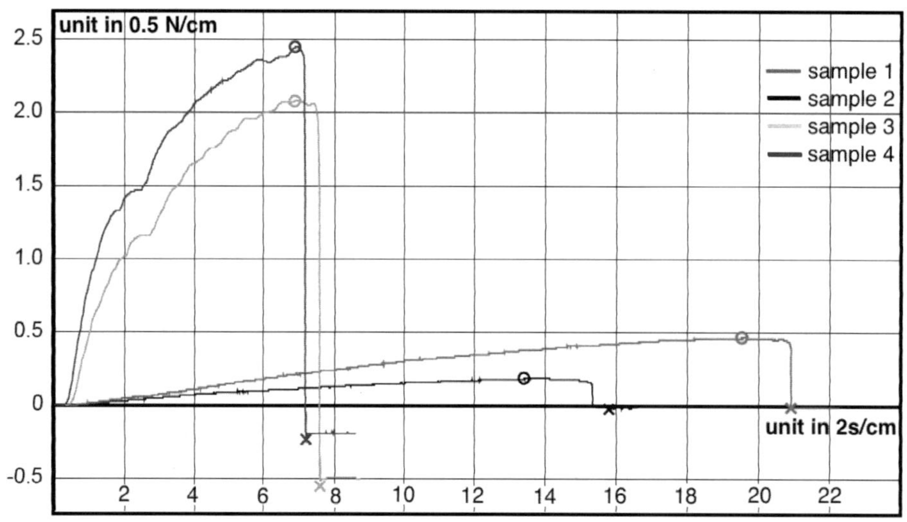

Figure 7. Tensil strength and elongation measurement of bioplastic prepared from cassava peel waste with and without kaffir lime essential oil (EO).

4. Conclusion

The cassava peel's waste can derive starch that is further applied for bioplastic. Combining kaffir lime essential oil (EO) into bioplastic create improvement in tensile strength stability and that increase elongation of bioplastic by 85% longer. This result can pave the way for further intensive research by improvement and optimization for cassava starch for starting material of bioplastic production.

5. Acknowledgement

All involved authors have an equal contribution to this paper; MSR have an idea, data analysis and writing paper, meanwhile AZA, IR and AF collecting the data. Part of the research is supported by directorate general for student innovation research grant 2017 and partly supported by MSR research grant 2017. Author thank to Mr Widji Sulistjo and Hadi Kurniawan from instrumentation unit of chemistry department of Brawijaya University.

Reference

[1] Raut J S and Karuppayil S M 2014 A status review on the medicinal properties of essential oils *Industrial Crops and Products* **62** 250–264

[2] Habsari R A, Warsito W and Noorhamdani N 2018 Chemical Composition of Oil Fraction Kaffir Lime (Citrus hystrix DC) as Antibacterial Activity of E. coli *The Journal of Pure and Applied Chemistry Research* **7** 32

[3] Warsito W, Noorhamdani N, Sukardi S and Suratmo S 2018 Assessment of antioxidant activity of citronellal extract and fractions of essential oils of Citrus hystrix DC *Tropical Journal of Pharmaceutical Research* **17** 1119–1125

[4] Abirami A, Nagarani, G and Siddhuraju P 2015 Hepatoprotective effect of leaf extracts from Citrus hystrix and C. maxima against paracetamol induced liver injury in rats *Food Science and Human Wellness* **4** 35–41

[5] Sun S, Phrutivorapongkul A, Dibwe D F, Balachandran C and Awale S 2018 Chemical constituents of Thai citrus hystrix and their antiausterity activity against the PANC-1 human pancreatic cancer cell line *Journal of Natural Products* **81** 1877–1883

[6] Croteau R, Kutchan T M and Lewis N G 2000 Natural products (secondary metabolites). In *Biochemistry and Molecular Biology of Plants* (2015th ed., Vol. 24, pp. 1250–1319) (Oxford, UK: American Society of Plant Physiologist)

[7] Krauskopf L 1993 Plasticizer structure/performance relationships *Journal of Vinyl Technology* **15** 140–147

[8] Vieira M G A, da Silva M A, dos Santos L O and Beppu M M 2011 Natural-based plasticizers and biopolymer films: A review *European Polymer Journal* **47** 254–263

[9] Susanti S, Al Karoma D, Mulyani D and Masruri M 2017 Physical Properties and Characterization of Cassava Peel Waste Modified by Esterification *The Journal of Pure and Applied Chemistry Research* **6** 255

Antioxidant activity of vacuum column chromatography fractions of ethanol extract of sarang banua (*Clerodenrum fragrans* vent willd) leaves

M Simorangkir[1]*, B Nainggolan[1] and S Silaban[1]

[1]Department of Chemistry, Faculty of Mathematics and Natural Sciences, Universitas Negeri Medan, Medan 20221, Indonesia

*Corresponding author's email: murniatysimorangkir@unimed.ac.id

Abstract. Sarang banua plants are found in Simalungun, North Sumatra, Indonesia, which have been used by the community as traditional medicinal plants. The result of plant determination, sarang banua plant is a type of *Clerodendrum fragrans* Vent Willd, including the family *Verbenaceae*. The antioxidant activity of vacuum column chromatography fractions of ethanol extract of *C. fragrans* leaves was completed. This study aims to find a pure fraction through vacuum column chromatography from the ethanol extract of *C. fragrans* leaves which has antioxidant activity. Purification of ethanol extract of *C. Fragrans* was carried out using vacuum column chromatography with silica gel 1.07733.1000 stationary phase and *n*-hexane, ethyl acetate and methanol as mobile phase.The antioxidant activity test of the fractions were carried out by DPPH free radical reduction method. The antioxidant test results using the DPPH free radical scavenging method showed that the F5 fraction had higher antioxidant activity than F4 and F3 fractions with IC_{50} values in a row 19.80; 24.65 and 84.02 ppm. The IC_{50} value of vitamin C as a positive control is 22.25 ppm. The F5 fraction has antioxidant activity higher than Vitamin C (positive control) and includes a very strong category of antioxidants.

Keywords: *Clerodendrum fragrans* Vent Willd, antioxidant activity, vacuum column chromatography fractions, ethanol extract.

1. Introduction

Sarang banua plants are found in Simalungun, North Sumatra, Indonesia, which have been used by the community as traditional medicinal plants. The result of plant determination, sarang banua plant is a type of *Clerodendrum fragrans* Vent Willd, including the family Verbenaceae [1].The *C.fragrans* plant is a semi-wood plant, typically shrubs and the community in Simalungun says the name of the plant is "sarang banua"(Figure 1). The genus Clerodendrum (Family Verbanaceae) has more than five hundred species that have been identified which are semi-woody shrub plants and are widespread in the tropics. Root, stems, leaves extracts several species of the genus Clerodendrum have been used for the treatment of asthma, cataracts, malaria, blood, skin and lung diseases traditionally [2].

Medicinal plants contain secondary metabolites such as alkaloids, flavonoids, steroids, saponins, tannins so that they can be used as pharmacological sources that are cheap and biodegradable [3].For example, *Clerodendrum volubile* plants contain saponins, anthraquinones, flavonoids, alkaloids, phenolic compounds, terpenes and glycosides have been used as analgesics[4].Ethanol extract of

Solanum blumei Nees ex Blume fruit which is a local Indonesian plant containing β-solanine glycoside alkaloid which is toxic and has anticancer activity for L_{1210} leukemia cells [5,6].

The secondary metabolites of alkaloids, triterpenoids, flavonoids, saponins, tannins and quinine were found in the ethanol extract of *C. fragrans* leaves [1] andhas antioxidant and antibacterial activity of *S. aureus* and *E. coli* [7]. Purification of ethanol extract from leaves of *C. fragrans* needs to be done to find active compounds that have potential as antioxidants. This study aims to find a pure fraction through vacuum column chromatography which has antioxidant activity.

Figure 1. "Sarang banua"- *Clerodendrum fragrans* Vent Willd.

2. Experimental Method

2.1. Preparation of ethanol extract of C. fragrans leaves

Fresh *C. fragrans* leaves are taken from Raya Usang village, Simalungun, Indonesia. The leaves of *C. fragrans* are dried and made into powder. Simplicia of *C. fragrans* leaf powder was extracted by maceration with solvents of different polarity[1].

2.2. Fractionation of ethanol extract of C. fragrans leaves

Purification of ethanol extract of *C. fragrans* was carried out using vacuum column chromatography with silica gel 1.07733.1000 stationary phase and the mobile phase of *n*-hexane, *n*-hexane: ethyl acetate (9: 1; 8: 2; 7: 3; 6: 4; 5: 5; 4: 6; 3: 7; 2: 8; 1: 9; 0:10), ethyl acetate: methanol (9: 1; 8: 2; 7: 3; 6: 4; 5: 5; 6: 4; 7: 3; 8: 2; 9: 1) and methanol. Column fractionation results were tested by TLC and fractions which had the same chromatogram combined.

2.3. Antioxidant test of the combined fraction of vacuum column chromatography

The antioxidant test of the combined fraction was carried out by the DPPH (1,1 diphenyl-2-picrylhydrazyl) free radical scavenging method using UV-Visible spectrophotometer at a wavelength of 517 nm.The concentration of the fraction sample solution was made in concentrations of 5, 10, 25 and 50 ppm and the concentration of DPPH solution was 0.4 mM in methanol p.a. As a positive control a solution of vitamin C with a concentration of 5, 10, 25 and 50 ppm in methanol p.a. is used.Percent of sample inhibition was calculated based on the difference in absorbance of DPPH solution with absorbance of sample solution divided by absorbance of DPPH solution multiplied by 100%.The amount of antioxidant concentration that can inhibit as much as 50% of free radicals is called the IC_{50} value.The IC_{50} value can be calculated using a linear regression equation, where Y equals 50 and X shows the IC_{50} value of the test sample[8, 9, 7, 10].

3. Result and Discussion

The results of vacuum column chromatography ethanol extract of leaves of *C. fragrans* obtained 20 fractions. The TLC test results for each column fraction are shown in Figure 2. Fractionation of vacuum column chromatography ethanol extract of *C. fragrans* leaves using column mobile phase solution which is different in polarity (increasingly polar). The next twenty column fractions are in TLC using the phase hexana solution: ethylacetate (7: 3) and hexana: ethylacetate (8: 2) to determine the purity of each fraction. Fractions that have the same chromatogram are then combined into a combined fraction.The fraction that has the same chromatogram is combined so that five combined fractions are obtained.

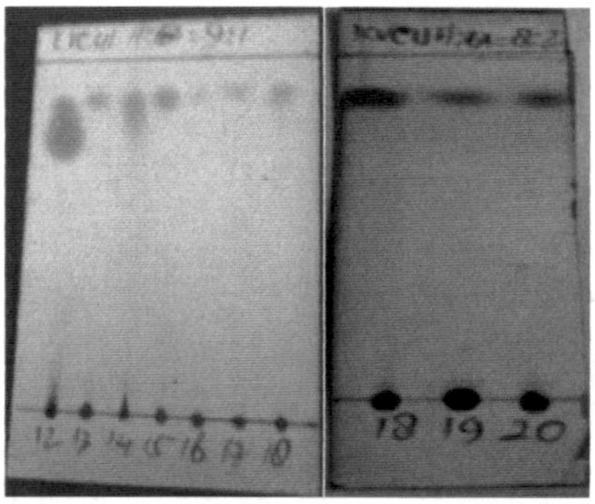

Figure 2. TLC chromatogram chromatography fraction column vacuum ethanol extract of *C. fragrans* leaf.

Results fractionation of vacuum column chromatography ethanol extract of *C. fragrans* leaves is presented in Table 1.

Table 1. Results of fractionation of chromatography of vacuum column ethanol extract of *C. Fragrans* leaf.

Fraction	No Vial	Column Mobile Phase	TLC Mobil Phase	Rf	Weight (g)
F1	1-2	Hex : EtilAc (9 : 1~ 8:2)	Hex:EtilAc (7:3)	0.93	0.148
F2	3	Hex:EtilAc (7:3)	Hex:EtilAc (7:3)	0.59;0.70;0.80	0.095
F3	4-5	Hex:EtilAc (1:9)	Hex:EtilAc (7:3)	0.42;0.59;0.70;0.82	0.397
F4	6-12	EtilAc:MeOH(8:2)	Hex:EtilAc (7:3)	0.19;0.42;0.59;0.70	1.819
F5	13-20	EtilAc:MeOH (6:4 ~ MeOH)	Hex:EtilAc (8:2)	0.88	3.850

Each combined fraction was then tested for antioxidant activity to determine the most active combined fraction. Because the weight of the combined fraction F1, F2 was obtained very little (0.148 and 0.095 g), the antioxidant test was not carried out on the two fractions. The antioxidant activity test was carried out on a combined fraction of F3, F4 and F5 using the DPPH free radical reduction method. The test results of the antioxidant activity of fractions F3, F4 and F5 are presented in Table 2.

Table 2. Antioxidant activity of chromatographic fraction column of leaves *Clerodendrum fragrans* Vent Willd with DPPH scavenging method.

Vacuum Column Chromatography Fraction	Concentration (ppm)	Absorbance	Scavenging Activity on DPPH (%)	IC_{50} (ppm)	Category of Antioxidant
Fraction F3	5	0.481	14.82	84.02	Strong antioxidants
	10	0.449	20.45		
	25	0.419	25.72		
	50	0.365	35.20		
Fraction F4	5	0.249	55.30	24.65	Very strong antioxidants
	10	0.250	60.15		
	25	0.132	76.20		
	50	0.095	83.17		
Fraction F5	5	0.194	65.00	19.80	Very strong antioxidants
	10	0.168	70.24		
	25	0.059	89.45		
	50	0.011	98.01		

Vitamin C	5	0.213	62.13	22.25	Very strong antioxidants
	10	0.129	77.10		
	25	0.007	86.45		
	50	0.015	97.34		

The F5 vacuum column chromatography fraction showed the highest percentage of (98.01%). The higher the percentage of DPPH free radical scavenging the sample, the higher the antioxidant activity.

The antioxidant test showed that results F5 vacuum column chromatography fraction is the highest antioxidant activity with an IC_{50} value of 19.80 ppm followed by F4 fraction (24.65 ppm) and F3 fraction (97.99 ppm). Antioxidant concentration (ppm) which can reduce 50% of free radicals is called IC_{50} value (inhibition concentration 50). Compounds that have IC_{50} values below 50 ppm are grouped as very strong antioxidants, 50-100 ppm are classified as strong antioxidants and 100-150 ppm are classified as weak antioxidants [11].

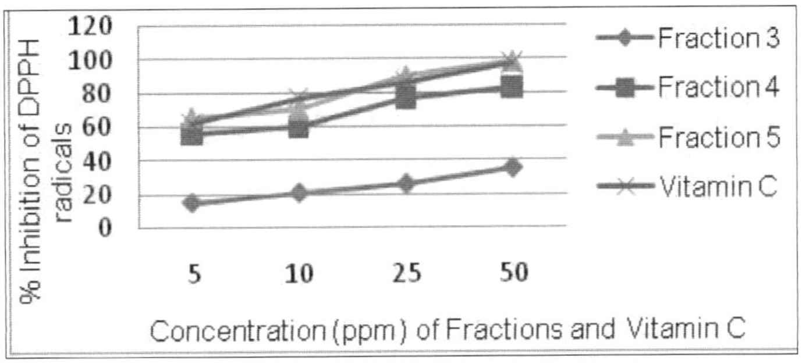

Figure 3. Scavenging activity of *C. fragrans* column fractions on DPPH radicals.

The F5 vacuum column chromatography fraction has antioxidant activity higher than Vitamin C (positive control) and includes a very strong category of antioxidants. The magnitude of antioxidant activity is caused by the ability of antioxidants to reduce DPPH free radicals (Figure 3).The ability to reduce free radicals F5 vacuum column chromatography fraction was the highest (98.01%), followed by vitamin C (97.43%), F4 fraction (83.17%) and F3 fraction (32.18%) (Figure 3). The F5 vacuum column chromatography fraction from the ethanol extract of *C. fragrans* leaves had the highest scavenging activity on DPPH (98.01%), followed by vitamin C (97.43%), F4 fraction (83.17%) and F3 fraction (32.18%).

4. Conclusion

The F5 vacuum column chromatography fraction had higher antioxidant activity than F4 and F3 fractions with IC_{50} values in a row 19.80; 24.65 and 84.02 ppm. The F5 fraction has antioxidant activity higher than Vitamin C (positive control) and includes a very strong category of antioxidants.

5. Acknowledgements

This work is supported by grant "Penelitian Dasar Unggulan Perguruan Tinggi (PDUPT with a contract **No.39/UN33.8/PL-DRPM/2019**) Year 2019" from Direktorat Jenderal Penguatan Riset and Pengembangan,Ministry of Technology Research and Higher Education, Republic of Indonesia.

References

[1] Simorangkir M, Nainggolan B and Silaban S, 2018 In Book of Pgrogram of The 2nd International Conference on Innovation in Education, Science and Culture (ICIESC), Medan City Indonesia

[2] Shrivastatava N and Patel T 2007 Clerodenrum and heathcare : an overview *Medicinal and Aromatic Plant Science and Biotechnology* **1** 142-150

[3] Udayan D, Nair S N, Padinchareveetil S K and Thumadath A K 2014 Evaluation of phytochemical constituents, proximate and fluorescence analysis of ethanolic extract and its fractions of *Clerodendrum philippinum* Schauer found in Wayanad region of Kerala, India *Research Journal of Chemical Sciences* **4** 1-6

[4] Senjobi C T, Fasola T R and Aziba P I 2017 Phytochemical and analgesic evaluation of methanol leaf extract of *Clerodendrum volubile* Linn *Ife Journal of Science* **19** 141-145

[5] Simorangkir M, Barus T, Surbakti R and Simanjuntak P 2016 Isolation and toxicity of steroidal alkaloid glycoside from fruits of ranti hitam (*Solanum blumei* Nees ex Blume) *Asian Journal of Chemistry* **28** 203-206

[6] Simorangkir M, Silaban S, Surbakti R, Barus T and Simanjuntak P 2017 Aktivitas antikanker ekstrak etanol buah ranti hitam (*Solanum blumei* Nees ex Blume) terhadap sel leukimia L_{1210}*ChimicaetNaturaActa* **5** 31-35

[7] Simorangkir M, Hutabarat W, Nainggolan B and Silaban S 2019 Antioxidant and antibacterial activities of non-polar to polar solvent extracts of sarang banua (*Clerodendrum fragrans*) leaves *Rasayan Journal of chemistry* **12** 959-965

[8] Molyneux P 2004 The use of the stable free radical diphenylpicrylhydrazyl (dpph) for estimating antioxidant activity Songklanakarin *J. Sci. Technol.* **26** 211-219

[9] Deore S L, Khadabadi S S, Patel Q R, Deshmukh S P, Jaju M S, Junghare N R, Wane T P and Jain R G 2009 In vitro antioxidant activity and quantitative estimation of phenolic content of Lagenaria siceraria *Rasayan Journal of chemistry* **2** 129-132

[10] Nerdy N and Manurung K 2018 Spectrophotometric method for antioxidant activity test and total phenolic determination of red dragon fruit leaves and white dragon fruit leaves *Rasayan Journal of Chemistry* **11** 1183–1192

[11] Blois M S 1958 Antioxidant determinations by the use of a stable free radical *Nature* **181** 1199–1200

The Difference of Amino Acid Profiling From Two Morphological Purple Sweet Potatoes From Kawi Mountain Cultivars, East Java, Indonesia

N Kurnianingsih[1,2,7*], R Ratnawati[2], F Fatchiyah[3,7], W Barlianto[4], M M Ali[5], A Safitri[6,7] and E Suyanto[3,7]

[1]Doctoral Program of Medical Science, Faculty of Medicine,Universitas Brawijaya, Indonesia
[2]Physiology Department, Faculty of Medicine,Universitas Brawijaya, Indonesia
[3]Biology Department, Faculty of Mathematics and Natural Science, Universitas Brawijaya, Indonesia
[4]Pediatric Department, Faculty of Medicine,Universitas Brawijaya, Indonesia
[5]Pharmacology Department, Faculty of Medicine, Universitas Brawijaya, Indonesia
[6]Chemistry Department, Faculty of Mathematics and Natural Science, Universitas Brawijaya, Indonesia
[7]Research Center of Smart Molecule of Natural Genetics Resource, Universitas Brawijaya, Indonesia

*Corresponding author's email: nia_fkub@ub.ac.id

Abstract. Indonesia is one of the main commercial producers of sweet potatoes. Previous study has presented well documented anthocyanin content of Purple Sweet Potatoes (PSP) from local cultivar in Indonesia. However, research about amino acids profiling of local cultivar PSP as high commodities in Kawi Mountain region has not undertaken. This study was explored amino acids profiles in two morphological PSP cultivated in Kawi Mountain region, Ngajum Village, Wonosari Subdistric, Malang Regency, East Java, Indonesia. Fresh storage roots were cleaned with running water and shade dried. The analysis was conducted at PT.Saraswanti Indo Genetech, Bogor, Indonesia. Method of UPLC was used for identification of Ser, Glu, Phe, Ile, Val, Ala, Arg, Gly, Lys, Asp, Leu, Tyr, Pro, Thr, His, Cys and Met; while tryptophan was analyzed using HPLC method. The five largest amino acids in dark-purple skin and flesh PSP were Glu 565.75 mg/kg, Asp 479.74 mg/kg, Arg 413.54 mg/kg, Ala371.87 mg/kg and Leu 336.67 mg/kg, whereas white flesh and purple skin PSP contained Glu 653.67 mg/kg, Asp 447.60 mg/kg, Arg 314.41 mg/kg, Lys 278.77 mg/kg and Leu 256.97 mg/kg. Future research is needed to investigate the health benefit of amino acids content in PSP from Kawi Mountain cultivar.

1. Introduction

Sweet potatoes (*Ipomoea batatas (L.) Lam*) is a Convolvulaceae family plants which are mainly originated from Central America countries. Currently, this plants widely cultivated around the world, both sutropical and tropical region including Indonesia [1]. Sweet potatoes are included into seven most important crops in the world following wheat, rice, maize, potato, barley, and cassava. Sweet potatoes is also a major food crop in anumber areas of Indonesia[2].

The sweetpotato is a perennial plant grown vegetatively either by storage roots or stem cuttings. Storage roots are the most commercial part from this plant. There are various color differences from storage roots of PSP, ranging from white to dark purple. This color difference can be seen in both the skin and tuber flesh, depending on the amount and distribution of anthocyanin pigments[2].

Purple sweet potatoes (PSP) is one of the sweet potato varieties that has a dark-purple color both on the skin and tuber flesh. The range of anthocyanin content in raw storage roots of PSP are 107.8 mg to 174.7 mg/100 fresh weight [3]. Anthocyanin from PSP has a stronger antioxidant activity than anthocyanin contained in purple corn, grape skin, red cabbage and elderberry. Even the components of peonidin and cyanidin from PSP show a DPPH radical-scavenging activity that is stronger than ascorbic acid [4].

Indonesia has many varieties of local PSP cultivars that contain anthocyanin and benefits as antioxidants, antiapoptosis and neuroprotective [5-7]. In East Java province, PSP is a high comodities in the Gunung Kawi region[8][10]. Exploration of the health benefits potential of the Kawi Mountain cultivars of PSP were resulted in increasing superoxide dismutase enzymes, decreasing NF-kB and preventing the formation of foam cells in atherogenic diet [9, 10]. The same cultivars were also reported to optimally reduce caspase-3 and apoptosis at doses of 20 mg/kgBW although it had not significantly improved spatial memory ability in diabetic rats [11].

Besides the health benefit properties of local PSP, for our knowledgelack of research that reported amino acids (AA) content in local PSP cultivars. Though AA is very important as a precursor of various substances that function as a regulator of metabolism, growth, development in order to maintain whole body homeostasis [12]. Thus, the aim of this study was to explore the AA contain in PSP from Kawi Mountain cultivar.

2. Methods

2.1. Material

Fresh storage roots of PSP were collected from local agricultural areas in Ngajum village, Wonosari District, in Kawi Mountain, Malang Regency, East Java, Indonesia. We analyzed two different tuber morphologies of PSP, purple skin-flesh and purple skin-white flesh. The storage roots were thencleaned with running water and shade dried.

2.2. Research procedure

2.2.1. Ultra Performance Liquid Chromatography (UPLC)

All the procedure was carried out at PT.Saraswanti Indo Genetech, Bogor, Indonesia. Ultra Performance Liquid Chromatography method was used for identification 17 of amino acids e.g L-serine (Ser), L-glutamic acid (Glu), L-phenylalanine (Phe), L-isoleucin (Iso), L-valine (Val), L-alanine (Ala), L-arginine (Arg), glycine (Gly), L-lysine (Lys), L-aspartic acid (Asp), L-leucine (Leu), thyrosine (Tyr), proline (Pro), threonine (Thr), histidine (His), cystine (Cys) and methionine (Met). In brief all procedure refered to Waters Acquity UPLC H Class and H Class Bio amino Acid Analysis System Guide [13]. The method was used column AccQ.Tag Ultra C18 1.7 µm (2.1 x 100 mm), with column temparature at 49^{0}C, flow rates of mobile phase was 0.5 mL/minute, 1µl of injection volume andphotohometric diode array (PDA) detector at 260 nm wavelength was used. Gradient composition systems of mobile phase were as followed A : Eluent A concentrate Amino Acid Analysis AccQ.Tag Ultra; B : Eluent B Amino Acid Analysis AccQ.Tag Ultra 10% in water; C : Aquabidest; D : Eluent B Amino Acid Analysis AccQ.Tag Ultra.

2.2.2. High Performance Liquid Chromatography (HPLC)

Tryptophan (Trp) was analyzed using High Performance Liquid Chromatography based on the AOAC official method. Lichrospher column RP-18, 250 mm x 4.0 mm, 5µm, Merck was used for separation, with ambient column temperatures. The mobile phase flow rate was 1.5 mL/min,isocratic pump

Journal of Physics: Conference Series **1374** (2019) 012017 doi:10.1088/1742-6596/1374/1/012017

system, injection volume was 15 μL and photohometric diode array (PDA) detectors at 280 nm wavelength were used. The mobile phase consists of A = Sodium Acetate 0.0085 M pH 4; B = Methanol (A: B = 95: 5).

3. Results and Discussion

The results of the analysis showed that although there were differences in the morphology of the color of tuber flesh, the two types of PSP both contained 12 AA, namely Glu, Asp, Arg, Ala, Leu, Thr, Lys, Val, Phe, Ile, Pro and Tyr. While the other 6 AAs namely Gly, His, Cys, Met, Trp and Ser were not detected in both samples. Nevertheless there are still differences in AA levels from both samples for Glu, Ala and Thr wherein the purple-skin and white-flesh have higher level of Glu but lower level in Ala and Thr (Table1).

Table 1. Comparison of AA content in two different morphological Purple Sweet Potatoes.

Amino acids	Level (mg/kg)	
	Purple skin and flesh	Purple skin-white flesh
L-Glutamic acid	565.74	653.67
L-Aspartic acid	479.74	447.60
L-Arginine	413.54	314.41
L-Alanine	371.87	<338.66
L-Leucine	336.67	256.97
L-Threonine	292.53	<239.36
L-Lysine	279.28	278.77
L-Valine	274.27	190.01
L-Phenylalanine	229.31	210.53
L-Isoleucine	224.62	162.93
L-Proline	223.25	159.00
L-Tyrosine	<222.88	<222.88
Glycine	n.d	n.d
L-Histidine	n.d	n.d
L-Cystine	n.d	n.d
L-Methionine	n.d	n.d
L-Tryptophan	n.d	n.d
L-Serine	n.d	n.d

*n.d = not detected

Compared to other plants containing anthocyanin, blueberries and strawberries have lower Glu content, 14.35 mg/g and 61.99 mg/g,respectively. Conversely, the strawberries have highest level of Trp whereas Cys is highest in blueberries [14]. Another study conducted on *Synsepalum dulcificum* pulp berry reported that essential amino acids (EAA) were detected with the highest values coming from leucine 2.35 g/100 g of protein and methionine has the lowest value of 0.31 g/100 g of protein. Glutamic acid is the highest value of NEAA in *Synsepalum dulcificum* pulp berry [15]. Free Amino Acid (FAA) was also detected in three Rosa roxburghii Tratt fruit genotypes, both amino acids and non-essential amino acids with various levels [16].

Amino acids have traditionally been classified into essential amino acids and non-essential amino acids according to nitrogen balance. Essential amino acids cannot be synthesized either in animals or humans body, thus they are necessary to be consumed from foods. Non-essential amino acids can be synthesized in the body[17]. However, this classification has conceptual limitation because when there is an insufficient of NEAA synthesis, the body requires amino acid intake from food [18].

Methionine, lysine, histidine, leucine, isoleucine, phenylalanine, valine, threonine, and tryptophan are classified as essential amino acids. Whereas aspartate, alanine, arginine, glycine, cysteine, glutamate, proline, tyrosine, and serine are amino acids that not essential for humans and most other animals[12]. However, some non-essential amino acids are classified as conditionally essential because the dramatically increase of demand, for example in pregnancy, infection, burns, injury, and breastfeeding [12]. Amino acids used by the body to synthesize various types of proteins that are important for physiological processes of the body, namely gene expression, cell growth, cell differentiation, hormone secretion, cell signaling, regulation of intestinal absorption, regulation of immune responses, etc[18].

The storage roots of PSP in the current study has glutamic acid as the highest value. Glutamic acid is well received as a conditional essential amino acid under conditions of stress and illness. Glutamic acid under physiological pH will be an anionic form called glutamate. Glutamate can also be synthesized into glutamine through the role of the glutamine synthetase enzyme[19]. Glutamate is an excitatory neurotransmitter in the central nervous system. Via glutamate receptors will mediate rapid synapse transmission. In addition glutamate is also needed in ammonia and protein metabolism, fetal-placental metabolism, maintenance of functions of the intestinal mucosal barrier, precursors of GABA neurotransmitters and others [20].

4. Conclusion

Current study revealed that purple sweet potatoes is a source of both essential or non-essential amino acid. In the future, further studies are needed to evaluate the health benefits of amino acids contained in purple sweet potatoes

Reference

[1] Meira M, da Silva E P, David J M and J P David 2012 Review of the genus Ipomoea: Traditional uses, chemistry and biological activities," *Brazilian J. Pharmacogn.* **22** 682–713

[2] Zosimo H 1999 *Sweetpotato Germplasm Management (Ipomoea batatas) Training manual*, H. Zosimo, Ed. I(Peru-International Potato Center) p 74–75.

[3] Steed L E and Truong V D 2008 Anthocyanin content, antioxidant activity, and selected physical properties of flowable purple-fleshed sweetpotato purees *J. Food Sci.* **73** 215–221

[4] Kano M, Takayanagi T, Harada K, Makino K and Ishikawa F 2005 Antioxidative Activity of Anthocyanins from Purple Sweet Potato, Ipomoea batatas Cultivar Ayamurasaki *Biosci. Biotechnol. Biochem.* **69** 979–988

[5] Wicaksono L A, Yunianta and Widyaningsih T D 2016 Anthocyanin extraction from purple sweet potato cultivar antin-3 (Ipomoea batatas L.) using maceration, microwave assisted extraction, ultrasonic assisted extraction and their application as anti-hyperglycemic agents in alloxan-induced wistar rats *Int. J. PharmTech Res.* **9** 181–192

[6] Dwiyanti G, Siswaningsih W and Febrianti A 2018 Production of purple sweet potato (Ipomoea batatas L.) juice having high anthocyanin content and antioxidant activity *J. Phys. Conf. Ser.* **1013**

[7] Adnyana I M O, Sudewi A R, Samatra D P and Suprapta D 2018 Neuroprotective Effects of Purple Sweet Potato Balinese Cultivar in Wistar Rats With Ischemic Stroke *Open Access Maced. J. Med. Sci.* **6** 1959–1964

[8] Ariadi B Y 2006 Comparison Among Various Sweet Potato Varieties Focused From Farm Income And Marketing At Malang Regency Jurnal Pembangunan Pedesaan **6** 104-113

[9] Suhardi C J, Ratnawati R and Khotimah H 2016 Effect Of Anthocyanin From Purple Sweet Potato (*Ipomoea Batatas L.*) Kawi Mountain Cultivarto Increase Sod Serum In Wistar Rat (Rattus Norvegicus) With Atherogenic Diet *Maalah Kesehatan* **3** 166–173

[10] Maharani T, Sargowo D, Tjokropranowo A and Ratnawati R 2014 Effect of extract purple Ipomoe Batatas cultivar kawi mountain chronic inflammation in Wistar Rats with atherogenic diet *International Journal of Science and Technology*. **3** 1–7

[11] Prabawati R K, Ratnawati R, Rahayu M and Prakosa A G 2019 Effect Anthocyanin of Purple Potato Gunung Kawi on Mda Levels, Expression of Caspase-3, and Spatial Memory Function on Diabetic Wistar Rats *Malang Neurology Journal*. **5** 34–41

[12] Wu G 2013 Functional amino acids in nutrition and health *Amino Acids* **45** 407–411
[13] Waters Corporation 2012 *Acquity UPLC H-class and H-class bio amino acid analysis* (Manchester-Waters Corporation)
[14] Hua Z, Zhen-Yu W, Xin Y, Hai-Tian Z, Ying-Chun Z, Ai-Jun D and Jing J J W 2014 Determination of free amino acids and 18 elements in freeze-dried strawberry and blueberry fruit using an Amino Acid Analyzer and ICP-MS with micro-wave digestion *Food Chem.* **147** 189–194
[15] Njoku N E, Ubbaonu C N, Alagbaoso S O, Eluchie C N and Umelo MC 2015 Amino acid profile and oxidizable vitamin content of Synsepalum dulcificum berry (miracle fruit) pulp *Food Sci. Nutr.* **3** 252–256
[16] Lu M, An H and Wang D 2017 Characterization of Amino Acid Composition in Fruits of Three Rosa roxburghii Genotypes *Hortic. Plant J.* **3** 232–236
[17] Wu G 2010 Functional Amino Acids in Growth *Advances.* **4** 31–37
[18] Hou Y, Yin Y and Wu G 2015 Dietary essentiality of 'nutritionally non-essential amino acids' for animals and humans *Exp. Biol. Med.* **240** 997–1007
[19] Tapiero H, Mathé G, Couvreur P and Tew K D 2002 II. Glutamine and glutamate.*Biomed. Pharmacother.* **56** 446–457
[20] Kulkarni C, Kulkarni K and Hamsa B 2009 L-Glutamic acid and glutamine: Exciting molecules of clinical interest *Indian J. Pharmacol.* **37** 148

Phytoconstituents Investigation on the Ethanolic Extract of *Azadirachta indica* var. Indonesian and Philippines

R Raissa[1], A Safitri[1], M A G Beltran[2] and A Aulanni'am[1*]

[1]Department of Chemistry, Faculty of Mathematics and Natural Science, Universitas Brawijaya, Malang, Indonesia
[2]College of Veterinary Medicine, Tarlac Agricultural University, Camiling, Tarlac, Philippines

*Corresponding author's email: aulani.fkhub@gmail.com

Abstract. In the present study, phytochemical screening of ethanolic extracts of *Azadirachta indica* leaves from Indonesia and the Philippines revealed the presence of different phytoconstituents. Preliminary qualitative chemical test, TLC and LC-MS were used. TLC for all the extracts showed bands in long UV 366 for presence of flavonoid, tannin, saponin, terpenoid. As a result of LC-MS analysis of ethanolic extract *Azadirachta indica* leaves from Indonesia and Philippines, 10 compounds from Indonesian varian and 7 compounds from the Philippines varian were detected using m/z value. In conclusion, phytochemical screening based on TLC and LC-MS/MS show diverse bioactive compounds in ethanolic extract *Azadirachta indica* leaves from Indonesia and the Philippines. These can be effective approach for selecting best quality of varian leaves and planting area.

1. Introduction

Characterize natural product for new drug discovery has been concerned lately. Some of herbal products prove to offer synthetic drug substance. There are several factors that influence the quality of the herbs. Variation, environment conditions, storage, processing can be those influence quality factors of the herbs. Characterization compounds of herbs extract can be a standard procedure to find out the quality of the herbs.

Azadirachta indica (Neem) leaves are native of dry areas (Rojas-Sandoval *et al.*, 2014). It was naturally distributed in Thailand, Malaysia, Philippines and Indonesia and has become one of the most widespread trees in tropical and subtropical areas. Neem plant as medicinal plant is reported to have antifungal (Lloyd *et al.*, 2005), hepatoprotective (Pingale, 2010), anti-inflammatory (Jagadeesh *et a*l., 2014), anthelmintic (Beltran *et al.*, 2019), anti-cancer, insecticidal. The chemical constituents of *Azadirachta indica* leaves have a very important role in medicinal applications and it is believed due to its biologically active components.

In this present study, analytical method TLC and LC-MS was undertaken for identifying phytoconstituent. Thin layer Chromatography (TLC) is a very commonly used technique for identifying compounds, is a method of analysis in which the stationary phase, a finely divided solid, is spread as a thin layer on a rigid supporting plate; and the mobile phase, a liquid is allowed to migrate across the surface of the plate (Gennaro, 2000). This analytical tool is used because of its simplicity, speed of separation, cost effectiveness and high sensitivity.

LC-MS/MS are used for characterization and quantitation of herbal medicines because full characterization of these product. The advantages of LC-MS/MS are its high sensitivity and high-throughput to confirm the identity of the components in complex herbal extract, along with the detection and identification of unknown and unexpected compounds (Krug *et al.*, 2008).

The aim of this research study was to assess the bioactive components present in the ethanolic extract of *Azadirachta indica* leaves varian Indonesia and Philippines using phytochemical screening and chromatographic analysis. So the result could be developed and applied to the pharmaceutical production and quality control of botanical product.

2. Material and Methods

2.1. Collection and Authentication of Plant Material

Fresh *Azadirachta indica* leaves were collected from Indonesia (Madura) dan Philippines (Camiling). The plant specimens were authenticated by Laboratorium of Plant Taxonomy, Universitas Brawijaya. Number identification of these plant is 0238/UN10.F09.42/03/2018.

2.2. Preparation of Plant Extract

The leaves were cleaned by washing with running water and shade dried and the milled to pass through 100-mesh sieve. The leaf powder was extracted by maceration for three days with 80% ethanol at room temperature. The extracts were concentrated at 45C using Rotary vacuum evaporator to yield 80% hydroethanolic fraction as brownish green viscous residue. The concentrated extracts were keep in refrigerator at 4°C until further use.

2.3. Preliminary Phytochemical Screening

Test for the presence and absence of phytochemical compounds using standard methods involves the addition of an appropriate chemical agent to all the extract in a test tube and shaken. The different qualitative chemical test were performed for establishing profile of given extract for its chemical composition. Phytochemical screening of ethanolic extract of *Azadirachta indica* varian Indonesia and Philippines were carried out for alkaloids, flavonoid, saponin, tannin, terpenoid.

2.3.1. Alkaloid. 500 µL of ethanolic extract of *Azadirachta indica* varian Indonesia and Philippines of 10,000 ppm put in a test tube, then added 0.5 mL 2% hydrochloric acid. The solution was divided into three tubes. Tube 1 solution was added 0.5 mL acid solution dilute as a comparison, tube 2 was added 2-3 drops of Dragendorff's reagent, and tube 3 was added 2-3 drops of Mayer's reagents. If tube 2 was formed orange precipitation and tube 3 was formed yellowish precipitation. It indicated the presence of alkaloids.

2.3.2. Flavonoid. 500 µL of ethanolic extract of *Azadirachta indica* var. Indonesia and Philippines of 10,000 ppm put in a test tube, then added 1-2 ml hot methanol 50%. The solution was mixed by Magnesium and 4-5 drops of hydrochloric acid concentrate. Formation of red or orange colour indicated the presence of flavonoid.

2.3.3. Saponin. 500 µL of ethanolic extract of *Azadirachta indica* var. Indonesia and Philippines of 10,000 ppm was diluted with 10 ml distilled water and this was shaken 1 minute. Formation of bubble was added 2 drops of HCl 1 N. The stable bubble indicated the presence of saponin.

2.3.4. Tannin. 500 µL of ethanolic extract of *Azadirachta indica* var. Indonesia and Philippines of 10,000 ppm was diluted with 1-2 ml distilled water and this was added 2 drops $FeCl_3$ solution.. Formation of green blackish colour indicated the presence of tannin.

2.3.5. Terpenoid. 500 µL of ethanolic extract of *Azadirachta indica* var. Indonesia and Philippines of 10,000 ppm was diluted with 0,5 ml chloroform and this was added 0,5 ml anhydrous acetic acid. Then 1-2 ml concentrated sulphuric acid. The appearance of reddish brown or violet ring indicated the presence of terpenoids.

2.4. Thin Layer Chromatography

The presence of number of phytoconstituents flavonoids, saponins, tannins, terpenoid in ethanolic extract of *Azadirachta indica* var Indonesia and Philippines, which further became the basis for the phytochemical investigations through TLC on analytical plates over silica gel (TLC grade; Macherey-Nagel; Germany). The plates were dried in hot air oven at 110°C for 30 mins and then stored in a dry atmosphere and used whenever required. Samples were prepared by diluting the crude extracts of ethanol with respective solvent and then applied usually 1-10μl volumes to the origins of a TLC plate 1 cm above its bottom with the help of capillary tubes. After the application of the sample on the plate the plates were kept in TLC glass chamber (solvent saturated) than mobile phase was allowed to move through adsorbent phase up to top of the plate. The developed TLC plates were air dried. They were later spray with different spraying reagent. Rf were observed under long UV 366 nm. Calculation of Rf value is done with this formula:

$$Rf = \frac{migration\ distance\ of\ substance}{migration\ distance\ of\ solvent\ front}$$

2.5. LC-MS Screening

LC-MS screening was performed using LC-MS Thermo Scientific Dionex Ultimate 3000 RSL Cnano with micro flow meter in Lembaga Sentral Ilmu Hayati Universitas Brawijaya, Malang, equipped with Electrospray ionization (ESI). The LC-MS analytical data were optimized using a background subtraction technique of chromatography with the Analyst version: Compound Discoverer with mzCLoud MS/MS Library. The principle of this method is to reduce background. Data containing more real m/z were observed. Each compound was then identified from reference compounds based molecular weight of the structure.

3. Result and Discussion

3.1. Phytochemical Screening

The presence and absence of the phytochemical in *Azadirachta indica* var. Indonesia and Philippines are listed in the table 1. Our result revealed that flavonoid, tannin, saponin, terpenoids are present in ethanolic extract of *Azadirachta indica* var. Indonesia and Philippines.

Table 1. Result of phytochemical screening.

Constituents	Test performed	Ethanolic extract	
		Indonesia	Philippines
Alkaloid	Dragendorff's Test	-	-
	Mayer's Test	-	-
Flavonoid	Hydrochloric acid Test	+	+
Saponin	Froth Test	+	+
Tannin	Ferric chloride Test	+	+
Terpenoids	Salkowski's Test	+	+

3.2. Thin Layer Chromatography Profiling

TLC profiling of all extracts gives result the presence of number of phytochemicals. Various phytochemicals gives different Rf values. Different Rf values of the compound reflect an idea about their polarity. Mixture of solvents can be used for separation of pure compound from plant extract. The presence of any significant bioactive natural product indicates the necessity of separation of the compound from the

Journal of Physics: Conference Series **1374** (2019) 012018 doi:10.1088/1742-6596/1374/1/012018

mixture of compounds through suitable chromatographic techniques. In the present study, flavonoids, saponin, tannin and terpenoid are confirmed to be present.

Table 2. Phytochemical analysis of *Azadirachta indica* leaves from Indonesia and the Philippines by Thin Layer Chromatography (TLC).

Chemical Name	Solvent System	Extract Neem Leaves	Rf Values	Spray Reagent
Flavonoid	ethyl acetate: methanol: aqua (5:1:5	Indonesia	0.03, 0.15	Ammonia
		The Philippines	0.06, 0.77	
Saponin	chloroform: methanol: aqua (13:7:2)	Indonesia	0.26, 0.87	Sulphuric acid 10%
		The Philippines	0.22, 0.85, 0.90	
Terpenoid	toluene: ethyl acetate (7:3)	Indonesia	0.19, 0.49	Liebermann burchard
		The Philippines	0.27, 0.32	
Tanin	n-butanol: acetic acid glacial: Aqua (2:0,5:1,1)	Indonesia	0.27	$FeCl_3$ 1%
		The Philippines	0.23	

(a) (b) (c) (d)

Figure 1. Photograph of TLC of flavonoid (a) Indonesian (b) Indonesian (after spray) (c) Philippines (d) Philippines (after spray).

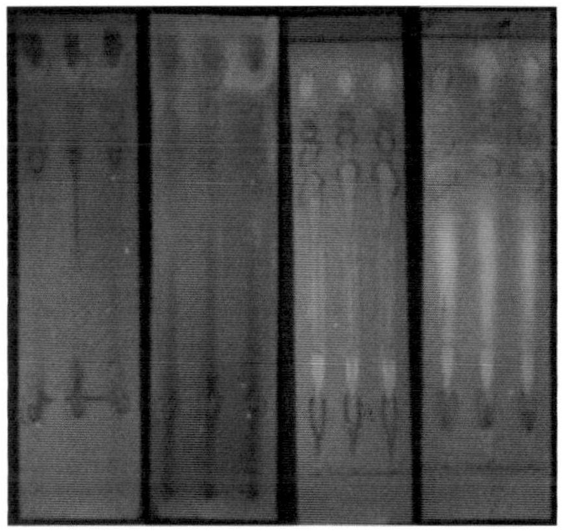

(a) (b) (c) (d)

Figure 2. Photograph of TLC of saponin (a) Indonesian (b) Indonesian (after spray) (c) Philippines (d) Philippines (after spray).

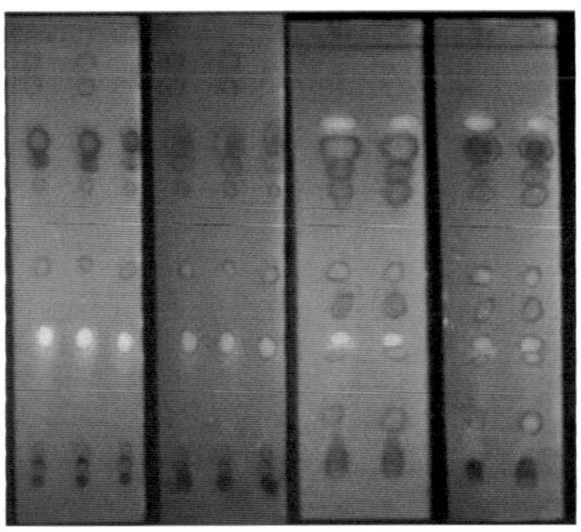

Figure 3. Photograph of TLC of terpenoid (a) Indonesian (b) Indonesian (after spray) (c) Philippines (d) Philippines (after spray).

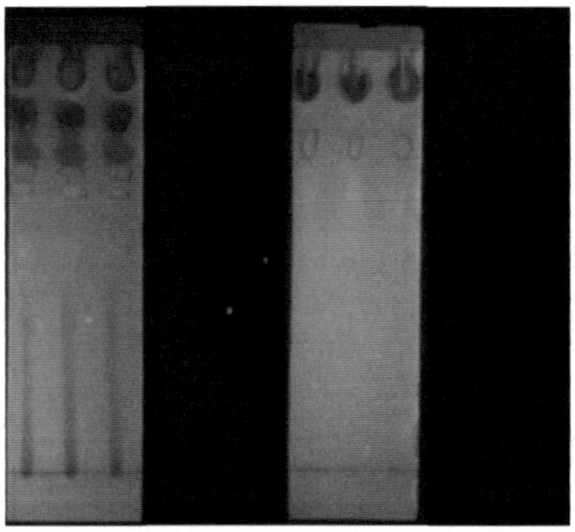

(a) (b) (c) (d)

Figure 4. Photograph of TLC of tanin (a) Indonesian (b) Indonesian (after spray) (c) Philippines (d) Philippines (after spray).

3.3. LC-MS Screening

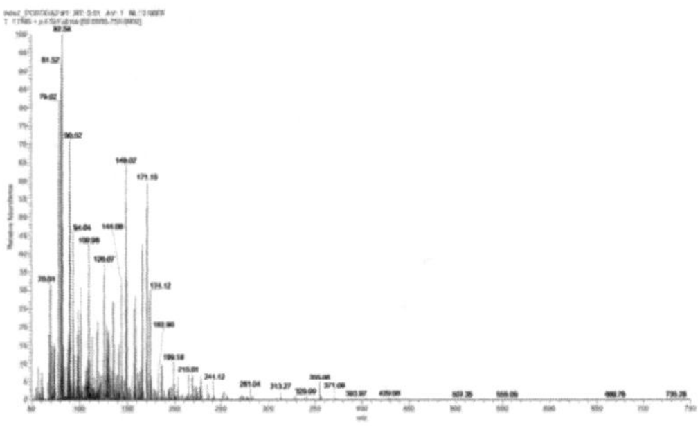

Figure 5. LC-MS chromatogram of ethanolic extract of *Azadirachta indica* varian Indonesia leaves.

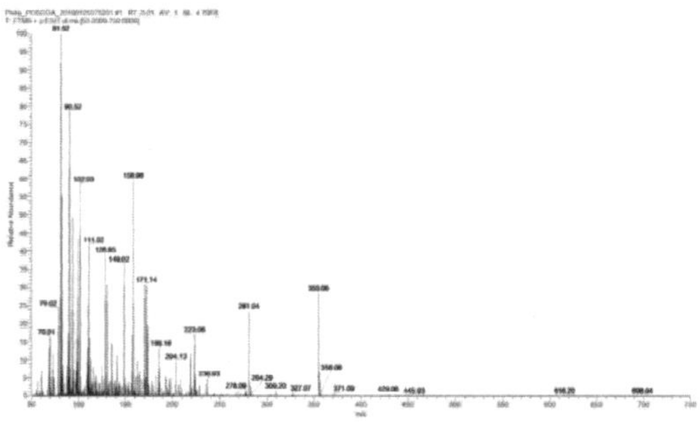

Figure 6. LC-MS chromatogram of ethanolic extract of *Azadirachta indica* varian Philippines leaves.

Table 3. Bioactive compounds in ethanolic extract of *Azadirachta indica* varian Indonesia leaves.

No.	Name of Compounds	Group	Molecular Formula	Molecular Weight	[M+HJ]+ (m/z) observed
1.	Betulin	Terpenoid	$C_{30}H_{50}O_2$	442.728	442
2.	Ginsenoside		$C_{42}H_{72}O_{13}$	785.025	785
3.	Caryophyllene oxide		$C_{15}H_{24}O$	220.356	220
4.	Soyasaponin I	Saponin	$C_{48}H_{78}O_{18}$	943.134	943
5.	Ecgonine	Alkaloid	$C_9H_{15}NO_3$	185.223	185
6.	Scutellarin	Flavonoid	$C_{21}H_{18}O_{12}$	462.363	462
7.	Epicatechin		$C_{15}H_{14}O_6$	290.271	290
8.	Icariin		$C_{33}H_{40}O_{15}$	676.668	676
9.	Sesamolin		$C_{20}H_{18}O_7$	370.357	370
10.	Lupeol		$C_{30}H_{50}O$	426.729	426

Table 4. Bioactive compounds in 80% ethanolic extract of *Azadirachta indica* varian Philippines leaves.

No.	Name of Compounds	Group	Molecular Formula	Molecular Weight	[M+HJ]+ (m/z) observed
1.	Betulin	Terpenoid	$C_{30}H_{50}O_2$	442.728	442
2.	Caryophyllene oxide		$C_{15}H_{24}O$	220.356	220
3.	Andrographolide		$C_{20}H_{30}O_5$	350.455	350
4.	10-Deacetylbaccatin III		$C_{29}H_{36}O_{10}$	544.597	544
5.	3-Acetyl-11-keto-β-boswellic acid		$C_{32}H_{48}O_5$	512.731	512
6.	O-chloroacetylcarbamoylfumagillol		$C_{19}H_{28}ClNO_6$	401.884	401
7.	Rutin	Flavonoid	$C_{27}H_{30}O_{16}$	610.521	610

On the basis of the LC-MS the known compounds, 10 compounds (4 terpenoids, 1 alkaloid, 2 flavonoids, 2 lignins, 1 saponin) were identified from 80% ethanolic extract of Azadirachta indica varian Indonesia leaves and 7 compounds (6 terpenoids, 1 flavonoid) were identified from 80% ethanolic extract of Azadirachta indica varian Philippines leaves. Betulin and caryophyllene oxide were identified in both varian.

4. Conclusion

In the present study, ethanolic extract of *Azadirachta indica* var. Indonesia and Philippines showed the presence of bioactive compound such as flavonoids, terpenoids, saponins, tannin. This study also leads to the further research in the way of isolation and identification of the active compound from the leave of *Azadirachta indica* var. Indonesia and Philippines using chromatographic and spectroscopic techniques.

Reference

[1] Beltran M A G, Robertson A and Martin R J 2019 The Effects of Plant Leaves Variants from The Philippines on Infective *Oesophagostomum dentatum* Larvae., Veterinary Biomedical and Clinical Journal 1

[2] Gennaro A R 2000 *Remington: The Science and Practice of Pharmacy*, 20th ed, Vol. II, Mack Publishing Company, Pennsylvania, 1016

[3] Jagadeesh K, Srinivas K and Revankar S P 2014 Anti Inflammatory Effect of Azadirachta Indica (Neem) In Albino Rats-An Experimental Study IOSR Journal of Pharmacy 4

[4] Krug D, Zurek G, Revermann O, Vos M, Velicer G J and Muller R 2008 Discovery the hidden secondary metabolome of Myxococcus xanthus: A study of intraspecific diversity *Appl. Environ. Microbiol.* **74** 3058-68

[5] Lloyd A C C, Menon T and Umamaheshwari K 2005 Anticandidal activity of Azadirachta indica, Indian Journal of Pharmacology **37**

[6] Pingale S S 2010 Hepatoprotection Study of Leaves Powder of *Azadirachta Indica* A. Juss., International Journal of Pharmaceutical Sciences Review and Research

[7] Rojas-Sandoval J and Acevedo-Rodriguez P 2014 *Azadirachta indica* (neem tree) Invasive Species Compendium (Wallingford, UK: CAB International)

Journal of Physics: Conference Series **1374** (2019) 012019 doi:10.1088/1742-6596/1374/1/012019

Growth of Sunu Grouper (*Plectropomus leopardus*) Larvae That Given Rotivera (*Bachionus rotundiformus*) Enriched with Taurine and Glutamine

R Rostika[1*], Iskandar[2], T Herawati[2], I Bangkit[2], G Banthani[2] and L P Dewanti[1]

[1]Study Center of Fisheries Development Faculty of Fisheries and Marine Sciences Universitas Padjadjaran
[2]Study Programe of Fisheries Faculty of Fisheries and Marine Sciences Universitas Padjadjaran

*Corresponding author's email: ritarostika_unpad@yahoo.com

Abstract. This study aims to determine the effect of taurine and glutamine enriched Brachionus on the growth and survival rate of Sunu Grouper (*Plectropomus leopardus*). This research was conducted in Marine Research and Fisheries Extension Center Gondol, Bali. The study was conducted for 1 month on September 2017. The research method used was using Completely Randomized Design (CRD) consisting of 4 treatments and 4 replication, namely treatment A (without Taurine and Glutamine; (B) 0,5 gram Taurine / 10 L media; (C) 0,5 gram Glutamine / 10 L media; (D) 0,5 gram Taurine and 0,5 Glutamine / 10 L media. The age of Sunu Grouper was D-2 until D-10, parameters observed were Survival Rate, Average Daily Length, Relative Growth Rate, number of Brachionus in the digestive tract, and water quality. Which included DO, temperature and pH. Data were analyzed using analysis of variance with F test with 95% confidence level and the difference between treatments was tested by Duncan multiple-range test, then water quality was analyzed descriptively by comparing to related literature and water quality standard according to SNI 1999. The results showed that the use of 0,5 gram Taurine and 0,5 gram Glutamine enriched Brachionus, had an effect on the growth of Sunu Kerapu.
Keyword: *Plectropomus leopardus*, brachionus, taurine, glutamine, growth, survival rate

1. Introduction

Sunu or leopard coral grouper (*Plectropomus leopardus*) is a grouper highly needed by Asian market and has important economic value for its relatively high price [1]. However, the development of sunu grouper farming in Indonesia is constrained by hardly continuous and adequate juvenile fish supply [2]. Aslianti [3] states that sunu grouper larvae survival rate is 0.15% - 0.27% and tends to die on day 14. This low survival rate requires appropriate natural feed supply at larva stadium, which is a very important initial phase for larva survival, with the appropriate feed is zooplankton Brachionus.

Brachionus is tiny with size of 50-250 μ [4][5], that it is suitable for the mouth opening of sunu grouper larva, with slow movement, easy to cultivate, easy to digest and easy to improve its nutrients with various fish oils [6]. The nutrients of Brachionus are inadequate to fulfill the nutritional need of sunu grouper larva, thus it needs to be enriched with various materials in support of rotifer's nutrients, such as *Chorella sp.*, *Nannocloropsis sp.*, baker's yeast and taurine [7].

The other attempts to improve rotifer's quality are with enrichment using *Chlorella* sp, vitamin C and vitamin B complex [4]. Enrichment using fatty acid [8] and enrichment using β-carotene [9]. Addition of β-carotene is evidently not able to maximally improve the survival rate of larva at stadium D2-D20. According to to Jusadi *et al.* [1], humpback grouper larva D16 fed with rotifer enriched with amino acid taurine may improve its survival rate. Rotifer enrichment with glutamine for humpback grouper may significantly enhance larva growth.

Taurine is non-essential amino acid which contains sulfur, but is not of protein group since it does not have carboxyl group (-COOH) needed to form peptide bond. Taurine is found in high content in skeletal muscle, heart and white blood cell and central nerve system. Taurine may be found in various sources of food like meat and fish [10]. This amino acid exists about 79.5% in total amino acid in abalone's muscle and in egg and larva of *Haliotis rubra* [6]. Taurine is known as non-essential amino acid which plays an important role in osmoregulation of marine invertebrates and a number of anaerobic energy metabolisms through formation of final product glycolysis, tauropine [11].

Glutamine is non-essential amino acid which is quantitatively free amino acid substantially found in blood plasma and muscle compared to other free amino acids. Glutamine is a source of energy which plays a role in biosynthesis of glucose, amino sugar and glutathione [12]. Additional glutamine in feed is expected to improve intestinal performance in feed digestion. This research aims at identifying the influence of enrichment and addition of taurine and glutamine to rotifer in support of sunu grouper larva's survival and growth.

2. Method

The instruments used in this research are graduated cylinder, Pasteur pipette, bucket, ruler, plankton net, thermometer, fiber vessel 1000 L, blender, analytical balance, microscope, light meter and hose. The materials are Rotifer (*B. rotundiformis*) in size of (80-120 μm) serving as natural feed with addition of chicken egg yolk, baker's yeast, fish oil, taurine and glutamine as enriching materials.

This research is conducted by employing a Completely Randomized Design (RAL) consisting of four treatments and four repetitions:

1. Treatment A : Larva is given with rotifer only (without taurine and glutamine).
2. Treatment B : Larva is given with rotifer enriched with 0.5 g taurine per 10 L of rotifer media.
3. Treatment C : Larva is given with rotifer enriched with 0.5 g glutamine per 10 L of rotifer media.
4. Treatment D : Larva is given with rotifer enriched with 0.5 g taurine and 0.5 g glutamine per 10 L of rotifer media.

The survival rate is the comparison between the number of live larvae at the end of research and the number of larvae at the beginning research.

$$SR(\%) = \frac{\sum \text{larvae at t1}}{\sum \text{larvae at t0}} \text{ x } 100\%$$

Larvae relative length-growth is the comparison between the addition of larvae length during the research to larvae initial length. According to Takeuchi [13], relative growth rate may be stated in the following formula.

$$RGR = \frac{Wt - Wo}{Wo \times t} \times 100\%$$

2.1. The Number of Rotifers in Fish Digestive Tract

The number of rotifers in fish larva's digestive tract is counted to predict the amount of consumed feed. The counting is performed one hour after feeding. Five of larvae are taken from each nurturing media and observed under microscope by putting them one by one onto microscope slide, in which the number of rotifers in digestion tract is counted.

2.2. Water Quality
Water quality test needs to be performed to observe the influence of enrichment materials on nurturing media, covering pH, temperature, Nitrite, and Ammonia. This water quality test is conducted at the beginning and at the end of research.

2.3. Data Analysis
A completely randomized design (RAL) is employed in this research with 4 treatments and 4 repetitions for each treatment. F test is conducted at level of 5% on the survival and larva's total final length parameters to examine whether the treatments influence the parameters, and in case of significant difference, the Duncan test will be conducted.

3. Result and Discussion
3.1. Survival Rate of Larvae Sunu Grouper
The research result shows that treatment A has survival rate of 2.12 ± 0.16 %, treatment B has survival rate of 2.29 ± 0.09 %, treatment C has survival rate of 2.40 ± 0.17 %, and treatment D has survival rate of 2.38 ± 0.16 %. Rotifer is enriched with addition of 0.5g taurine, 0.5g glutamine and mixture of 0.5g taurine and 0.5g glutamine. The survival rate data are presented in Table 1.

Table 1. Survival Rate of Sunu Grouper Larvae.

Treatment	Taurine and Glutamine (gr)	Survival Rate (%)
A	Control	2.12 ± 0,16
B	0.5 Glutamine	2.29 ± 0,09
C	0.5 Taurine	2.40 ± 0,17
D	0.5 Glutamine + 0.5 Taurine	2.38 ± 0,16

The analysis of variance results show that amino acid taurine and glutamine additions do not significantly influence (Fhit < F table) in Duncan test at level 5% on sunu grouper larvae's survival. The reason is the function of taurine and glutamine as the neurotransmitter in the central nerve system. Neurotransmitter accelerates message delivery to the nerve system in the brain and facilitates communication between brain cells [14]. The chemical compound neurotransmitter and adequate energy supply from taurine or glutamine and other important roles of taurine and glutamine in larva's body may expectedly speed message delivery through nerve cells to the nerve system in the brain, thus the organogenesis process controlled by the central nerve system occurs more smoothly. In addition, the other important roles of taurine in larva's body function like vision, brain growth and heart function supposedly make larva has better body growth which may eventually improve the larva's survival. This conforms to the statement of Russheim [15] that taurine also plays an important role in vision, brain, nerve system, heart function and also serving as bile acid conjugator.

The lowest result is of treatment A (control) without taurine or glutamine addition. According to Guffron and Kordi [16], that a farming biota's survival is low is caused by some factors, one of which is inappropriate feed nutrition. This may be caused by no supplementation of amino acid taurine or glutamine to rotifer, thus larva's low amino acid absorption will make nutrient availability for protein synthesis and energy production in treatment A lower than that of treatment B, C, and D.

3.2. Average of Larvae Sunu Grouper Length
Taurine and glutamine supplementation to rotifer may result in different growth rate in sunu grouper larvae between the treatments during the 10 days of nursery, ranging between 12.07% - 8.40%. The highest growth of leopard coral grouper larva is presented by treatment A (0.5g taurine and 0.5g glutamin) of 12.07%, treatment C (0.5 taurine) of 11.57%, and treatment B (0.5 glutamin) of 10.17%.

Meanwhile, treatment D (control), which is not supplemented with taurine and glutamine, has the lowest growth rate of 8.40%. The relative growth rate data are presented in Figure 1.

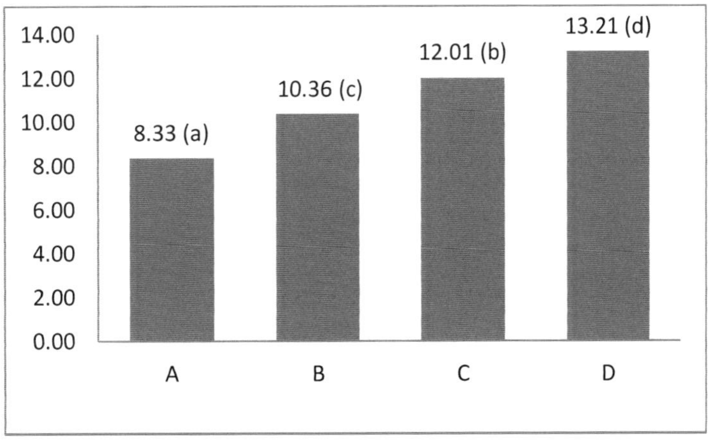

Figure 1. The Relative Growth Rate of Sunu Grouper Larvae.

Based on the research results, rotifer enriched with taurine and glutamine may result in sunu grouper larvae's different growth from treatment without taurine and glutamin supplementation. Table 2 show average of larvae sunu grouper length, treatments A, B, and C have significant growth compared to the control treatment (D).

Table 2. Average of Larvae Sunu Grouper Length.

Treatment	Taurine and Glutamine (gr)	Length (cm)
A	Control	2.84 ± 0.02^a
B	0.5 Glutamine	3.01 ± 0.10^c
C	0.5 Taurine	3.15 ± 0.06^b
D	0.5 Glutamine + 0.5 Taurine	3.20 ± 0.08^d

Factor which influences growth is good farming management, such as high stocking density, feed quality, water quality, parasite or disease. In addition, fish growth is influenced by internal and external factors. Internal factor includes heredity, resistance to disease, age, and capability to exploit feed, while external factor includes temperature, feed quality and quantity, and movement space [16]. Fish will have slow growth and small size in case of inadequate feed. When body organ's functions and body physiological functions operate well, larvae body's metabolism rate will increase, thus larvae digestion speed also increases. Free amino acid does need hydrolysis since it can be directly absorbed into larvae's digestion tract. Larvae's increased digestion speed makes larvae absorbs nutrients faster for its body. This makes larvae of treatments B, C, and D stronger so that they can find feed faster and more than larvae of treatment A. The higher content of amino acid in rotifer because of enrichment process with taurine and glutamine makes the source of energy in larvae body available faster since the free amino acid absorbed in the digestion tract is transferred 3.5 times faster to larvae body's tissue than in protein form.

3.3. Rotifer in Digestive Tract

Rotifers eaten in larvae's digestive tract are observed using microscope one hour after rotifers are given, which at 13.00. The number of rotifers in digestive tract of larvae data are presented in Figure 2.

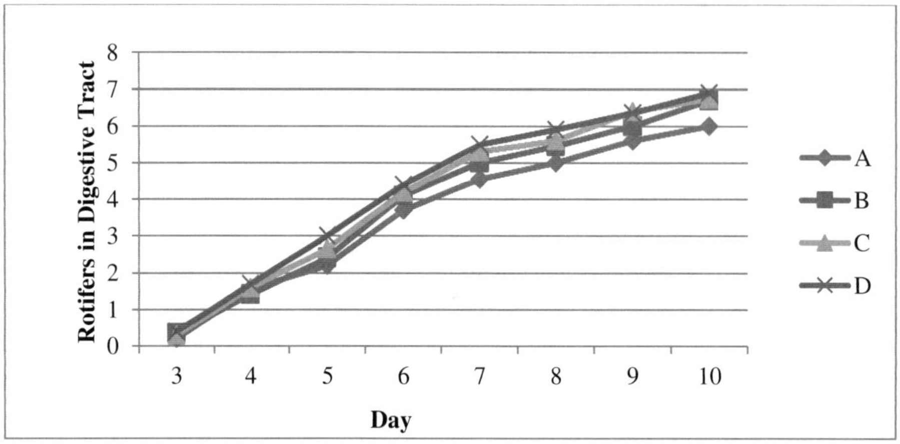

Figure 2. Number of Brachionus in Digestive Tract's Larvae.

The results of observation of the number of Brachionus consumed by larvae one hour after feeding show an increase of the number of rotifers in digestion tract each day. However, from day 6 and 7, the number of rotifers eaten in treatments B, C and D tends to be more than that in treatment A. The mean number of rotifers consumed by larvae of treatments D, C and B is higher than that of treatment A. This is expectedly related to taurine and glutamine supplementation to the enrichment media which may improve larva's appetite. This conforms to the research conducted by Aprilia [17], that rotifer added with taurine and glutamine is able to improve larva's appetite compared to control treatment. The high amino acid absorption by larva will result in nutrient availability for protein synthesis and energy production with treatments B, C and D is higher than that of treatment A. The correlation with source of energy is that marine fish absorbs amino acid more than protein [18]. The reason is that amino acid may be directly absorbed by larva digestion system and does not require enzyme to break peptide bonds.

3.4. Water Quality

The temperature observation result shows that during the research taken in three different times, there is a range of temperature between 26.3°C – 28.6°C, as, for clarity, presented in Table 3. There is an increase of temperature during research for 2°C – 3°C. Meanwhile, according to Aslianti [3], the change in minimum-maximum temperature is 1°C - 1.5°C. This value does not conform to the standard quality of temperature for leopard coral grouper larvae of 28-30°C. The reason is that the research location is semi-outdoor, thus the optimal temperature cannot be achieved. The unstable ambient temperature makes the larvae grow abnormally and their survival cannot be enhanced. It is this change in temperature which makes it difficult for larvae to adapt. The temperature observation data are presented in Table 3.

Table 3. Water Quality during Research Period (°C).

Treatment	Time		
	07.00	12.00	17.00
A (Control)	26.2 – 27.7	26.5 – 28.2	28.6 – 27.7
B (Glutamine 0,5gr)	26.3 – 27.2	26.5 – 28.1	27.7 – 28.8
C (Taurine 0,5 gr)	26.3 – 27.1	26.5 – 28.1	27.6 - 28.8
D (Glutamine 0,5gr + Taurine 0,5 gr)	26.4 – 27.4	26.3 - 28.1	28.6 – 27.6

Based on the observation results of light parameter during research taken in three different times, the range is 101 – 4803 Lux, as for clarity, presented in Table 4. Light is an important factor in the survival of larvae, especially of grouper. The light parameter value obtained in this research ranges about 101 -

4803 Lux. The light parameter in this research has different value from the standard quality of light of 500-700 lux. The reason is that the research location has high light intensity.

Excessive light intensity will get the fishes stressed, since as nocturnal animal, sunu grouper are used to doing activities, including eating in dim condition. Consequently, an environment with high light intensity will disrupt their eating activity and result in stress. The light observation data on the nurturing media are presented in Table 4.

Table 4. Light Parameter During Research (Lux).

Treatment	Time		
	07.00	12.00	17.00
A (Control)	50 - 1651	391 - 4803	52 - 160
B (Glutamine 0,5gr)	53 - 630	391 - 2270	41 - 182
C (Taurine 0,5 gr)	55 - 596	116 - 2830	38 - 192
D (Glutamine 0,5gr + Taurine 0,5 gr)	50 - 550	144 - 4040	39 - 186

Based on the observation, pH parameter is 7.82 - 8.19mg/L, nitrite parameter is 0.75 - 1mg/L and ammonia parameter is 0 - 0.75mg/L. The measurement of pH, nitrite, and ammonia parameters in sampling is conducted before changing the water, thus the value is not too far different from the standard quality. The data of pH, Nitrite and Ammonia observation are presented in Table 5.

Table 5. Observation Result of pH, Nitrite, Ammonia During Research (mg/L).

Treatment	pH	Nitrite	Ammonia
A (Control)	7.84 – 7.86	0.88 - 1	0 – 0.563
B (Glutamine 0.5gr)	7.98 – 8.20	1	0 – 0.625
C (Taurine 0.5 gr)	7.88 – 7.92	0.75 - 1	0 – 0.625
D (Glutamine 0.5gr + Taurine 0.5 gr)	7.83 – 7.88	0.75 - 1	0 – 0.563

4. Conclusion

Based on the research results, we may conclude that giving rotifers enriched with glutamine and taurine does not enhance the survival of sunu grouper larvae, but giving rotifers enriched with 0.5g taurine and 0.5g glutamine may enhance larvae's growth during the 10 days of nursery.

References

[1] Jusadi D, Ruchyani S, Mokoginta I and Ekasari J 2011 Peningkatan kelangsungan hidup dan perkembanganlarva udang putih melalui pengayaan Rotifera dengan taurin *J. Akuakultur Indonesia* **10** 131-136

[2] Sunyoto P and Mustahal 2002 Pembenihan Ikan Laut Ekonomis; Kerapu, Kakap, Baronang (Jakarta: Penebar Swadaya)

[3] Aslianti T, Suwirya K and Asmanik 2008 Teknologi Pemeliharaan Larva Kerapu Sunu (*Plectropomus leopardus*) Secara Massal *J. Ris. Akuakultur* **3** *1-11*

[4] Mudjiman A 1987 Makanan Ikan Cetakan ke-3 (Jakarta: Penebar Swadaya)

[5] Lubzens E, A Tandler and G Minkeff 1989 Rotiferas as Food in Aquaculture. National Center for Mariculture, Israel Oceanography and Limnological Research. Israel *Hydrobiology 2* 1861187: 387-400

[6] Watanabe T, Kitajima C and Fujita S 1983 Nutritional values of live organisms used in Japan for mass propagation of fish: a review *Aquaculture* **34** 115-143

[7] Martinez J B, Chatzfolis S, Divanach P and Takeuchi T 2004 Effect of Dietary Taurine Suplementatation on Growth Performance and Feed Selection of Sea Bass *Dicentrachlrs labrax* Fry Feed With Demand-Feeders *Fish Sci.* **70** 74-79

[8] Indah D 2001 Pengaruh Pemberian Rotifera *Brachionus* sp. yang Diperkaya dengan Beta Karoten terhadap Kelangsungan Hidup Larva Kerapu Bebek *Cromileptus altivelis*. Institut Pertanian Bogor.xxx

[9] Santoso D 2011 Taurin untuk performa mental dan atletik yang optimal http://www.dennysan-toso.com [8 januari 2017]

[10] Sato M, Taekuchi M, Kanno N, Nagasiha E and Sato Y 1991 Characterisation and physiological role of tauropine dehydrogenase and lactate dehyrogenase from muscle of abalone, *Haliotis discus hannai*. Tohoku *J. Agr.Res.* **41** 83-95

[11] Bartell S M and Batal A B 2007 The effect of supplemental glutamine on growth performance, development of the gastrointestinal tract and humoral immune response of broilers. *Poultry Science.* **86** 1940–1947

[12] Esparza D P 2006 Neurotransmitter. Amino Acids & Metal Health-I: The Role of L-Glutamine. http://www.naturalhealthweb.com/articles/esparza.html . [9 Januari 2017]

[13] Russheim Ch M 2000 Taurine www.serve.com/BatonRouge/taurine_chmr.htm. [27 April 2017]

[14] Ghufran M H and Kordi K 2005 Budidaya Ikan Laut di Karamba Jaring Apung (Jakarta: Rineka Cipta)

[15] Gusrina 2008 Budidaya Ikan Jilid 2 Direktorat Pembinaan Sekolah Menengah Kejuruan Direktorat Jenderal Manajemen Pendidikan Dasar dan Menengah Departemen Pendidikan Nasional Jakarta

[16] Aprilia T 2008 Aplikasi Pengkayaan Rotifera dengan Asam Amino Bebas Untuk Larva Kerapu Bebek Cromileptes altivelis. Skripsi. Program Studi Teknologi dan Manajemen Akuakultur Fakultas Perikanan dan Ilmu Kelautan Institut Pertanian Bogor hal 11

[17] Ronnestad I, Thorsen A and Finn R N 1999 Fish larval nutrition: A review of recent advances in the roles of amino acids. *Aquaculture* **177** 201-216

Antioxidant potential ethanolic extract of *Glycine max* (l.) Merr. Var. Detam and daidzein

S Prahastuti[1*], M Hidayat[1], S T Hasianna[1], W Widowati[1*], A Amalia[2], D T Yusepany[2], R Rizal[2] and H S W Kusuma[2]

[1]Faculty of Medicine, Maranatha Christian University, Jl. Prof. Drg. Surya Sumantri 65, Bandung 40163, West Java, Indonesia
[2]Aretha Medika Utama, Biomolecular Research Center, Jl Babakan Jeruk II No. 9, Bandung 40163, West Java, Indonesia

*Corresponding author's email: wahyu_w60@yahoo.com

Abstract. Antioxidants in chemical terms are electron donors and antioxidants are biologically compounds that can overcome the negative effects of oxidants in the body such as damage to vital cells of the body. The balance between oxidants and antioxidants is very important. Antioxidant from natural sources are safer than synthetic antioxidants. Soybeans are functional food source that contains essential amino acids, vitamin e, saponins and are rich in antioxidants such as flavonoids, isoflavones and anthocyanins. The aim of this research is to investigate antioxidant potential of ethanolic extract of black soybean (EEBS) and its compound daidzein. The phytochemical screening assay evaluated by modified Farnsworth methods and to find out phenolic and flavonoid content, total phenols and flavonoids were tested. DPPH (2,2-diphenyl 1-pichylhydazy), ABTS (2,2'-Azinobis-(3-ethylbenzothiazoline-6-sulfonic acid) reducing activity, FRAP (ferric reducing antioxidant power) activity and H_2O_2 scavenging activity assay were used to measure antioxidant activity. The analysis of phytochemical of EEBS exhibit the presence of saponins, alkaloids, tannins, steroids and triterpenoids and terpenoids. Total phenol and flavonoid assay showed the presence of phenols and flavonoids. Four assay of antioxidant activity showed that EEBS at highest concentration exhibit higher activity (%), and from three assays EEBS has higher antioxidant activity (expressed as IC_{50}) than daidzein. Overall, three of the four antioxidant tests performed can be concluded that EEBS has better antioxidants activity than daidzein. Further research is needed regarding black soybean as a promising antioxidant resource.
Keywords : antioxidant, oxidant, black soybean, daidzein, phytochemical

1. Introduction

Antioxidants in chemical terms are electron donors and antioxidants are biologically compounds that can overcome the negative effects of oxidants in the body such as damage to vital cells of the body. The balance between oxidants and antioxidants is very important because it is related to the work function of the body's immune system, especially to maintain the integrity and functioning of lipid membranes, cell proteins, and nucleic acids, and control signal transduction and gene expression in immune cells [1,2]. Free radicals are able to oxidize biomolecules, leading to mutagenic changes, tissue damage and cell death. They play a significant pathological role in atherosclerosis, cancer, arthritis, cirrhosis, emphysema and various other degenerative diseases [3-5].

The production of antioxidants in the human body occurs naturally to compensate for the production of free radicals. The antioxidants function as a defence system against free radicals, but increased production of free radicals that are formed due to stress factors, UV radiation, air pollution and the environment results in inadequate defence systems, so that additional antioxidants from outside are needed [6,7].

Antioxidants exogen can be obtained in synthetic and natural forms. Synthetic antioxidants such as buthylated hydroxy toluene (BHT), buthylated hydroxyanisol (BHA) and tert-butylhydroquinone (TBHQ) can effectively inhibit oxidation. However, the use of synthetic antioxidants is limited by government rules because if its use exceeds the limit it can cause toxic in the body and carcinogenic effect, so it needs safe natural antioxidants. One promising potential source of natural antioxidants is plants because they contain flavonoids, tannins, carotenoids, terpenoids, vitamins C and E and polyphenols [8,3,4].

Soybeans are functional food source that contains essential amino acids, vitamin E, saponins and are rich in antioxidants such as flavonoids, isoflavones and anthocyanins. One important aspect of soybeans as a functional food source is that isoflavones are secondary metabolites produced by plants through the synthesis of 2hydroxyisoflavone synthase (IFS). This compound is not synthesized by microorganisms, therefore, this plant is the main source of isoflavones in nature. Soybeans must have a fairly high isoflavone content [9].

Isoflavones in soybean consist of malonil-glycosides, acetyl-glycosides, glycosides, and aglycones. Diophysis of the four forms of isoflavones, the highest biological activity by isoflavones aglycones, especially genistein (5,7,4'-trihydroxy isoflavones), daidzein (7,4'-dihydroxy isoflavones) and glycitine (6-methoxy-7,4' dihydroxy isoflavones) [10].

The active compounds in soybeans that are responsible for biological activity were requested by daidzein and genistein. Several studies have shown that genistein, isoflavones in soybeans can prevent cancer [11], breast cancer [12]. Daidzein genistein compounds can prevent osteoponia in postmenopausal women [13], antimetastatic prostate cancer in humans [14], anti-inflammatory [15].

The types of soybeans that are mostly studied are black and yellow soybeans [16,17]. Previous study stated that black soybean has a better ability to prevent oxidation-related diseases compared to yellow soybeans [17]. Black soybeans have a higher antioxidant content than yellow soybeans [18]. The use of black soybeans in Indonesia is still limited compared to other kind of soybeans even though black soybeans have a lot of potential to be developed [16]. The potential for black soybean as a therapeutic is very promising, therefore it is necessary for further research on black soybeans especially as antioxidants because of its rich compounds.

2. Material and Method

2.1. Preparation of EEBS (Ethanol Extract of Black Soybean)

Fresh Black Soybeans obtained from Unit Pengelolaan Benih Sumber (UPBS) Balai Penelitian Tanaman Aneka Kacang dan Umbi, Malang, East Java. The plants were identified by herbarium staff, Department of Biology, School of Life Science and Technology, Bandung Institute of Technology, Bandung, West Java, Indonesia. The dried soy bean then grinded, mashed into powder. The powder of *Glycine max* (L.) Merr. var. Detam (250g) were extracted with distilled ethanol 70% by a maceration methods. Ethanol filtrate was filtered, and wastes were remacerated until colourless filtrate in triplicates. The pasta form is obtained by condensation using 50°C evaporapor (Zhengzhou Well-known, RE-201D). The yield of EEBS was 2.63 g stored at 20°C [3,19-23,24]. EEBS was used for next experiment. Daidzein (Chengdu Biopurify Phytochemical Ltd, BP0445) were used as standard compound.

2.2. Phytochemical Assay

2.2.1. Flavonoids. 10 mg EEBS was dissolved in 2 N HCl [Merck 1003171000] in the test tube. Added enough Mg [Merck EM105815, USA], then heated for 5-10 minutes, cooled and filtered and add 1 ml of amyl alcohol. The sample contains flavonoid compounds if it produces red/orange [20,21,24].

2.2.2. Saponins. 10 mg EEBS was dissolved using ddH_2O in the test tube, boiled in a bath for 5 minutes then filtered and shaken vigorously and added 1 N. HCl. The sample contained saponin compounds if the foam remained stable and still after dropping HCl 1 N [19-24].

2.2.3. Phenols. Dissolved 10 mg EEBS in 5 ml ddH_2O, added 1% $FeCl_3$ [Merck 103943] solution as much as 500 µl. Sample contains phenol compounds if it produces green/red/purple/blue/ black [19-24].

2.2.4. Tannins. 10 mg EEBS was dissolved in 2 ml 2 N HCl [Merck 1003171000] in a test tube, then heated in a waterbath for 30 minutes, wait until it's cold then add 500 µl amyl alcohol [Merck 10979, USA]. The sample contains tannin compounds if the amyl alcohol layer is orange/red [19-24].

2.2.5. Alkaloids. 10 mg EEBS dissolved in 5 ml ddH_2O was evaporated in the waterbath. The resulting residue dissolved with 5 ml 2N HCl. The solution obtained was divided into 2 test tubes. The first tube is added 3 drops of HCN 2N which functions as a blank. The second tube solution was transferred as much as one drop to the drip plate, then added 3 drops of Dragendorff reagent. The orange sediment formed indicate the presence of alkaloids [19-24].

2.2.6. Steroids/Triterpenoids. 10 mg EEBS put in a drop plate, add glacial acetic acid until submerged, let stand for 10-15 minutes then adding one drop of H_2SO_4 [Merck 109073, USA] concentrated. If it produces blue green, the sample contains steroid class compounds, whereas if it produces purple/red/orange, the sample contains a triterpenoid compound [19-24].

*2.2.7. Terpenoids***.** 10 mg of EEBS put into a drop plate, added vanillin sufficiently, then drip H_2SO_4 [Merck 109073, USA] concentrated one drop and homogenized. The sample contains terpenoid compounds if it produces purple [19-24].

2.3. Total Phenol Assay
Briefly 15 µL standard solution in 6 concentration level (50; 25; 12,5; 6,25; 3,125; 1,5625 µg/mL) of gallic acid [Sigma 398225] and sample (EEBS) in concentration of 50 µg/mL were prepared for total phenol assay. Standard and sampled was mixed with 75 µL of Folin-Ciocalteu's reagent 2.0 M [Merck 1.090.010.500] and 60 µL of Na_2CO_3 7,5 % [Merck A897992745] in microplate. The reaction then incubated at 50°C for 10 min. The absorbance was measured in 760 nm of wavelength using Multiskan Go Reader [Thermo Fisher Scientific 1510]. The linear regression equation ($y = ax + b$) was created based on the standard (gallic acid) absorbance value. The analysis of phenol content of sample was performed based on the each of standard linear regression equation [19-25].

2.4. Total Flavonoids Assay
Briefly 15 µL standard solution in 6 concentration level (50; 25; 12,5; 6,25; 3,125; 1,5625 µg/mL) of quercetin [Sigma Q4951-10G] and sample (extract from black soybean) in concentration of 50 µg/mL were prepared for total flavonoid assay. Standard and sampled was mixed with 75 µL $AlCl_3$ 2% [Merck 449598] into the well plate containing the sample (well sample) and add 150 µl sample solvent (DMSO [Merck1029522500]) to well blank. The absorbance was measured in 415 nm of wavelength using Multiskan Go Reader. The linear regression equation ($y = ax + b$) was created based on the standard (quercetin) absorbance value. The analysis of flavonoids content of sample was performed based on the each of standard linear regression equation [26].

2.5. DPPH scavenging activity assay
Method from Widowati et al. was used for the DPPH assay [20-22]. The method is based on the formation of non-radical DPPH-H results from the addition of hydrogen from an antioxidant characterized by a reduction in alcoholic DPPH solution [19-24]. Briefly 50 µl various concentration of samples (EEBS, daidzein), was introduced in 96-well microplate followed by addition of 200 µl of

0,077 mmol/L DPPH [Sigma Aldrich D9123, USA] into the well. The mixture incubated at room temperature for 30 min in the dark. After that, absorbance was measured using a microplate at 517 nm. The scavenging activity (%) calculated as below :
Scavenging Activity (%) = (Ac - As)/Ac x 100
Ac = negative control absorbance (without samples)
As = samples absorbance
The scavenging activity (%) was the continued to be calculated as median inhibitory concentration (IC_{50}) [19-23,27,28].

2.6. ABTS-reducing activity assay

The antioxidant capacity of EEBS and daidzein was evaluated by ABTS (2,2'-Azinobis-(3-ethylbenzothiazoline-6-sulfonic acid) [Sigma A1888-2G, USA] diammonium salt-free radical assay.Two microlitres of various level samples (EEBS, daidzein) were added to the sample well, then ABTS*+ solution (198 µl) was added to each well (96 well-plate), incubated for 6 min at 30 °C. Absorbance was measured at 745 nm wavelengths using microplate spectrophotometer [19-24].

2.7. FRAP assay

FRAP assay was performed using previous modified method [19-24]. Briefly 7.5 µl of various level samples (EEBS, daidzein) was added with 142.5 µl FRAP reagent into each well in a 96 well-microplate and then incubated at 37 °C for 30 min. Absorbance was measured at 593 nm wavelength using microplate spectrophotometer.

2.8. H_2O_2 scavenging activity

The mixture of 12µL ferrous ammonium sulphate 1mM [Sigma 7783859], 60µL of various level samples (EEBS, daidzein), 3 µL of H_2O_2 5 mM Merck 1.08597.1000] was incubated at dark room temperature for 5 min. After incubation, 75 µL of 1,10-phenanthroline 1 mM [Sigma 131377] was added to the well plate containing samples (well samples) and blank, the plate was incubated for 10 min in a dark place at room temperature. The absorbance of scavenging activity was measured at 510 nm wavelength [27,29]. The formula used to measured H_2O_2 scavenging activity:

$$H_2O_2 \text{ scavenging activity } (\%) = \frac{Abs\ control - Abs\ sample}{Abs\ control}\ x\ 100$$

Abs sample= Sample absorbance
Abs control= Control absorbance

3. Results and Discussions

Soybean is one of the most important plants for human and animal consumption, and the most important organic component of soybean seeds is protein (about 40%) and oil (about 20%). With the exception of isoflavonoids, few studies have been done on other phenolic classes present in soybean [30]. Black soybeans contain essential amino acids, vitamin E, saponins and are rich in antioxidants, for example with flavonoids, isoflavones and anthocyanins. However, black soybeans have a 4-fold tannin content compared to yellow soybeans. Tannins is one of the anti-nutrients because of its ability to bind proteins so that proteins are difficult to digest by the protease enzyme. From the results of phytochemical screening that has been done, it is known that black soybeans contain saponins, tannins, steroids/triterpenoids, terpenoids and alkaloids. Total phenol and flavonoids show the presence of phenols and flavonoids [9,31].

3.1. Phytochemical Assay of EEBS

Phytochemical screening of EEBS aimed to evaluate qualitatively presence of flavonoids, saponins, phenols, tannins, steroids/triterpenoids, terpenoids and alkaloids. The result of EEBS phytochemical screening showed in Table 1. Phytochemical screening shows that only saponins, tannins, steroids/triterpenoids and alkaloids were detected.

Table 1. The result of qualitative phytochemical screening of EEBS.

Phytochemical content	EEBS
Flavonoids	-
Saponins	+
Phenols	-
Tannins	+
Steroids/Triterpenoids	+/+
Terpenoids	+
Alkaloids	+

+ : detected; - : Not detected

3.2. Total Phenol and Flavonoids Assay
Total phenol assay was to find out phenol content contained in EEBS. Total phenol of EEBS based on gallic acid equivalent (GAE) was performed using linear regression equation (y = 0,0374 x + 0.1452 with R^2 = 0.99) [19-23]. The total phenol obtained from EEBS was 5.60±0.18 µg GAE/mg EEBS. Total flavonoid assay is to find out flavonoid content contained in EEBS. Total flavonoids of EEBS based on quercetin equivalent (QE) was performed using linear regression equation (y = 0.0174 x + 0.0368 with R^2 = 0.99) [26,27]. This study shows that EEBS has total flavonoids content is 2.03±0.12 µg QE/mg EEBS.

In the phytochemical screening assay, phenol and flavonoids were not detected in EEBS, according to the results of total phenol (5.60±0.18 GAE µg/mg EEBS) and flavonoid (2.03±0.12 QE µg/mg EEBS) which showed low results when compared to previous studies which reported total phenol from black soybean (80% ethanol) was 17.75 ± 0.39 (GAE mg/g extract) and total flavonoids 5.73 ± 0.26 catechin equivalent (CE mg/g extract), so that they are not detected qualitatively by phytochemical assay [32].

Other ingredients in black soybeans are essential amino acids, vitamin E, saponins and are rich in antioxidants such as flavonoids, isoflavones and anthocyanins. The several foods that have been analyzed, it is known that soybeans rank first, containing isoflavones and their derivatives. Isoflavones and their derivatives are compounds known to function as antioxidants, antitumor, antiosteroclerosis [33,34]. Isoflavones are flavonoid antioxidants that have aglycone, namely genistein (5,7,4'-trihydroxy isoflavones), daidzein (7,4'-dihydroxy isoflavones) and glycitine (6-methoxy-7,4' dihydroxy isoflavones)[10] that exert antioxidant action [3,35]. Total isoflavone content in soybean seed was around 558.2 µg/g - 1716.9 µg/g [36]. When compared with other types of soybeans, black soybeans contain than yellow soybeans. The total polyphenols in yellow soybeans are 0.02 mg/g, while black soybeans have a total polyphenol content of 0.56 mg/g [9].

3.3. Antioxidant Assay
3.3.1. DPPH scavenging activity. The antioxidant capacity of EEBS and daidzein was evaluated using DPPH free radical scavenging activity. This method is based on the formation of non-radical DPPH-H results from the addition of hydrogen from antioxidants which is characterized by a reduction in the DPPH alcohol solution [37,20-23]. Antioxidant molecules extinguished by DPPH free radicals are marked in purple from the change in DPPH sample to colourless [19-23,24]. The inhibitory concentration (IC_{50}) of EEBS and daidzein to scavenge DPPH can be seen in Table 4.

Table 2. The IC_{50} value of DPPH scavenging activity of EEBS and daidzein.

Samples	Equation	R^2	IC_{50} µg/mL	IC_{50} (µM)
EEBS	y = 0.2463x + 22.069	0.99	113.40	-
Daidzein	y = 0.2052x + 27.376	0.99	110.25	433.66

*Linear equations, coefficient of regression (R^2) and IC_{50} of each sample were calculated. IC_{50} of EEBS was presented in µg/mL, while daidzein were presented in µM and µg/mL.

As shown in Table 2, the IC_{50} value of daidzein (110.25 µg/mL) was lower than EEBS (113.40 µg/mL). These results indicate low scavenging activity of EEBS compared to daidzein. Nevertheless, at the highest concentration (200 µg/mL) EEBS has higher scavenging activity (70.18 ± 0.8%) than daidzein (67.34 ± 1.16%) (shown in Figure 1).

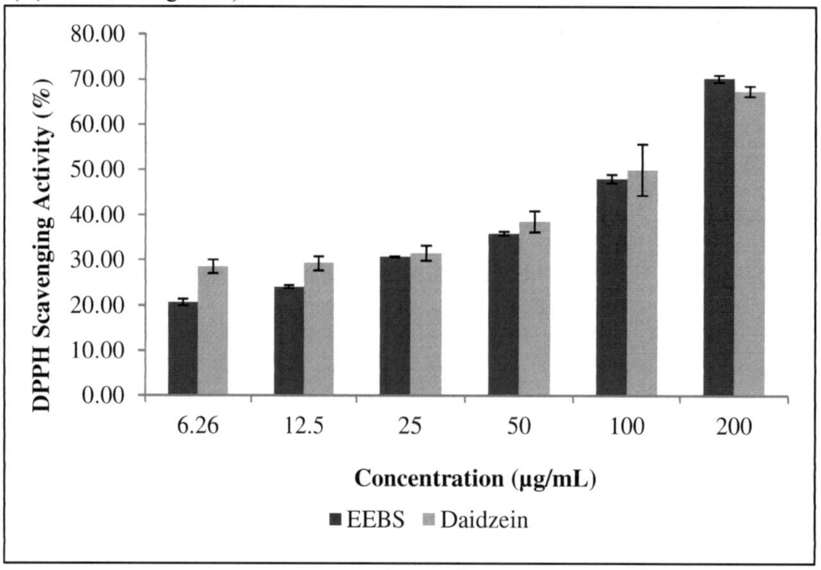

Figure 1. Effect various concentration of EEBS and daidzein toward DPPH scavenging activity.
EEBS and daidzein were diluted in DMSO to reach the final concentration of 6.26; 12.5; 25; 50; 100; 200 (µg/mL

Research conducted by Takahata et al. (2001) showed that DPPH radical scavenging activity of black soybeans was the strongest among the three types of soybeans associated with seed peels, which also included reddish brown, chocolate, and black beans, and were highly dependent on the content of phenolic compounds [38].

According to Table 4. the most active extract in the DPPH scavenging activity of the two tested compounds was daidzein extract which was shown by a lower IC_{50} value compared to EEBS. Daidzein has higher antioxidant activity through DPPH scavenging activity and IC_{50} value compared to EEBS. This is because daidzein is a polyphenolic compound that has a broad spectrum of physiological and pharmacological functions and is known to act as an antioxidant in vivo and in vitro [39].

3.3.2. ABTS-reducing activity assay. The ABTS reduction activity assay measures the relative ability of antioxidants to scavenge the ABTS generated. The reaction of a strong oxidizing agent (potassium permanganate / potassium persulfate) with ABTS salt will produce ABTS. Reduction of the blue-green ABTS radical solution by antioxidants that donate hydrogen is measured by 745 nm long wave absorption spectrum [19-23,24,28]. ABTS-reducing activity of EEBS and daidzein based on IC_{50} values can be seen in Table 3.

Table 3. The IC_{50} value of ABTS reducing activity of EEBS and daidzein.

Samples	Equation	R^2	IC_{50} µg/mL	IC_{50} (µM)
EEBS	y = 0.5008x + 12.441	0.99	75.00	-
Daidzein	y = 0.4736x + 8.9021	0.99	86.78	341.34

*Linear equations, coefficient of regression (R^2) and IC_{50} of each sample were calculated. IC50 of EEBS was presented in µg/mL, while daidzein were presented in µM and µg/mL.

EEBS has higher ABTS-reducing activity as indicated by lower IC_{50} (75.00 µg/mL) compared to daidzein (86.78 µg/mL). Along with reducing activity at the highest concentration (50 µg/mL) EEBS have higher activity (36.97 ± 0:09%) than daidzein (32.30 ± 1.87%) shown in Figure 2. The results show that EEBS has high antioxidant activity compared to standard compounds.

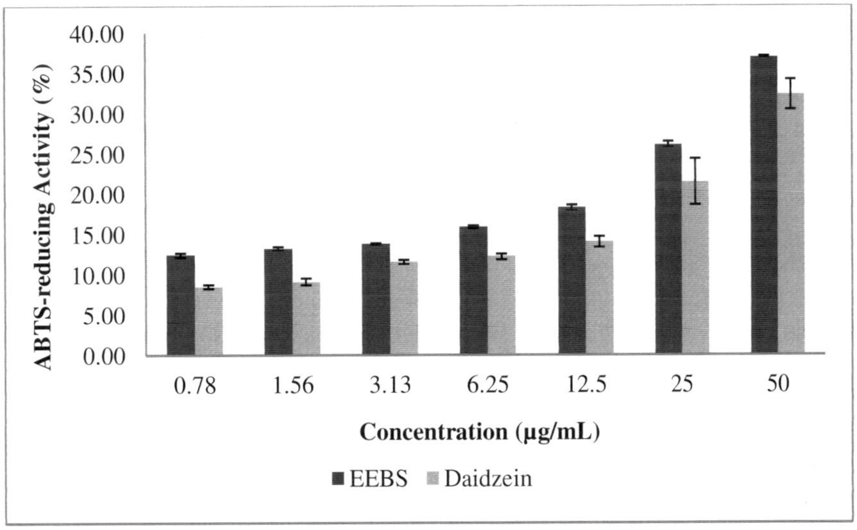

Figure 2. Effect various concentration of EEBS and daidzein toward ABTS-reducing activity. EEBS and Daidzein were diluted in DMSO to reach the final concentration of 0.78; 1.56; 3.13; 6.25; 12.5; 25; 50 (µg/mL).

IC_{50} EEBS value in ABTS reducing activity assay is lower than daidzein, this shows that EEBS has higher antioxidant potential than daidzein. Some studies reported that black soybeans have the potential to possess antioxidants.

3.3.3. FRAP activity. FRAP activity can be used to determine the EEBS antioxidant capacity and daidzein. FRAP activity of the EEBS and daidzein can be seen in Figure 3. The method of FRAP activity based on the reduction Ferroin analog, complex tripyridyltriazine Fe^{3+} of Fe $(TPTZ)^{3+}$ into Fe^{2+} complex dark blue coloured Fe (TPTZ) $^{2+}$ by antioxidants in acidic medium [5,19-23,28]. In this study, FRAP activity of EEBS and daidzein at the highest concentration (50 µg/mL) was 148.89 ± 4.16 and 98.36 ± 4.57 µM Fe (II)/µg respectively. These results indicate that EEBS has higher FRAP activity than daidzein [21,23].

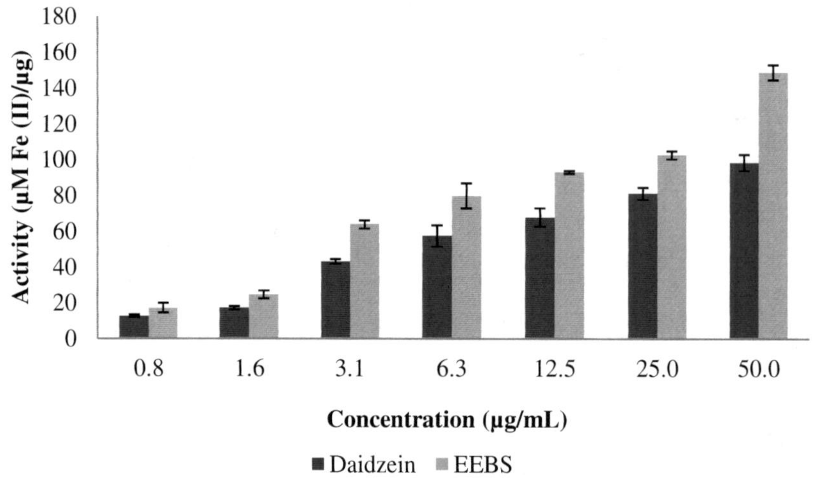

Figure 3. Effect various concentration of EEBS and daidzein toward FRAP activity.
EEBS and daidzein were diluted in DMSO to reach the final concentration of 0.8; 1.6; 3.1; 6.3; 12.5; 25; 50 (µg/mL).

The content of flavonoids in EEBS, especially isoflavones and their derivatives (daidzein and genistein), has an important role as an antioxidant in black soybeans. As one of the flavonoid groups, isoflavone (daidzein is one of them) bioactive compounds containing phenolic groups have been reported to have ability as an antioxidant and prevent the occurrence of free radical damage through two mechanisms, namely donating hydrogen ions, and acting as a direct free radical scavenger [40].

3.3.4. H_2O_2 scavenging activity. The reaction of ferrous ammonium sulphate and phenanthroline could form Fe^{2+}-tri-phenanthroline complex with the color of orange, but if H_2O_2 exists in that reaction, Fe^{2+}-tri-phenanthroline complex would not be formed, thus scavenger of H_2O_2 might not form Fe^{2+}-tri-phenanthroline complex [29]. The H_2O_2 scavenging activity of EEBS and daidzein exhibited in Table 6. The present data showed that EEBS has higher H_2O_2 scavenging activity with an IC_{50} value 284.61 µg/mL, than daidzein with an IC_{50} value 363.60µg/mL. EEBS has higher antioxidant activity compared to daidzein.

Table 4. The IC_{50} value of H_2O_2 scavenging activity of EEBS and Daidzein.

Samples	Equation	R^2	IC_{50} µg/mL	IC_{50} (µM)
EEBS	y = 0.0946x + 23.076	0.99	284.61	-
Daidzein	y = 0.1284x + 3.3139	0.99	363.60	1,430.201

*Linear equations, coefficient of regression (R^2) and IC_{50} of each sample were calculated. IC50 of EEBS was presented in µg/mL, while Daidzein were presented in µM and µg/mL.

In line with reducing activity (shown in Figure 4) at the highest concentration (300 µg/mL) EEBS has a higher activity (50.87 ± 1.4%) than daidzein (41.56 ± 0.40%).

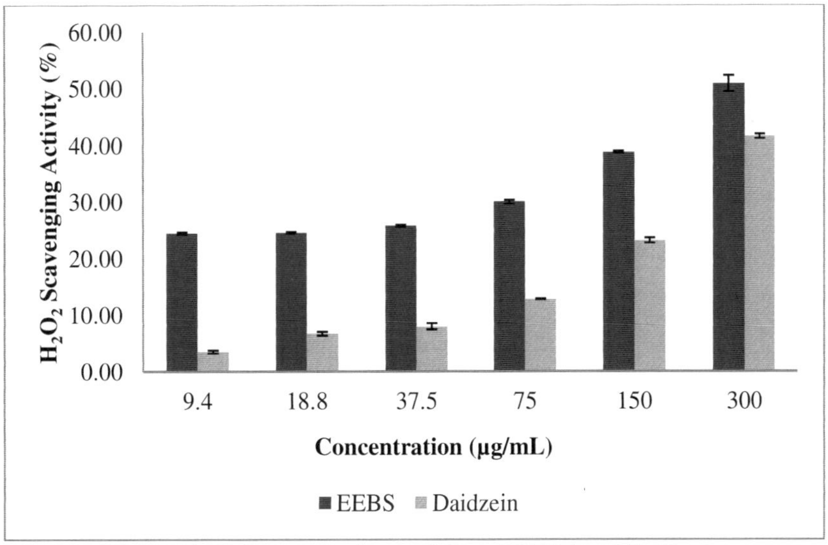

Figure 4. H_2O_2 scavenging activity of EEBS and Daidzein.
EEBS and daidzein were diluted in DMSO to reach the final concentration of 9.4; 18,8; 37.5; 75; 150; 300 (µg/mL).

Meta 5,7-dihydroxyl structure on ring A shows the ability of isoflavones to act as hydrogen ion donors so that more stable compounds are formed and peroxyl radicals are less reactive, while 4'-hydroxyl groups on ring B Isoflavone compounds act as ROS compound scavenger. The hydroxyl group configuration in ring B flavonoid compounds has been reported to act as a compound scavenger of ROS. It was further stated that the hydroxyl group at ring B can donate hydrogen ions by donating an electron to hydroxyl and peroxyl radicals; stabilize these two radicals, and form flavonoid radicals that are relatively more stable [40].

Daidzein included in the isoflavones group has a hydroxyl group responsible as a positive group for its antioxidant properties. The amount of hydroxyl groups possessed by daidzein can be a factor that can cause free radical scavenging activities. The more hydroxyl groups possessed by active compounds, the more free radicals can be scavenged [41]. Different from DPPH, the results of the ABTS reducing activity assay, FRAP activity and H_2O_2 scavenging the activity show that EEBS has a better antioxidant activity than daidzein. As previously explained, black soybeans contain essential amino acids, vitamin E, saponins and are rich in antioxidants, isoflavones and anthocyanins [33,34], also contain saponins, tannins, steroids/triterpenoids and alkaloids detected in the phytochemical test. The synergistic reaction of various compounds contained in EEBS causes EEBS to have a higher antioxidant activity than just one compound, daidzein.

The IC_{50} value is defined as the concentration of the test compound which can reduce free radicals by as much as 50%. The smaller the IC_{50} value the higher the free radical reduction activity. According to the standards that classified antioxidant capacity based on IC_{50} values such as very strong (IC_{50} value less than 50 ppm), strong (IC_{50} value between 50-100 ppm), moderate (IC_{50} between 100-150 ppm), and weak (IC_{50} value between 151-200 ppm or more) (1 ppm equal to 1 µg/mL) [20]. DPPH showed the results of moderate antioxidant activity, ABTS showed the results of strong antioxidant activity and H_2O_2 showed low/weak antioxidant activity. The three tests (ABTS, FRAP and H_2O_2) showed better EEBS antioxidant activity than daidzein.

4. Conclusion

Overall, EEBS has better antioxidant activity capability than daidzein as evidenced by three assay results, ABTS reducing activity, FRAP activity and H_2O_2 scavenging activity.

5. Acknowledgements
This research was financially supported by the Ministry of Research, Technology and Higher Education of the Republic of Indonesia and for research grant 2019 (Penelitian Dasar Unggulan Perguruan Tinggi). This research was also supported by Aretha Medika Utama Biomolecular and Biomedical Research Center, Bandung, Indonesia for the research methodology and laboratory facilities. We are thankful to Ni Luh Wisma Eka Yanti, Satrio Haryo Benowo Wibowo, Ubaydillah Zedd Munshy, Dwi Surya Artie, Rizki Amalia from Aretha Medika Utama Biomolecular and Biomedical Research Center, Bandung, Indonesia for their valuable assistance.

References
[1] Krinsky N I 1992 Mechanism of action of biological antioxidants *Proc. Soc. Exp. Biol. Med.* **200** (2) 248-54

[2] Lobo V, Patil A, Phatak A and Chandra N 2010 Free radicals, antioxidants and functional foods: an health *Pharmacogn. Rev.* **4** (8) 118

[3] Fritz K L, Seppanen C M, Kurzer M S and Csallany A S 2003 The in vivo antioxidant activity of soybean isoflavones in human subjects *Nutr. Res.* **23** (4) 479-87

[4] Oteiza P I, Erlejman A G, Verstraeten S V, Keen C L and Fraga C G 2005 Flavonoid-membrane interactions: a protective role of flavonoids at the membrane surface? *J. Immunol. Res.* **12** (1) 19-25

[5] Yang J H, Mau J L, Ko P T and Huang L C 2000 Antioxidant properties of fermented soybean broth *Food Chem.* **71** (2) 249-54

[6] Johnson P 2002 Antioxidant enzyme expression in health and disease: effects of exercise and hypertension. *Comp. Biochem. and Physiol Part C: Toxicol & Pharm.* **133** (4) 493-505

[7] Blois MS 1958 Antioxidant determinations by the use of a stable free radical *Nat.* **181** (4617) 1199

[8] Jin L, Zhang Y, Yan L, Guo Y and Niu L 2012 Phenolic compounds and antioxidant activity of bulb extracts of six *Lilium* species native to China *Mol.* **17** (8) 9361-78

[9] Sulistyowati E, Martono S, Riyanto S and Lukitaningsih E 2018 Analisis Daidzein Dan Genistein Pada Kedelai (*Glycine max* L. Merril) Varietas Anjasmoro, Argomulyo Dan Dena 2 Menggunakan Metode Kckt *Media Farmasi Indonesia* **13** (1) 1299-304

[10] Tipkanon S, Chompreeda P, Haruthaithanasan V, Suwonsichon T, Prinyawiwatkul W and Xu Z 2010 Optimizing time and temperature of enzymatic conversion of isoflavone glucosides to aglycones in soy germ flour *J. of Agri. and Food Chem.* **58** (21) 11340-5

[11] Ravindranath MH, Muthugounder S, Presser N and Viswanathan S 2004 Anticancer therapeutic potential of soy isoflavone, genistein Complementary and Alternative Approaches to Biomedicine (Boston, MA:Springer) pp 121-65

[12] Kang X, Jin S and Zhang Q 2009 Antitumor and Antiangiogenic Activity of Soy Phytoestrogen on 7, 12-Dimethylbenz [α] anthracene-Induced Mammary Tumors Following Ovariectomy in Sprague–Dawley Rats *J. of Food Sci.* **74** (7) 237-42

[13] Herwana E, Setiabudy R, Soegondo S, Baziad A and Hidayat A 2015 Soy isoflavone supplementation increases equol-producing capability in postmenopausal women with osteopenia. *UnivMed* **31** (2) 120-30

[14] Pavese J M, Krishna S N and Bergan R C 2014 Genistein inhibits human prostate cancer cell detachment, invasion, and metastasis *The American J. of Clinical Nut.* **100** (supp_1):431S-6S

[15] Parida S, Singh T U, Thangamalai R, Reddy C E, Panigrahi M, Kandasamy K, Singh V and Mishra S K 2015 Daidzein pretreatment improves survival in mouse model of sepsis *J. Surg. Res.* **197** (2) 363-73

[16] Krisna A, Ika W and Wardani R 2014 Exploration of Black Soybean For Functional Food Production Improve the Public Health **2** (4) 58–67

[17] Takahashi R, Ohmori R, Kiyose C, Momiyama Y, Ohsuzu F and Kondo K 2005 Antioxidant activities of black and yellow soybeans against low density lipoprotein oxidation *J. of Agri. and Food. Chem.* **53** (11) 4578-82

[18] Dajanta K, Janpum P and Leksing W 2013 Antioxidant capacities, total phenolics and flavonoids in black and yellow soybeans fermented by *Bacillus subtilis*: A comparative study of Thai fermented soybeans (thua nao) *Int. Food Research J.* **20** (6) 3125

[19] Nurhayati B, Rahayu I G, Rinaldi S F, Zaini W S, Afifah E, Arumwardana S Kusuma H S W, Rizal R and Widowati W 2018 The Antioxidant and Cytotoxic Effects of *Cosmos caudatus* Ethanolic Extract on Cervical Cancer *Indones Biomed J.* **10** (3) 243–9

[20] Widowati W, Rani A P, Hamzah R A, Arumwardana S, Afifah E, Kusuma H S W, Rihibiha D D, Nufus H and Amalia A 2017 Antioxidant and antiaging assays of *Hibiscus sabdariffa* extract and its compounds *Nat. Product Sci.* **23** (3) 192-200

[21] Widowati W, Fauziah N, Herdiman H, Afni M, Afifah E, Kusuma H S W, Nufus H, Arumwardana S and Rihibiha D D 2016 Antioxidant and anti-aging assays of *Oryza sativa* extracts, vanillin and coumaric acid *J. Nat. Remed* **16** (3) 88-89

[22] Widowati W, Janeva B W, Nadya S, Amalia A, Arumwardana S, Kusuma H S W and Arinta Y 2018 Antioxidant and antiaging activities of *Jasminum Sambac* extract, and its compounds *J. of Reports in Pharm. Sci.* **7** (3) 270

[23] Rusmana D, Wahyudianingsih R, Elisabeth M, Balqis B, Maesaroh M and Widowati W 2017 Antioxidant Activity of *Phyllanthus niruri* Extract, Rutin and Quercetin *Indones. Biomed. J.* **9** (2) 84

[24] Widowati W, Wargasetia T L, Afifah E, Mozef T, Kusuma H S W, Nufus H, Arumwardana S, Amalia A and Rizal R 2018 Antioxidant and antidiabetic potential of *Curcuma longa* and its compounds *Asian J Agri Biol.* **6** (2) 149–61

[25] Ivanova D, Gerova D, Chervenkov T and Yankova T 2005 Polyphenols and antioxidant capacity of Bulgarian medicinal plants *J. of Ethnopharmacology* **96** (1-2) 145-50

[26] Atanassova M and Georgieva S (2011) Total phenolic and total flavonoid contents, antioxidant capacity and biological contaminants in medicinal herbs. *Ivancheva J. Univ. Chem. Technol.Metall.* **46** (1) 81–8

[27] Mukhopadhyay D, Dasgupta P, Roy D S, Palchoudhuri S, Chatterjee I, Ali S and Dastidar S G 2016 A Sensitive In vitro Spectrophotometric Hydrogen Peroxide Scavenging Assay using 1, 10-Phenanthroline *Free Rad. Antiox.* **6** (1) 124-32

[28] Miller N J and Rice-Evans C A 1997 Factors influencing the antioxidant activity determined by the ABTS+ radical cation assay *Free Radic. Res.* **26** (3) 195–9

[29] Utami S, Adityaningsari P, Sosiawan I, Endrini S, Sachrowardi Q R, Laksono S P, Nafik S, Arrahmani B C, Afifah E, Widowati W 2017 Antioxidants and Anticholinesterase Activities of the Characterized Ethanolic of ripe Sesoot (*Garcinia picrorrhiza* Miq.) Fruit Extract (GpKar) and Xanthone *Trad. Med. J.* **22** (3) 160-5

[30] Malenčić D, Popović M and Miladinović J 2007 Phenolic content and antioxidant properties of soybean (*Glycine max* (L.) Merr.) seeds *Molecules* **12** (3) 576–81

[31] Hasler C M 1998 Functional foods: their role in disease prevention and health promotion *Food Technol-Champaign Then Chicago* **52** 63-147

[32] Juan M Y and Chou C C 2010 Enhancement of antioxidant activity, total phenolic and flavonoid content of black soybeans by solid state fermentation with *Bacillus subtilis* BCRC 14715 *Food Microbiol.* **27** (5) 586-91

[33] Dixon R A and Steele C L 1999 Flavonoids and isoflavonoids–a gold mine for metabolic engineering *Trends in Plant Sci.* **4** (10) 394-400

[34] Yuan D, Pan Y, Chen Y, Uno T, Zhang S and Kano Y 2008 An improved method for basic hydrolysis of isoflavone malonylglucosides and quality evaluation of Chinese soy materials *Chem. Pharm. Bull.* **56** (1) 1-6

[35] Andlauer W, Martena M J and Fürst P 1999 Determination of selected phytochemicals by reversed-phase high-performance liquid chromatography combined with ultraviolet and mass spectrometric detection *J. Chromatogr. A.* **849** (2) 341-8

[36] Sakthivelu G, Akitha Devi M K, Giridhar P, Rajasekaran T, Ravishankar G A, Nikolova M T, Angelov G B, Todorova R M and Kosturkova G P 2008 Isoflavone Composition, Phenol Content, and Antioxidant Activity of Soybean Seeds from India and Bulgaria *J. Agric. Food Chem.* **56** (8) 2090–5

[37] Sohn DH, Kim YC, Oh SH, Park EJ, Li X, Lee BH 2003 Hepatoprotective and free radical scavenging effects of *Nelumbo nucifera Phytomedicine* **10** (2-3) 165-9

[38] Takahata Y, Ohnishi-Kameyama M, Furuta S, Takahashi M and Suda I 2001 Highly polymerized procyanidins in brown soybean seed coat with a high radical-scavenging activity *J. Agric. Food Chem.* **49** (12) 5843-5847

[39] Dwiecki K, Neunert G, Polewski P and Polewski K 2009 Antioxidant activity of daidzein, a natural antioxidant, and its spectroscopic properties in organic solvents and phosphatidylcholine liposome *J. Photochem Photobiol B Biol.* **96** (3) 242–8

[40] Heim K E, Tagliaferro A R and Bobilya D J 2002 Flavonoid: chemistry, metabolism and structure-activity relationship *J. Nutr. Biochem.* **13** (10) 572-584

[41] Pietta P G 2000 Flavonoids as antioxidants *J. Nat. Prod.* **63** (7) 1035-42

The antibacterial activities of bacteriocin *Pediococcus acidilactici* of breast milk isolate to against methicillin-resistant *Staphylococcus aureus*

S D Lestari[1], A L O Sadiq[1], W A Safitri[2], S S Dewi[3] and M E Prastiyanto[3*]

[1]DIV Medical Laboratory Technology, Faculty of Nursing and Health Secience Muhammadiyah University of Semarang, Indonesia
[2]DIII Medical Laboratory Technology, Faculty of Nursing and Health Secience Muhammadiyah University of Semarang, Indonesia
[2]Department of Medical Laboratory Technology, Faculty of Nursing and Health Sience Muahmmadiyah University of Semarang, Indonesia

*Corresponding author's email: evy_prastiyanto@unimus.ac.id

Abstract. Resistance has occurred if an antibiotic cannot kill pathogenic bacteria therefore an antibiotic therapeutic effect cannot be achieved. Broad and free use of antibiotics effected in health problems, one of the health problems is bacteria are experienced the resistance such as Methicillin-Resistant Staphylococcus aureus (MRSA). Therefore an antibiotic alternative is needed from natural ingredients have the potential as anti-MRSA such as bacteriocin. This research was conducted to determine the antibacterial activity of bacteriocin from Pediococcus acidilactici of breast milk isolate (ASI) against MRSA. Bacteriocin isolation was carried out by growing of P.acidilactici on the media of Rogosa Sharpe (MRS) broth. Antibacterial activity test was carried out by diffusion method are to see the inhibition zone and dilution to determine the value of Minimum Inhibitory Concentration (MIC) and Minimum Bactericidal Concentration (MBC). The test results shown that bacteriocin P.acidilactici strains A11 and C12 strains had antibacterial activity against MRSA. The A11 strain inhibition zone is 15 mm, MIC and MBC values are 25%. While the C12 strain inhibition zone produced 18 mm, the value of MIC and MBC was 12.5%.

1. Introduction

Extensive and inappropriate use of antibiotics has caused bacteria to experience resistance. Bacterial resistance is a condition where the bacteria can survive or not die when attacked by antibiotics, even have the ability to adapt and develop when exposed to antibiotics [1]. According to the CDC U.S, The Department of Health and Human Services in 2013, 23,000 people has died each year from disease infections accompanied by antibiotic resistance. Resistance events such as the discovery of Methicillin-resistant *Staphylococcus aureus* (MRSA) in acute infections are a public health threat. MRSA is a strain of *Staphylococcus aureus* that is resistant to the methicillin group. The high rate of bacterial resistance encourages efforts to utilize natural ingredients that have antibacterial activity which is used as an antibiotic agent. One natural product such as bacteriocin lactic acid bacteria (LAB) has the potential as an antibacterial in the pharmaceutical field [2].

BAL is a group of bacteria have a Gram positive, not have spores, do not have cytochrome, aerotolerant, anaerobic, have microaerophilic, spherical or stem which produces lactic acid as the main metabolic end product during carbohydrate fermentation, H_2O_2 anticancer and antimicrobial in bacteriocin [3]. Bacteriocin is an extracellular enzyme as a probiotic are contributes to suppressing pathogenic bacteria in the digestive tract [4]. The advantage of this compound compared to other antibacterial compounds is works selectively, is safe and is able to prevent resistance [5]. Bacteria is can produce bacteriocin are LAB groups, one of which is isolated from Breast Milk.

Breast Milk is a source of nutrition for babies containing LAB [6]. Breast milk contains bifidogenic factors, an oligosaccharide called acetylglucosamine and glycoprotein which can support the growth of LAB especially bifidobakteria [7]. Previous study was able to isolate probiotic bacteria from breast milk is namely the genus Lactobacillus which could potentially reduce cholesterol [8]. This study aims to determine the antibacterial activity of LAB bacteriocin isolates breast milk against MRSA.

2. Research methods

This study uses a type of experimental test research with the well diffusion method. The research object is bacteriocin obtained from LAB breast milk isolates. MRSA bacterial samples were obtained from the research sites of the Microbiology Laboratory of the Faculty of Nursing and Health Sciences of the University of Muhammadiyah Semarang. MRSA bacteria strain from diabeticus ulcer base swabs and resistant to antibiotics Methicillin, Benzylpenicillin, Oxacillin, Gentamicin, Ciprofloxacin, Levofloxacin, Moxifloxacin, Trimethoprim/Sulfamethoxazole.

2.1. Tools and materials

The tools has used are autoclave (Huramaya), incubator (Memmert), centrifuge (Gemmy), refrigerator (Frimed), freezer (Electrolux), laminar air flow (Mascotte), petri dish (disposible), micropipette. The ingredients used are breast milk, media by Roger Rogosa Sharpe (MRS) broth, Man of Rogosa Sharpe (MRS) so that the media Muller Himton Agar (MHA), physiological NaCl, bacteria test MRSA, aquadest, 70% alcohol, St. Mc. Farland, and antibiotics (Amphicilin, Eritromycin, Gentamycin, Tetracycline).

2.2. Isolation of Bacteriocin

Bacteriocin was isolated from the LAB of *Pediococcus acidilactici* strains A11 and C12 which were then cultured again in the media DeMan, Rogosa, Sharpe (MRS) broth, and would produce a large amount of LAB which was characterized by turbidity. After being cultured on the media then incubated for 24 hours then centrifugation in cold conditions to get the bacteriocin content in BAL [9]. Next part of the supernatant was taken and 1N of NaOH was added and therefore bacteriocin was obtained in a neutral state.

2.3. Bacterial Inoculation

MRSA are available at the Microbiology Laboratory of the Faculty of Nursing and Health Sciences of the University of Muhammadiyah Semarang was inoculated into the media of Blood Plate Agar (BAP) at a temperature of 35 ± 2 °C for 24 hours. And the next step, the colony with positive catalase was selected and continued with the Fire Staph test. Furthermore, sensitivity was tested for antibiotics gentamicin, ciprofloxacin, levofloxacin, clindamycin, tetracycline, rifampicin and trimethoprim.

2.4. Antibacterial Activity Test

Antibacterial activity test was carried out by the well diffusion method. The test bacteria were sub-cultured on BAP media for 24 hours at a temperature of 35 ± 2°C. Then turbidity 0.5 of the McFarland standard (1.5 × 108) is made. Using sterile cotton is blocked throughout the MHA media. Then has made wells by using a cork borer with a diameter of 0.5 cm. The antibacterial activity of bacteriocin was carried out by bacteriocin isolated from 200 μL to each well. Then it incubated at 35 ± 2oC for 16-20 hours. Antibacterial activity is determined by measuring the inhibitory zone [10].

2.5. Minimum inhibitory concentration (MIC)
The antibacterial activity of bacteriocin was carried out by microplate using microplate. The method of microdilution is a method recommended by the Clinical and Laboratory Standards Institute (CLSI). Microplate consists of 96 wells, 12 columns and 8 rows. The wells used are 24 wells, each well filled with 100 μL MHB. The first column was added 100 μL bacteriocin strain A11, while the second column was added 100 μL bacteriocin strain C12. Then each bacteriocin is resuspended to the next column until column 7. Then each column is filled with a suspension of the MRSA test with turbidity of 0.5 McFarland standard (1.5 × 108) as much as 100 μL. Microplate is incubated at 35 ± 2oC for 16-20 hours. MIC is the smallest concentration where there is no bacterial growth in the well which is shown in clear color and carried out by visual observation.

2.6. Minimum bactericidal concentration (MBC)
MBC is done by taking a solution from the microplate well which shows the MIC value and from all other wells that are above the MIC value using the ose. Then it was scratched on the surface of the prepared BAP media. Then BAP was incubated at 35 ± 2oC for 16-20 hours. BAP which shows the visualization of clarity and non-bacterial overgrowth was determined as the MBC value.

3. Result and Discussion

3.1. Regeneration of Pediococcus acidilactici
The bacteriocin-producing *P. acicilactici* isolate used must always be regenerated so that the bacteria are in the growth phase. Bacterial rejuvenation is an important thing that must be done to get new bacterial cultures so that they can reproduce well when used [11]. According to [12], regeneration of *P. acicilactici* using MRSA media for 48 hours to provide nutrition and grow bacteria, at the 48th hour *P. acicilactici* was ready to be harvested and inoculated into the MRSB media to make bacteriocin.

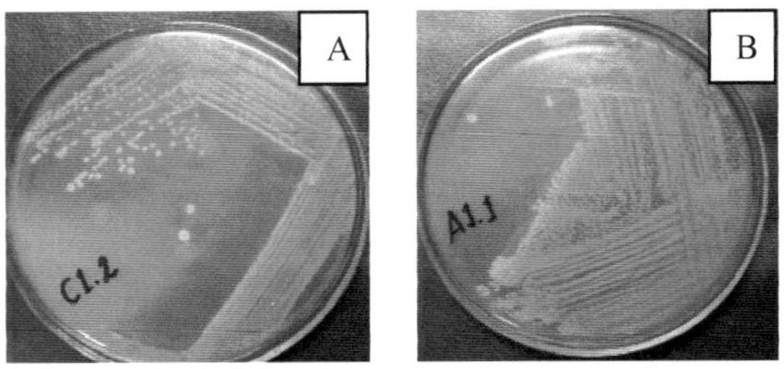

Figure 1. Isolates of *Pediococcus acidilactici* strains C12 (A) and A11 (B).

3.2. Bacteriocin Production
Bacteriocin isolation was made by inoculating 2 groups of P.acicilactici isolates into 200 mL MRSB media for 48 hours at 37°C. Then centrifugation at a speed of 12000 rpm at 4°C for 10 minutes. Then the part of the supernatant is separated and add NaOH 1 N to get bacteriocin in neutral.

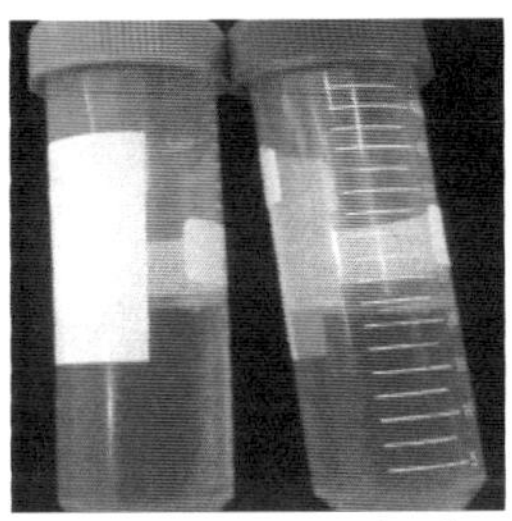

Figure 2. Bacteriocin.

3.3. Antibacterial Activity of Bacteriocin

The antibacterial activity of bacteriocin in the *P. acidilactici* strains A11 and C12 was seen with the resulting inhibition zone. Testing is done with three repetitions. The inhibitory zone that is formed is measured using a caliper.

Table 1. Average Diameter of Bacteriocin Inhibitory Zone.

Bacteria	Inhibition zone of bacteriocin (mm)	
	A11	C12
MRSA	15	18

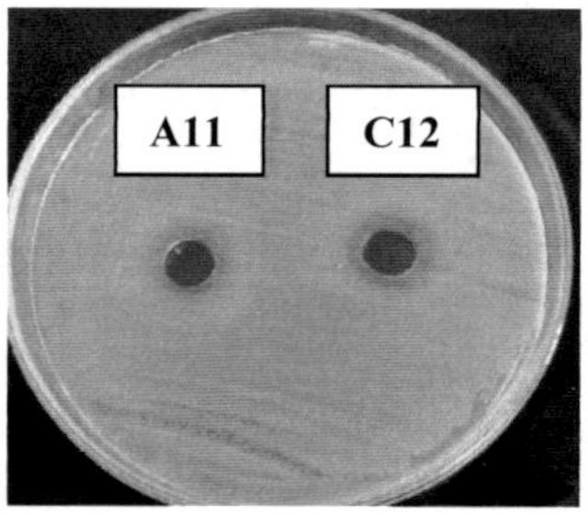

Figure 3. Inhibition zone of bacteriocin.

In Table 1. shows that bacteriocin is able to inhibit the growth of MRSA in the presence of the resulting clear zone of 15 mm and 18 mm. According to [13], the diameter of the inhibition zone with a value of 0-3 mm is included in the weak category, the inhibition zone between 3-6 mm is included in the medium category and the inhibition zone is more than 6 mm in the strong category. So it can be concluded that bacteriocin antibacterial activity of *P. acidilactici* ASI isolates against MRSA is included in the strong situation. These results further confirm that bacteriocin is a safer antibacterial agent and prevents resistance.

Table 2. Differences between MRSA treatment with bacteriocin and antibiotics.

No	Treatment of MRSA	
	Antibiotics (mm)	*Bacteriocin* (mm)
1	Amphicilin (0) Eritromycin (0) Genatamycin (0) Tetracycline (18)	A11P (15)
2	Amphicilin (0) Eritromycin (0) Genatamycin (0) Tetracycline (18)	C12P (18)

As for bacteriocin testing and antibiotics control of MRSA in Table 2. the results of the Kriskal Wallis test showed a significance value of 0.735 ($p \geq 0.05$). This means that there is no difference between the treatment of MRSA and bacteriocin and antibiotics.

Table 3. The bacteriocin MIC values for MRSA.

Bacteriocin	MIC *Bacteriocin* Against MRSA						
	100%	50%	25%	12,5%	6,25%	3,12%	1,5%
P. acidilactici A11	-	-	-	+	+	+	+
P. acidilactici C12	-	-	-	-	+	+	+

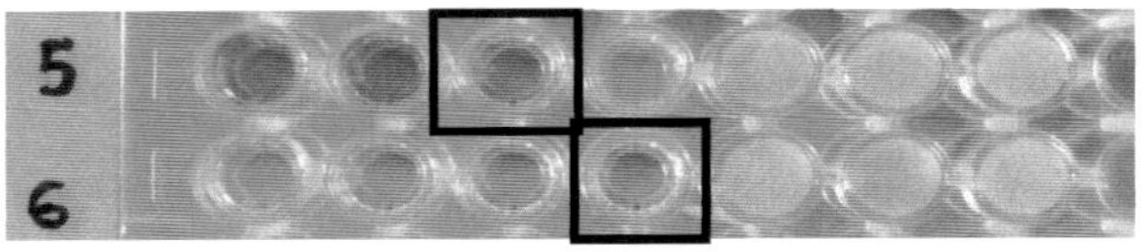

Figure 4. The bacteriocin MIC values of A11 (5) dan C12 (6) against MRSA.

Table 4. MBC bacteriocin values against MRSA.

Bacteriocin	MBC *Bacteriocin* Against MRSA						
	100%	50%	25%	12,5%	6,25%	3,12%	1,5%
P. acidilactici A11	-	-	-	+	+	+	+
P. acidilactici C12	-	-	-	-	+	+	+

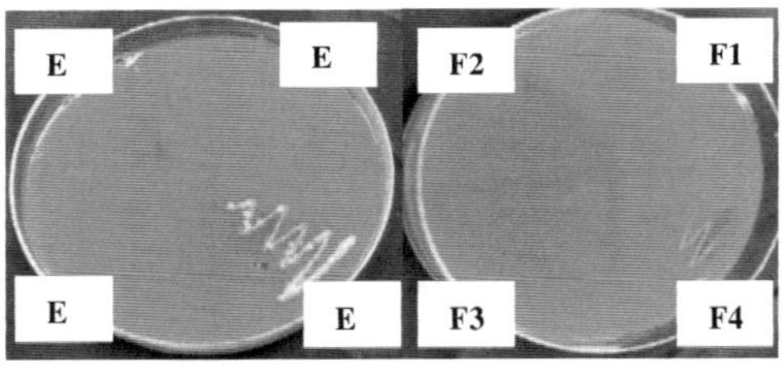

Figure 5. MBC value of *bacteriocin* A11 (E) dan C12 (F) againts MRSA.

The results of testing the determination of MIC and MBC from each bacteriocin against MRSA can be seen in Table 2. and Table 3. The MIC and MBC values of bacteriocin A11 and C12 were respectively 25% and 12.5%. The MBC value will always be higher or equal to the MIC value.

The target of bacteriocin from LAB is the sensitive cytoplasmic membrane of bacterial cells because the initial bacteriocin reaction is to damage membrane permeability and eliminate proton motive force (PMF) thereby inhibiting energy production and biosynthesis of proteins or nucleic acids. Bacteriocin inhibition activities require cell surface specific receptors, for example in pediocin AcH. Besides that it results in cell lysis. This is a secondary effect of pediocin AcH activity through depolymerization of the peptidoglycan layer, so that it can indirectly activate the cell autolysis system [14].

The mechanism of bacteriocin bactericidal activity is as follows: (1) bacteriocin molecules are in direct contact with cell membranes, this contact process is able to disrupt membrane potential in the destability of cytoplasmic membranes so that cells become not strong, and (3) membrane instability capable of producing holes cell membrane through the process of interference with PMF (Proton Motive Force). Leaks that occur due to the formation of holes in the cytoplasmic membrane are indicated by the presence and exit of cellular molecules. This leakage results in a decrease in cellular pH gradient. The effect of forming cytoplasmic holes is the impact of bacteriocin which causes changes in gradient potensal membrane and the release of intracellular molecules as well as the entry of extracellular substances (environment). The effect causes cell growth to be inhibited and produce a death process in cells that are sensitive to bacteriocin [15].

Bacteriocin as a bacterial metabolite has the potential to be utilized in therapeutic effects. But there are fundamental differences between bacteriocin and antibiotics. Bacteriocin is generally a peptide or peptide complex with intraspecific antagonistic effects on the strain of the product and relatively narrow bactericidal spectrum, this is what distinguishes it from antibiotics.

One of the advantages of using bacteriocin LAB as an antimicrobial is its ability to eliminate pathogenic microbes and decompose food from milk and meat. The possibility of further therapeutic applications is closely related to the characteristics of the antimicrobial substance, which is not toxic, can inhibit low levels and is produced by bacteria classified as GRAS (generally recognized as safe), which is a microbe that is not at risk to health. Thus, it provides an opportunity for its use as an antimicrobial agent against MRSA.

4. Conclusion

The test results can be concluded that the two bacteriocin strains of *P. acidilactici* has shown potential antimicrobial activity against MRSA with inhibitory zone values of 15 mm and 18 mm, respectively. While the value of MIC and MBC are 25% and 12.5% respectively. So it is confirmed that bacteriocin *P. acicilactici* isolates ASI has the potential as a natural antibacterial agent that can prevent resistance and is not toxic. The advantages of using bacteriocin, especially from LAB as an antimicrobial substance, are able to inhibit low levels and are produced by bacteria classified as GRAS (generally recognized as safe), which is a microbe that is not at risk to health.

References

[1] Giedraitiene A, Vitkauskiene A, Naginiene R and Pavilonis A 2011 Antibiotic Resistance Mechanisms of Clinically Important Bacteria *Jurnal Medicina (Kaunas)* **47** (3) 137-46

[2] Sifour M, Tayeb L, Haddar H O, Namous H and Aissaoi S 2012 Production and Caracterization of Bacteriocin of Lactobacillus plantarum F12 with Inhhibition Activity Againt Listeria monocytogenes *TOJSAT* **2** 55-61

[4] Dobson A, Cotter P D, Ross R P and Hill C 2012 Bacteriocin production: a probiotic trait? *Appl. Environ Microbiol.* **78** 41

[5] Marshall S H 2003 Antimicrobial peptides: as natural alternative to chemical antibiotics and potential for applied biotechnology *Electron J. Biotech* **3** 6

[6] Setianingsih S 2010 *Kajian Senyawa Antimikroba Bakteri Asam Laktat Homofermentatif Isolat Asi* (Bogor: Fakultas Teknologi Pertanian Institut Pertanian Bogor)

[7] Salminen S, Wright A V and Ouwehand A 2004 *Lactid Acid Bacteria* (New York: Marckel Dekker)

[8] Djide M N and Wahyudin E 2008 Isolasi Bakteri Asam Laktat dari Air Susu Ibu, dan Potensinya dalam penurunan kolesterol secara in vitro *Majalah Farmasi dan Farmakologi* **12** 3-Nopember 2008 (ISSN: 1410-7031)

[9] Dewi S S and Anggraini H 2012 *Viabilitas Bakteri Asam Laktat Asal ASI terhadap pH Asam Lambung dan Garam Empedu* Seminar Hasil-hasil Penelitian LPPM UNIMUS Semarang

[10] Pratiwi S 2008 *Mikrobiologi Farmasi* (Jakarta: Penerbit Erlangga)

[12] Khoiriyah H, Puji Ardiningsih and Afgani Jayuska 2014 Penentuan Waktu Inkubasi Optimum Terhadap Aktivitas Bakteri *Lactobacillus* sp. **3** (1) 7-14

[13] Pan X, Chen F, Wu T, Tang H and Zhao Z 2009 The acid, bile tolarance and antimicrobial property of *Lactobacillus acidophilus* NIT *Food Control* **20** 598-602

[14] Salvadogo A 2006 Bacteriocins and Lactic Acid Bacteria-a miniriview *African Journal of Biotechnology* **5** 678-683

[15] Luc De Vuyst 2007 Bakteriosin from Lactic Acid Bacteria: Production, Purification, and Food Apllication *Journal of Molecular Micribiology and Biotechnology* **13** 194-1999

Biocellulose isolated from the waste of pinecone flower (*Pinus merkusii* Jungh Et De Vriese)

S E Lusiana[1], M M Aisiyah[1], U Z 'Uyunin[1], Z D Nasihin[1], A Srihardyastutie[1], M F Rahman[1] and M Masruri[1]*

[1]Chemistry Department, Faculty of Mathematics and Natural Sciences, Brawijaya University, Indonesia
Jl. Veteran 65145 Malang

*Corresponding author's email: masruri@ub.ac.id

Abstract. A bio-cellulose isolated from pinecone flower waste has been isolated by two type procedures. The first method is commonly used in industrial protocol. It is undergone via delignification and bleaching processes. In the delignification process provide a black-liquid of lignin waste, meanwhile the bleaching process afford neutralized clear liquid. The bio-cellulose is the main product. On the hand, the second method is initiated by soxhlet extraction using three different solvents i.e. n-hexane, ethanol and toluene, respectively. The extraction process produces organic hexane extract, ethanol extract, and toluene extract. All these extract have different appearance in color and prospective. Further delignification and bleaching process provide black lignin waste and neutralized liquid waste. The bio-cellulose still also become a major product.

1. Introduction

Bio-cellulose is a cellulose derived from biological sources, such as plants, animals, and microorganism [1]. The cellulose is classified as polysaccharide that is composed of the monomers of glucose. To be precise is D-glucopyranose which is the glucose that has a cyclic structure (Figure 1). The connection inter-glucose molecule is attached by glycoside bond. It occurs between hydroxyl group at C4 and hydroxyl-group attached in C1 by dehydrating of water molecule in enzymatic reaction. In biological sample, cellulose mixes to the other chemicals inside the cell such as water, lipid, lignin, and protein. And to isolate a pure cellulose, the procedure has to be able to separate cleanly the other matrix from cellulose [2].

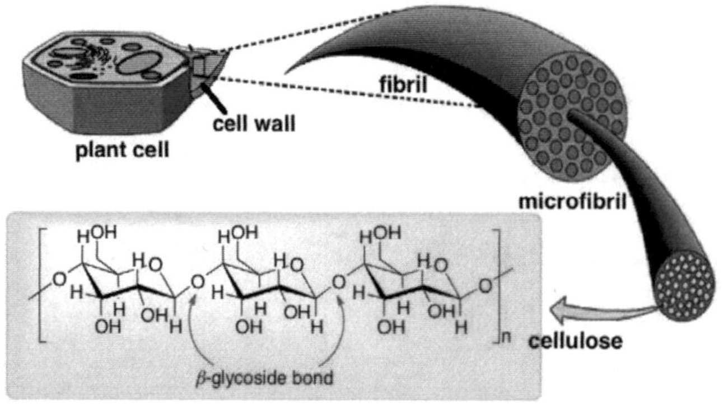

Figure 1. Molecular structure of cellulose. Glycoside bond connect the glucose monomers.

The common procedure applies a base such as sodium hydroxide and potassium hydroxide solution to the mixture for dissolving lignin, protein, chlorophyll and other water soluble material. A black-soluble chemical mostly contain lignin generally extracted, and cellulose is located as residue. The bleaching process oxidized the other material that are soluble in basic conditions [3]. However, in term of "green chemistry", this procedure disposes many organic chemicals that composes in the cell, such as terpenoid, fatty acid, phenolic, tannin, and alkaloid group of the secondary metabolites [4].

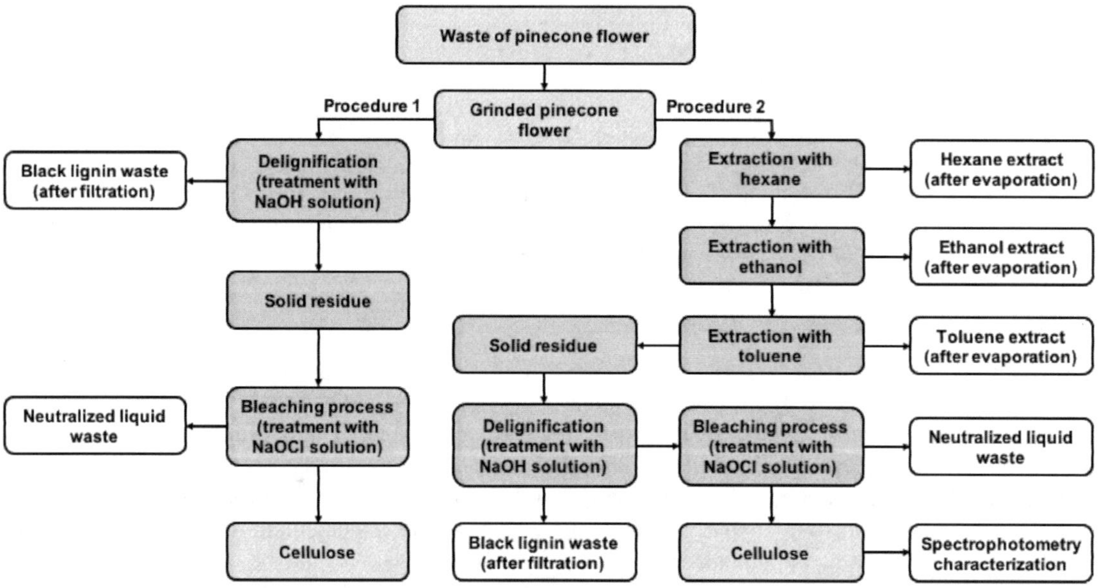

Figure 2. Two type methods for cellulose isolation from pinecone flower waste.

2. Experiment section

2.1. Sample and chemicals

The pinecone flower waste is collected from local industrial pine forest in Malang, Indonesia. Pine species is *Pinus merkusii* Jungh Et De Vriese and the taxonomy is previous reported by the same group [5, 6]. The sample is further washed and dried in air for 2 days before grinding to powder. Meanwhile the chemicals used for research includes sodium hydroxide (Merck), sodium hypochlorite (Sigma), hydrochloric acid (Smart Lab), iron trichloride (Merck), gallic acid (Merck), formic acid (Merck), acetic acid (Merck), n-hexane (Merck), ethanol (Merck), toluene (Smart Lab), and potassium bromide (Merck).

2.2. Instruments

The instruments used for the research includes FTIR spectrophotometer (Shimadzu FTIR), laboratory grinder, analytical balance (Mettler), and mechanical stirrer.

2.3. Experiment procedure

The schematic of procedure for isolation of cellulose from pinecone flower waste is depicted in Figure 2. There are 2 type procedure for isolation. Namely procedure 1 and procedure 2.

2.3.1. Procedure 1. A dried powder of pinecone flower waste (50 g) in Erlenmeyer-1000 mL is added 500 mL sodium hydroxide solution 6.0%. This mixture is heated at 70 °C and stirred for 4 hours. Then, it is filtered off and the solid part is undergone washing with water to neutral. The neutral solid is further undergone bleaching process by addition 300 mL of sodium hypochlorite solution 6%. It is stirred for 2 h at 70 °C and the solid part is separated from the liquid, and is further bleached for 6 time, consecutively. The product is washed with water exhaustively to afford neutral cellulose. Then, it is separated by filtration, and the solid product is cellulose.

2.3.2. Procedure 2

2.3.2.1. Soxhlet extraction. A dried powder of pinecone flower waste (150 g) in the timble is set to the soxhlet apparatus. The n-hexane (650 mL) is added and the process is undertaken in several cycles for 6 hours. The liquid extract is evaporated under reduced pressure to afford a brown-black concentrated extract of hexane. The solid residue retained in the soxhlet timble is further undergone extraction using toluene as solvent (650 mL). The process is proceeded for several cycle for 6 hours. After concentration in rotary evaporator affords the toluene extract as a concentrated of black-brown extract of toluene. The residue retained in the timble is further extracted using ethanol (650 mL). It is also proceeded for several cycles for 6 hours. The ethanol extract is afforded as a clear-brown liquid after concentration in vacuo using rotary evaporator. The solid residue in the soxhlet timble is undergone further process for cellulose isolation.

2.3.2.2. Cellulose isolation. The dried of solid residue in above procedure (50 g) is basified with 500 mL of sodium hydroxide solution 6.0%. This mixture is heated at 70 °C and stirred for 4 hours. Then, it is washed with water until neutral pH. The solid residue is further added with 300 mL of sodium hypochlorite 6.0%, heated at 70 °C and stirred for 2 hours. The product is separated and the solid residue is added with sodium hypochlorite. The process is repeated for 4 times to afford a white cellulose. Then, is is washed untio neutral pH and dried at room temperature.

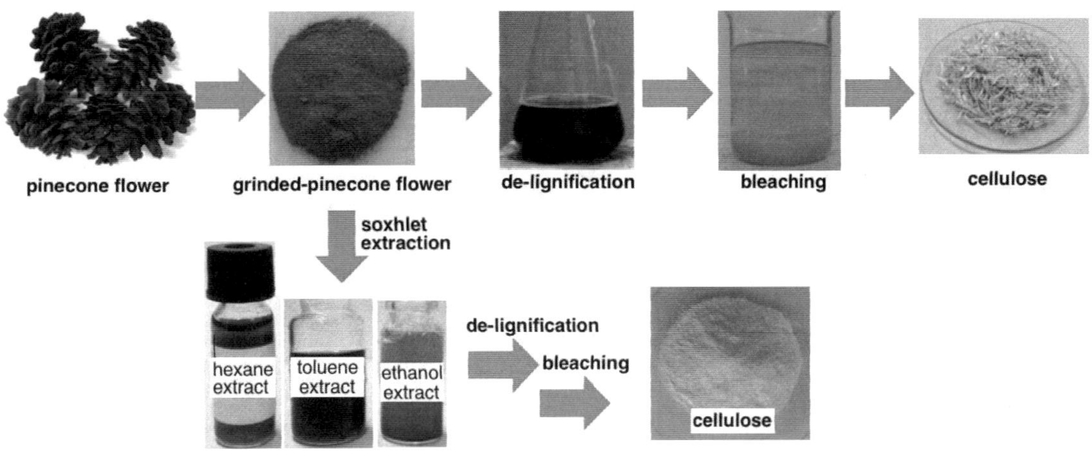

Figure 3. Cellulose results is produced following 2 type procedure.

3. Result and discussion

The cellulose composes a part of cell life in plants. It sticks and mixes with other chemicals contains in the cell wall. Protein, secondary metabolite products, and other biological matrix has to be opened and separated from cellulose. For example, lignocellulose binds strongly to cellulose, and cleavage the lignin from cellulose requires a strong base. Including other polar of phenolic groups such as tannin, phenylpropanoid compounds and the alkaloids [7].

Grinding process of sample pinecone flower waste can increase the surface area interaction between reagent and chemicals (Figure 3). The direct delignification of the pinecone flower powder affords a black aqueous lignin after filtration. The residue, further bleaching process can afford cellulose in coarse size. A yellowish-white cellulose is afforded.

The second procedure give a better result. Some organic solvent extracts derived from n-hexane, toluene, and ethanol is afforded. It is predicted secondary metabolite composes these extract. Generally, non polar group such as sesquiterpene steroid, and triterpenoid class of terpenoid composed the n-hexane extract. Meanwhile, toluene generally applied for tannin, lignin, and other phenolic groups extraction process. And, ethanol can extract all of fatty acid and polar groups of compound include protein and polyketide [8]. The quantity of cellulose isolated from that procedure is reported in Table 1. The first procedure is able to provide better yield. Cellulose is isolated in 44.06%. Meanwhile the second procedure give a slightly lower yield (38.45%). However, cellulose appearance and potential additional product given in the second procedure opens the more prospective.

Table 1. Yield of cellulose isolated.

Method	Cellulose isolated (%)	Cellulose appearance	Potential additional products
1	44.06%	coarse powder white-brown	Lignin
2	38.45%	fine powder white-brown	Lignin n-Hexane extract (6.9%) Toluene extract (8.6%) Ethanol extract (5.1%)

Molecular characterization of the cellulose isolated is undertaken using FTIR spectrophotometry (Figure 4 and Figure 5). Finger printing analysis is methodologically applied for qualitative analysis. The similarity of bands wavelength of the sample is compared to the band wavelength from reference, especially in finger printing region. This region correlates to the similarity in the structure and composition [9].

Table 2. Functional group and fingerprint bands of cellulose.

Characterized of functional groups	Bands absorption wavelength (1/cm)		
	Cellulose 1	Cellulose 2	Commercial cellulose
Hydroxyl (O-H)	3400.0	3420.0	3391.0
Alkyl (CH_3; CH_2-)			
Stretching band	2901.0	2909.0	2906.7
Bending band	1267.9-1509.9	1267.9-1510.9	1300.0-1430.0
Carbonyl (C=O)	-	-	1760.0
C=C alkene	1634.4	1640.1	1654.4-1635.2
C-O-C ether group	1130.7-1161.8	1130.6-1161.8	1061.7-1162.4

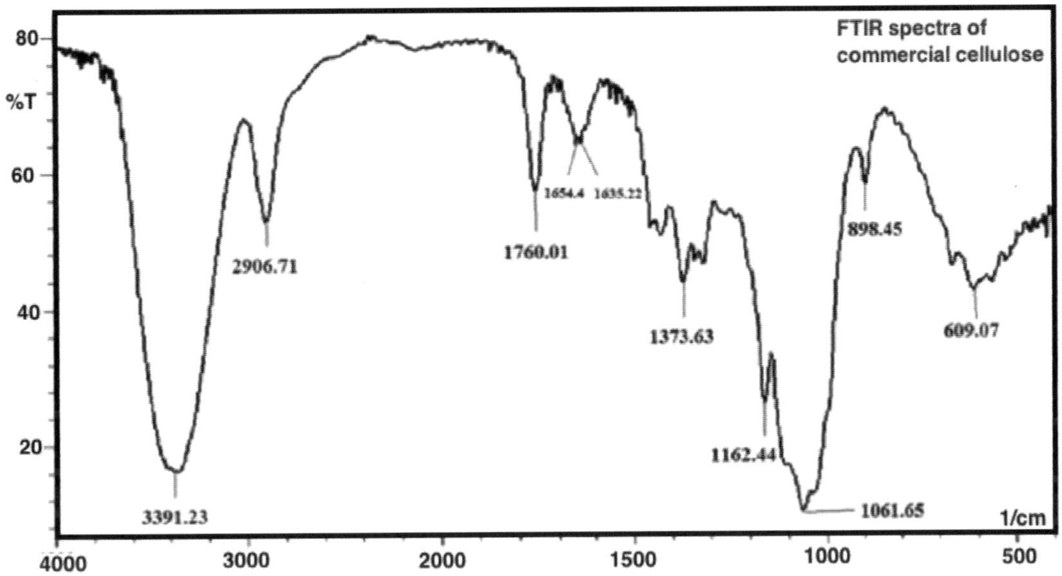

Figure 4. The FTIR spectra of the commercial cellulose as the reference.

The hydroxyl- functional group is detected in about 3400 cm^{-1}. The reference spectra from the commercial cellulose come up at 3391 cm^{-1}. In addition, the corresponding band stretching for this group, C-O, is recorded in between 1000 cm^{-1} and 1265 cm^{-1}. For the alkyl stretching band, such as C-H and H-C-H, is detected in 2900 cm^{-1} in all sample cellulose. And also their bending bands (1200-1500 cm^{-1}). Carbon-carbon double bond group is recorded in all samples (1634-1654 cm^{-1}). However, the carbonyl- group is only detected in commercial cellulose (1760 cm^{-1}) (Table 2). This finding indicate the high purity and no structural changing of cellulose isolated from pinecone flower waste.

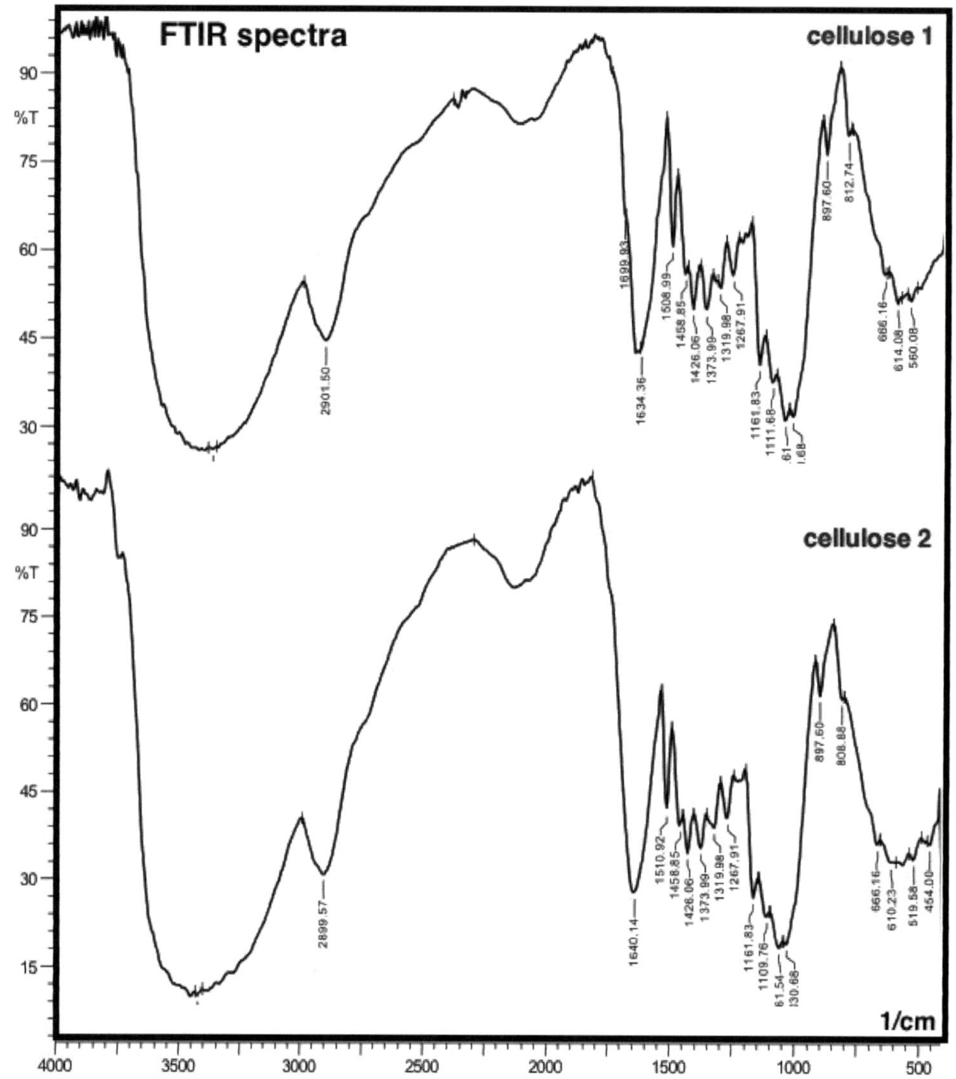

Figure 5. The FTIR Spectra of cellulose afforded from 2 type procedure.

4. Conclusion

To summarized, cellulose can be isolated in high yield from pinecone flower waste. The two type procedure can be applied to afford cellulose. The second procedure can generate another potency for prospecting of secondary metabolite extracted using n-hexane, toluene and ethanol as solvents. Further stage has to be undertaken to analyze the composition these extract, and also conversion of the isolated cellulose into prospective products or starting materials.

5. Acknowledgement

The research is funded by Directorate for Research & Community Development (DRPM) from Ministry of Research, Technology and Higher Education under contract number 167/SP2H/LT/DRPM/2019.

Reference

[1] Wang L, Peng W, Zhang M and Lin Z 2013 Separation characteristics of lignin from Eucalyptus lignin cellulose for medicinal biocellulose preparation *Journal of Pure and Applied Microbiology* **7** 59-66

[2] Gartaula G, Dhhital S, Netzel G, Flanagan B M, Yakubov G E, Beahan C T, Collins H M, Burton R A, Bacic A and Gidley M J 2018 Quantitative structureal organisation model for wheat endosperm cell walls: Cellulose as an important constituent *Carbohydrate Polymer* **196** 199-208

[3] Rajinipriya M, Nagalakshmaiah M, Robert M and Elkoun S 2018 Importance of agricultural and industrial waste in the field of nanocellulose and recent industrial developments of wood based nanocellulose: a review *ACS Sustainable Chemistry & Engineering* **6** (3) 2807–2828

[4] Guerriero G, Berni R, Munoz-Sanchez J, Apone F, Abdel-Salam E, Qahtan A, Alatar A, Cantini C, Cai G, Hausman J F and Siddiqui K 2018 Production of plant secondary metabolites: Examples, tips and suggestions for biotechnologists *Genes* **9** (6) 309

[5] Azkiya N, Masruri M and Ulfa S 2018 Green Synthesis of Silver Nanoparticles using Extract of Pinus merkusii Jungh & De Vriese Cone Flower (Vol. 299, p. 012070) Presented at the IOP Conference Series: Materials Science and Engineering, IOP Publishing

[6] Masruri M, Pangestin D N, Ulfa S M, Riyanto S, Srihardyastutie A and Rahman M F 2018 A Potent Staphylococcus Aureus Growth Inhibitor of A Dried Flower Extract of *Pinus Merkusii* Jungh & De Vriese And Copper Nanoparticle (Vol. 299, p. 012072) Presented at the IOP Conference Series: Materials Science and Engineering, IOP Publishing

[7] Starmans D A and Nijhuis H H 1996 Extraction of secondary metabolites from plant material: a review *Trends in Food Science & Technology* **7** (6) 191–197

[8] Azmir J, Zaidul I S M, Rahman M M, Sharif K M, Mohamed A, Sahena F, Jahurul M H A, Ghafoor K, Norulaini N A N and Omar A K M 2013 Techniques for extraction of bioactive compounds from plant materials: A review *Journal of Food Engineering* **117** (4) 426-436

[9] Oh S Y, Yoo D I, Shin Y and Seo G 2005 FTIR analysis of cellulose treated with sodium hydroxide and carbon dioxide *Carbohydrate Research* **340** (3) 417–428

Journal of Physics: Conference Series **1374** (2019) 012023 doi:10.1088/1742-6596/1374/1/012023

Cellulose nanocrystal (CNC) produced from the sulphuric acid hydrolysis of the pine cone flower waste (*Pinus merkusii* Jungh Et De Vriese)

S E Lusiana[1], A Srihardyastutie[1] and M Masruri[1]*

[1]Chemistry Department, Faculty of Mathematics and Natural Sciences, Brawijaya University, Jl. Veteran 65145 Malang Indonesia

*Corresponding author's email: masruri@ub.ac.id

Abstract. Cellulose nanocrystal (CNC) has been produced up on the hydrolysis of the cellulose isolated from the waste of pine flower (*Pinus merkusii* Jungh Et De Vriese). The hydrolysis process is undergone using 10%, 30% and 60% of sulfuric acid solution. the hydrolysis process is also performed using normal heating at 70 °C under stirring with magnetic stirrer for 4 hours. The CNC afforded is characterized by means of infra red spectrophotometry (FTIR), x-ray diffraction spectrometry (XRD) and transmission electron microscope (TEM) analysis. The TEM analysis afford the dimension of CNC in about 15 nm x 185 nm. This finding open the way for further prospective application.

1. Introduction

Nanocellulose is a cellulose in a nanometer dimension, which can be isolated from biological sources, such as plants, animal and microorganism. Nano size (1-100 nm) of cellulose has many industrial applications, such as gel antibacterial, functional biodegradable plastic for food packaging, material for medical tool and device, and drug delivery materials [1, 2]. Previously, it has been reported that the waste of Indonesian pine industry can come from the cone flower dropped around the tree [3, 4]. It is about 40-60% cellulose contained in the cone flower. The direct cellulose isolation and purification is already reported elsewhere recently, and the strategy to convert cellulose into nanometer in size is reported in this paper. Previous research reported the methods for producing nanocellulose, such as hydrolysis in some acid media [5, 6], hydrolysis-assisted sonication [7] and microwave [8, 9], oxidative-hydrolysis using some oxidation reagents [5]. This strategy applies sulfuric acid in different concentration by conventional stirring to afford cellulose nanocrystal (CNC) from cone flower waste of *Pinus merkusii* Jungh Et De Vriese.

2. Experiment section

2.1. Sample and chemicals

A pine cone flower's sample was collected from local forest in Malang Region, East Java. Then, it was washed by immersion in hot water for 2 h, and further dried in oven at 100 °C for 24 h. The dried sample was undergone milling to afford a dried powder.

2.2. Cellulose isolation
Cellulose preparation is conducted following previous reference [10] with modification. A 50 g of pine cone powder was mixed with 500 mL sodium hydroxide 6%. This mixture was heated at 70 °C under magnetic stirring for 4 h. The solid part was separated by filtration and further washed with distillated water until neutral pH. From this product, was undergone bleaching process by mixing with 300 mL of 6% sodium hypochlorite solution. The mixture was stirred at 70 °C for 2 h and filtered off. This process was repeated for 4 times.

2.3. Cellulose nanocrystal preparation
The procedure follows the previous reference [5] with modification. A 1.0 g of dried cellulose in Erlenmeyer, which has been afforded in the previous step is added with 20 mL of 10% sulfuric acid solution. The mixture is warmed at 45 °C and stirred for 60 minutes. A cold distillated water is added to stop the reaction. The mixture is washed and undergone centrifugation until neutral pH. The solid product precipitate in bottom is collected as nanocellulose and dried before further characterization. Similar procedure is undertaken using different sulfuric acid concentration, i.e. 30% and 60%. The nanocellulose afforded is also characterized using similar methods.

2.4. FTIR analysis
The FTIR analysis of cellulose nanocrystal is undertaken using FTIR spectrophotometer Shimadzu. The sample is homogeneously mixed with potassium bromide and pressed to form a clear and transparent disk with 1.5 cm in diameter. The spectra analysis is scanned in between 4000 and 400 cm^{-1} of wavelength.

2.5. XRD analysis
Crystallinity analysis is performed in x-ray diffraction spectrometer (PanAnalytical, Type E'xpert Pro). CuKα radiation 10-90 (2-theta), scanning rate 5 °/min, the operation voltage 40 kV and current 25 mA.

2.6. TEM analysis
The morphology analysis using transmission electron microscope is recorded using JEOL JEM-1400 operated in 120 V. The sample is coated in carbon prior to analysis.

3. Result and discussion
Cellulose in the plant mixes with other cell matrix. Protein, amino acid, organic acid, lignin, and other chemical components. Isolation from its mixture requires separation by destructing of the other chemicals. A basic hydrolysis generally undergone to dispose protein, lignin, organic acid, and other chemicals. Table 1 is displayed cellulose isolated after delignification and bleaching process [1]. In average 22.22 ± 1.07 g of cellulose is able to isolated from 50.0 g of a dried powder of cone flower waste.

Table 1. Cellulose isolation from pine cone flower.

Entry	Cellulose code	Yield (g)
1	C1	21.05
2	C2	21.37
3	C3	22.03
4	C4	23.32
5	C5	23.35
	average yield	22.22 ± 1.07

Hydrolysis process of the cellulose breaks the poly-glucose into smaller chain to afford nanosize of cellulose. The sulfuric acid hydrolysis, using 10%, 30% and 60%, respectively increase glycoside bonds. Moreover, applying 60% of sulfuric acid intensify the carbonyl band absorption in FTIR (Figure 1). The increasing of carbonyl band means the opening of glucofuranose chain. Sulfuric acid 10% and 30% give in about 35% absorption unit, however sulfuric acid 60% give the carbonyl peak in 50% of absorption unit. Furthermore, the rest of functional groups bands remain similar; such as the hydroxyl (O-H) band in 3485 cm^{-1}, C-H alkyl groups in between 2950 cm^{-1}and 2800 cm^{-1}, the C-O-C band absorption in 1300-950 cm^{-1}.

The opening of glucofuranose chain also affects the crystallinity of the cellulose nanocrystal (Figure 2). An amorphous nanocellulose is produced from hydrolysis using 60% of sulfuric acid. Conversely, 10% and 30% sulfuric acid provide a similar crystallinity of nanocellulose. The peak at 2-theta 15° (101) and 18° (10i) is detected for nanocellulose from 10% and 30% sulfuric acid hydrolysis. The sharp peak 22° (002) is recorded, includes the peak at shoulder in 21° (021). Similar reference is reported for nanocellulose [11]. However, for nanocellulose afforded by 60% sulfuric hydrolysis provide a broad peak 22° (002).

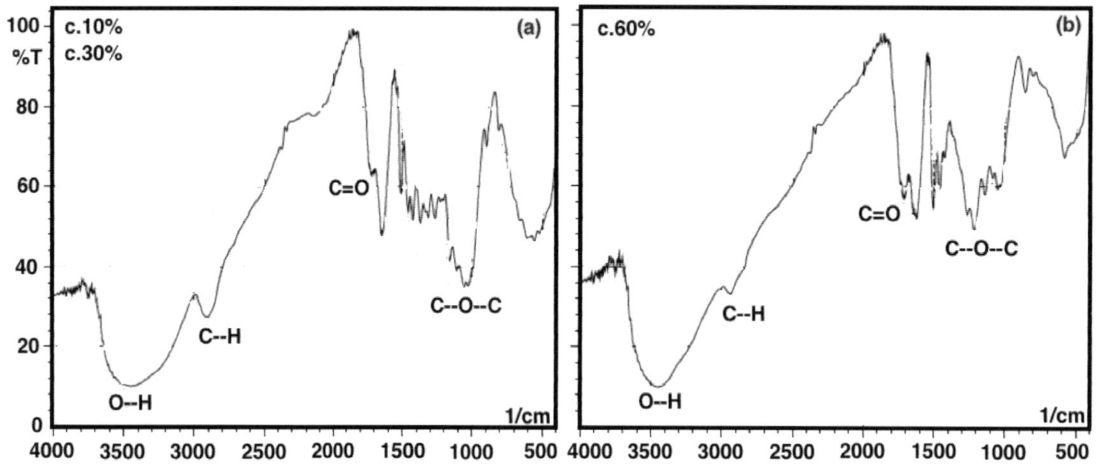

Figure 1. The FTIR spectra of cellulose nanocrystal from pine cone flower waste. The nanocellulose produced by hydrolysis using 10% and 30% of sulfuric acid (a), and 60% sulfuric acid (b).

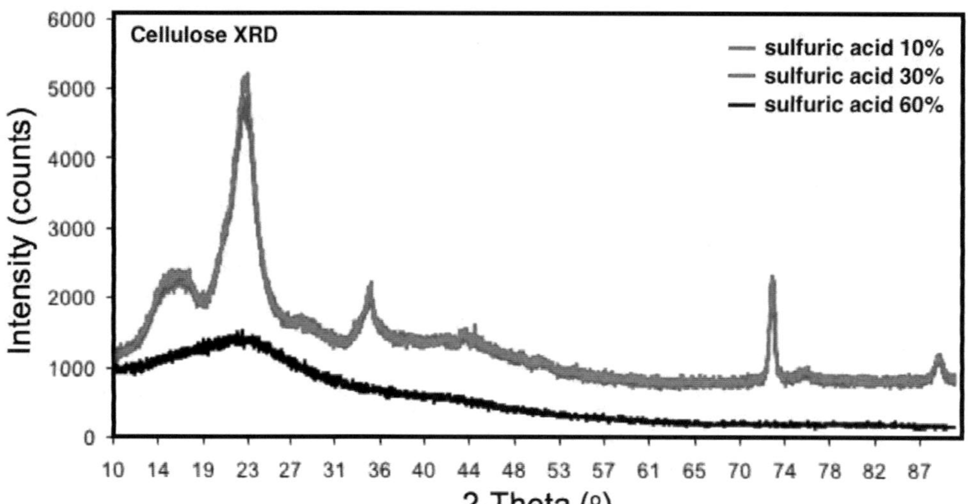

Figure 2. The XRD diffractogram of cellulose nanocrystal from pine cone flower waste.

The nanocrystal size of cellulose is determined based on the direct measurement from TEM image (Figure 3). cellulose nanocrystal produced from sulfuric acid hydrolysis using 10% and 30% give similar dimension in size. It is about 15 nm in wide and 185 nm of length (Figure 3a). However, nanocellulose afforded from 60% sulfuric acid hydrolysis give a variation in dimension. A shorter nanosize than that in 10% and 30% hydrolysis products. It is about 110-125 nm in length and 15-35 nm in wide.

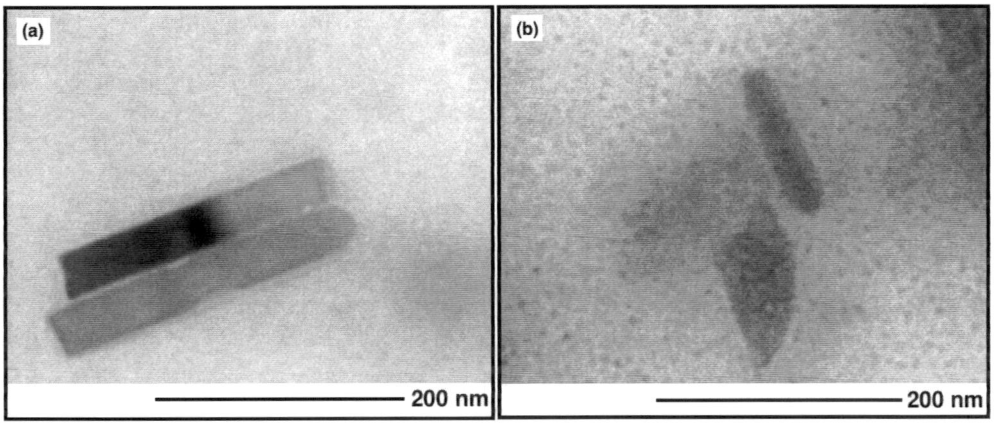

Figure 3. The transmission electron microscope (TEM) image of bionanocellulose from pine cone flower waste. Bionanocellulose produced by hydrolysis using 10% and 30% of sulfuric acid (a), and 60% sulfuric acid (b).

4. Conclusion

Cellulose nanocrystal is able to be isolated from the hydrolysis of pine cone flower waste-rich of cellulose. The sulfuric acid concentration affects the nanocrystalinity and nano size of the cellulose. A 10% and 30% sulfuric acid give a high crystallinity with nano dimension 15 nm (wide) and 185 nm length. However, amorphous nanocellulose is afforded from 60% hydrolysis of sulfuric acid..

5. Acknowledgement

Authors thank to Directorate General for Research and Community development (DRPM) of Ministry of Research, Technology and Higher Education for Penelitian Unggulan Dasar Perguruan Tinggi (2019) under contract number 167/SP2H/LT/DRPM/2019.

Reference

[1] Dufresne A 2019 Nanocellulose Processing Properties and Potential Applications *Current Forestry Reports* **5** (2) 76–89

[2] Oun A A, Shankar S and Rhim J-W 2019 Multifunctional nanocellulose/metal and metal oxide nanoparticle hybrid nanomaterials *Critical reviews in food science and nutrition* 1–26

[3] Masruri M, Pangestin D N, Ulfa S M, Riyanto S, Srihardyastutie A and Rahman M F 2018 A Potent Staphylococcus aureus Growth Inhibitor of A Dried Flower Extract of Pinus Merkusii Jungh & De Vriese And Copper Nanoparticle (Vol. 299, p. 012072) Presented at the *IOP Conference Series: Materials Science and Engineering*, IOP Publishing

[4] Azkiya N, Masruri M and Ulfa S 2018 Green Synthesis of Silver Nanoparticles using Extract of Pinus merkusii Jungh & De Vriese Cone Flower (Vol. 299, p. 012070) Presented at the *IOP Conference Series: Materials Science and Engineering*, IOP Publishing

[5] Liu C, Li B, Du H, Lv D, Zhang Y, Yu G, Mu X and Peng H 2016 Properties of nanocellulose isolated from corncob residue using sulfuric acid, formic acid, oxidative and mechanical methods. *Carbohydrate polymers* **151** 716–724

[6] Rambabu N, Panthapulakkal S, Sain M and Dalai A 2016 Production of nanocellulose fibers from pinecone biomass: evaluation and optimization of chemical and mechanical treatment conditions on mechanical properties of nanocellulose films. *Industrial Crops and Products* **83** 746–754

[7] Tang Y, Yang S, Zhang N and Zhang J 2014 Preparation and characterization of nanocrystalline cellulose via low-intensity ultrasonic-assisted sulfuric acid hydrolysis *Cellulose* **21** (1) 335–346

[8] Kos T, Anžlovar A, Kunaver M, Huskić M and Žagar E 2014 Fast preparation of nanocrystalline cellulose by microwave-assisted hydrolysis *Cellulose* **21** (4) 2579–2585

[9] Singh S, Gaikwad K K, Park S-I and Lee Y S 2017 Microwave-assisted step reduced extraction of seaweed (Gelidiella aceroso) cellulose nanocrystals *International journal of biological macromolecules* **99** 506–510

[10] Taflick T, Schwendler L A, Rosa S M, Bica C I and Nachtigall S M 2017 Cellulose nanocrystals from acacia bark–Influence of solvent extraction. *International journal of biological macromolecules* **101** 553–561

[11] Park S, Baker J O, Himmel M E, Parilla P A, and Johnson D K 2010 Cellulose crystallinity index: measurement techniques and their impact on interpreting cellulase performance *Biotechnology for biofuels* **3** (1) 10

Comparative Analysis of Wharton's Jelly Mesenchymal Stem Cell (WJ-MSCs) Isolated Using Explant and Enzymatic Methods

W Widowati[1*], R F Gunanegara[1], R Rizal[2], W S Widodo[2], A Amalia[2], S H B Wibowo[2], K Handono[3], M Marlina[4], I N E Lister[5] and L Chiuman[5]

[1]Faculty of Medicine, Maranatha Christian University, Jl Surya Sumantri No. 65, Bandung 40164 West Java, Indonesia
[2]Biomolecular and Biomedical Research Center, Aretha Medika Utama, Jl Babakan Jeruk II No. 65, Bandung 40163, West Java, Indonesia
[3]Faculty of Medicine, Brawijaya University, Jl Veteran Malang 65145, East Java, Indonesia
[4]Faculty of Pharmacy, Andalas University, Padang 25163, West Sumatera, Indonesia
[5]Universitas Prima Indonesia Jl. Belanga No. 1, Medan 20118, North Sumatera, Indonesia

*Corresponding author's email: wahyu_w60@yahoo.com

Abstract. Wharton's Jelly is one of the best sources for mesenchymal stem cells. Human Wharton's Jelly Mesenchymal Stem Cells (hWJ-MSCs) have high proliferation, multi-lineage differentiation potential, and do not produce any teratogen or carcinogen. These characteristics make hWJ-MSCs become suitable for regenerative medicine. Some methods were developed to isolate hWJ-MSCs from umbilical cord, such as explant method and enzymatic method. This study aims to characterize hWJ-MSCs which are isolated by two different methods, explant attachment method and enzymatic method. hWJ-MSCs isolation was performed through explant method and enzymatic method using trypsin, hyaluronidase and collagenase type 1 with certain ratio of concentration. Isolated hWJ-MSCs was characterized using flow cytometer to detect the expression of CD44, CD90, CD105, CD73 and negative lineage. MSCs differentiation assay was performed to analyze adipogenic, chondrogenic and osteogenic cells lineage. We successfully isolated hWJ-MSCs from umbilical cord through enzymatic and explant methods. Immunophenotyping assay through flow cytometry analysis showed high purity of WJ-MSCs. The isolated hWJ-MSCs from both methods showed positive expression of CD44, CD90, CD105, and CD73. The isolated hWJ-MSCs exhibited capacity to differentiate into adipocyte, chondrocyte, and osteocyte cells. hWJ-MSCs isolated through explant and enzymatic method have high proliferation capacity and be able to differentiate into three different lineage cells. Both methods explant attachment and enzymatic methods are efficiently produced hWJ-MSCs.
Keywords : Wharton's Jelly Mesenchymal Stem Cell, adipogenic, chondrogenic, osteogenic

1. Introduction

Cell therapy and tissue engineering have been increasingly used and promising therapies [1]. Human mesenchymal stem cells (hMSCs) are the powerful tools because self-renewing, multi potent differentiation capability to differentiate into adipocytes, osteoblasts, chondrocytes [2][3], strong immunosuppressive properties and can be used for autologous and allogeneic therapy [4][5], and possible differentiation potential into hepatocytes, cardiomyocytes, skeletal myocytes and neurons [3][6]. The hMSCs are easy to acquire, harvesting process and culture, fast in vitro expansion, feasible for autologous and allogenic therapy, and paracrine function [1].

The hMSCs were successfully isolated from adult tissue including bone marrow, peripheral blood, adipose tissue, skeletal muscle, synovium, dental pulp [7], hair follicle (HF) [8][9], cornea [9][10], and neonatal birth-associated tissues including umbilical cord blood, umbilical cord, placenta, decidua, chorion villi, chorion membrane, amniotic fluid and Wharton's Jelly [7][11]. Ontologically, stem cells that derived from earlier phases of life possess a higher proliferative capacity as foetal MSCs replicate much faster than adult BM-MSCs [12][13].

Umbilical cord tissue is one of the major sources of MSCs since isolation of umbilical cord MSCs (UCMSCs) is easier, does not require an invasive procedure and has a higher proliferation potential compared with other sources [3], has a higher proliferation potential compared with other sources [3][15]. Moreover, umbilical cord is usually discarded after birth so the collection of cells without any ethical conflict or moral concerns [3][7][13][15][16], low risk of infection, non-carcinogenesis, multipotency [7], exhibit low immunogenicity and low immunity after cyto-therapy [5]. Wharton's jelly (WJ) was proven as the best MSC source among various compartments of UC, WJ is perivascular regions surrounding the vessels, arteries, veins, the media and the adventitia compartment of the walls of UCB vessels [18][19]. WJ-MSCs have a higher proliferation rate and self-renewal capacity than adult tissue-derived MSCs [2][20], and favourable for tissue engineering purpose [21].

Clinical therapies require a large number of cells, so many strategies are used to improve the quality and quantity of stem cells [2][22]. Isolation methods were categorized into two major methods; enzymatic method and explant culture method. The procedure of enzymatic method are proteolytic enzymes for cells separation from tissue; the single cell suspension is then cultured in appropriate medium for further cell proliferation [9]. Most studies used collagenase digestion alone or in association with different proteases to accelerated degradation of extracellular matrix proteins [21]. The studies used a panel of enzymes to induce the release of cells (collagenase/trypsin or collagenase/hyaluronidase/trypsin) [21][24]. The explant method is based on the migratory ability of the cells [21], the original tissue is cut into smaller pieces which are placed in culture dishes or flasks, and cells start to migrate out of tissue and adhere to the culture surface [9]. These two techniques are different by many aspects such as time to reach isolated cells and economy, however in the case of MSCs isolation, it seems that both of them provide acceptable cell yield [9][23]. The choice method is critical for cells isolation which it can influence cells yield, viability, integrity, and function [21][23]. The cells isolation method was affected by multiple parameters including tissue source, isolation method and medium composition [24].

This research was done to compare between enzymatic and explant method for human Wharton's Jelly-derived Mesenchymal Stem Cells (hWJ-MSCs) including immunophenotyping, differentiation capacities, population doubling capacity from various sources and passages.

2. Methods

2.1. hWJ-MSCs Isolation with Explants Procedure

Fresh human umbilical cords (UC; n=3) were collected from women aged 25-40 years after normal vaginal delivery, have signed the informed consent using the guidelines approved by Ethics Committee in collaboration between Maranatha Christian University, Bandung, Indonesia and Immanuel Hospital Bandung, Indonesia [2][5][7]. The UC was washed with PBS (Biowest, X0515-500) supplemented with antibiotic and transferred to laboratory using transport medium containing phosphate buffer saline (PBS) (Biowest, X0515-500), 1% Antibiotic-Antimycotic (Ab-Am) (Biowest,

L0010-100), 1% Amphotericin B (Amp B) (Biowest, L0009-100), 0.1% Gentamicin (Gibco, 15750060), 1% Nanomycopulitin (Biowest, LX16-100), and transferred immediately to the laboratory to be processed.

UC was washed again by PBS, and the vessel was removed from the UC. After vessel removal, Wharton's jelly was dissected into small segments (1-2 mm) and transferred onto 100 mm petri dishes. The explants were cultured in MEM-α (Biowest, L0475-500) supplemented with 20% fetal bovine serum (FBS) (Biowest, S1810-500). Cell cultures were incubated in a humidified atmosphere with 5% CO_2 at 37^{0}C for 3 weeks after explantation, when fibroblast like and spindle-shaped adherent cells migrated from the tissue fragments, the adherent cells and tissue fragments were detached using Trypsin-EDTA solution (Biowest, L0931-500) followed by washing with basal medium. The cells were harvested and re-plated at a density 8 x 10^3 cells/cm^2 when cells reached 80-90% confluence. The culture flask was left undisturbed for 3 - 4 days and maintained at 37°C in a humidified atmosphere containing 5% CO_2. The medium changed every 2-3 days thereafter [2][5][7].

2.2. hWJ-MSCs Isolation with Enzymatic Digestion
Wharton's Jelly fragments were washed again with PBS 1x three times and cut into 1-2 mm, and then crushed using sterile blade. The crushed tissues then were inserted into a 50 ml falcon tube (TPP, 91050) containing the mixed enzyme solution contains 4 mg/ml Collagenase Type I (Gibco, 10114532), 1 mg/ml Hyaluronidase (Sigma, 515397), Trypsin 0,1% (Biowest, L0931-500), and then rocked 3 h, 300 rpm, 37 °C, and 5% CO_2 condition followed with overnight incubation, 37 °C, 5% CO_2. Sample was filtered using 70 µm cell strainer (Corning, 431752) then centrifuged at 1600 rpm for 10 min (MPW-250R). Supernatant was discarded and the pellet was washed with PBS 1x and centrifuged at 1600 rpm for 5 min. Washed cells were cultured in MEM-α containing 10% FBS and 1% Ab-Am, 1% Amp B, 0.1% Gentamicin, 1% Nanomycopulitin in 100 mm dish in a humidified 37 °C, 5% CO_2 incubator. After 3-week culture, we harvested the P0 cells. The medium was changed twice a week [25].

2.3. hWJ-MSCs Markers Detection using Fluorescence Activated Cell Sorting (FACS)
The hWJ-MSCs were evaluated using surface marker detection at Passage 3 (P3) and P8 to confirm MSCs characterization, hWJ-MSCs at 80% confluence were harvested and dissociated with trypsin-EDTA and centrifuged at 300 x g for 10 min. The pellet was resuspended with PBS 1x, FBS, and cells were counted with a hemocytometer. Between 10x10^4 – 20x10^4 cells in 500 µL PBS were introduced into FACS (Macsquant Analyzer 10, Miltenyi). The cells were stained with CD90, CD105, CD4, CD73 and their isotype control in hWJ-MSCs analysis kit and the procedure was according to manufacture protocol (BD stem flowTM kit, 562245). The experiments and measurement of surface marker were performed in triplicate [2][5][7].

2.4. Morphology Analysis of hWJ-MSCs
The hWJMSCs cultures were routinely visualized using an inverted research microscope and the phenotype and growth of primary and passaged cells were monitored Photomicrographs were taken with a digital camera model of Inverted Microscope (Olympus, CKX41-F32FL) under inverted phase-contrast optics [2][5][7][26].

2.5. hWJ-MSCs Differentiation
For osteogenic differentiation, hWJ-MSCs were seeded at density 5x10^3 cells in 24 well plate (Nunc, 72296-18) using StemPro Osteogenesis Differentiation Kit (Gibco A10072-01) for 3 weeks. Calcium deposits were visualized using Alizarin red S (Sigma, A5533). For chondrogenic differentiation, hWJ-MSCs were seeded at density 5x10^3 cells in 4 well plate (Nunc, 176740) using StemPro Chondrogenesis Differentiation Kit (Gibco, A10071-01) for 3 weeks. Chondrocytes were visualized using Alcian blue (Sigma, A5268). Adipogenic differentiation of hWJ-MSCs was done using StemPro

Adipogenesis Differentiation Kit (Gibco, A10070-01) for 3 weeks. Oil Red O (Sigma, O0625) to confirm lipid droplets [5][7].

2.6. Passaging hWJ-MSCs Cell Proliferation Analysis
Passaging hWJ-MSCs were counted and passaged at a confluence of 80%. Briefly, cultured cells were detached by trypsin (Biowest, L0931-500), then incubated for 1-3 min at 37°C, completed medium consisting of 80% MEM-α, FBS 20%, 1% Antibiotic and anti-mycotic and 1% heparin were added to stop trypsin and centrifuged (MPW-2000) at 1600 rpm, 5 min at 240 C. The pellet of cells was resuspended with trypan blue solution (Sigma, 25200072) and diluted 1:1 dilution. Then, cells were counted by a haemocytometer (Neubauer,17849). Population Doubling (PD) was counted at every passage with the formula:

$$\mathrm{PD} = [\log_{10}(\mathrm{NH}) - \log_{10}(\mathrm{NI})]/\log_{10}$$

where NI is the inoculum cell number and NH the cell harvest number. Then, to determine cumulative PD data by added the PD at the previous passage. The PD time (PDT) was determined by the formula [2][27]:

$$\text{PD time} = (\text{t (time)})/(\text{PD (in days)})$$

3. Results and Discussion

3.1. Morphology of Isolated MSC
MSCs from Wharton's Jelly was successfully isolated through both explant and enzymatic methods. The isolated cell from both methods showed Spindle-shaped and fibroblast-like MSCs and clearly observed under inverted microscope. The homogenous monolayer hWJ-MSCs attached in plastic flask and cultured until passage 8. The morphological of primary hWJ-MSCs changed in the higher passage, resulted in big Spindle-shaped MSCs.

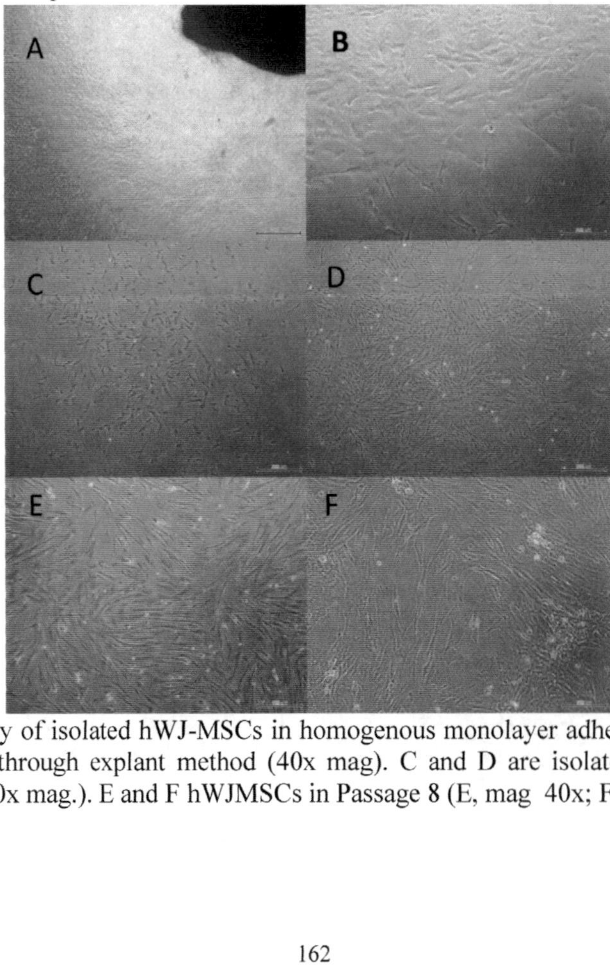

Figure 1. Morphology of isolated hWJ-MSCs in homogenous monolayer adherent flask. A and B are isolated hWJ-MSCs through explant method (40x mag). C and D are isolated hWJ-MSCs through enzymatic method (40x mag.). E and F hWJMSCs in Passage 8 (E, mag 40x; F, mag.. 100x).

3.2. hWJ-MSCs Markers Detection

CD90, CD105, CD73 and CD44 were used to characterize the immunophenotypes of isolated hWJ-MSCs from both methods. The immunophenotypes of hWJ-MSCs were analysed by flow cytometry. The results showed in figure 2. In general, the expression of all markers was higher in explant method than enzymatic one. The expression of isolated hWJ-MSCs P 3 through enzymatic method were 92.03% for CD90, 91.91 % for CD44, 90.76 % for CD105 and 96.64 % for CD73, when the expression of isolated hWJ-MSCs (P3) through explant method 96.75 % for CD90, 96.20 %for CD44, 98.3 % for CD105, and 96.53 % for CD73. In higher passage (P8) the isolated MSCs still exhibit MSCs characteristics (Figure 2, Tabel 1.). The isolated MSCs from both methods showed a very little expression (0-1.5%) of negative lineage (CD34, CD11b, CD19, CD45, HLA-DR) See Figure 3.

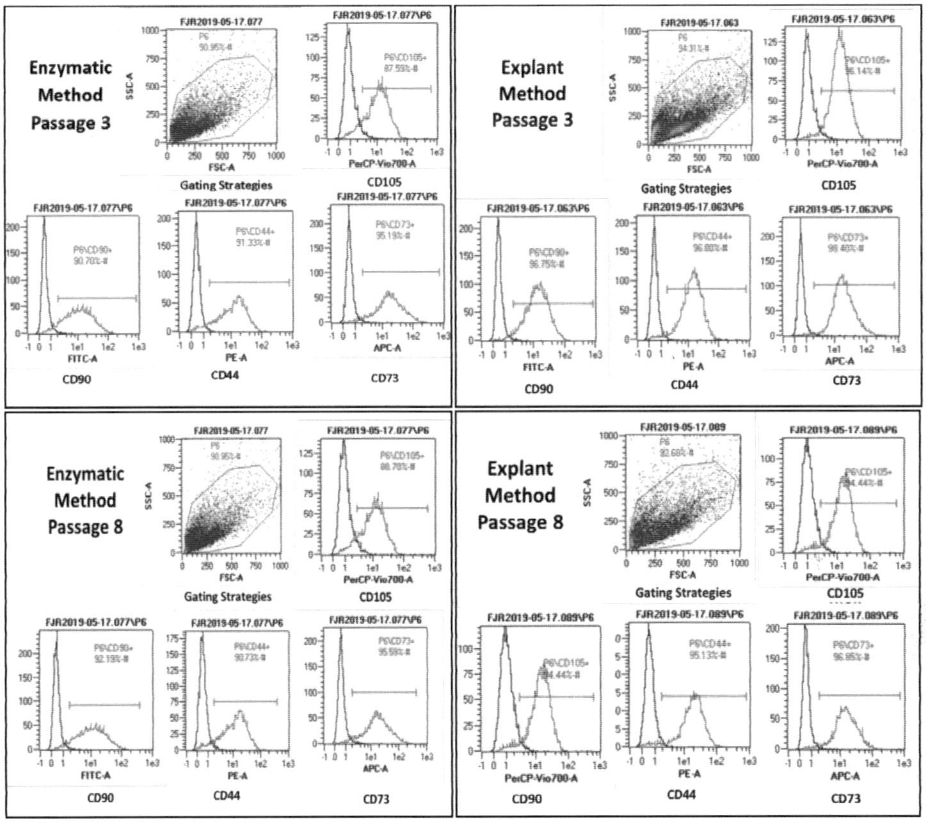

Figure 2. Immunophenotyping of isolated hWJ-MSCs in both methods at passage 3 and 8. The isolated hWJ-MSCs showed positive expression of CD90, CD105, CD44 and CD73 (green line). To avoid positive false, the isotype of each markers was analysed (black line).

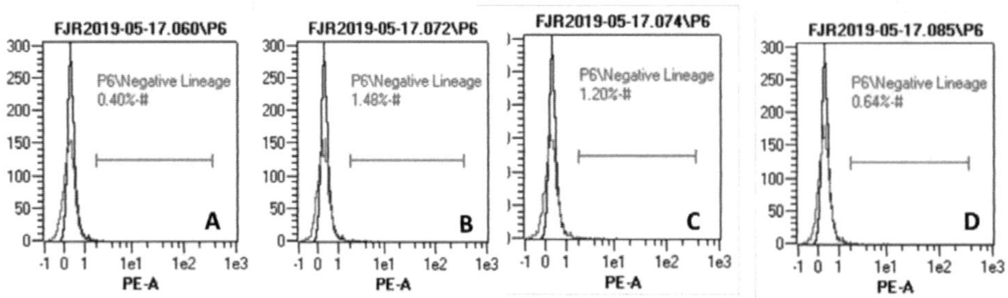

Figure 3. Negative Lineage Cocktail (Immunophenotyping of isolated hhWJ-MSCs in both methods in passage 3 and 8. The isolated hWJ-MSCs showed negative CD34, CD11b, CD19, CD45, LA-DR expression (green line). To avoid positive or negative false, the negative isotype cocktail was analysed (black line). A=Explant P3, B=Enzymatic P3, C=Explant P8, D=Enzymatic P8.

Table 1. Marker expression of isolated hWJMSCs.

Isolation Methods and Passage	Marker Expression (%)				
	FITC-CD90	PE-CD44	PerCP-CD105	APC-CD73	Neg. Lin.
Enzymatic P3	92.03±0.47[a]	91.91±0.51[a]	90.76±0.52[a]	96.64±0.67[a]	1.21±0.28[a]
Explant P3	96.91±0.26[b]	96.51±0.29[b]	96.81±0.35[b]	98.36±0.21[b]	0.40±0.12[a]
Enzymatic P8	91.42±0.81[a]	91.23±0.94[a]	90.56±0.66[a]	95.92±0.46[a]	1.29±0.29[a]
Explant P8	95.42±1.64[b]	94.39±1.61[b]	94.46±1.64[b]	96.65±1.20[a]	0.68±0.11[a]

*The data was shown in mean±standard deviation. Different subscript alphabet in the same column was statistically significant differences among treatment (isolation method and passage) through ANOVA and Tuckey Analysis with $P<0.05$

Statistical analysis of marker expression showed that the expression of CD90, CD105, and CD73 was higher in the MSCs isolated through explant method than enzymatic one. But, the expression of those marker still high (>90%) in enzymatic method-isolated MSCs. Other research shows that the expression of MSCs positive marker (CD90, CD105, CD73 and CD90) were between 85-99% [16][28][29].

CD90 or Thy-1, is a glycosylphosphatidylinositol (GPI)-anchored glycoprotein. This protein has a number of activities such as neuritis outgrowth, T-cell activation, apoptosis in carcinoma cells, tumour suppression, wound healing. Recent studies also indicate that this protein regulates fibroblast focal adhesion, cytoskeleton organization, and cell migration. In MSCs, this protein participated in self-renewal and differentiation [30]. Endoglin (CD105) is a component of the transforming growth factor-beta receptor (TGF-βR) complex. This protein expressed in primitive hematopoietic cells indicate that this molecule has effect in cell cycle and cell survival and differentiation [31]. CD44 antigen is a glycoprotein that has function in cell–cell interactions, cell migration and adhesion [32]. And the last, CD73 or ecto-5′-nucleotidase (ecto-5′-NT, EC 3.1.3.5) is a glycosyl-phosphatidylinositol (GPI)-linked well-known as lymphocyte differentiation antigen. The protein had a function as a co-signaling molecule on T lymphocytes and as an adhesion molecule. This protein help lymphocyte to bind to endothelium [33]. The different expression of this marker in both enzymatic and explant method may affect the proliferation of MSCs. The expression of negative lineages (CD34 PE, CD11b PE, CD19 PE, CD45 PE, HLA-DR PE) was measly expressed. This means that isolated MSCs has high purity.

3.3. Multilineage Differentiation

The assessment of differentiation capacity of hWJ-MSCs toward three lineages (adipocyte, chondrocyte, and osteocyte) was done. All isolated hWJ-MSCs can be differentiate toward all three lineages. The result showed in Figure 4.

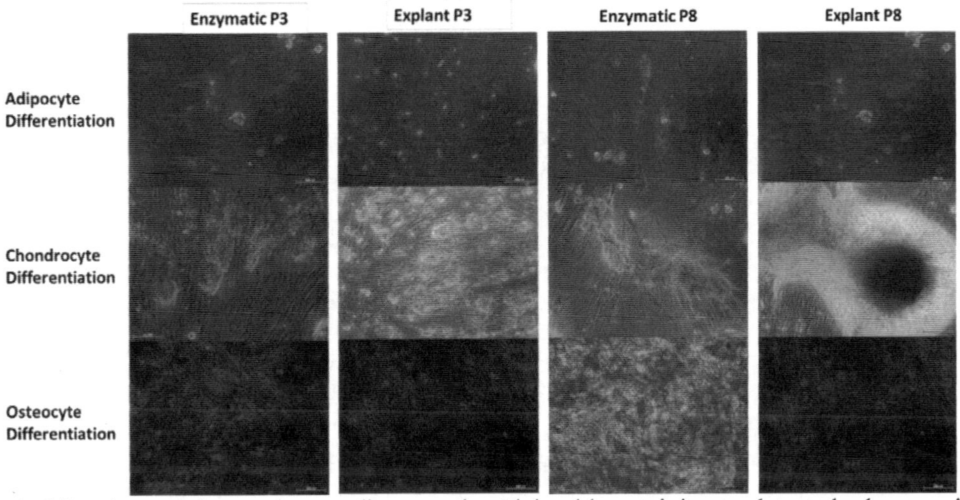

Figure 4. Oil-red-o staining to detect adipogenesis, Alcian blue staining to detect chodrogenesis and, Alizarin-red staining to detect osteogenesis (Mag.100x) in multilineage differentiation of Isolated MSCs from both enzymatic and explant method.

3.4. Population Doubling Time (PDT)

To evaluate the self-renewal capacity of both isolated hWJ-MSCs we used population doubling time (PDT) assay and the result showed in Figure 4. PDT is time for cells to divide become two individuals. The lower PDT value means the higher proliferation capacity of the cells. The result of this assay showed in Figure 4.

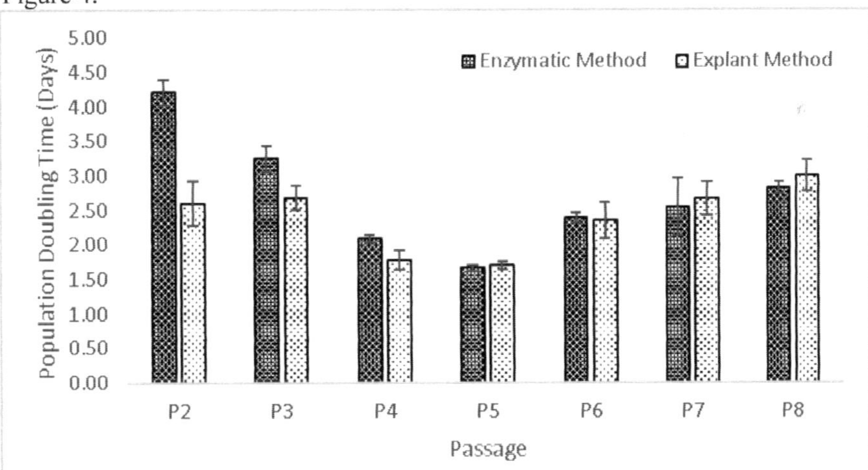

Figure 5. PDT of isolated hWJ-MSCs through explant and enzymatic methods. The PDT of hWJ-MSCs were presented as median±standard deviation. The experiment was done in triplicates. The early passage (P2-P3) was high PDT value. The lowest PDT was P4-5, the older passage increased the PDT value.

The result of PDT assay showed that in early passage, the explant method has lower PDT. It is mean, in early passage the cells isolated through explant methods has higher proliferation capacity. The lower proliferation capacity in isolated hWJ-MSCs through enzymatic methods may be due to the enzyme treatment. The best proliferation capacity of MSCs was in Passage 5. The increasing passage in line with the lower proliferation capacity of MSCs (Figure 5).

Proliferation assay showed that in early passage, isolated hWJ-MSCs from explant procedure had higher self-renewal capacity. This may be caused by enzymatic digestion in the hWJ-MSCs isolation process. The cells that isolated may be contains other cells in Wharton's Jelly tissue. After passage 4, the PDT of hWJ-MSCs increase in the increasing of passage. This result in line with the research conducted by Widowati et al., (2014) and Antonius et al., (2015) that the rising passage caused the decreasing proliferation capacity of primary hWJ-MSCs [2][27].

Our results show that both methods can be used for isolation of MSCs from Wharton's Jelly tissues. hWJ-MSCs is one of the best sources of MSCs. It can substitute the Bone Marrow-Mesenchymal Stem Cells (BM-MSCs) since it has similar characteristics. hWJ-MSCs are neither embryonic stem cells nor adult stem cells. hWJ-MSCs have the characteristics of both embryonic stem cells and mesenchymal stem cells. Then, that properties can be used for stem-cells based therapies especially for regenerative diseases [28][34][35].

This study was aimed to compare the characteristics of two different methods that commonly used to isolate hWJ-MSCs, explant and enzymatic methods. The enzymatic method is fastest method, but it uses some enzyme to digest Wharton's Jelly tissues and extract the MSCs efficiently. The enzymes that usually used to digest Wharton's Jelly matrix is Collagenase 1, Hyaluronidase and Trypsin [36]. Collagenase is an enzyme isolated from Clostridium histolyticum and can degrade collagen tissue [37]. Trypsin is a proteolytic enzyme that degrade adherent protein, so the cells can be extracted from the tissue [38]. Hyaluronidase is an enzyme that catalyse the degradation of hyaluronic acid, one of components in Wharton's Jelly Matrix [39]. The combination of that enzymes provided great activity to digest extracellular matrix of Wharton's Jelly and extract MSCs efficiently.

Other methods to isolate MSCs from WJ tissue is Explant Method. The Tissue was cropped and chopped into small pieces., about 1 mm2. The figure was easiest method to isolate MSCs. The MSCs will outgrow from attached tissue. The disadvantage of this method is the inability to determine the number of the cells isolated from the tissue, so the determination of population doubling time at Passage 1 can be measured [36].

We suggest further works to obtain a quick method to obtain high purity of MSCs with xenofree medium. In this study, the isolation and characterization of WJ-MSCs was achieved with high purity and obtained at a more suitable time, but it still used FBS, which was xenogeneic, and that xeno genic medium may be has bad effect in hWJMSCs culture. In our study, the morphology of hWJ-MSCs in higher passage (P8) showed large shape hWJMSCs and had low self-renewal capacity (Figure 1 and Figure 4). That was caused by the use of FBS in the medium. Tonti and Manello (2008) stated that the use of FBS can cause immune reactions and transmission of pathogen [40].

4. Conclusion

In conclusion, the enzymatic and explant methods can be used for isolation of hWJMSCs resulted in high purity of hWJ-MSCs. The explant method is the easiest one, and resulting higher purity of MSCs than the enzymatic method.

5. Acknowledgements

This research was financially supported by the Ministry of Research, Technology and Higher Education of the Republic of Indonesia and for research grant 2019 (Penelitian Terapan Unggulan Perguruan Tinggi). This research was also supported by Aretha Medika Utama Biomolecular and Biomedical Research Center, Bandung, Indonesia for the research methodology and laboratory facilities. We are thankful to Ni Luh Wisma Eka Yanti, Hanna Sari Widya Kusuma, Ika Adhani Sholihah, Dewani Tediana Yusepany, Dwi Surya Artie, Anisa Siwianti, Rizki Amalia from Aretha Medika Utama Biomolecular and Biomedical Research Center, Bandung, Indonesia for their valuable assistance.

References

[1] Noverina R, Widowati W, Ayuningtyas W, Kurniawan D, Afifah E, Laksmitawati D R, Rinendyaputri R, Rilianawati R, Faried A, Bachtiar I and Wirakusumah F F 2019 Growth factors profile in conditioned medium human adipose tissue-derived mesenchymal stem cells (CM-hATMSCs) Clinical Nut. Exper. **224** 34 – 44

[2] Widowati W, Wijaya L, Bachtiar I, Gunanegara R F, Sugeng S U, Irawan Y A, Sumitro S B, Widodo M A 2014 Effect of oxygen tension on proliferation and characteristics of Wharton's jelly-derived mesenchymal stem cells *Biomark. Genom. Med.* **6** 43 – 48

[3] Hassan G, Kasem I, Soukkarieh C and Aljamali M A 2017 A Simple method to isolate and expand human umbilical cord derived mesenchymal stem cells: using explant method and umbilical cord blood serum *Int. J. Stem Cells.* **10** (2) 184 – 192

[4] Jones B J and McTaggart S J 2008 Immunosuppression by mesenchymal stromal cells: from culture to clinic *Exp. Hematol.* **36** 733-741

[5] Widowati W, Murti H, Jasaputra D K, Sumitro S B, Widodo M A, Widyastuti H, Agustina D, Laksmitawati D R, Fauziah N, Sumitro S B, Widodo M A and Bachtiar I 2015 Conditioned medium from normoxia (WJMSCs-norCM) and hypoxia-treated WJMSCs (WJMSCs-hypoCM) in inhibiting cancer cell proliferation *Biomark. Genom. Med.* **7** 8– 17

[6] Chua S J, Bielecki R, Wong C J, Yamanaka N, Rogers I M and Casper R F 2009 Neural progenitors, neurons and oligodendrocytes from human umbilical cord blood cells in a serum- free, feeder-free cell culture *Biochem. Biophys. Res. Commun.* **379** 217 – 221

[7] Widowati W, Murti H, Jasaputra D K, Sumitro S B, Widodo M A, Fauziah N, Maesaroh M and Bachtiar I 2016 Selective cytotoxic potential of IFN-γ and TNF-α on breast cancer cell lines (T47D and MCF7) *Asian J. Cell .Biol.* **11** (1) 1 – 12

[8] Ohyama M, Zheng Y, Paus R and Stenn K S 2010 The mesenchymal component of hair follicle neogenesis: background, methods and molecular characterization *Exp Dermatol.* **19** 89 – 99

[9] Hendijani F 2017 Explant culture: An advantageous method for isolation of mesenchymal stem cells from human tissues *Cell Proliferation.* **50**:e12334: 1 – 14

[10] Vereb Z, Poliska S, Albert R, Olstad O K, Boratko A, Csortos C, Moe M C, Facsko A and Petrovski G 2016 Role of human corneal stroma-derived mesenchymal-like stem cells in corneal immunity and wound healing *Sci. Rep.* **6** 26227

[11] Gjorgieva D, Zaidman N and Bosnakovski D 2013 Mesenchymal stem cells for anti-cancer drug delivery *Recent Pat. Anticancer Drug Discov.* **8** (3) 310 – 318

[12] Hass R, Kasper C, Böhm S and Jacob R 2011 Different populations and sources of human mesenchymal stem cells (MSC): A comparison of adult and neonatal tissue-derived MSC *Cell Commun. Signal.* **9** 12

[13] Yusoff Z, Maqbool M, George E, Hassan R M D and Ramasamy R 2016 Generation and characterisation of human umbilical cord derived mesenchymal stem cells by explant method *Med. J. Malaysia* **71** (3) 105-110

[14] Can A and Karahuseyinoglu S 2007 Concise review: human umbilical cord stroma with regard to the source of fetus-derived stem cells *Stem Cells.* **25** 2886 – 2895

[15] Pirjali T, Azarpiral N, Ayatollahi M, Aghdaie M H, Geramizadeh B and Talai T 2013 Isolation and characterization of human mesenchymal stem cells derived from human umbilical cord Wharton's jelly and amniotic membrane *Int. J. Org. Transplant. Med.* **4** (3) 111 – 116

[16] Beeravolu N, McKee C, Alamri A, Mikhael S, Brown C, Perez-Cruet M and Chaudhry G R 2017 Isolation and characterization of mesenchymal stromal cells from human umbilical cord and fetal placenta *Jo.VE.* **122** e55224

[18] Bongso A and Fong C Y 2013 The therapeutic potential, challenges and future clinical directions of stem cells from the Wharton's jelly of the human umbilical cord *Stem Cell Rev.* **9** 226–240

[19] Bharti D, Shivakumar S B, Park J-K, Ullah I, Subbarao R B, Park J-S, Lee S-L, Park B-W, Rho G-J 2018 Comparative analysis of human Wharton's jelly mesenchymal stem cells derived from different parts of the same umbilical cord *Cell Tissue* **372** 51–65

[20] Troyer D L and Weiss M L 2008 Wharton's jelly-derived cells are a primitive stromal cell population *Stem Cells.* **26** 591-599

[21] Margossioan T, Reppel L, Makdissy N, Stoltz J F, Bensoussan D and Huselstein C 2012 Mesenchymal stem cells derived from Wharton's jelly: comparative phenotype analysis between tissue and in vitro expansion *Bio-med. Materials Engineering.* **22** (4) 243 – 254

[22] Nekanti U, Mohanty L, Venugopal P, Balasubramanian S, Totey S, Ta M 2010 Optimization and scale-up of Wharton's jelly-derived mesenchymal stem cells for clinical applications *Stem cell Res.* **5** (3) 244 – 254

[23] Salehinejad P, Alitheen N B, Ali A M, Omar A R, Mohit M, Janzamin E, Samani F S, Torshizi Z and Nematollahi-Mahani S N 2012 Comparison of different methods for the isolation of mesenchymal stem cells from human umbilical cord Wharton's jelly *In Vitro Cellular Develop. Biology-Animal.* **48** (2) 75 – 83

[24] Mushahary D, Spittler A, Kasper C, Weber V, and Charwat V 2018 Isolation, cultivation, and characterization of human mesenchymal stem cells *Cytomet.* Part A **93** (1) 19 – 31

[25] Yoon J H, Roh E Y, Shin S, Jung N H, Song E Y, Chang J Y, Kim B J and Jeon H W 2013 Comparison of explant-derived and enzymatic digestion-derived mscs and the growth factors from Wharton's jelly *Bio. Med. Res. Int.* **2013** 1 – 8

[26] Azandeh S, Orazizadeh M, Hashemitabar M, Khodadadi A, Shayesteh A A, Nejad D B, Gharravi A M and Allahbakhshi E 2012 Mixed enzymatic-explant protocol for isolation of mesenchymal stem cells from Wharton's jelly and encapsulation in 3D culture system *J. Biomed. Sci. Engineering.* **5** 580 – 586

[27] Antoninus A A, Widowati W, Wijaya L, Agustina D, Puradisastra S, Sumitro SB, Widodo M A and Bachtiar I 2015 Human platelet lysate enhances the proliferation of Wharton's jelly-derived mesenchymal stem cells *Biomar. Genom. Med.* **7** (3) 87 – 97

[28] Ranjbaran H, Abediankenari S, Mohammadi M, Jafari N, Khalilian A, Rahmani Z, Marzei Amiri M M and Ebrahimi P 2018 Wharton's jelly derived-mesenchymal stem cells: Isolation and characterization *Acta Med. Iran.* **56** (1) 28 – 33

[29] Mennan C, Wright K, Bhattacharjee A, Balain B, Richardson J and Roberts S 2013 Isolation and characterisation of mesenchymal stem cells from different regions of the human umbilical cord *Bio. Med. Res. Int.* **2013** 916136

[30] Moraes D A, Sibov T T, Pavon L F, Alvim P Q, Bonadio R S, Da Silva J R, Pic-Taylor A, Toledo O A, Marti L C, Azevedo R B and Oliveira D M 2016 A reduction in CD90 (THY-1) expression results in increased differentiation of mesenchymal stromal cells *Stem cell Res Ther* **7** (1) 97

[31] Pierelli L, Bonanno G, Rutella S, Marone M, Scambia G and Leone G 2001 CD105 (endoglin) expression on hematopoietic stem/progenitor cells *Leukemia Lymphoma* **42** (6) 1195 – 1206

[32] Misumi Y, Ogata S, Ohkubo K, Hirose S and Ikehara Y 1990 Primary structure of human placental 5′-nucleotidase and identification of the glycolipid anchor in the mature form *European J. Biochem.* **191** (3) 563 – 569

[33] Zhang B 2010 CD73: a novel target for cancer immunotherapy *Cancer Res.* **70** (16) 6407 – 6411

[34] Battula V L, Treml S, Abele H and Bühring H J 2008 Prospective isolation and characterization of mesenchymal stem cells from human placenta using a frizzled-9-specific monoclonal antibody *Differentiation* **76** (4) 326 – 336

[35] Carlin R, Davis D, Weiss M, Schultz B and Troyer D 2006 Expression of early transcription factors Oct-4, Sox-2 and Nanog by porcine umbilical cord (PUC) matrix cells *Reprod. Biolog. Endocrinolog.* **4** (1) 8

[36] Seshareddy K, Troyer D, Weiss ML 2008 Method to isolate mesenchymal-like cells from Wharton's Jelly of umbilical cord *Methods Cell Biol.* **86** 101 – 119

[37] Breite A G, McCarthy R C and Dwulet F E 2011 Characterization and functional assessment of Clostridium histolyticum class I (C1) collagenases and the synergistic degradation of native collagen in enzyme mixtures containing class II (C2) collagenase. In *Transplantation Proceedings* **43** (9) 3171 – 3175

[38] Huang H L, Hsing H W, Lai T C, Chen Y W, Lee T R, Chan H T, Lyu P C, Wu C L, Lu Y C, Lin S T and Lin CW 2010 Trypsin-induced proteome alteration during cell subculture in mammalian cells *J. Biomedic. Sci.* **17** (1) 36

[39] Stern R, Kogan G, Jedrzejas M J and Šoltés L 2007 The many ways to cleave hyaluronan *Biotechnolog. Advances.* **25** (6) 537 – 557

[40] Tonti G A and Mannello F 2002 From bone marrow to therapeutic applications: different behaviour and genetic/epigenetic stability during mesenchymal stem cell expansion in autologous and foetal bovine sera? *Int. J. Develop. Biol.* **52** (8) 1023 – 1032

Genotype and Serotype Identification of Hepatitis B Virus in Chronic Hepatitis B Patients Treated with Telbivudine

Y Purnamasari[1]*, Tien[2], Arimaswati[3], A N K Syarifin[2], A Ali[4] and N Massi[5]

[1] Microbiology Department, Medical Faculty of Halu Oleo University, Kendari, Indonesia
[2] Biochemistry Department, Medical Faculty of Halu Oleo University, Kendari, Indonesia
[3] Parasitology Department, Medical Faculty of Halu Oleo University, Kendari, Indonesia
[4]Anesthesiology and Intensive Care Department, Medical Faculty of Halu Oleo University, Kendari, Indonesia
[5]Microbiology Department, Medical Faculty of Hasanuddin University, Makassar, Indonesia

*Corresponding author's email: yenti.purnamasari@ymail.com

Abstract. Hepatitis B is an inflammatory liver disease caused by hepatitis B virus (HBV). In Indonesia, the prevalence of Hepatitis B Surface Antigen (HBsAg) is 9.4%, categorizing the country as endemic hepatitis B. HBV has been classified into at least ten genotypes and four serotypes. Each genotype has different clinical significance and virologic characteristics, which can be an independent risk factor of Hepatocellular Carcinoma in addition to the male sex, older age, and positive Hepatitis B Envelope Antigen status. Telbivudine antiviral therapy suppressed the virus, but in case of resistance mutations, the mutant might grow continuously due to drugs inefficiency. This study aimed to detect the genotypes and serotypes of HBV in hepatitis B chronic patients after 12 weeks of telbivudine treatment. The subject of this study involved 26 patients with chronic hepatitis B, receiving 12 weeks treatment of telbivudine in the gastrohepatology division of the RSUP Wahidin Sudirohusodo hospital. The HBV genotype was identified by analyzing the HBV P gene, while the serotype was detected by deducing the nucleotide of the HBV S gene that overlapped with the P gene. Nine samples (75%) were genotype b, and three samples (25%) were genotype c. for serotype, six (50.0%), three (25.0%), and three (25.0%) samples were *ayw*, *adw*, and *adr,* respectively. genotype b is generally associated with less progressive liver disease than genotype c. genotypes b and c are prevalent in highly endemic areas in which the perinatal or vertical transmission play an important role in spreading the virus.
Keywords: Genotype, Hepatitis B Virus, Serotype, Telbivudine Treatment

1. Introduction

An estimation of 4.0 to 20.3% of hepatitis B cases in a healthy population of Indonesia has ranked this country with moderate to high endemicity of hepatitis B infection [1]. In 2007, the prevalence of Hepatitis B surface antigen (HBsAg) was 9.7% in men and 9.3% in women [2]. Meanwhile, the prevalence of people infected with hepatitis B virus-hepatitis B core antibody (HBc) was 34%, indicating that a third of Indonesia's people have been infected with HBV. The highest hepatitis B virus infection in Indonesia found in various Provincial, including East Nusa Tenggara, Papua, South Sulawesi, Central Sulawesi, Maluku, Southeast Sulawesi, North Sulawesi, Aceh, West Nusa Tenggara, Central Kalimantan, North Sumatra, and South Kalimantan [2].

At least ten hepatitis B virus (HBV) genotypes (A to J) with distinct geographic distributions and several HBV mutants, including precore/core promoter mutations and pre-S/S deletion mutations, have been recognized to be not only predictive of liver disease progression but also associated with response to antiviral therapy. HBV genotypes and variants may serve as viral genetic markers in predicting disease progression as well as helping physicians for optimizing individualized antiviral therapy in clinical practice [3]. Genotypes B and C are found mostly in the Asia Pacific countries, including Indonesia [4,5].

One of chronic hepatitis B (CHB) treatments is the lifelong oral administration of nucleos(t)ide analogs such as Telbivudine. The antiviral nucleos(t)ide analogs targeted the HBV polymerase protein-translated from the HBV P gene, which is responsible for the reverse transcription process during HBV replication. However, HBV drug-resistant mutants often arise from such treatment. Currently, three Telbivudine-resistant mutants have been recognized, namely M204I, L80I/V and/or L180M [3]. The antiviral will suppress HBV titer to an undetectable level, unless in the presence of drug-resistant mutants, which can still replicate even with the antiviral pressure [6,7]. According to various international guidelines, the primary therapy for CHB is considered failed if it could not decrease the HBV DNA titer more than 1 $\log_{10}$ IU/mL after 12 weeks of compliant administration. Therefore, monitoring of antiviral therapy in CHB patients should be performed at week 12 [8-10].

The pathogenic differences among various HBV genotypes have been partially clarified and influenced the clinical outcomes of HBV infection [3]. Hence, the identification of HBV genotype is crucial for many reasons. Furthermore, monitor developments are necessary for those who received antiviral therapy based on infecting vhb genotype. This study aimed to detect the genotypes and serotypes of HBV in hepatitis b chronic patients after 12 weeks of Telbivudine treatment.

2. Materials and Methods

As many as 26 CHB patients received 600 mg daily dose of Telbivudine for 12 weeks were recruited from the RSUP Wahidin Sudirohusodo hospital. Venous blood samples (5 mL) were taken at week 12 of treatment. The HBV DNA was extracted, and P gene region amplified with nested polymerase chain reaction (PCR) method. The PCR method began with a pre-denaturation step at 94°C for 5 minutes, followed by denaturation (94°C for 30 s), annealing (55°C for 30 s), and extension steps (72°C for 1 m) for the first 35 cycles and second 25 cycles of PCR phase, and ended with a final extension step at 72°C for 7 min. Sequences of the oligonucleotide primers are listed in Table 1. Amplification products were visualized using ethidium bromide-stained 1.5% agarose gel under ultraviolet light. All experiments were carried out with the Kwok and Higuchi's rules, to avoid any cross-contamination [11].

Positive PCR products were purified using a Gel/PCR DNA Fragments Extraction Kit (Geneaid Biotech Ltd., Taipei, Taiwan) and sequenced using a Bigdye Terminator Cycle Sequencing Ready Reaction kit (Applied Biosystems, Foster City, CA) on an automatic sequencer (Applied Biosystems 337 DNA, Perkin Elmer). The resulting nucleotide sequences were aligned and compared with reference sequences from the GenBank (HBV genotype B, Acc no. M54923 and HBV genotype C, Acc. no. GQ358153) using the BioEdit v.7.0.5 software. Phylogenetic analysis with 1.000 bootstrapping and neighbor-joining method using MEGA v 5.1 software was used to classify the HBV isolates into genotypes based on 4% nucleotide heterogeneity of the overlapping HBV S gene region (12). HBV serotypes were determined using the HBsAg analysis (which is deduced from nucleotide S gene), based on amino acid variations in positions 122 and 160 [12-14].

Table 1. HBV DNA-specific primers used in nested PCR and sequencing methods.

Primer	Nucleotide sequence (5'→3')	Nt position[a]	Direction
006F	CCTGCTGGTGGCTCCAGTTC	56-75	Sense
995R	CCTATTGATTGGAAAAGTATGTCAA	970-993	Anti-sense
850R	ATACATTTAAACCCTAACAAAC	826-847	Anti-sense
S2-1	CAAGGTATGTTGCCCGTTTG	455-474	Sense
S88	TGTTGCCCGTTTGTCCTCTA	462-481	Sense

[a]Nucleotide numbering based on EcoRI restriction site

3. Results and Discussion

The P gene amplification results showed 12 positive samples (46,2%) and 14 negative samples (53,8%). In the positive sample, VHB DNA was detected. The test continued with the direct sequencing method to determine nucleotide sequences, followed by the genotype analysis. The 12 positive samples consisted of 9 (75%) HBV genotype B and 3 (25%) HBV genotype C (Figure 1). This result was in agreement with a previous study by Thedja and co-workers. The study reported the HBV genotype distribution in Makassar, Indonesia, which is dominated by HBV genotypes B and C [15]. Both genotypes are commonly found in regions of East Asia and Southeast Asia, the Pacific islands, and Pakistan. Meanwhile, genotypes A and D are found in India and the Philippines, and the genotype D is common in the Pacific islands. HBV genotypes B and C are often found in areas with high vertical transmission, while genotypes A, D, E, F, G, and H occurred in areas with the horizontal transmission [16]. In addition to genotypes, VHB serotypes can also be determined by conducting HBsAg analysis (which is deduced from nucleotide S gene), based on amino acid variations in positions 122 and 160 [12-14]. Six samples (50.0%) obtained were the final subtypes, three samples (25.0%) were adw subtypes, and three samples (25.0%) were adr subtypes. Identification of HBV genotype is crucial for many reasons. The pathogenic differences among various HBV genotypes have been partially clarified and influenced the clinical outcomes of HBV infection. HBV genotypes in viral factors are not only predictive of clinical progression but also related to interferon (IFN)-α treatment response [17].

```
                120       130       140       150       160       170
           ....|....|....|....|....|....|....|....|....|....|....|....|
VHB C  116 TSTGPCKTCTIPAQGTSMFPSCCCTKPSDGNCTCIPIPSSWAFARFLWEWASVRFSWLSL 175
297    116 XX.......................................................... 175
99     116 XXX......................................................... 175
206    116 XXX.......T................................................. 175
40.12  116 -------X..T................................................. 168

                180       190       200       210       220
           ....|....|....|....|....|....|....|....|....|....|..
VHB C  176 LVPFVQWFAGLSPIVWLSVIWMMWYWGPNLYNILNPFLPLLPIFFCLWVYI* 227
297    176 .....H..E.F..S..............S...............L......* 227
99     176 ........V....T..............S.......................* 227
206    176 ........V....T..............S.....K.................* 227
40.12  169 ........V....T..............S.......................* 220
```

Figure 1. Bioedit analysis of HBsAg protein amino acid VHB genotype.

Clear HBV genotype-related associations exist between clinical outcomes and treatment efficacy in patients with CHB [18]. A study conducted in China has investigated the reasons for the longer immune clearance period in HBV patients infected with genotype C as compared to genotype B. These including a higher level of viral replication, a high hepatic histological activity, recurrent or persistently high ALT levels and IFN, nucleos(t)ide analogs, and low response to treatment. The possible relationship among genotypes B and C and peripheral blood follicular helper T (Tfh) cells in CHB patients under treatment has been investigated. Tfh cells play a major role in spreading signals that affect cellular division, helping the activation of B cells and the regulation of humoral response. In addition, Tfh cells secreted specific cytotoxic T lymphocyte (CTL) interleukin (IL)-21, to sustain long-acting, effective, and antiviral immunity in chronic infection. High serum HBV DNA and ALT ratios in patients with genotype C might be related to lower peripheral blood Tfh cell levels. The lowering levels also decreased the levels of IL-21 as compared to the genotype B. The study also reported low levels of HBV-specific CTL [19].

Chronic hepatitis B patients can be successfully treated using nucleos(t)ide analogs (NAs), which was the choice of therapy for patients in this study. The NAs work mainly by inhibiting the hepatitis B virus (HBV) DNA polymerase activity and thus suppress the HBV replication [20].

However, among 26 samples studied, 12 samples showed a positive HBV DNA. These cases emerged a necessity for continuous observation with a long period of treatment. Persistent HBV replication may trigger strong and continued immune responses against the virus, resulting in even more severe liver damages [21].

4. Conclusion

Genotypes B and C were dominant genotypes found in patients of this study, while the dominant serotypes were ayw, adw, and adr. Further prospective research is needed to evaluate the treatment outcomes of patients with more numbers, to provide more evidence of the relationship between genotype and treatment development.

References

[1] Perhimpunan Peneliti Hati Indonesia 2012 *Konsensus Nasional Penatalaksanaan Hepatitis B.* (Jakarta: PPHI)

[2] Riset Kesehatan Dasar 2007 (Jakarta: Badan Penelitian dan Pengembangan Kesehatan Depekes RI)

[3] Lin C-L and Jia-Horing K 2015 Hepatitis B Virus Genotypes and variants. *Cold Spring Harb Respect Med.* **5** (5) A021436

[4] Siburian M D, Utama A, Dhenni R, Arnelis N, Fanany I, Intan M D, Kurniasih T S, Andriani F, Afadlal S, Julianto E B, *et al.* 2013 High prevalence of hepatitis B virus genotype C/C1 in the Minangkabau ethnic group in Indonesia *Virol J.* **10** 27

[5] Prasetyo A A, Dirgahayu P, Sari Y, Hudiyono H, Kageyama S 2013 Molecular epidemiology of HIV, HBV, HCV, and HTLV-1/2 in drug abuser inmates in central Javan prisons, Indonesia *J Infect Dev Ctries.* **7** 453–467

[6] Zoulim F and Locarnini S 2009 Hepatitis B Virus Resistance to Nucleos(t)ide Analogues *Gastroenterology* **137** 1593–1608

[7] Coleman P F 2006 Detecting Hepatitis B Surface Antigen Mutants *Emerg Infect Dis* **12** (2) 198-293

[8] Lok A S and McMahon B J 2009 Chronic Hepatitis B Update 2009. *Hepatology* **50** (3) 661-662

[9] Liaw Y F *et al.* 2012 Asian-Pacific Consensus Statement on the Management of Chronic Hepatitis B: a 2012 Update *Hepatol Int* **6** (3) 531-561

[10] Seto W K *et al.* 2011 Significance of HBV DNA Levels at 12 Weeks of Telbivudine Treatment and the 3 Years Treatment Outcome *J Hepatol* **55** (3) 522–528

[11] Kwok S and Higuchi R 1989 Avoiding false positives with PCR *Nature* **339** (6221) 237-238

[12] Le Bouvier G L 1971 The Heterogeneity of Australia Antigen *The Journal of Infectious Disease* **123** 671-675

[13] Bancroft W H *et al.* 1972 Detection of Additona Antigenic Determinants of Hepatitis B Antigen *The Journal of Immunology* **109** 4

[14] Yamanaka T *et al.* 1990 Hepatitis B Surface Antigen Particles with All Four Subtypic Determinants: Point Mutations of Hepatitis B Virus DNA Inducing Phenotypic Changes or Double Infection with Viruses of Different Subtypes *Molecular Immunology* **27** 443-449

[15] Thedja M D *et al.* 2011 Ethnogeographical structure of hepatitis B virus genotype distribution in Indonesia and discovery of a new subgenotype, B9 *Arch Virol* **156** (5) 855-868

[16] Asian Pacific Association for the Study of the Liver (APASL) 2012 *Asian-Pacific Consensus Statement on the Management of Chronic Hepatitis B*: a 2012 Update.
[17] Liu C J and Kao J H 2013 Global perspective on the natural history of chronic hepatitis B: role of hepatitis B virus genotypes A to J *Semin Liver Dis.* **33** 97–102
[18] Lin S, Liu C, Shang H, Chen H, Yang B, Chen J, Chen Y, Chen D and Ou Q 2013 HBV serum markers of 49164 patients and their relationships to HBV genotype in Fujian Province of China *J Clin Lab Anal.* **27** 130–136
[19] Xibing G, Xiaojuan Y, Juanhua W and Zhong H 2013 Relationship between HBV genotypes B, C and follicular helper T cells in patients with chronic hepatitis B and its significance. *Hepat Mon.* **13** e6221
[20] Fung J, Lai C L, Seto W K and Yuen M F 2011 Nucleotide Analogues in The Treatment of Chronic Hepatitis B *J Antimicrob Chemother.* **66** (12) 2715-25
[21] Perrillo R P 2001 Acute flares in chronic hepatitis B: the natural and unnatural history of an immunologically mediated liver disease *Gastroenterology* **120** 1009–1022

Molecular Docking Study from Lunacridine, Scopoletin and Skimmianine as Antidiabetes through α-Glucosidase Inhibitor

Adriani[1,2*], Noorhamdani[3], S Winarsih[3] and T Ardyati[4]

[1] Doctoral Program in Medical Science, Faculty of Medicine, Brawijaya University, Malang, Indonesia
[2]STKIP Pembangunan Indonesia Makassar, Indonesia
[3]Department of Microbiology Faculty of Medicine, Brawijaya University, Malang, Indonesia
[3]Department of Microbiology Faculty of Medicine, Brawijaya University, Malang, Indonesia
[4]Department of Biology, Faculty of Mathematics and Science, Brawijaya University, Malang, Indonesia

*Corresponding author's email: adrimarsya@gmail.com

Abstract. Diabetes is a metabolic syndrome disease characterized by hyperglycemia in patients. The use of medicinal plants for the treatment of diabetes can control blood levels because it contains anti-diabetic active substances. The medicinal plant as an anti-diabetic through inhibition of the α-glucosidase enzyme thereby reducing the absorption of glucose in the small intestine. Lunacridine, skimmianine and scopoletin are found in the Rutaceae family but there is no information about them as α-glucosidase inhibitors. The purpose of the study to determine the ability of lunacridine, skimmianine and scopoletin as inhibitors of α-glucosidase enzymes based on docking molecular studies. The research method is ligand and receptor preparation using Pymol and docking. The docking process uses Autodoct vina in Pyrx and using acarbose as controls. The docking results are visualized using Ligplot and Discovery studio software. The results showed that lunacridine, skimmianine, scopoletin interacted with α-glucosidase and various binding affinity value. The lunacridine binding affinity is close to the acarbose control and can cross cell membranes based on Lipinski rules. Lunacridine has an anti-diabetic ability through inhibition of α-glucosidase enzyme with the inhibitory value close to acarbose control
Keywords: Lunacridine, scopoletin, skimmianine, α-glucosidase inhibitor, docking

1. Introduction

Diabetes mellitus, the metabolic syndrome, is characterized by postprandial hyperglycemia which results in increased blood glucose levels. The cause of diabetes is insulin resistance or pancreatic beta-cell damage. Another trigger is the less active style and consumption of unhealthy foods can increase the prevalence of diabetes. Monitoring and controlling blood glucose levels is the most common method in the treatment of type 2 diabetes currently. One of the targets for the treatment of diabetes is the inhibition of α-glucosidase enzyme activity. α-glucosidase is an enzyme that plays a role in breaking down the 1.4 glycosidic bonds in carbohydrates into glucose so that facilitate absorption in the small intestine [1, 2]. Inhibition of this enzyme can control blood glucose levels of diabetic patients shortly after eating.

α-glucosidase inhibitors are produced by microorganisms and medicinal plants, one of which is from Rutaceae family. Previous phytochemical research on this family reported various types of compounds, including lunacridine [3], skimmianine [4] steroid, saponin[5], scopoletin[6] and polyphenol [5,7]. Lunacridine, skimmianine, and scopoletin have an anticancer effect [8-10] but its potential as an α-glucosidase inhibitor is unknown.

Screening of active compound bioactivity from natural materials based on molecular docking is widely used. Molecular docking is a method of exploring new compounds that have the potential to be drug candidates based on bioinformatics modeling. In general, molecular docking is used to determine the interaction of ligands and receptors and the energy produced by these interactions. This method considered efficient because it provides fast results and does not differ greatly from laboratory results [11, 12]. The purpose of this study was to determine the activity of lunacridine, skimmianine, and scopoletin as α-glucosidase inhibitors through a molecular docking approach.

2. Methods

2.1. Receptor and ligands preparation

The 3D structure of the receptor is downloaded from the Protein Data Base (RCSB) GDP ID: 2QMJ [13] then prepared using PyMOL software. The preparation process involves removing water molecules and ligands attached and adding hydrogen atoms. The 3D structure of ligands (lunacridine, skimmianine and scopoletin) was downloaded from Pubchem, then the ligand structure was minimized using PyRx 0.8 software.

2.2. Molecular docking

The docking process begins with the preparation of ligands and receptors. the next step is setting the gird box on the active site of the receptor to get the best binding position between the ligand-receptor. Docking results are stored in PDB format while affinity binding values are saved in Microsoft Excel format. The ligand and receptor interactions formed were visualized using LigPlot +, PyMOL and Discovery studio.

2.3. Pharmacokinetics analysis

The pharmacokinetic analysis includes the test of the human intestine absorbs (HIA) and toxicity test. The ligand structure stored in the molfile format (* mol) is inputted to the online PreADMET software (https://preadmet.bmdrc.kr/adme/), then submitted [14,15]. The ability of ligands to pass through cell membranes and interact with target proteins was analysis based on the Lipinski Rule of Five. The 3D ligand structure is inputted to the SCFBIO Server (Http: //www.scfbio ittd.res.in/software/drugdesign/lipinski.jsp) and then run to get the data [16].

3. Result and Discussion

3.1. Activity of α-glucosidase inhibitors from the active compound Rutaceae

The docking results showed that lunacridine, scopoletin, and skimmianine were able to inhibit the α-glucosidase enzyme activity. Lunacridine has the highest inhibitory activity compared to scopoletin and skimmianine (Table 1). This can be seen from the lunacridine binding affinity value of -6.1 and still under the acarbose control (-7.8). The binding affinity value shows the lowest energy used by receptors to interact with ligands. The more negative the affinity binding value is formed, the more stable the interaction ligand and receptor because the bonds formed are stronger [11,17].

Table 1. Binding affinity value, prediction of HIA and toxicity lunacridine, scopoletin and skimmianine from Rutaceae.

Ligand	Binding affinity value	HIA (%)	Toxicity
Lunacridine	-6,1	95,91	Negative
Scopoletin	-5,8	93.92	Negative
Skimmianine	-5,4	97,89	Negative
Acarbose	-7,8	0,00	Negative

The type and number of bonds formed between the ligand and the receptor affect the binding affinity value. Hydrogen bonds are factors that affect protein stability because it strengthens the interaction effect between proteins and ligands [18]. The more hydrogen bonds, the binding affinity value is higher. Lunacridine interacts with 2 amino acids namely Asp 203 and Asp 443 through hydrogen bonds. Skimmianine only interacts with 1 amino acid (Asp A542) and scopoletin not interact. Hydrogen bonds have more energy than hydrophobic bonds. The interaction will be stronger if the ligand interacts with key amino acid residues from α-glucosidase namely Asp A203, Asp A443, Tyr A299, Trp A406, Met A444, Phe A450, Lys A480, Asp A542, Phe A575, Gln A603 [19]. Acarbose as a controlled drug bound to Asp A542 key amino acid with a distance of 3.43A and Asp A443.

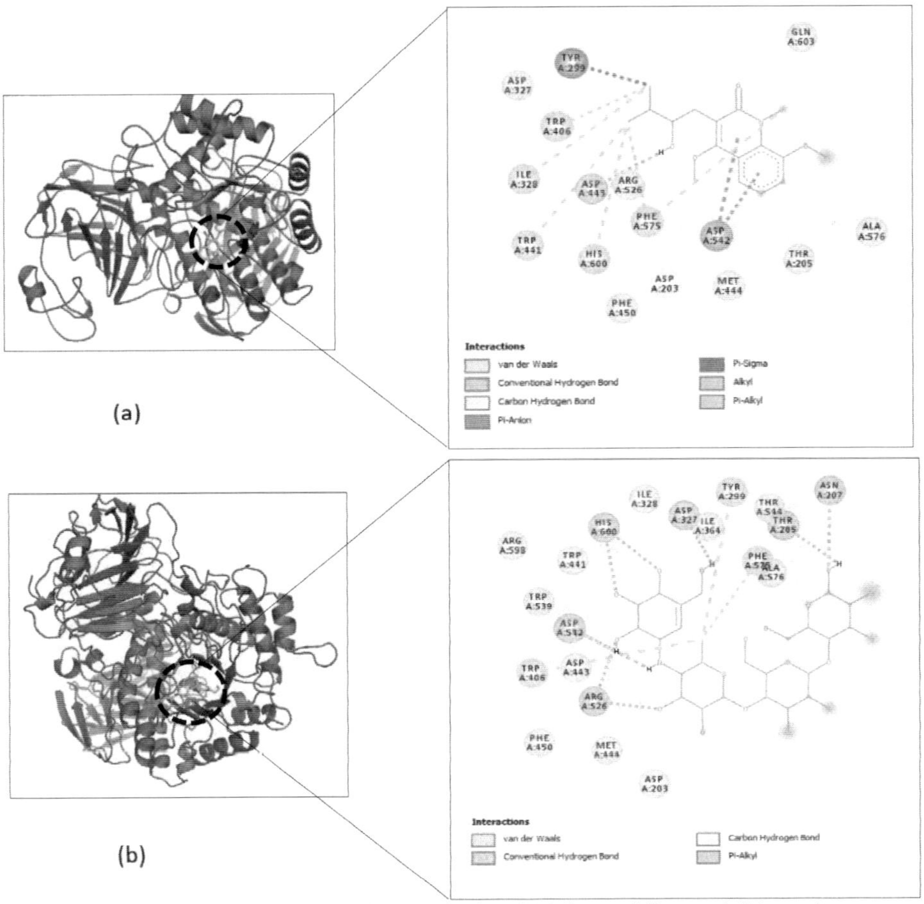

Figure 1. Molecular interaction analysis of (a) lunacridine and (b) acarbose.

3.2. Pharmacokinetic analysis of lunacridine, skimmianine and scopoletin as drug candidate

The pharmacokinetic analysis of lunacridine, skimmianine and scopoletin using PreADMET software states that all three are well absorbed by the small intestine (Table 1). The value obtained above 70% indicates that active compounds can be absorbed high in the intestine [20]. The ability of a drug to pass through cell membranes and interact with target proteins is important in the treatment of a disease. The results of the analysis based on the Lipinski Rule of Five states that lunacridine, skimmianine, and scopoletin can pass through cell membranes. They have a molecular weight of ≤ 500 Da, the number of H-donors ≤ 5, the number of H-acceptors ≤ 10, and Log P ≤ 5 (Table 2). Acarbose as a controlled drug cannot pass through the cell membrane because it is not in aby following with the provisions of the Lipinski rule of five. Ligand not in by following accordance with Lipinski rule parameter, it is most likely to cause problems when it enters the body [21].

Table 2. Lipinski rule of five (Ro5) of Lunacridine, Scopoletin and Skimmianine from Rutaceae.

Ligand	Molecule weight (Da)	H-donor	H-acceptor	Log P
Lunacridine	305	1	5	2.436100
Scopoletin	192	1	4	1.333000
Skimmianine	259	0	0	3.006799
Acarbose	645	14	19	-8.564498

Molecular weight ≤ 500 Da, H-donor ≤ 5, H-acceptor ≤ 10, Log P ≤ 5 [22]

Ligands with a molecular weight of less than 500 Da are easier to cross cell membranes because they do not interfere with the diffusion process. The number of H-donors and H-acceptors is related to the hydrophobicity and absorption of water. Many H-donor bonds cause the solubility of ligands in water to increase so that absorbance in cell membranes is limited, thus inhibiting ligands from interacting with targets [23]. Log P is also related to ligand hydrophobicity in crossing the cell membrane. The higher the hydrophobicity of a drug the ability to exit across the cell membrane is reduced. Thus ligands will be toxic because they are longer in cells [24].

4. Conclusion

Based on the results of the study can be concluded that the compound lunacridine from the family *Rutaceae* is a promising candidate as an α-glucosidase inhibitor. The active compound from *Rutaceae* has potential as an anti-diabetic drug candidate.

Acknowledgements

We thank Mr. Didik Huswo Utomo and Inbio team for his help and guidance for *in silico* study during the manuscript preparation.

References

[1] Shihabudeen M S H, Priscilla D H and Thirumurugan K 2011 Cinnamon extract inhibits α-glucosidase activity and damps postprandial glucose excursion in diabetic rats. *Nutr. Metab.* **8** 1-11

[2] Van D L 2008 A-glucosidase inhibitors in the early treatment of type 2 diabetes *Vasc Health Risk Manag* **4** 1189-95

[3] Prescott T, Ian H S, Robert K and Sutherland K M 2007 Lunacridine from *Lunasia amara* is a DNA intercalating topoisomerase II inhibitor *Journal of Etnopharmacology* **109** 290-4

[4] Macabeo A P G and Alicia M 2008 Chemical and phytomedicinal investigations in *Lunasia amara Phcog Rev.: Plant Review* **2** 317-25

[5] Hasnaeni and Aminah 2019 Uji aktivitas antioksidan dan profil fitokimia ekstrak kayu beta-beta (*Lunasia amara* Blanco) *Galenika Journal of Pharmacy* **5** 101-7

[6] Hasnaeni, Sudarsono, Arief N and Sitarina W 2017 Identification of active anti-inflammatory principles of beta-beta wood (*Lunasia amara* Blanco) from Siawung Barru-South Sulawesi, Indonesia *Trop. J. Pharm. Res.* **16** 161-4

[7] Rafiq S, Rajkumari K, SA Sofi, Nadia B, Fiza N and Gulzar A N 2018 Citrus peel as a source of functional ingredient: a review *Journal of The Saudi Society of Agricultural Science* **17** 351-8

[8] Zubair M S and Subehan 2010 Molecular docking of lunacridine from *Lunasia amara* to DNA: its inhibition and interaction study correlated with the cytotoxic activity on P388 murine leukemia cells, *Indonesian Journal of Cancer Chemoprevention* **1** 108-17

[9] Molnar J, Imre O, Laszlo P, Tannaz G, Judit H and Istvan A 2013 Investigation of the antiproliferative action of the quinolone alkaloids kokusagine and skimmianine of human cell lines. *Current Signal Transduction Theraphy* **8** 148-155

[10] Shi W, Hu J, Bao N, Li D, Chen L and Sun J 2016 Design synthesis and cytotoxic activities of scopoletin-isoxazole and scopoletin-pyrazole hybrids *Europe PMC* **27** 147-151

[11] Zafar M, Khan A, Rauf A, Khan A and Lodhi M A 2016 In silico study of alkaloids as α-glucosidase inhibitors : hope for the discovery of effective lead compounds *Front. Endocrinol.***7** 1-17

[12] Thipparapu G, Ajumeera R and Venkatesan V 2017 Novel dihydropyrimidine derivates as potential HDAC inhibitors: in silico study *Insilico pharmacol.***5** 1-9

[13] Mohapatra S, Prasad A, Haque F, Ray S, De B and Ray S S 2015 In silico investigation of black tea component α-amylase and α-glucosidase and lipase *Journal of Applied Pharmaceutical Science* **5** 042-7

[14] Nisha C M, Kumar A, Nair P, Gupta N, Silakari C, Tripathi T and Kumar A 2016 Molecular docking and in silico ADMET study reveals acylguanidine 7a a potential inhibitor of β-secretase *Advances in Bioinformatics* **2016** 1-6

[15] Luo H, Liang D, Bao M, Sun R, Li Y, Li J, Wang X, Lu K and Bao J 2017 In silico identification of potential inhibitors targeting *Streptococcus mutans* sortase A *International Journal of Oral Science* **9** 53-62

[16] Fitria A, Kholifah H, Umiyatus S, Fitriah and Didik HU 2018 In silico approach for revealing anti-breast cancer and estrogen receptor α inhibitory activity of *Artocarpus altilis* AIP Conf. Proc. *The 8 th Annual Basic Sains International Conference* (Malang) **2021** 1-7

[17] Mohan R and Venogopal S 2013 In silico molecular interaction studies of gamma-hemolysin of *Staphylococcus aureus* with flavonoid compounds *Trends in Bioinformatic* **6** 91-100

[18] Fu Y, Zhao J and Chen Z 2018 Insights into the molecular mechanisms of protein-ligand interactions by molecular docking and molecular dynamics simulation: a case of oligopeptide binding protein *Computational and Mathematical Methods in* Medicine **2018** 1-12

[19] Bharathi A, Selvaraj M R, C S Vasavi, Punnagai M, G A Gayathri and M Gayathri 2014 In silico molecular docking and in vitro antidiabetic studies of dihydropyrimidine [4,5-α] acridin-2-amines *BioMed Research International* **2014** 1-10

[20] Nerkar G, Shena A K, Poorvashree J E and Hemant U C 2012 In silico screening synthesis and pharmacological evaluation of novel quinazolines as NMDA receptor inhibitors for anticonvulsant activity *Inter. J. Pharm. Pharm. Sci.* **4** 449-53

[21] Mendhis S, Puska P and Norrving B 2011 Global atlas on cardiovascular disease prevention and control (Switzerland : World Health Organization)

[22] Lipinski C A, Lombardo F, Dominy B W and D and Feeney P J 2001 Eksperimental and computational approaches to estimate solubility and permeability in drug discovery and development settings *Advance Drug Delivery Reviews* **64** 1-17

[23] Patrick G 2001 *Instant Notes In Medicinal Chemistry* (Oxford: BIOS Scientific Publisher)

[24] Ekins S, Mestres J and Testa B 2007 In Silico Pharmacology for drug discovery: applications to targets and beyond *British Journal Pharmacology* **152** 21-37

Preparation of Nanoparticles from *Curcuma longa* L. and *Cosmos caudatus* Extracts

A S Putri[1], T D Octavianty[1], N T Wahyudi[1] and A Safitri[1,2*]

[1]Chemistry Department, Brawijaya University, Malang, 65145, Indonesia
[2]Research Centre for Smart Molecules of Natural Genetic Resources, Brawijaya University, Malang, 65145, Indonesia

*Corresponding author's email: a.safitri@ub.ac.id

Abstract. This study aims to prepare nanoparticles from *Curcuma longa L* and *Cosmos caudatus* extracts, and to investigate their physico-chemical characterizations. The first step in this research was maceration extraction of *Curcuma longa L* and *C. caudatus,* using ethanol (96%), and formulated into nanoparticles. Nanoparticles were characterized using spectrophotometry UV-Vis, FTIR spectrometry, and SEM (scanning electron microscopy). UV-Vis spectra confirmed that formation of nanoparticles emerged in the UV-vis region around 420 nm. FTIR analysis revealed the existence of functional flavonoids compounds from the extracts, showing at wavenumber 1440-1420 cm^{-1}. From SEM analysis, the formation of nanoparticles results in mostly flake-like morphology with particles size found within the range of 15-30 nm. The biologically synthesized nanoparticles could be of immense use in medical field for their potential as anti-cancer drugs.
Keywords: *Curcuma longa L*, *Cosmos caudatus*, nanoparticles, SEM, FTIR

1. Introduction
Cancer is the leading cause of death worldwide by donating 7.9 million deaths in the year 2007. Most of the medications for cancer treatment available today have limited potential, because it is toxic, inefficient in treating cancer, or very expensive. Effective method with the use of plant extracts as anti-cancer drug, due plant-based products have natural properties, multi-targeted capabilities, as well as affordable and safe compared with synthetic materials, have been proposed [1-2].

Curcumin is a natural yellow pigment isolated from Curcuma herbaceous rhizome which is a mixture of curcuminoids. Curcumin is known to be an antioxidant, anti-inflammatory, anti-fatigue, antiparasitic, anti-allergic, anti-microbial, anti-mutagenic and anticancer. Curcuminoid is a group of phenolic compounds in the rhizome of the family of zingiberaceae, among others: *Curcuma longa* Linn. syn. *Curcuma domestica* Val. (turmeric) and *Curcuma xanthorhiza rox* (*temulawak*) [3]. Curcuminoids are beneficial to prevent the onset of infections of various diseases [4].

Carotenoids are subclasses of phytonutrients that stand out in fruits and vegetables, one of which is *Cosmos caudatus* (*kenikir*). Lutein in *C. caudatus* is one of the types of xanthophyll with zeaxanthin, it has been a concern to be examined regarding what substances are contained therein [1].

The transmission of nanoparticles is described as a formulation of particles dispersed at the nanometer size or scale per thousand micrometer [5]. The nanoparticles drug should be contained in a drug with adequate quantities in the matrix on each particle grain. Nanoparticles in pharmaceutical preparations can be systems of drugs in matrices such as nanosphere and nano capsules nano liposomes, nano emulsions, and as a system combined in scaffold and transdermal transmission [6-7].

In this study, the mixture of turmeric extract and *kenikir* was designed into nanoparticles. The formation of a nanoparticle requires a capping agent and cross-linker agent. Chitosan and TPP (tripolyphosphate) were chosen as capping and cross-linker agents. The nanoparticles, then characterized by Scanning Electron Microscopy (SEM), Fourier Transform Infrared (FTIR), to determine the physical and chemical properties of the nanoparticles produced.

2. Materials and Methods

2.1. Chemicals and instrumentation

The plant specimens of *Curcuma domestica* L. and *Cosmos caudatus* were collected from Materia Medica, Batu, East Java. The plant was identified and authenticated by a plant taxonomist of the herbarium unit, UPT Materia Medica, Batu. The fresh parts of this species were washed under running tap water, shade dried at room temperature and crushed into powder. Other materials were purchased from Sigma-Aldrich: ethyl alcohol (pure, d = 0.789 g/mL), glacial acetic acid (pharmaceutical secondary standard), HCl (37%, analytical grade), NaCl (powder, ≥99.5%, analytical grade), $FeCl_2{\cdot}4H_2O$ (≥99.0%, analytical grade), low-molecular-weight chitosan, and Na-TPP. Water used was demineralized water (Hydrobatt).

Instruments used was UV-Vis spectrophotometer (1601, Shimadzu), FTIR spectrophotometer (Shimadzu FTIR 8400S, sample was analyzed using KBr plate) in the 4000-400 cm^{-1} and Scanning Electron Microscopy (SEM) (FEI, Type: Inspect-S50).

2.2. Extract preparation

Each of the *kenikir* leaves powder and turmeric rhizome powder were macerated with 96% ethyl alcohol. The solvent was completely removed by rotary evaporator at 80 °C ,120 rpm to obtain concentrated extracts. The extracts were collected in dark glass bottles, and stored at 4 °C for further use. *Kenikir* extracts and turmeric extracts were each weighed at 200 mg and dissolved in 10 mL of 96% ethyl alcohol ethanol. The mixture of *kenikir*-turmeric extracts in the ratio of 2:1; 1:1, and 2:1 were prepared.

2.3. Nanoparticles preparation

The nanoparticles were prepared from *kenikir* and turmeric extracts, in the variations of ratio as mentioned above. The mixture of extracts was subsequently dissolved in 35 mL 96% ethyl alcohol, mixed with 15 mL of water, then 100 mL of chitosan solution, then, 1% glacial acetic acid solution was added. A 350 mL of Na-TPP solution was added slowly, accompanied by stirring at a steady speed for 2 h. All the mixture was continued to stir for approximately 2 h at a steady speed. Nanoparticles form mixture of *kenikir* and turmeric extracts were then separated by centrifugation. Then, the nanoparticles were dried using a spray dryer. The resulting nanoparticles were analyzed using SEM, UV-Vis and FTIR.

2.4. Phytochemical screening

Kenikir and turmeric extracts, and mixture of *kenikir* and turmeric extracts were subjected to phytochemical screening to determine the presence of active secondary metabolites using standard procedures, for detection of terpenoids, steroids, flavonoids, phenolic compounds, and tannins [8-10].

3. Results and Discussion

The phytochemical test aims to determine secondary metabolite compounds contained in the turmeric and *kenikir* extracts, and mixture of those extracts. Based on Table 1, the *kenikir* extracts contains flavonoids, terpenoids and steroids; while the turmeric extract contains phenolics, tannins and flavonoids. These results were in agreement with the literatures indicating that *kenikir* contains terpenoids compounds and steroids, whereas turmeric contains phenolic groups and tannins [11-12].

Table 1. Phytochemical screening results of *kenikir* leaves extracts, turmeric extracts, and mixture of *kenikir* and turmeric extracts.

	kenikir Extract	turmeric extract	Mixture of kenikir and turmeric extracts		
			2:1	1:1	1:2
Flavanoid	v	v	v	v	v
Steroid	v	-	v	v	v
Terpenoid	v	-	v	v	v
Phenolic	-	v	v	v	v
Tannin	-	v	v	v	v

- = undetected; v = detected

All samples were measured their absorbance at a wavelength range of 200-800 nm. The UV-Vis spectra for the *kenikir* and turmeric extracts, also the mixture of turmeric and *kenikir* extracts is presented in Figure 1. Based on Figure 1, the maximum wavelength for all samples show the same numbers, at around 420 nm. This is categorized as visible light, and at around 420-430 nm light renders a substance yellow, as shown in all sample solutions that have yellow colour. Interestingly, the absorbance of each sample in this wavelength are different. The *kenikir* extract shows highest absorption at 0.420, the turmeric extract has the lowest absorption at 0.145. The mixture of extracts, however, still indicating that the mixture retains their extract properties. For example, *kenikir* and turmeric extract in 2:1 ratio has the maximum absorption closed to *kenikir* extract only; while *kenikir*:turmeric extract in 1:2 shows maximum absorption that close to turmeric extract only. The 1:1 mixture of extract displays maximum absorption of 0.213, this is in between the maximum absorption of *kenikir* and turmeric extract only.

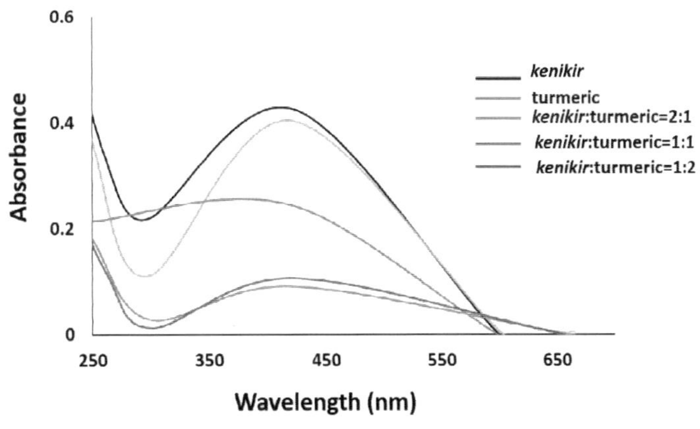

Figure 1. UV-Vis spectra of *kenikir* and turmeric extracts, and mixture of turmeric and *kenikir* extracts on different ratios.

Results of FTIR spectra of *kenikir* and turmeric extracts, and mixture of those extracts can be seen in Figure 2, and peak assignments are presented in Table 2. In turmeric extract only, there were functional groups of alcohol, alkane, alkenes and aromatics; in turmeric and *kenikir* mixture there were similar functional groups, nonetheless, in *kenikir* extract, there was indication of group of terpenes. It can be seen that the mixture of *kenikir* and turmeric extracts showing more intense of FTIR spectra. The broad and intense absorption peak at around 3500-3400 cm^{-1}, corresponds to the O-H stretching vibrations of polyphenolic compounds. Peaks in region 5 (1430-1450 cm^{-1}), also correspond to organic, aromatic compounds, and derivations of these compounds including polyphenols, alcohol or terpenoids, contained in the extracts. Another peak is observed in region no 10 within the range of 980–960 cm^{-1}, in accordance with C-H bending from aromatic compounds, and or/ alkenes.

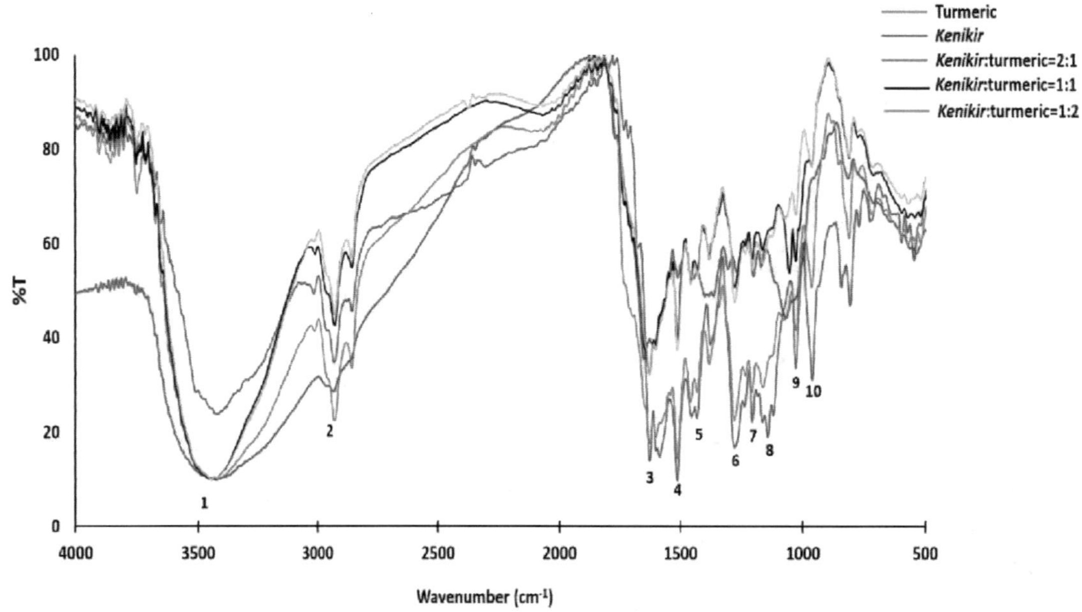

Figure 2. FTIR spectra of *kenikir* and turmeric extract, and mixture of turmeric and *kenikir* extracts on different ratios.

Table 2. Interpretation of FTIR spectra of *kenikir* and turmeric extract, and mixture of turmeric and *kenikir* extracts on different ratios.

Label	Wavenumber (cm^{-1})	Assignment [13-15]	Compound indicated
1	3500-3400	-OH stretching	polyphenols
2	2960-2850	-C-H stretching	aromatics
3	1680-1670	-C=C stretching	alkenes
4	1560-1590	-C-C bending	aromatics ring
5	1430-1450	-CH_3 bending -C-C stretching	alkenes aromatics ring
6	1350-1330	-C-H stretching	alkanes, alcohol, carboxylic acid
7	1250-1230	-C-O- stretching	carbonyl
8	1150-1090	-C-O-C- bending	alcohol
9	1090-1080	-C-O-C asymmetrical stretching	carbonyl, alkenyl
10	980-960	-C-H bending	alkenes, aromatics ring

The morphology of turmeric, *kenikir* extracts, and mixture of *kenikir* and turmeric extracts are presented in Figure 3. From Figure 3, the resulting images of turmeric and *kenikir* extracts only were rough and uneven, with some irregular and large sheet shapes (Figures 3a and 3b). The mixture of *kenikir* and turmeric extract showed that the nanoparticles retained the original extract surface shapes. The morphology of nanoparticles from 2:1 variation (Figure 3c) was more dominant and there are grains-shape attached to the sheet. In the variation of 1:1, composition of the nanoparticles of the sheet and granules look more balanced, thus, the grain attached to the sheet looks more solid (Figure 3d). As in Figure 3e, the morphology of nanoparticles was also similar to that in Figure 3a. Overall, the SEM images of the resulted nanoparticles of *kenikir* and turmeric extracts indicated that the nanoparticles were not optimally formed. This may cause by the process of freeze dry in test samples and coating process in nanoparticles synthesis that was not ideal.

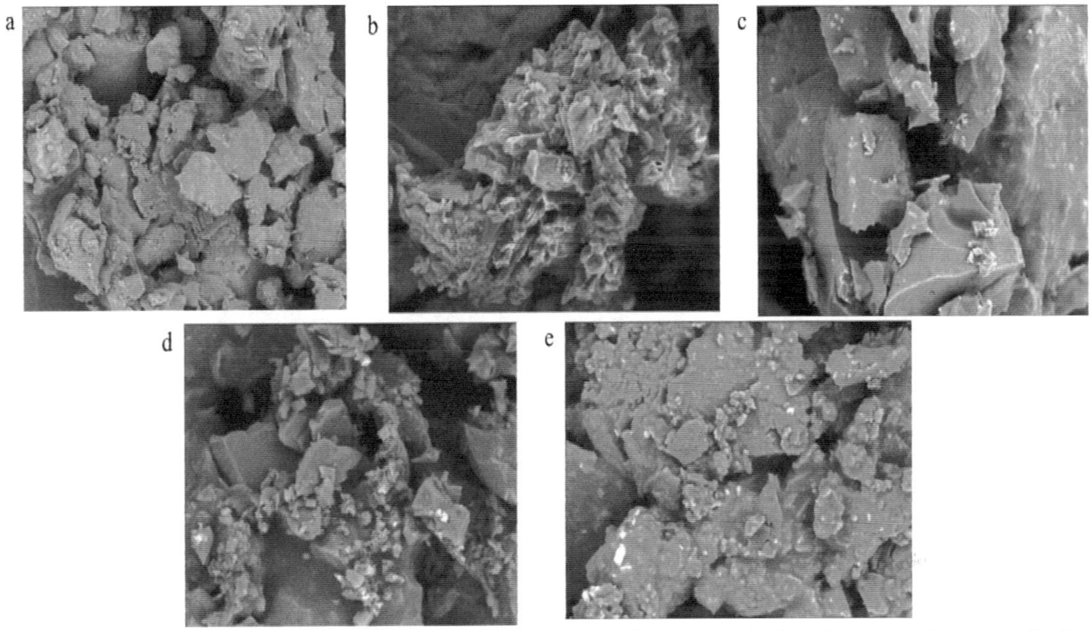

Figure 3. SEM images of the: (a) kenikir extract; (b) turmeric extract; (c) *kenikir*:turmeric extract in 2:1 ratio; (d) *kenikir*:turmeric extract in 1:1 ratio; and (e) *kenikir*:turmeric extract in 1:2 ratio, magnification was 10,000×.

4. Conclusion

Results from this study have shown nanoparticles of mixed *kenikir*:turmeric contains antioxidant, anti-inflammatory, anti-microbial, anti-mutagenic and anticancer compounds. In this study, three variations of the composition of ethanol extract between kenikir and turmeric extract were made by making comparisons of 2:1, 1:1 and 1:2. After characterization, results that the mixing of two plant extracts of *kenikir* and turmeric is able to produce better physical characteristics, compared to *kenikir* or turmeric extract alone.

5. Acknowledgement

This research was supported by the Doctorate Grant from Brawijaya University year of 2019, grant number 39/UN10.F09.01/PN/2019 for Anna Safitri.

References

[1] Ayyadurai N, Valarmathy N, Kannan S, Jansirani D and Alsenaidy A 2013 *Afr. J. Pharm. Pharmacol.* **7** 736-739

[2] Safitri A, Roosdiana A, Rosyada I, Evindasari C A, Muzayyana Z and Rachmawanti R 2019 *IOP Conf. Series: Mat. Sci. Eng.* **509** 012017 (9 pages)
[3] Fadus M C, Lau C, Bikhchandani J and Lynch H T 2017 *Trad. Complement. Med.* **7** 339-346
[4] Boyanapalli S S S, Huang Y, Su Z, Cheng D, Zhang C, Guo Y, Rao R, Androulakis I P and Kong A-N 2018 *Biopharm. Drug. Dispos*. **39** 289-297
[5] Banerjee A, Qi J, Gogoi R, Wong J and Mitragotri S 2016 *J. Control. Release* **238** 176−185
[6] Cho K, Wang X, Nie S, Chen Z and Shin D M 2008 *Clin. Cancer Res.* **14** 1310–1316
[7] De Jong W H and Borm P J 2008 *Int. J. Nanomed.* **3** 133-149
[8] Ismail A M, Mohamed E A, Marghany M R, Abdel-Motaal F F, Abdel-Farid I B and El-Sayed M A 2016 *J. Saudi Soc. Agric. Sci.* **15** 112-117
[9] Evans W C 2009 Trease and Evans' Pharmacognosy E-Book: Elsevier Health Sciences
[10] Harborne J B 2012 Phytochemical Methods: A Guide to Modern Techniques of Plant Analysis: Springer Science & Business Media
[11] Tandi J, Claresta J A, Ayu G and Irwan 2018 *Ind. J. Pharm. Sci. Technol.* **1** 70-80
[12] Pawar H, Karde M, Mundle N, Jadhav N and Mehra K 2014 *Med. Chem.* **4** 588-519
[13] Vargas-León E A, Díaz-Batalla L, González-Cruz L, Bernardino-Nicanor A, Castro-Rosa J, Reynoso-Camacho R and Gómez-Aldapa C A 2018 *Ind. Crops Prod.* **116** 201-208
[14] Weng X, Huang L, Chen Z, Megharaj M and Naidu R 2013 *Ind. Crops Prod.* **51** 342-347
[15] Grasel F.dos S, Ferrão M F and Wolf C R 2016 *Spectrochim. Acta A Mol. Biomol. Spectrosc*. **153** 94-101

The Effect of Topical Toman Fish (*Channa micropeltes*) Extract on Macrophages and Lymphocytes in Diabetes Mellitus Wound Healing

A N Carabelly[1]*, I W A K Firdaus[1], P C Nurmardina[2], D A Putri[2] and M L Apriasari[3]

[1]Department of Oral Biology, Faculty of Dentistry, University of Lambung Mangkurat, Banjarmasin - Indonesia
[2]Faculty of Dentistry, University of Lambung Mangkurat, Banjarmasin - Indonesia
[3]Department of Oral Medicine, Faculty of Dentistry, University of Lambung Mangkurat, Banjarmasin - Indonesia

*Corresponding author's email: amy.carabelly@ulm.ac.id

Abstract. Oral administration of Toman fish extract is proven to accelerate wound healing in rat model of Diabetes Mellitus. Toman fish extract contains albumin which may elevate antioxidant level and reduce inflammation. These changes will later enhance cell proliferation and inflammatory cell activity in wound healing process. This study was aimed to determine the topical application effect of 20% toman fish extract on macrophages and lymphocytes in Diabetes Mellitus wound healing. It was an experimental research using post-test only with control group design. This study was performed on 36 diabetic male wistar rats which was comprised of negative control group without given any treatment, 20% Toman fish extract ointment group and 10% Haruan fish (*Channa striata*) extract ointment group. Rats were sacrificed on day 4, 8, and 14 to investigate wounded area with biopsy technique which later obtained histological preparation using Haematoxylin-Eosin staining. The preparation was then observed using microscope to count macrophages and lymphocytes number. One-Way ANOVA test demonstrated a significant difference in macrophage number on day 4 ($p=0.002$), day 8 ($p=0.000$), day 14 ($p=0.000$). The result of the test also indicated a significant difference in lymphocyte number on day 4 ($p=0.001$), day 8 ($p=0.001$), and day 14 ($p=0.001$). Furthermore, Post Hoc LSD test resulted in a significant difference ($p<0.05$) of macrophage and lymphocyte in day four, eight, and fourteen among all treatment groups. Toman fish extract ointment at 20% concentration for topical application may raise the number of macrophage and lymphocyte than those of negative control and 10% haruan fish extract ointment on day 4 and 8 with gradual decrease on day 14.
Keywords: Albumin, Diabetic wound healing, Lymphocyte, Macrophage, Toman fish extract ointment.

1. Introduction

Diabetes Mellitus is the precipitating factor which induce complication in various organs including oral cavity in the form of xerostomia, gingivitis, periodontitis and candidiasis [1,2]. Hyperglicemic condition promote the decline of cell ability for phagocytosis entailing longer wound healing time [1,2,3]. Indonesia ranks 6th in the world for its high Diabetes Mellitus prevalence and 38.113 South Kalimantan populations were reported by Basic Health Survey to suffer from Diabetes Mellitus in 2013 with gradual increase in 2018 [2,4]. Empirically, Haruan fish (*Channa striata*) extract has been utilized by health practitioner as a basic ingredient in supplement product thus Haruan fish (Channa striata) demand elevates significantly. Alternative compound is mandatory to prevail over the shortage in Haruan fish (*Channa stria*ta) population [5,6].

Haruan fish (*Channa striata*) and Toman fish (*Channa micropeltes*) are categorized in the same genus [6]. Both species contain albumin to accelerate body tissue or cell rehabilitation which undergo destruction [6,7,8]. Albumin is the main protein required for wound healing process [9]. Albumin in inflammation stage acts as an osmotic pressure regulator in circulatory system to prevent the worsening of edema [8]. In diabetic wound inflammation, the number of macrophages increase in day 4 and then reduce in number in day 8 with gradual decrease to day 14 [8,9,11,12]. Notwithstanding, the number of lymphocytes in diabetic wound elevate in day 4 [13,14]. Lymphocytes in diabetic wound healing is essential for antigen binding which activates and produces lymphokine. Phagocytosis process reveals that lymphocyte and macrophage interchangeably stimulate each other in a persistent way to eliminate antigen [13]. Wound healing phase subsequently enters proliferative and maturation phase after phagocytosis until the final achievement of wound closure [15,16].

Haruan fish and Toman fish may accelerate wound healing process in diabetic Wistar Rat. Topical application of 10% Haruan fish (*Channa striata*) extract depicted an acceleration in incised wound healing of diabetic Wistar Rat [17], while topical Toman fish extract at 20% concentration is proven to enhance incised wound closure on the back of Wistar Rat induced with STZ in day 9 [18]. Topical application of Toman fish extract is conducted in the form of ointment. Toman fish extract ointment contains adeps lanae and vaselin flavum as bases to prolong drug contact on the skin so that drug absorption ability will increase [6,18]. According to that, this study is conducted to analyze the effect of Toman fish (*Channa micropeltes*) ointment application at 20% concentration toward macrophages and lymphocytes in diabetic wound on the back of Wistar Rat (*Rattus novergicus*) on day 4, 8 and 14 in vivo.

2. Material and Methods

This research had obtained Ethical Clearance letter No.079/KEPKG-FKGULM/EC/XII/2018 from Faculty of Dentistry, Universitas Lambung Mangkurat, Banjarmasin, Indonesia. This study was performed in several steps which were comprised of fish extract manufacturing and animal treatment at Biochemical Laboratory, Faculty of Medicine, Universitas Lambung Mangkurat, Banjarmasin-Indonesia. Haruan and Toman fish extract oinments were fabricated at Pharmacy Laboratory, Faculty of Mathematics and Science, Universitas Lambung Mangkurat, Banjarbaru-Indonesia. Paraffin blocks were crafted at Anatomical Pathology Laboratory RSUD Ulin, Banjarmasin-Indonesia. Histopathological staining and interpretation were conducted at Histopathology Laboratory, Faculty of Medicine Research Centre, Surabaya-Indonesia. The sample of this study was male Wistar-strain rat, age 2-3 months, 250-300 g body weigh (BW). Inclusion criteria was comprised of healthy rat (active with good appetite) while exclusion criteria was comprised of death or unhealthy rat (limp, weigh loss more than 10%). It was an experimental research using post-test only with control group design. This study was performed on 36 diabetic male Wistar rats comprised of negative control group without given any treatment, 20% Toman fish extract ointment group and 10% Haruan fish (Channa striata) extract ointment group applied topically. Rats were sacrificed on day 4, 8, and 14. Wounded areas were investigated with biopsy technique to produce histological preparation using

Haematoxylin-Eosin staining. The preparation was then observed using microscope to count the number of macrophages and lymphocytes.

2.1. Formulating Haruan and Toman Fish Extract

Toman and Haruan fish were obtained from Ahad Km 7 Market, Banjarmasin, South Kalimantan. Toman and Haruan fish utilized in this research possessed 600-1000g BW [18]. Utilized part was the flesh of the fish. Each sample was scaled and followed by the removal of its digestive organ. Afterwards, the flesh was steamed inside pan for 25-35 minutes under 60oC temperature [19]. Toman fish and Haruan fish flesh was subsequently wrapped with flannel and inserted in hydraulic press instrument for pressing procedure. Derived extract was inserted in reaction tube and was centrifuged for 15 minutes under 6000 rpm. Centrifugation result was separated from contaminating compound. Toman and Haruan fish was kept inside isolated glass bottle and covered using aluminum foil and clean pack [20].

2.2. Manufacturing Haruan and Toman Fish Extract Ointment

The fabrication of Toman and Haruan fish extract ointment was initiated by weighing adeps lanae and vaselin flavum using analytic balance. As much as 16.875 gram adeps lanae and 23.125 gram vaselin flavum were obtained for Toman fish extract while 16.875 gram adeps lanae and 28.125 gram vaselin flavum were prepared for Haruan fish extract. Toman fish extract at 20% concentration and Haruan fish extract at 10% concentration was previously mixed with adeps lanae and gradually poured with the whole volume of Toman fish or Haruan fish extract respectively to the extent that the extract was absorbed by vaselin flavum and digerum homogenously [18].

2.3. Storing Haruan Fish and Toman Fish Extract Ointment

After the mixture of Toman fish and Haruan fish extract to the base homogenously, each extract was poured inside 50 gram ointment pot and stored inside refrigerator.

2.4. Diabetes Mellitus Induction of Wistar Rat

Diabetic rat model was obtained from STZ injection in Wistar rat under 35 mg/kg dosage. Glucose level in rats was examined using glucometer before induced with STZ. Rats were only fed twice a day and glucose level was examined in the respective first three days after STZ induction. Rat was diagnosed with Diabetes Mellitus when glucose level was over 126 mg/dL. Diabetic wistar rat will display limpness and inactivity [8].

2.5. Wound Construction on the Back of Experimental Animal

Trial animal used was white male rat adapted for a week in laboratory environment. Rat was taken and wound design on rat's back was measured prior to the incision. Procedure was commenced by hand-washing and the application of sterile gloves. Sedative procedure was conducted using 5 ml dosage of diethyl ether inhalation that rat was asleep. Fur around the back was shaved in 3 cm diameter and cleaned using 70% ethanol. Incision wound was achieved in 1 cm length and depth limited to dermal layer on the back of Wistar rat using disposable scalpel and blade no 11. Overflowing blood was cleaned using aquadest [8].

2.6. Animal Treatment

Rats were randomly divided into nine groups with each group was given treatment as follow [4]: negative control groups (1-2) were comprised of 4 diabetic rats given feed twice a day only. Positive control groups (4-6) were comprised of 4 Wistar rats given BR2 feed and 10% concentration of Haruan fish extract applied topically using cotton bud twice a day on the back wound of diabetic Wistar rats. Treatment groups (7-9) were comprised of four rats given BR2 feed and 20% concentration of Toman fish extract topically using cotton bud twice a day on the back wound of diabetic Wistar rats. Back wound was subsequently covered by sterile gauze which was moistened with NaCl. Additional dry sterile gauze were presented on it.

2.7. Rat Sacrificing Using Diethyl Ether

Rats from each group were sacrificed on day 4, 8 and 14 to observe macrophage and lymphocyte on wound healing process using inhaled anesthesia of 5 ml diethyl ether. Inhalation process was embarked by putting white rats into covered Beaker bottle so that ethyl evaporation would not likely to occur. The process was preceded until white rat was sentenced death. All sacrificed rats were later buried [21].

2.8. Tissue Collection

Tissues for observation were obtained using excisional biopsy technique. Biopsy was conducted by collecting wound tissue and normal tissue around incised back wound of Wistar rat [20].

2.9. Management of Experimental Animal after Tissue Extraction

Unutilized experimental animal organ were buried. Burying of experimental animal organ was performed by cleaning experimental animal organ which was not deployed. Later on, the organ was covered with a fabric and buried in the depth of ± 25 – 50 cm [8].

2.10. Fabricating Preparation

Each biopsy specimen from each treatment group was collected to fabricate histopathological preparation of back wound with the following step [22]: incised area and normal tissue were fixated in 10% Buffer Neutral Formalin (BNF) less than 24 hours. Fixating result was trimmed to ± 1 cm and put in embedding cassette. Organ put in embedding cassette was inserted into a bucket and was rotated using tissue processing machine ± 18 hours with the following steps: 10% formalin (I) for one hour, 10% formalin (II) for one hour, 85% alcohol for one hour, 90% alcohol (I) for 1 hours, 90% alcohol (II) for one hour, absolute alcohol (I) for two hours, absolute alcohol (II) for two hours, xylol (I) for two hours, xylol (II) for two hours, liquid paraffin (I) for two hours, and liquid paraffin (II) for three hours. Tissue was relocated into base mold or a mold containing liquid paraffin and then attached to embedding cassette until it was cold. The preparation was later stored in the refrigerator. Block was placed in microtome and then cut into 5 microns width. Cutting result was selected carefully and put into a water bath with 37oC-47oC temperature. It was kept until no shrinkage observed and obtained using object glass while dried.

2.11. Haematoxyllin Eosin (HE) Staining

Fabricated preparation was stained using Haematoxyllin Eosin (HE) with respective steps [22]: 1) xylol (I) 5 minutes, 2) xylol (II) 5 minutes, 3) xylol (III) 5 minutes, 4) absolute alcohol (I) 3 minutes, 5) absolute alcohol (II) 3 minutes, 6) 80% alcohol 3 minutes, 7) 70% alcohol 3 minutes, 8) Aquabidest/flowing water till clean, 9) Mayer Haematoxyllin 2-5 minutes, 10) Aquabidest/flowing water till clean, 11) Eosin 5-10 minutes, 12) 70% Alcohol 3 minutes, 13) 80% alcohol 3 minutes, 14) absolute alcohol (I) 3 minutes, 15) Absolute alcohol (II) 3 minutes, 16) xylol (I) 5 minutes, 17) xylol (II) 5 minutes, 18) xylol (III) 5 minutes, 19) xylol (IV) 5 minutes, 20) mounting.

2.12. Observation of Histopathological Preparation

Histopathological preparation was observed using light microscope (IMERCO IM-910B, made in Germany). Macrophage and lymphocyte was calculated under 400 magnifications in 5 section fields observed with inflammatory cell. Identified macrophage and lymphocyte was counted and compared in each negative control given feed only, treatment groups given 20% concentration of Toman Fish extract topically, positive control given 10% Haruan fish extract topically in diabetic rats [21].

Journal of Physics: Conference Series **1374** (2019) 012028 doi:10.1088/1742-6596/1374/1/012028

2.13. Statistical analysis
Statistic data in this study was analyzed using Saphiro-wilk normality test ($p>0.05$) and variance homogeneity of Levene's test. Since data was normally distributed and homogenous, data was then analyzed using One Way ANOVA and Post Hoc LSD through SPSS program.

3. Results and Discussion

Table 1. Mean value and standart deviation.

Groups	**Day-4**		**Day-8**		**Day-14**	
	Macrophages	Lymphocytes	Macrophages	Lymphocytes	Macrophages	Lymphocytes
Toman	12.0 ± 0.81	11.7 ± 0.9	16.0 ± 8.1	15.2 ± 1.2	6.0 ± 0.8	6.2 ± 0.9
Haruan	10.25 ± 0.95	9.7 ± 0.9	13.7 ± 0.95	13.0 ± 0.8	8.75 ± 0.95	8.2 ± 0.9
Control	8.75 ± 0.95	8.0 ± 0.8	11.5 ± 0.81	11.0 ± 0.8	10.25 ± 0.95	10.0 ± 0.8

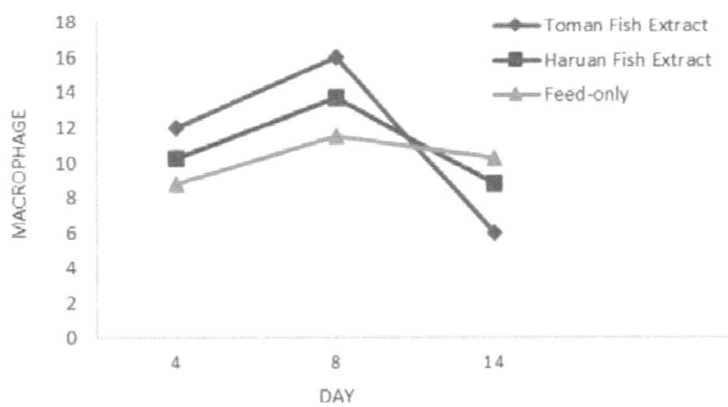

Figure 1. Average number of macrophage cell on back wound of Wistar rats for fourteen days in each treatment.

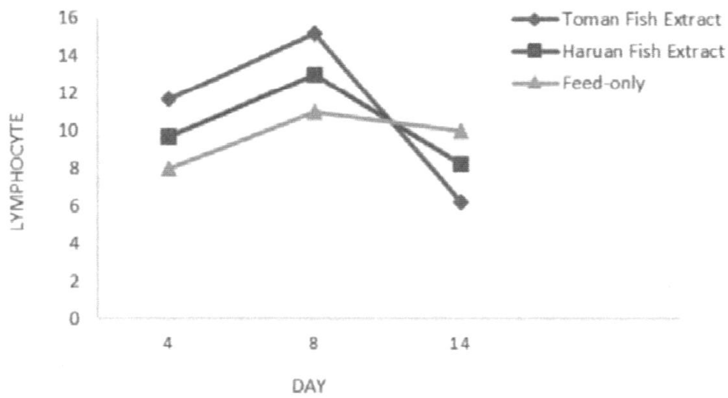

Figure 2. Average number of lymphocyte on back wound of Wistar rats for fourteen days in each treatment.

One Way ANOVA test demonstrated a significant difference in macrophage number on day 4 ($p=0.002$), day 8 ($p=0.000$), day 14 ($p=0.000$). The result of the test also indicated a significant difference in lymphocyte number on day 4 ($p=0.001$), day 8 ($p=0.001$), and day 14 ($p=0.001$). Furthermore, Post Hoc LSD test resulted in a significant difference ($p<0.05$) of macrophage and lymphocyte in day four, eight, and fourteen among all treatment groups.

Macrophages are the primary producers of pro-inflammatory cytokines in wound. Recent wound healing studies have focused on studying macrophage dysfunction in chronic wounds of diabetic humans and mice. Classically activated or M1-like macrophages are known for killing microorganisms and producing pro-inflammatory cytokines. In contrast, the alternatively activated or M2-like macrophages produce anti-inflammatory factors [23]. Lymphocyte in the form of T lymphocyte and B lymphocyte act for antigen paghocytosis in wound healing process [24].

Diabetes mellitus is a condition produced from the elevation of blood glucose level or hyperglycemia. Hyperglycemia may induce disturbance in wound healing process thus wound healing will be delayed [23]. Acceleration of diabetic wound healing process requires an exogenous material, such as albumin. Albumin is proven to accelerate wound healing process and is presented in Haruan and Toman fish ointment. The level of albumin in Toman fish is higher than Haruan where 5.53% unit of albumin were detected in 20% Toman fish extract ointment while 4.53% unit in 10% Haruan fish extract [12]. This induces an increase in macrophage and lymphocytes with the highest number to be found in 20% Toman fish extract ointment group in day 4 and 8 of this study. Albumin is attested to elevate Nuclear Factor Kappa Beta (NF-kβ) thus increase the number and the activity of macrophage [26]. NF-kβ is a transcription factor to regulate pro-inflammatory process. NF-kβ binds inhibitory protein in cytoplasm to prevent its binding with deoxyribonucleic acid (DNA) which is essential in optimizing NF-kβ function through extracellular stimulation. Albumin is an immunomodulator contained in Toman and Haruan fish extract ointment. This compound shall be trapped by macrophage protein receptor and activate NF-kβ which may elevate the number of macrophage [4,27,28].

Albumin may also increase the level of Prostaglandin E2 (PGE2) and prostaglandin D2 (PGD2). PGE2 is increased during initial inflammatory cascade for insult clearance and PGD2 is increased during the active event of inflammatory resolution and healing [29]. Prostaglandin plays a role in activating M1-like macrophages cell [30]. Activated M1-like macrophages will produce interleukin-1 (IL-1) mediator which subsequently stimulates interleukin-6 (IL-6) to increase lymphocyte differentiation and proliferation [31,32]. Albumin has been proven to increase IL-1 and IL-6 in open heart surgery [32]. IL-6 has important regulatory function in the immune system as a mediator of acute-phase response in wound healing process.

It was early recognized as an important cytokine for T and B lymphocyte differentiation. It is involved in early steps of T cell activation via the promotion of T cells proliferation and/or survival, in B cells terminal differentiation via the secretion of immunoglobulins and down-regulation of MHC class II expression, and it is a differentiation factor for cytotoxic T lymphocytes (CTL) [34].

Diabetic wound is confined in a persistent inflammatory state with elevated levels of pro-inflammatory cytokines and proteases, thus resulting in prolonged inflammation. The transition of macrophages from a pro-inflammatory M1-like phenotype to an alternative M2-like phenotype has been suggested as a pre-requisite for the switch from inflammatory to proliferative phase in the healing of the wound [23]. Albumin elevates the number of Transforming Growth Factor-beta (TGF-β) which acts as M1-like macrophages, T lymphocytes, and B lymphocyte growth inhibitor decreasing the number of macrophage and lymphocyte in 20% Toman fish extract ointment group compared to 10% Haruan fish extract application in day 14 [24,34]. TGF-β may also stimulate fibroblast migration, stimulate and inhibit fibroblast proliferation, stimulate fibroblast collagen synthesis and inhibit endothelial cell proliferation which may accelerate proliferative phase on diabetic wound healing process [24]. Supplementation of Channa striatus (Haruan Fish) capsule can also significantly decreased Tumor Necrosis Factor alpha (TNF-α) in the twelfth weeks [35]. TNF-α is a product of innate cells or CD4 and CD8 T cells effector [36]. Bearing a role as a transporter in proliferative stage, albumin carries not only nutrient but also oxygen required for new tissue formation in wound healing so that diabetes mellitus wound healing acceleration can be achieved [37].

4. Conclusion

Toman fish extract ointment at 20% concentration for topical application may raise the number of macrophage and lymphocyte than those of negative control and 10% haruan fish extract ointment on day 4 and 8 with gradual decrease on day 14.

References

[1] Walukow W G 2013 Gambaran Xerostomia pada Penderita Diabetes Melitus Tipe 2 di Poliklinik Endokrin RSUP Prof dr. R. D. Kandou Manado *Journal e-Gigi* **1** (2) 1-5

[2] Latifah N L 2017 Hubungan Durasi Penyakit dan Kadar Gula Darah dengan Keluhan Subyektif Penderita Diabetes Melitus *Jurnal Berkala Epidemiologi* **5** (2) 232-234

[3] Lede M J, Hariyanto T and Ardiyani V M 2018 Pengaruh Kadar Gula Darah Terhadap Penyembuhan Luka Diabetes Melitus Di Puskesmas Dinoyo Malang *Nursing News* **3** (1) 21-24

[4] Audina N, Carabelly A N and Apriasari M L 2018 The Effect Toman (Channa micropeltes) Fish Extract on Epithelial Thickness in Diabetes Mellitus Wound Healing *Dentino Jurnal Kedokteran Gigi* **3** (1) 22-28

[5] Chasanah E, Nurilmala M, Purnamasari A R and Fithriani D 2015 Komposisi kimia, kadar albumin dan bioaktivitas ekstrak protein ikan gabus (Channa striata) alam dan hasil budidaya *Jurnal Pascapanen dan Bioteknologi Kelautan dan Perikanan* **10** (2) 123-132

[6] Andrie M and Sihombing D 2017 Efektivitas Sediaan Salep yang Mengandung Ekstrak Ikan Gabus (Channa striata) pada Proses Penyembuhan Luka Akut Stadium II Terbuka pada Tikus Jantan Galur Wistar *Pharm. Sci Res.* **4** (2) 88-101

[7] Soepribadi I 2013 Regenerasi dan Penyembuhan Untuk Kedokteran Gigi. Sagung Seto, Jakarta p. 11-15

[8] Fajriani N, Carabelly A N and Apriasari M L 2018 The Effect Of Toman Fish Extract (Channa Micropeltes) On Neuthropil In Diabetes Meliitus Wound Healing (In Vivo Study in the Back of Male Wistar Mice (Ratus Novergicus) *Dentino Jurnal Kedokteran Gigi* **3** (1) 15-19

[9] Kusumaningrum G A, Alamsjah M A and Marsitah E D 2014 Uji Kadar Albumin dan Pertumbuhan Ikan Gabus (Channa striata) Dengan Kadar Protein Pakan Komerisal Yang Berbeda *Jurnal Ilmiah Perikanan dan Kelautan* **6** (1) 25-26
[10] Winarsih W, Wientarsih I and Sutardi L N 2012 Aktivitas Salep Ekstrak Rimpang Kunyit dalam Proses Penyembuhan Luka pada Mencit yang Diinduksi Diabetes Melitus *Jurnal Veteriner* **13** (3) 242-250
[11] Sura G M, Carabelly A N and Apriasari M L 2013 Aplikasi Ekstrak Haruan (Channa striata) 100% Pada Luka Punggung Mencit (Mus musculus) *Jurnal PDGI* **62** (2) 44-48
[12] Firlianty 2016 Vacuum Drying Albumin Powder of Snakehead (Channa micropeltes) Potential for Wound Healing from Central Kalimantan, Indonesia *International Journal of ChemTech Research* **9** (5) 263-269
[13] Izzaty A, Dewi N and Pratiwi D I V 2014 Ekstrak Haruan (Channa Striata) Secara Efektif Menurunkan Jumlah Sel Limfosit Fase Inflamasi Dalam Penyembuhan Luka *Dentofasial* **13** (3) 176-179
[14] Apriasari M L and Puspitasari D 2018 Effect of Channa Micropeltes for Increasing Lymphocyte and Fibroblast Cells in Diabetic Wound Healing *Journal of Medical Sciences.* **18** (4) 205-210
[15] Landen N X, Li D and Stahle M 2016 Transition From Inflammation to Proliferation: a Critical Step During Wound Healing *Celuller and Molecular Life Sciences* **73** 3861-3865
[16] Destri C, Sudiana I K and Nugraha J 2017 Potensi Jatropha Multifida Terhadap Jumlah Fibroblast Pada Apthous Ulcer Mukosa Mulut Tikus *Jurnal Biosains Pascasarjana* **19** (1) 6-12
[17] Prastari C, Yasni S and Nurilmala M 2017 Karakteristik Protein Ikan Gabus Yang Berpotensi Sebagai Antihiperglikemik *Jurnal Pengolahan Hasil Perikanan Indonesia* **20** (2) 413-423
[18] Karuniawan A M, Andrie M and Riza H 2016 Uji Efek Penyembuhan Luka Sayat Salep Ekstrak Ikan Toman (Channa Micropeltes) Secara Topikal Pada Tikus Yang Diinduksi Streptozotocin *Jurnal Mahasiswa Farmasi UNTAN* **1** (1) 1-13
[19] Nugroho M 2012 Pengaruh Suhu dan Lama Ekstraksi Secara Pengukusan Terhadap Rendemen dan Kadar Albumin Ikan Gabus (Ophiocephalus striatus) *Jurnal Teknologi Pangan* **3** (1) 64-75
[20] Royyana A, Carabelly A N and Aspriyanto D 2018 The Influence of Toman Fish (Channa Micropeltes) Extract on The Number of Neovascular In Diabetes Mellitus Wound Healing In Vivo Study on The Back Of Male Wister Rat (Rattus Novergicus) *Dentino Jurnal Kedokteran Gigi* **3** (2) 101-107
[21] Khairunnida, Carabelly A N and Apriasari M L 2018 The Effect Of Giant Snakehead (Channa micropeltes) Extract On The Number Of Macrophage In Diabetes Melitus Wound Healing (In Vivo Study on the Back of Male Wistar Rat (Rattus novergicus) *Dentino Jurnal Kedokteran Gigi* **3** (2) 189-193
[22] Harijati N, Samino S, Indriyani S and Soewondo A 2017 Mikroteknik Dasar (Malang: UB Press) p. 85-94
[23] Salazar J J, Ennis W J and Koh T J 2016 Diabetes Medications: Impact on Inflammation and Wound Healing *Diabetes Complications* **30** (4) 746–752
[24] Schaffer M and Barbul A 1998 Lymphocyte function in wound healing and following injury *British Journal of Surgery* **85** 444–460
[25] Andrie M and Sihombing D 2017 Efektivitas Sediaan Salep yang Mengandung Ekstrak Ikan Gabus (Channa striata) pada Proses Penyembuhan Luka Akut Stadium II Terbuka pada Tikus Jantan Galur Wistar *Pharm. Sci. Res.* **4** (2) 88-101
[26] Wheeler D S, Giuliano J S, Lahni P M, Denenberg A, Wong H R and Zingarelli B 2011 The Immunomodulatory Effects of Albumin In Vitro and In Vivo Hindawi Publishing Corporation Advances in Pharmacological Sciences p.1-7

[27] Akrom, Widjaya A and Armansyah T 2015 Ekstrak Etanol Biji Jintan Hitam (Nigella Sativa) Meningkatkan Aktivitas Fagositosis Makrofag Mencit Swiss yang Diinfeksi Lysteria monocytogene *Jurnal Kedokteran Hewan* **9** (2) 94-100
[28] Apriasari M L, Puspitasari D and Carabelly A N 2018 Anti Infalammatory Effect of Toman Fish (Channa Micropeltes) Extract in Wound Healing Process *Jurnal Berkala Kedokteran* **14** (2) 84-88
[29] Frederick E D, Hausburg M A, Thomas G W, Rael L T, Brody E and Bar-or D 2016 The Low Molecular Weight Fraction of Human Serum Albumin Upregulates COX2, Prostaglandin E2 and Prostaglandin D2 Under Inflammatory Conditions in Osteoarthritic knee synovial Fibroblast *Biochemistry and Biophysics Reports* **8** 68-74
[30] Suhartono E and Djati M S 2014 Radikal Bebas dan Intoksikasi Kadmium (Banjarmasin: Pustaka Banua) p 21-23
[31] Macleod A S and Mansbridge J N 2014 The Innate Immune System in Acute and Chronic Wounds *Wound Healing Society* **2** (2) 71-72
[32] Simsek E, Karapinar K, Bugra O, Ulus A T and Sarigul A 2014 Effects of Albumin and Synthetic Polypeptide-coated Oxygenators on IL-1, IL-2, IL-6, and IL-10 in Open Heart Surgery *Asian Journal of Surgery* **37** 93-99
[33] Horn F, Henze C and Heidrich K 2000 Interleukin-6 Signal Transduction and Lymphocyte Function *Immunobiology* **201** 151-167
[34] Wolf G, Schroeder R, Ziyadeh F N and Stahl R A K 2014 Albumin up-regulates the type II transforming growth factor-beta receptor in cultured proximal tubular cells *Kidney International* **66** 1849–1858
[35] Paliliewu N, Datau E A, Matheos J C and Surachmanto E E 2013 Channa striatus capsules induces cytokine conversion in pulmonary tuberculosis patients *Journal of Experimental and Integrative Medicine* **3** (3) 237-242
[36] Mehta A K, Gracias D T and Croft M 2016 TNF activity and T Cell *Cytokine* **101** 14-18
[37] Murdani O J, Andrie M and Taurina W 2016 Uji Efek Penyembuhan Luka Sayat Ekstrak Ikan Toman (Channa Micropeltes) Secara Oral Pada Tikus Jantan Galur Wistar Yang Diinduksi Streptozotocin *Jurnal Mahasiswa Farmasi UNTAN* **1** (1) 1-2

The preventive effect of activated charcoal on HDL levels and aorta histopathological profiles in hypercholesterol rat models

A Roosdiana[1*], D Vidiastuti[2] and H Herenda[2]

[1]Department of Chemistry, Faculty of Mathematics and Natural Sciences, University of Brawijaya, Malang, Indonesia
[2]Faculty of Veterinary Medicine, University of Brawijaya, Malang, Indonesia

*Corresponding author's email: aroos@ub.ac.id

Abstract. Hypercholesterolemia is a condition which the cholesterol levels in the blood exceeds the normal value. The use of hypercholesterolemia drugs or synthetic drugs in a long term will cause severe side effects. Therefore, it is important to explore an alternative preventive agent of hypercholesterolemia derived from natural materials, such as activated charcoal. The activated charcoal can absorb substances thousands times of its own weight and might be an effective and non-toxic material. The research aimed to determine the preventive effect of activated charcoal towards HDL level and aorta's histopathology. This experiment used 24 male rats in randomized design that were divided into six groups: negative control (1); positive control (2), and four groups of treatments with activated charcoal doses of 2.250 mg/kg BW (3), 4.950 mg/kg BW (4), 6.750 mg/kg BW (5) 4.950 mg/kg BW (6), respectively, for 14 days. Simultaneously, all groups were received high cholesterol diet except group 1 and group 6. High Density Lipoprotein (HDL) level was measured by the spectrophotometry method and the data of HDL were analyzed using One-Way ANOVA and followed by the Tukey test ($p<0.05$), while aorta's histopathology were analyzed descriptively. Therapeutic dose of 4.950 mg/kg BW (group 4) showed the best dose in increasing HDL levels by 42% compared to positive control. Activated charcoal reduced the inflammation cells and the defect of aorta's histopathology. Based on the results, activated charcoal prevents the decreasing of HDL level and a defect of aorta's histopathology.

1. Introduction

Hypercholesterolemia is a metabolic disease, which is caused by an elevated total cholesterol level in blood circulation. The adverse effects of an increase in total cholesterol have associated with several life threatening diseases, such as hypertension, atherosclerosis, cardiovascular diseases, metabolic syndrome and obesity. About 40-70% of all human being suffer from hypercholesterolemia world widely [1]. This disease can also occured in pet animal such as dog, 32.8% from 192 American dogs [2] and cats 13% [3].

Hypercholesterolemia can decline of HDL level, because of continuous feeding of high cholesterol diet increase the free radicals in the body. The free radicals inhibit Apo-A1 activity as an apolipoprotein which is required in the formation of HDL. Consequently, the inhibition of apoA-1 activity can decrease the formation of HDL [4]. Another effect of hypercholesterolemia is a damage of aorta. The damage of aortic endothelial cells can be indicated by the formation of random and rough tunica intima, and the appearance of inflammation cells. The occurrence of endothelial cells damage as an inflammation respond to oxidized-low density lipoprotein(Ox- LDL). The Ox-LDL can damage endothelial cell layers and the Ox-LDL infiltration in endothelia are phagocytes by monocytes and macrophages [5].

Currently available treatments for hypercholesterolemia is to regulate a diet that maintains normal body weight and reduce plasma lipid levels. However, in many cases, diet alone will not reduce blood lipid levels. In addition, 75-85% of serum cholesterol is endogenous and the change of diet will only reduce 10-30% of total cholesterol [6]. Another treatment is using a synthetic drug such as statin, the use of this drug in long-term will cause side effects including joint pain and liver damage. In order to overcome this problem, it is necessary to explore alternative treatment based on preventive hypercholesterolemia medication derived from natural product such as activated charcoal.

Activated charcoal is an adsorbent and commonly used as oral medication to treat poisoning and reduce intestinal gas in human. The activated charcoal is not toxic and safe to be used in oral administration because it cannot be digested in gastrointestinal tract [7] Activated charcoal is expected to be able to absorb the excess of cholesterol in the intestine before entering the bloodstream. To date, there have been no reports of the treatment effect of activated charcoal to aorta condition of hypercholesterolemia animal. Therefore, the present study designed to determine the precautionary influence of activated charcoal towards HDL level and aorta's histopathology.

2. Materials and Methods

2.1. Chemicals and Instrumentations

The activated charcoal used in the present study was purchased from PT. Haycarb. Eight to ten weeks old, male Wistar rats at weight range of 100-150 g were acclimatized for 7 days prior to the assay. The rats were housed in fully controlled conditions (room temperature of 25± 2 °C and 12 h of light and dark cycle) The research ethics committee of Brawijaya University approved the study protocol for animals (718-KEP-UB). Twenty-four rats were divided into six groups, each group consisted of 4 rats. Group 1 was the negative control received normal feed, while group 2 was the positive control given high cholesterol diet. Group 3, 4, 5, and 6 were chosen as treatments groups which received activated charcoal at doses of 2.250 mg/kgBW, 4.950 mg/kgBW, 6.750 mg/kgBW, and 4.950 mg/kgBW, respectively, for 14 days. Simultaneously, all treatment groups were received high cholesterol diet except in group 6. At the end of the experiment, all rats were sacrificed, aortic organ and serum were saved for assay examination.

2.1.1. Measurement of High Density Lipoprotein(HDL). Measurement of HDL were carried out by spectrophotometry method. Serum HDL was determined by precipitation methods using various reagents [8].

2.1.2. Aortic histopathological observation. Histopathologic analysis of the rats' liver aortic were prepared using the HE staining method. Aortic histopathological images were captured using the Olympus BX51 microscope with 400× magnification to observe the damage of endothelia and the occurrence of inflammation cells. A digital camera was used to capture aorta histopathology pictures.

2.1.3. Statistical Analysis. Quantitative HDL data levels was analyzed using One-way Analysis of variance (ANOVA), continued to Tukey test multiple comparison tests. The level for statistical significance was set at a p value<0.05. The analysis used SPSS 23.0 for windows; and the aortic histopathologic profiles were analyzed descriptively.

3. Results and Discussion

Hyper cholesterol rat models have been produced successfully by force feeding rats with high cholesterol nutrition. All hypercholesterolemia rats of groups 2 showed total cholesterol levels in blood above 54 mg/dL that were 151 mg/dL, while the healthy rats showed 43.25 mg/dL. It is known that at range of 10-54 mg/dL is categorized as normal value of total cholesterol level in rat blood [9]. The high total cholesterol level is linked to free radical such as reactive oxygen species (ROS). The decrease levels of HDL result from the imbalance of ROS and endogenous anti-oxidant. Groups 3, 4, 5, 6 were preventive treated with activated charcoal and the HDL levels of hypercholesterolemia rats are displayed at **Table 1**.

Table 1. Average HDL Levels in Groups 1, 2, 3, 4, 5, 6 after 14 days of therapy with activated charcoal (AC).

Group	HDL levels (mg/dL)*
1. Negative control (healthy group)	40.74 ± 1.42^{c}
2. Positive control (hypercholesterolemia group)	28.08 ± 0.75^{a}
3. Therapy AC 2.250 mg/kg bw	37.20 ± 0.82^{b}
4. Therapy AC 4.950 mg/kg bw	39.86 ± 0.98^{c}
5. Therapy AC 6.250 mg/kg bw	36.06 ± 0.69^{b}
6. AC 4.950 mg/kg bw	43.02 ± 0.85^{d}

*different letters (a-d) show significant statistical different effect in each group p<0.05

As shown in Table 1, HDL level of hypercholesterolemia rats dropped significantly compared to healthy rat. The standard normal of rat HDL level is ≥ 35mg/dL. The decline of HDL level due to receiving hypercholesterol diet. The high cholesterol intake triggers an increase levels of total cholesterol and LDL due to inadequate HDL level to bring cholesterol ester back to liver. In hypercholesterolemia condition, the body tries to balance plasma cholesterol level by converting cholesterol to bile acids. The synthesis of bile acid involves 7 α-hydroxylase, O_2, NADPH and cytochrome P_{450}; ROS as by product is resulted from this process.

The oral administration of activated charcoal may decrease the high level of total cholesterol and increase HDL level. These were significantly different ($p<0.05$) between HDL level of hypercholesterolemia and therapeutic hypercholesterolemia rats. At the dose of activated charcoal 0.4950 mg/kg body weight is effective dose. This is because the HDL level in group 4 was not different significantly compared to group 1 (healthy rats).

Feeding rats with high cholesterol nutrition, and activated charcoal therapy changed the histopathologic image of rats' aorta. Variations in the histopathologic image of aorta can be seen in **Figure 1,** particularly, the changes of tunica intima. The aorta layer comprised of Tunica intima, tunica media and tunica adventitia. The tunica intima condition of negative control group (Group 1) shows that normal layer of endothelial cells which are neatly arranged, flat shapes with the long axis of the cells parallel to the bloodstream. Tunica media consists of elastic fiber, fibroblast, collagen fiber and smooth muscle, while tunica adventitia is composed of fibro elastic connective tissue and smooth muscle cells [10]. That negative control groups shown as the healthy rats. The aorta of group 2 indicates hypercholesterolemia condition. In tunica intima shows that the damaged endothelial cells release to lumen, while in tunica adventitia occur the inflammation cells amongst fibroblast. Enhancement of histopathologic features of the aorta can be shown in the therapy groups. Group 3 shows a decline of endothelial cell erosion in tunica intima compared to Group 2. In addition, group 4 shows a more reduction of endothelial cells erosion and inflammation cells in tunica adventitia. Group 5 shows that aorta histopathology does not different with group 3. Group 6 received 4.950 mg/kg bw without hypercholesterol diet, that shows no damage in tunica intima and inflammation cells, this condition is similar to negative control groups. This group is carried out to investigate the effect of activated charcoal to aorta histopathology and HDL levels in healthy rat.

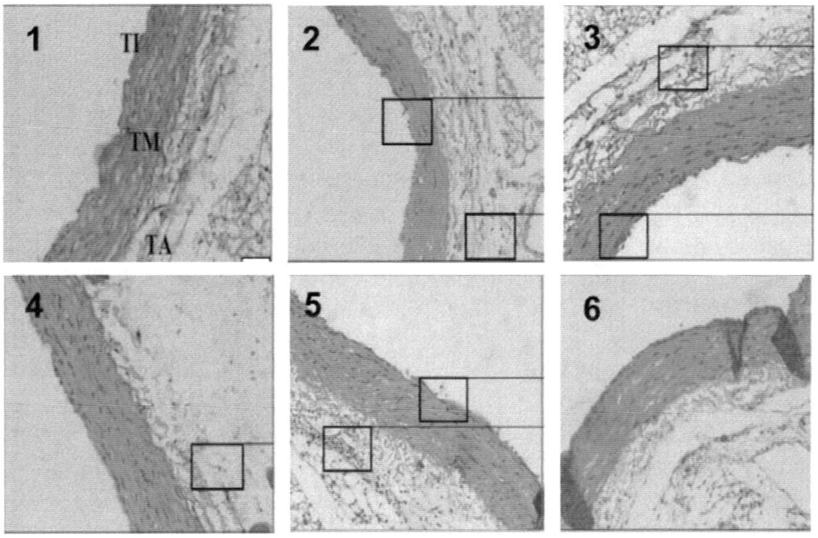

Figure 1. Comparison of histopathologic images of rats aorta from groups 1, 2, 3, 4, 5, 6 after 14 d treatments with activated charcoal with HE staining (400× magnification), TI: Tunica intima, TM: Tunica media, TA: Tunica Adventitia.

Decreases of HDL levels and repair in aorta histopathology of hypercholesterolemia rats treated with activated charcoal can be connected to physical properties of activated charcoal as an adsorbent. Activated charcoal function to inhibit intestine from fully absorbing of cholesterol diet, because The porous of activated charcoal can absorb large molecules such as cholesterol before being absorbed in intestine. The adsorption mechanism of cholesterol takes places through an internal diffusion of cholesterol from a boundary layer to the surface of adsorbent. Some of them are adsorbed on outer surface and most of its further diffuse internally inside adsorbent pore. The activated charcoal cannot be absorbed by intestine or digestive tract; the activated charcoal is only bound to molecules in gastrointestinal lumen. Therefore, activated charcoal are not metabolized in the body and excreted together with feces in the unchanged form [11]. The absorbed cholesterol in activated charcoal will be wasted with feces that can prevent the damage of aorta. The result of SEM shows that activated charcoal in this study has diameter pore of 4.23 µm, while diameter pore of chylomicron of 75-1200 nm or 0.075-1.2 µm [12], that shows activated charcoal has ability to adsorb cholesterol. Based on the results of aorta histopathology observation, activated charcoal can be used as a preventive medication of hypercholesterolemia at effective doses of 4.950 mg/kg body weight.

4. Conclusion

This paper confirmed that activated charcoal at dose of 4.950 mg/kg bw is effective to prevent the decrease of HDL level on hypercholesterolemia animal models using *Rattus norvegicus*. Moreover, the activated charcoal can improve histopathology of its aorta which shown by reducing endothelial cell erosion and inflammation cells. Further study should be conducted in order to investigate the highest and nontoxic dose of activated charcoal.

References

[1] Niho N M, Mutoh M, Takahashi M, Tsutsumi K, Sugimura T and Wakabayashi K 2005 Concurrent Suppression of Hyperlipidemia and Intestinal Polyp Formation by NO-1886, Increasing Lipoprotein Lipase Activity in Min Mice. Tokyo: National Cancer Center Research Institute

[2] Xenoulis P G and Steiner J M 2010 Lipid Metabolism and Hyperlipidemia in Dogs. T*he Veterinary Journal* **183** 12-21

[3] Tapan T 2010 Degenerative Diseases Alex Media Komputido (In Indonesian)

[4] Fielding C J 2007 *High-Density Lipoproteins: From Basic Biology to Clinical Aspects* Wiley-VCH

[5] Davis 2005 Atherosclerosis an Inflammatory Process *Journal Insur Med.* **37** 72

[6] Mahley R W and Bersot T P 2008 Goodman and Gilman's The Pharmacological Basis of Therapeutic

[7] Olson K R 2010 Activated Charcoal for Acute Poisoning: one toxicologist's journey *J. Med. Toxicol.* **6** 190-198

[8] Warnick G R, Benderson J C and Albers J J 1982 DExtran –sulfat Mg2+ Precipitation Procedure for Quantitation of High Density Lipoprotein Cholesterol *Clin.Chem.* **28** 1379-88

[9] Riesanti D G, Padaga M C and Herawati 2012 The level of HDL, LDL and Histopathologic Aorta of Hypercholesterolemia Rattus norvegicus therapies by Hydro Extract of *Dendrophthoepentandra,* Faculty of Veterinary Medicine *(*Malang: University of Brawijaya)

[10] Moini J 2013 *Phlebotomy: Principles and Practice* (Burlington: Jones & Bartlett)

[11] Lehmann J and Joseph J 2009 *Biochar for Environmental Management: Science and Technology* (London: Earthscan)

[12] Feingold K R *and* Grunfeld. C 2015 *Introduction to Lipid and Lipoprotein* https://www.ncbi.nlm.nih.gov/books/NBK305896/

Anti-diabetic properties of root extracts of *Ruellia tuberosa* L: effects on serum enzyme activity

A Safitri[1,2*], A Roosdiana[1], N Arrochmah[1] and S S Nur'Adya

[1]Chemistry Department, Brawijaya University, Indonesia
[2]Research Centre for Smart Molecules of Natural Genetic Resources, Brawijaya University, Indonesia

*Corresponding author's email: a.safitri@ub.ac.id

Abstract. This study aims to determine the anti-diabetic properties of the root extracts of *Ruellia tuberosa L* to enzymes (amylase, protease, and lipase) activities in serum of diabetic rats. Rats were divided into five groups: control, diabetic, and treatment groups with doses of 250, 375, and 500 mg/kg body weights. All treatment groups received root extracts of *R. tuberosa L* for 21 days by oral administration. Results from LC-MS study revealed that the flavonoids compounds including sorbifolin, cirsimaritin, cirsimarin, and cirsiliol 4'-glucoside, were detected in the extracts. The animal study results showed decreases in blood glucose levels by 54.56%, 37.70%, and 16.79%, for treatment doses of 250, 375, and 500 mg/kg body weights, respectively. All enzymes activities increased in diabetic rats, and after treatment the enzyme activities decreased. The decreases in protease activity were 52%, 36%, and 20% for doses of 250, 375, and 500 mg/kg body weights, respectively. In addition, the amylase and lipase activities also showed similar trends with decreasing activities to 63% and 51%, 42% and 37%, 16% and 21%, for treatment doses of 250, 375, and 500 mg/kg body weights, respectively.
Keywords: *Ruellia tuberosa* L, diabetes, serum, protease, amylase, lipase

1. Introduction

Based on our previous research, *Ruellia tuberosa* L n-hexane root extracts had anti-diabetic activity by conducting an *in vivo* study [1]. These included reduce blood glucose levels, reduce malondialdehyde (MDA) levels, and repair on the liver of histopathologic profiles [1]. Moreover, *R. tuberosa* L. hydroethanolic root extracts also had anti-diabetic activity from their positive effects on the pancreatic of diabetic rats [2]. Phytochemical investigation of this study revealed that *R. tuberosa* L. roots extracted with ethanol and water contained mostly flavonoid compounds [3].

In the current work, *R. tuberosa* L. roots are extracted using maceration technique, followed by identification and characterization of the resulted extracts, as previously mentioned in our paper [3]. The anti-diabetic properties of root extracts were determined by their effects on serum enzyme activity. The enzymes investigated were digestive enzymes: amylase, lipase, and protease.

Amylase plays a role in catalyzing the process of starch hydrolysis producing simple molecules such as glucose so that the body can be converted into energy [4]. Lipase is an enzyme that hydrolyzes ester bonds such as triglycerides to free fatty acids and glycerol [4]. Pancreatic damages in diabetic condition may lead to exocrine and endocrine inflammation or pancreatitis [5]. Progressive pancreatic inflammation causes irreversible pancreatic damage and results in exocrine and endocrine dysfunction.

In addition, pancreatic inflammation is also associated with an increase and decrease in amylase and or/ lipase activity [6, 7].

Increased protease enzyme activity can also an indicator of inflammation in people with diabetes. Increased protease activity can be affected by advanced glycosilation end products (AGEs) and release of cytokines. Release of cytokines due to increase of protease activity causes the formation of an immune response activate the inflammatory system; whereas AGEs are formed through the replication process. Increased AGEs damages cells because of disturbing intracellular and extracellular proteins [8, 9].

2. Materials and Methods

2.1. Chemicals and Instrumentation

Chemicals were purchased from Sigma-Aldrich: ethyl alcohol (pure, d = 0.789 g/mL), glacial acetic acid (pharmaceutical secondary standard), HCl (37%, analytical grade), H_2SO_4 (99.99%, analytical grade), $FeCl_3$ (97%, reagent grade), NaCl (powder, ≥99.5%, analytical grade), $KMnO_4$ (≥99.0%, reagent grade), streptozotocin (Bioworld), PBS solution (Phosphate Buffer Saline), 10% trichloroacetic acid (TCA), PBS-Tween, PMSF (Poly Methyl Sulfonil Fluoride), buffer Tris-HCl pH 6.5, phenolphthalein indicator, starch, DNS reagent, tyrosine, casein. The *in vivo* study used male rats (*Rattus norvegicus*), Wistar strain, purchased from Institute of Biosains, Brawijaya University, Malang. The study has obtained a certificate of ethics from the Research Committee of Universitas Brawijaya No: 873-KEP-UB. The plant simplisia of *R. tuberosa* L. were obtained from Materia Medica, East Java. The plant was identified and authenticated by a plant taxonomist of the herbarium unit, UPT Materia Medica, Batu. Instruments used were UV-Vis spectrophotometer (1601, Shimadzu).

2.2. Extract Preparation and Characterization

Extract preparation was reported previously [10]. The identification and characterization of the extract was published in our preceding paper [3].

2.3. In Vivo Study

Rats were divided into five groups: (1) control group; (2) diabetic group; (3) treatment group 1; (4) treatment group 2; and (5) treatment group 3. Rats were kept in the animal house, in Biosains Laboratory, Brawijaya University. Streptozotocin in a dose of 20 mg/kg of body weight was injected intraperitoneally to rats in groups 2 to 5, for 5 days consecutively. Rats in groups 3, 4, and 5 were treated with *R. tuberosa* L. extracts in doses of 250; 375; and 500 mg/kg body weights per day, respectively, for 21 days. Changes in blood glucose levels were monitored per week during the treatment. At the end of the experiment (day 21), rats were sacrificed, and the blood serum was collected and stored at 4 °C for subsequent process.

2.4. Measurement of Serum Enzyme Activity

2.4.1. Enzyme Isolation. PBS-Tween:PMSF (9:1) solution was added to serum, 5× sample volume. The mixture was mixed, then centrifuged for 10 min at 6000 rpm (4 °C). The supernatant was separated, ethanol solution was added to the supernatant with a ratio of 1:1 and left overnight until the precipitate formed. The precipitate then re-centrifuged for 15 min at a speed of 10,000 rpm. The precipitate was dried until all solvent evaporated. The Tris-HCl buffer solution pH 6.5 was added to the precipitate, in 1:1 ratio of volume.

2.4.2. Enzyme Activity Determination. For amylase serum activity, 1% starch was used as enzyme substrate. A 1% starch (0.5 mL) was mixed with serum (0.05 mL), free-reductant water (0.5 mL), and then incubated for 10 min at 42 °C. The DNS reagent was added to the mixture, and boiled for 5 min. The absorbance of the final solution was determined using UV-Vis spectrophotometer at λ 490 nm. The amylase activity was calculated using this equation:

$$\text{enzyme activity } \left(\frac{\mu g}{mL}.\text{min}\right) = \frac{[glucose]}{mol\ weight\ of\ glucose} X \frac{v}{p\ x\ q} x\ dilution\ factor$$

with: v is sample total volume (mL); p is enzyme volume (mL); and q is incubation time (min).

For lipase activity, olive oil was used as substrate. A 2.5 mL of olive oil was transferred into a Erlenmeyer flask. Mixture of 22.5 mg of Arabic gum, 15 mL of $CaCl_2$, 10 mL of 3 M NaCl, and phosphate buffer, pH 7. From this mixture, a 2.5 mL of solution mixture was mixed with 0.5 mL sample, and shaken for 15 min. The mixture was incubated for 6 min, at 30 °C, and boiled for 1 min. The solution was titrated with 0.05 M NaOH, with phenolphthalein indicator. The lipase activity was calculated using this equation:

$$\text{enzyme activity } \left(\frac{\mu g}{mL}.\min\right) = \frac{(Vs - Vb)NaOH \text{ x } M\ NaOH \text{ x } 1000}{\text{enzyme volume x min}}$$

with: Vs is volume of sample titration; and Vb is volume of blank titration

For protease activity, casein was used as substrate. A 500 ppm of casein was mixed with phosphate buffer and sample solution, and incubated for 60 min at 37 °C. A 4% of TCA solution was added and incubated for 30 min at room temperature. The solution mixture was centrifuged at 4000 rpm, for 10 min, at room temperature. Supernatant was separated and diluted 3× sample volume with phosphate buffer, and the sample absorbance was measured at λ 275 nm.

$$\text{enzyme activity } \left(\frac{\mu g}{mL}.\min\right) = \frac{[tyrosine]}{mol\ weight\ of\ tyrosine} X \frac{v}{p\ x\ q} x\ dilution\ factor$$

with: v is sample total volume (mL); p is enzyme volume (mL); and q is incubation time (min).

3. Results and Discussion

One unit of amylase activity is defined as μmol glucose that resulted from hydrolysis of glycoside bond in starch, due to amylase action. Enzyme activity is determined by converting samples absorbance to glucose concentration. Based on Table 1, on diabetic rats (group 2), amylase activity increased 229.46%, with the average of amylase activity was 2.142±0.027 μmol/mL.min. In treated groups, 3 to 5, amylase activity has decreased. In group 3, treated group with 250 mg/kg body weight, has shown the highest decreasing effect, to 3.466±0.047 μmol/mL.min (51.16%). At doses of 375 and 500 mg/kg body weights, amylase activity decreased to 4.477±0.042 (36.65%) and 5.5.79±0.030 (20.94%) μmol/mL.min, respectively.

Table 1. Amylase activity on serum of rats in groups 1 to 5, after treatment with *R. tuberosa* L. root extracts.

No	Group	Amylase activity* (μmol/mL. min)
1	1 (healthy rats)	2.142 ± 0.027[a]
2	2 (diabetic rats)	7.057± 0.046[b]
3	3 (treated rats, 250 mg/kg body weight)	3.466 ± 0.047[c]
4	4 (treated rats, 250 mg/kg body weight)	4.477± 0.042[d]
5	5 (treated rats, 250 mg/kg body weight)	5.579± 0.030[e]

*different letters (a-e) indicate significant statistical effects in each group, at $p<0.05$

Lipase activity on rats from groups 1-5 is presented in Table 2. The lipase activity in diabetic rats group increased up to 381.4%, with the average of 43.125±1.2 μmol/mL. min. In the groups 3, 4, and 5 or treatment groups, there were decline in the lipase activity. Again, at a dose of 250 mg/kg body weight of *R. tuberosa* L. extract, resulted in the highest decrease of lipase activity. At doses of 375 and 500 mg/kg body weights, also had effects in lowering lipase activity, at 41.55% and 15.46%, to 25.208±1.423 and 36.458±2.295 μmol/mL. min, respectively.

Table 2. Lipase activity on serum of rats in groups 1 to 5, after treatment with *R. tuberosa* L. root extracts.

No	Group	Lipase activity* (µmol/mL. min)
1	1 (healthy rats)	8.958 ± 1.8[a]
2	2 (diabetic rats)	43.125± 1.2[b]
3	3 (treated rats, 250 mg/kg body weight)	16.042± 1.5[c]
4	4 (treated rats, 250 mg/kg body weight)	25.028± 2.2[d]
5	5 (treated rats, 250 mg/kg body weight)	36.048± 2.1[e]

*different letters (a-e) show significant statistical effects in each group, at $p<0.05$

Table 3 lists protease activity on rats on groups 1 to 5. Similar trends in protease serum activity with those in amylase and lipase activity have been obtained. In the diabetic conditions, as in group 2, protease activity increased up to %, with the average of 0.068±0.007 µmol/mL. min. Treatments with *R. tuberosa* L. root extracts had effects in decreasing protease activity. These have been shown in the protease activity in rats group 3, 4, and 5. Dose of 250 mg/kg body weight of *R tuberosa* L extract, resulted in the highest decrease of protease activity, at 0.029±0.002 µmol/mL. min or decreased to 52.83%. Doses of 375 and 500 mg/kg body weights, also affected to the protease activity. These doses had decreased protease activity to 36.23% and 20.16%, or to 0.041±0.004 µmol/mL and 0.054±0.001 µmol/mL, respectively.

Table 3. Protease activity on serum of rats in groups 1 to 5, after treatment with *R. tuberosa* L. root extracts.

No	Group	Protease activity (µmol/mL. min)
1	1 (healthy rats)	0.019 ± 0.008[a]
2	2 (diabetic rats)	0.068 ± 0.007[b]
3	3 (treated rats, 250 mg/kg body weight)	0.029 ± 0.002[c]
4	4 (treated rats, 250 mg/kg body weight)	0.041 ± 0.004[d]
5	5 (treated rats, 250 mg/kg body weight)	0.054 ± 0.001[e]

*different letters (a-e) show significant statistical effects in each group, at $p<0.05$

In these results, all digestive serum enzymes activity increased dramatically in diabetic conditions, and decreased after rats treated with root extracts of *R. tuberosa* L. Pancreatic β-cell damage after streptozotocin induction caused decrease insulin production. It is known that insulin as a hormone that facilitates glucose uptake to the cells. In addition, this hormone is also important for the liver, since it helps the liver to absorb glucose and store glucose in the form of glycogen. In the case of glucose deficiency, this induces acinar cells to secrete the amylase enzyme, which catalyzes the hydrolysis of starch into glucose, thus the glucose requirement in the body is still adequately fulfilled [11]. As a result, amylase activity in diabetic rats without any treatments (group 2) increased and had the highest amylase activity among others. Other digestive enzymes, lipase and protease, also have shown the highest activity in diabetic rats. In the diabetic states, lack of insulin secretion activates lipase to hydrolyze triglycerides into fatty acids and glycerol, hence, they produce energy. In addition, increasing blood glucose levels triggers the formation of advanced glycosylation products (AGEs), which are formed in the process of non-enzymatic glycosylation [12]. In AGEs pathway, neutrophils are activated as an indication of inflammation. The formation of AGEs in various biological pathways also lead to the synthesis of cytokines and free radicals. Protease activity is a sign of increased inflammation in cells.

Decreases in enzyme serum activity after treatment with root extracts of *R. tuberosa* L, can be related to the secondary metabolites compounds that contained in the extracts. Extracts of *R. tuberosa* L. contained mostly phytosterol and flavonoids compounds [3, 10]. The phytosterol compounds were including stigmasterol, campesterol, and β-stigmasterol [3]; whereas the flavonoids in the extracts were sorbifolin, cirsimaritin, cirsimarin dan cirsiol 4-glucoside [10].

Phytosterol and flavonoid compounds are suggested act as free radical scavengers. In the molecular structure of flavonoids, the hydroxyl (-OH) groups can release H atoms and bind to free radicals, forming R-H bond, subsequently produce phenoxyl flavonoid radicals (FIO*). These flavonoid radicals are stabilized by C=C double bonds in the flavonoid rings [3]. The proposed mechanism of action of flavonoids compound acting as free radical scavenger is illustrated in Figure 1. In addition, phytosterol compounds (stigmasterol, β-sitosterol, and campesterol) also contributed to the anti-diabetic activity. Stigmasterol improves GLUT 4 translocation which causes an increase in glucose uptake, therefore, glucose levels are reduced and insulin increases [13]. β-sitosterol is also useful in stimulating insulin secretion from the pancreatic-β cells.

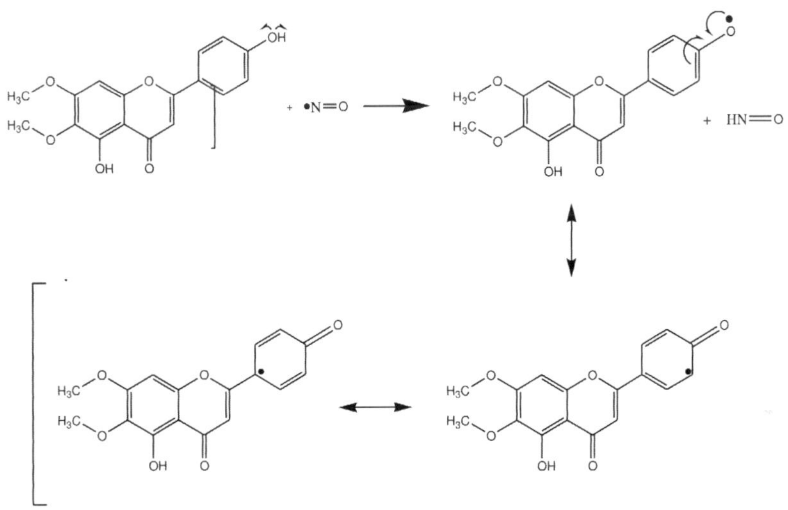

Figure 1. Proposed mechanism of actions of flavonoid as free radical scavengers.

4. Conclusion

The current work has investigated effects of *R. tuberosa* L. root extracts in serum enzymes activity. The root extracts contained steroids, flavonoids, ascorbic acids, phytosterol, and flavonoid compounds. The root extracts of *R. tuberosa* L. have reduced enzymes (lipase, protease, and amylase) activity on the serum of diabetic rats. In summary, root extracts of *R. tuberosa* L. have the potential to be used as a natural cure for diabetes.

5. Acknowledgement

This research was supported by *Hibah Penelitian Unggulan*, Universitas Brawijaya, grant number 697.1/UN10.C10/PN/2019.

References

[1] Kurniawati A N L, Aulanni'am A, Srihardyastutie A and Safitri A 2018 *IOP Conf. Series: Mat. Sci. Eng.* **299** 012022 7 pages

[2] Roosdiana A, Sutrisno S, Mahdi C and Safitri A 2019 *IOP Conf Series: Earth Environ. Sci.* **217** 012041 6 pages

[3] Safitri A, Roosdiana A, Rosyada I, Evindasari C A, Muzayyana Z and Rachmawanti R 2019 *IOP Conf. Series: Mat. Sci. Eng.* **509** 012017 9 pages
[4] Martinez-Gonzalez A I, Díaz-Sánchez A G, de la Rosa L A,Vargas-Requena C L, Bustos-Jaimes I and Alvarez-Parrilla E 2017 *Molecules* **22** (4) 669
[5] Srihardyastutie A, Djoko W S, Fatchiyah F and Aulani'am A 2015 *Biomed. Res.* **26**
[6] Sosenko J M, Krischer J P and Palmer J P 2008 *Diabet.Care* **31** (3) 528-533
[7] Grunnet L G and Mandrup-Poulsen T 2011 *Diabetes* **60** (3) 697-699
[8] Kaur J 2014 *Cardiol. Res. Prac.* **2014**
[9] Forbes J M and Mark E C 2013 *Amer. Physiol. Soc.* 137-188
[10] Ramadhan M, Sabarudin A and Safitri A 2019 *IOP Conf. Series: Earth. Environ. Sci.* **239** 012028 9 pages
[11] Aiyer P V 2005 *African J. Biotechnol.* **4** (13) 1525-1529
[12] Purnomo Y, Djoko W S, Sutiman B S and Widodo M A 2015 *Asian Pas. J. Trop. Biomed.*1-5
[13] Wang J, Mi H, Jie Y, Xinhua M, Sijian Z, Shihao D, Yun H, Xinzhou Y and Ping Z 2017 *Food Nutr. Res.* **61** (1)

Total and Differential Cells of Leukocyte Mice (*Mus musculus*) On Evaluation In Vivo Anticancer Extracts Ethanol Marine Sponges *Aaptos suberitoides*

A P D Nurhayati[1*], R Pratiwi[2] and Soekardiman[3]

[1]Biology Department, Faculty of Mathematic and Natural Science, Sepuluh November Institute of Technology Surabaya (ITS)
[2]Biology Faculty, Gadjah Mada University, Yogyakarta
[3]Pharmacy Faculty, Airlangga University, Surabaya

*Corresponding author's email: awiknurhayati@gmail.com

Abstract. The aim of this research is to know activity of ethanol extract from marine spons *Aaptos suberitoides* in total number and differentiation leukocyte cells which induce by carcinogenic agent Benzo(a)pyrene. METHODS. Mice (*Mus musculus*) were grouped in 6 group, group I (healthy mice), group II (induce by CMC Na), group III (treatment by cyvlophospamide), group IV (administered by spons extract 500 mg/kg BW), group V (administered by spons extract 1000 mg/kg BW) and group VI (administered by spons extract 1500 mg/kg BW). Benzo(a)pyrene induced for 10 days (5 times). The doses that given is 0,3 gram / 0,2 ml CMC Na. Mice were killed and the blood were taken. The sacrificed and blood sampling were taken. The total number and differentiation of leukocyte cells is made by smear method. RESULT AND DISCUSSION.The result of this examination is there were average total number of leukocyte cells in group I was 10,800 cells/μl, II and VI group were 3,700-5,500 cells/μl, the groups III, IV, V of 2,100-5,500 cells μl. The number of neutrophils and lymphocytes showed a significant difference ($P < 0.05$) while the monocyte was no significant difference ($P > 0.05$).
Keywords: *Aaptos suberitoides*, anticancer, total number of leukocytes, leukocyte differentiation

1. Introduction

Indonesia is one of the country who have the most rich of natural resources. Unfortunately, the natural resources in Indonesia was not used properly. One of the natural resources is a sessile benthic organism that has pharmacological activities as anti-microbial, anti-viral, anti-inflammatory and anti-cancer [1]. One of the benthic organism that has high bioactive activity is a sponge. Sponges have the ability to produce high secondary metabolites and synthesize a variety of organic components such as polyketide, alkaloid, peptide and terpene [2].

Sponges also produce toxins and have the prospect of being used in medicine. It has also been reported that some of the compounds isolated from the sponge have high toxic activity to living organisms [3]. *Spongosorites* sp. is a sponge that lives in tropical waters. Sponges *Spongosorites* sp. reported to have bioactive potential [4]. Sponges that taken from Pasir Putih of Situbondo have potential bioactive compounds as anti-cancer *A. suberitoides* is the most toxic sponge with an LC50 value of

134.1362 ± 36.611 [5]. This research is a basic development in the field of health and as a pharmaceutical product for anticancer and immune stimulant.

2. Materials and Methods

2.1. Test Animal Preparation

The experimental animals used were male mice (*Mus musculus*) 3-month-old B albino clone (BALB/c) strain of 24 individuals, grouped randomly in 6 groups. Before the treatment, mice are acclimated in the treated cage for 1 week and were given the same feed and drinking water, namely Par G produced by Comfeed and aquades as drinking water.

2.2. Carcinogenic Induction of Animal Test with Benzo (a) pyrene

The carcinogenic effect treatment is performed by giving a benzo (a) pyrene solution (representing a type of chemical compound that is carcinogenic) to the subcutaneous tissue in the cervical part of the mouse (*M. musculus*). Benzo (a) 0.3 gram powder dissolved in 0.2 ml CMC Na.

2.3. Test of Sponges A. Suberitoides against Animal Test

After the onset of cancer, at week 15 the mice were given anticancer therapy / treatment with sponge extract *A. suberitoides*. Treatment is done orally every day for 2 weeks. The concentration of sponge extract of *A. suberitoides* was 500, 1000, and 1500 mg / kg BW. The dose is based on the dose of drug worthiness in humans. Mice (*M. musculus*) were grouped randomly in 6 groups. Each group received the following treatment:

Group I (healthy mice)
Group II (induce by CMC Na)
Group III (treatment by cyvlophospamide)
Group IV (administered by spons extract 500 mg/kg BW)
Group V (administered by spons extract 1000 mg/kg BW)
Group VI (administered by spons extract 1500 mg/kg BW)

2.4. White blood cell count (Leucocytes)

Blood sampling was performed at week 17. The number of leukocytes was calculated using the modified Klonz [6]. Calculation of leukocyte amount was performed by using Thoma leukocyte pipette. Blood samples treated with anti-coagulant were smoked with pipettes until the "0.5" mark. The pipette then immersed in the Rees-Ecker solution inhaled until the "11" sign so as to obtain 1:20 dilution. The pipette is flipped for about 3 minutes by forming a quarter of the circle, then the first 2-3 drops of blood are removed. Furthermore, the blood drops on the side of the counting room with an angle of 30 °. Room count is allowed one minute which aims to lyse erythrocytes and give leukocyte chance to occupy count room. The leukocyte amount was performed with a 40x magnification microscope on four large boxes of count chambers. The number of leukocytes per cubic millimeter (mm^3) is the number of calculated cells multiplied by 50.

2.5. Type/Differential Types of White Blood Cells (leukocytes)

Blood samples without EDTA were dripped on the object glass and made smear using right hand placed another object in front of the blood drops at an angle of 30-40 ° C. The second object glass was pushed forward to form a thin smear. After drying the smear is fixed with methanol for 3-5 minutes, allowed to dry in the air. The preparations are then stained with giemza solution with 1: 9 dilution for 30 minutes (phosphate buffer pH 6.8-7.2). The preparations were then washed with aquades and allowed to dry on the shelf. After dry preparations are examined under a microscope with 100x magnification calculated each type of leukocytes by battlement method, using a blood counter tabulator. Interrupted count of at least 100 cells and calculated the percentage of leukocyte types. The number obtained is the relative amount of each type of leukocyte from all types of leukocytes.

3. Results and Discussion

3.1. Number of white blood cells (Leukocytes) Mice (M. musculus) on In Vivo Evaluation Anticancer from Sea Sponges Aaptos suberitoides

The total number of leukocyte cells in the Group I treatment remained within the total number of normal leukocyte cells of 5,100 - 11,600 cells / µl) [18]. The total number of leukocyte cells control (Group I) differed significantly (P <0.05) by treatment of Group I, II, III, IV and VI. In control, leukocyte cell amount was greater than other treatments. The total number of leukocyte cells at treatment Group II and K6 ie 3,700-5,500 cells / µl did not differ significantly (Table 1).

Table 1. Total Number of Leukocyte Mice Cells (*Mus musculus*) on In Vivo Evaluation of Anticancer from Sea Sponges *A. suberitoides*.

Repetition	Group (cell/µl)					
	I	II	III	IV	V	VI
1	11.950	4.400	2100	2.250	3.150	3.800
2	10.700	5.500	2000	3.700	3.250	4.150
3	9.750	3.800	4650	3.100	2.750	3.700
Average	10.800^{cd} ± 1.103	4.566,67^{ab} ± 862	2916,67^{a}± .502	3.016,67^{a}± 29	3.050^{a}±2 65	3.883,33^{ab}±23 6

(Description: different superscripts in the same column show significantly different (P <0.05)

3.2. Differentiation White blood cells (Leukocytes) Mice (M. musculus) on In Vivo Evaluation Anticancer from Sea Sponges A. Suberitoides

In leukocyte cell differentiation, the writer only found neutrophils, lymphocytes and monocytes, whereas eosinophils and basophils were not found (Table 2).

The average percentage of neutrophil cells in Group I is 59.67%, still close to the normal range of 60-70% [7]. But statistically, Group I did not differ significantly with Group III and VI (59-73%) (Table 2).

Table 2. Percentage of Mice Neutrophils (*M. musculus*) on In Vivo Evaluation Anticancer from Sea Sponges *Aaptos Suberitoides*.

Repetition	Group (%)					
	I	II	III	IV	V	VI
1	55	36	73	51	50	59
2	64	42	59	42	33	66
3	60	48	65	47	41	71
Average	59.67 ^{cd}± 4.509	42^{a}± 6.00	65.67 ^{cd}±7.024	46.67 ^{ab}± 4.509	41.33 ^{ab}± 8.505	65.33^{cd}± 6.028

(Description: different superscripts in the same column show significantly different (P <0.05

The average percentage of K1 lymphocyte cells was 27.67%, and the value still in the normal lymphocyte cell percentage range of 25-33%. The percentage of lymphocyte cell amount in Group III and VI were 7-17% did not differ significantly. Group II and V do not differ significantly (Table 3).

Table 3. Presentation of Mice Lymphocytes Mice (*M. musculus*) in Differentiation Leucocyte Evaluation In Vivo Anticancer from Sea Sponges *A. suberitoides*.

Repetition	Group (%)					
	I	II	III	IV	V	VI
1	34	56	9	42	44	15
2	28	49	17	42	55	7
3	21	45	8	42	55	11
Average	27.67^{b} ±6.506	50^{de} ±5.568	11.33^{a} ±4.933	42^{cd} ±0.00	51.33^{de} ±6.351	11^{a} ±4.00

(Description: different superscripts in the same column show significantly different ($P < 0.05$)

The percentage of lymphocyte cell amount in the Group III and VI of 7-17% did not differ significantly. Group III and VI have decreased lymphocyte amount (Table 3). On average the percentage of monocyte cells did not differ significantly between treatments (Table 4).

Table 4. Presentation of Monocyte Cells Mice (*M. musculus*) on Differentiation Leucocyte Evaluation In Vivo Anticancer from Sea Sponges *A. suberitoides*.

Repetition	Group (%)					
	I	II	III	IV	V	VI
1	11	8	9	7	6	15
2	8	9	17	16	12	7
3	17	7	8	11	4	11
Average	12	8	11.3333	11.3333	7.3333	11

In cancer-infected mice (Group II) had the same total number of leukocyte cells as mice treated with *A. suberitoides* sponge extract at the highest dose of 1500 mg / kg BW (Group VI). This means that sponge *A. suberitoides* extract with the highest dose has not been effective in the treatment of cancer. While the mice treated with *A.suberitoides* sponge extract at doses of 500 mg / kg and 1000 mg / kg body weight have the total number of leukocyte cells that are relatively the same as the mice induced by cyclophosphamide cancer drugs so that it can be said that the extract of sea sponges *A.suberitoides* At doses of 500 mg / kg and 1000 mg / kg body weight has the same effectiveness as anti-cancer drugs. It is suspected that the extract of *A. suberitoides* sponge is a more effective and useful type of traditional medicine when used at appropriate doses and appropriate use times [8].

In mice occurred damage to the blood-producing system that is bone marrow. All blood cells are formed in the bone marrow so that if the bone marrow is damaged then the production of blood cells will decrease including leucocyte cells. In addition, there are other factors that cause low levels of leukocytes, namely nutrition and stress. Stress over a long period of time will cause weight decreased so bad impact on the immune system. This is because the immune system requires sufficient leptin levels to function properly. Thus a deficiency in fatty deposits can lead to immune deficiency. Poor nutrition can lead to decreased leukopenia and phagocytosis [9].

Sponge activity of *A. suberitoides* at low doses (500 mg/Kg BW and 1000 mg / Kg BW) as anticancer drugs is suspected to contain bioactive compounds. The most commonly found compound in the marine sponge *A. suberitoides* is the alkaloid bioactive compound [10]. There are three alkaloids that can be isolated from sponges *A. suberitoides* are aaptamine, 9-dimetylaaptamine and isoaaptamine [11]. The aaptamine compound can induce the expression of the p21 protein that serves to withstand the cell cycle in G2 / M phase. Protein p21 is a member of the family of kinase inhibitor proteins (Cyclin Dependent Inhibitor Kinase / CDIK) which inhibits cell cycle and its expression is controlled by p53-independent [12]. Protein p53 acts as a regulator of cell proliferation and mediators in apoptosis [13], so that uncontrolled cancer cells may experience apoptosis. If cancer cells experience cell death (apoptosis) then the number of cancer cells will decrease so that the number of leukocytes that respond to the presence of cancer cells will be lower.

In leukocyte cell differentiation, only neutrophils, lymphocytes and monocytes are discovered, whereas eosinophils and basophils are not found. There are no infection or inflammation caused by bacteria. Neutrophils play a major role in the early defense of non-specific immunity against bacterial infections [9]. Similarly, eosinophil cells were not observed. And mice is expected not experience allergies or parasitic diseases. Eosinophils cells play a role against parasitic and allergic diseases [14]. While the absence of basophile cells because the number of basophile cells in the blood circulation is relatively small (0.07%). Basophil cells are precursors to mast cells. In the basophile cell contained heparin (anticoagulant). This heparin is released in an inflammatory area to prevent freezing and static blood and lymphs [15].

Lymphocytes play a role in the immune response, responsible for the presence of antigens or foreign bodies by forming circulating antibodies in the blood or in the cellular immune system [15]. In mice infected with cancer (K2) and treated with A.suberitoides sponge extract at dose 1000 mg / Kg BW (K5) showed no significant difference (44-56%). In cancer-infected mice, an increase in the number of lymphocyte cells is caused by a specific immune response to cancer cells. Cancer cells are not familiar with the cell death program known as apoptosis. Apoptosis (programmed cell death) is a process that runs physiologically in the cell's life through specific molecular signals that play a role to regulate and determine the process of cell death itself [16] [17]. Protein p53 serves as a regulator of cell proliferation and mediators in apoptosis. Loss of p53 gene function or mutation of the gene makes the cell protected from DNA damage, uncontrolled cell growth and death, cancer cell division occurs continuously without apoptosis [13]. So it can encourage the production of lymphocytes to respond to cancer cells.

On average the percentage of monocyte cells did not differ significantly between treatments. It is expected there is no bacterial infection in mice, so there is no phagocytosis. Monocytes are the largest leukocytes that are 15 to 20 μm in diameter and account for 3 to 9% of all white blood cells. Monocytes are formed in the bone marrow, entering the circulation in immature form and undergoing the maturation process into macrophages after entering the tissues [7]. In addition, there is also difficulty in the identification of monocytes in the presence of transitional forms between small and large lymphocytes.

4. Conclusion

The conclusions of this paper are sponge ethanol extract *A. suberitoides* did not significantly affect the number of leukocyte cells and the percentage of basophil cells, eosinophils and monocytes. Extract ethanol sponge *A. suberitoides* effect on lymphocyte and neutrophil cells.

References

[1] Suryati E, Parenrengi A, dan Rosmiati 2000 Penapisan Serta Analisis Kandungan Bioaktif Sponge Clathria sp. Yang efektif sebagai Antibiofouling pada teritif (*Balanus amphitrit*) *Jurnal Penelitian Perikanan Indonesia* **5**

[2] Sjorgen M 2006 Antifouling Activity of Synthesized peptide analogs of the sponge metabolite barettin *Peptides* **27** 2058-2064

[3] El Sayed K A, Hamann M T, Hashish N E, Shier W T, Kelly M and Khan A A 2001 Antimalarial, antiviral, and antitoxoplasmosis norsesterterpene peroxide acids from the Red Sea sponge *Diacarnus erythraeanus J. Nat. Prod.* **64** (4) 522-4

[4] Bao B, Sun Q, Yao X, Hong J, Lee C-O, Sim C J and Jung J H 2007 Cytotoxic Bisindole Alkaloids from a Marine Sponge *Spongosorites* sp *Journal of Natural Products* **68** (5) 711–715
[5] Nurhayati A P D 2009 Skrining, Isolasi, dan Evaluasi In Vivo Antikanker dari Spons Laut LPPM ITS
[6] Klonz A, Wonigeit K, Pabst R and Westermann J 1996 The marginal blood pool of the rat contains not only granulocytes, but also lymphocytes, NK-cells and monocytes: a second intravascular compartment, its cellular composition, adhesion molecule expression and interaction with the peripheral blood pool *Scand. J. Immunol.* **44** 461–469
[7] Handayani W 2008 Asuhan Keperawatan pada Klien dengan Gangguan Sistem Hematologi (Jakarta: Salemba Medika)
[8] Pramono S 2002 Kontribusi bahan obat alam dalam mengatasi krisis bahan obat di Indonesia *Jurnal Bahan Alam Indonesia* **1** (1) 18-20
[9] Baratawidjaja K G 2004 Imunologi Dasar (Jakarta: Fakultas Kedokteran Universitas Indonesia)
[10] Larghi E, Obrist B and Kaufman T 2008 A formal total synthesis of the marine alkaloid aaptamine *Tetrahedron.* **64** (22)
[11] Rombang W A R, Rymond J R, Ponis T, Anthony J H and Lewis N M 2004 Chemical Constituents of Indonesia Marine Sponges: Bisdemethylaaptamine, a Napthyridine Alkaloid from *Aaptos* sp *Buletin of The Indonesian Society of Natural Products Chemistry* **4** (1)
[12] Aoki S, Dexin K, Hideaki S, Yoshihiro S, Toshiyuki S, Andi S and Motomasa K 2006 Aaptamin, a Spongean Alkaloid, Activates p21 Promoter in a p53-Independent Manner *Biochemical and Biophysical Research Communications* **342** 101-106
[13] Seitz C S, Lin Q, Deng H and Khavari P A 2010 Alterations in NF- B function in transgenic epithelial tissue demonstrate a growth inhibitory role for NF-B *Proc. Natl. Acad. Sci. USA* **95** 2307–2312
[14] Hoffbrand V 2006 At a Glance Hematology (Jakarta: EMS)
[15] Frandson R D 1992 Anatomi dan Fisiologi Ternak Edisi ke-4 (Yogyakarta: Gadjah Mada University Press)
[16] Grodzicky T and Elkon K B 2002 Apoptosis: a case where too much or too little can lead to autoimmunity *Mt Sinai J. Med.* **69** 208-219
[17] Wyllie A H 1993 Thermocyte apoptosis induced by p53-dependent and independent pathways *Nature* **362** (6423) 849-852
[18] Heumann D 1983 Human Large Granular Lymphocytes contain an Esterase Activity usually Considered as Specific for The Myeloid Series *Eur. J. Immunol.* **13** 254-258

The Effects of Isolate of Goat Casein Yoghurt Milk to MDA Level and Gastric Histopathology of Rats Exposed to *2,3,7,8-Tetrachlorinedibenzo-P-Dioxin* (TCDD)

C Mahdi[1*], A E P Haskito[2] and M H Robbie[2]

[1]Department of Chemistry, Faculty of Mathematics and Natural Science, Brawijaya University, Malang 65145, Indonesia
[2]Faculty of Veterinary Medicine, Brawijaya University, Malang 65145, Indonesia

*Corresponding author's email : chanifmahdi@gmail.com

Abstract. Dioxins are toxic compounds belonging to organochlorine compounds are highly reactive and produce free radicals that can damage stomach cells. 2,3,7,8-*Tetrachlorinedibenzo-p-dioxin* (TCDD) is the type of dioxin that has the highest toxicity. Casein goat milk yogurt has the potential as an antioxidant. This study aims to determine the preventive effect of goat milk yoghurt on exposure to dioxin observed levels of *malondialdehyde* (MDA) and gastric histopathology. Experimental research using Completely Randomized Design (RAL). The white rats (*Rattusnorvegicus*) were divided into 6 groups, the normal group, casein control (goat milk yoghurt dose 600 mg / kg BW), positive control (TCDD dose 100 ng / kg BW), treatment 1 (goat milk yoghurt 300 mg / kg BW and TCDD 100 ng / kg BW), treatment 2 (goat milk yoghurt 600 mg / kg BW and TCDD 100 ng / kg BW), and treatment 3 (goat milk yoghurt 900 mg / kg BW and TCDD 100 ng / kg BW). Gastric MDA levels were measured using the Thiobarbituric Acid (TBA) assay method with 530 nm wavelength UV-Vis spectrophotometric measurement and gastric histopathology using Hematoxylin-Eosin (HE) staining observed in the form of gastric mucosal cell damage. Gastric MDA levels were analyzed using one way ANOVA statistical test, $\alpha = 5\%$ and gastric histopathology analysis were descriptively processed. The results showed that goat milk yoghurt with dose of 600 and 900 mg / kgBW was the best dose in preventing the increase of MDA level of gastric from white rat (*Rattusnorvegicus*) exposed TCDD significantly ($p < 0{,}01$), and dose 900mg / kgBB prevented erosion of gastric mucosal cells. The conclusion of this research is giving goat milk goat yoghurt able to prevent the increase of MDA level and erosion of gastric mucosal cells caused by TCDD.
Keywords : Dioxin, Casein goat milk yoghurt, Gastric, MDA, Gastric histopathology

1. Introduction

Dioxins are air pollutants that can arise due to various activities such as industrial by-products or chlorine-burning [1]. 2,3,7,8-tetrachlorinedibenzo-p-dioxin (TCDD) is the most toxic type of dioxin among other types of dioxins.

TCDD can cause a variety of health problems because of its persistent nature and can accumulate in body fat, such as heart problems. TCDD is a chemical that produces radical oxygen in the cell nucleus. Free radicals as reactive compounds that can damage cells in the body. If the body is unable to neutralize existing free radicals, these free radicals can interfere with tissue or organ cells [2]. Tissue cells that can be disturbed are the stomach. Malondialdehyde (MDA) is one indicator to measure oxidative stress caused by the accumulation of free radicals [3]. The higher levels of free radicals in an organ, the higher the level of MDA organs [4].

Yoghurt is the product produced from milk fermented with a starter of Lactic Acid Bacteria [5]. The main protein component in milk is casein which is a source of peptides and has a health effect as an antioxidant. Fermented milk (yogurt) has been shown to have higher antioxidant activity than pure milk [6]. The benefit of goat milk casein yogurt is expected to be an alternative to prevent an increase or reduce the number of free radicals due to TCDD exposure.

2. Method

2.1. Chemical material

Preventive provided in the form of casein goat's milk yogurt, which is made from Etawagoat's milk fresh from Surabaya Valenta Goat Milk with *starter* yoghurt (*Yógourmet Yoghurt Starter, LYO-SAN. INC 500 Aéroparc, C. P. 598, Lachute, QC.Canada, J8H 4G4)*, it's containing *Lactobacillus bulgaricus*, *Lactobacillus acidophilus,* and *Streptococcus thermophilus*. Induction of cardiac toxic TCDD as inductors used *2,3,7,8- tetrachloro-dibenzo-p-dioxin (2,3,7,8-TCDD Sigma 48599)*. Giving TCDD diluted with corn oil for every 100 ng / ml TCDD in 100 ml of corn oil.

2.2. Casein manufacture of goat's milk yoghurt

Goat's milk yoghurt was centrifuged at a speed of 12000rpm for 10 minutes at a temperature of 5°C and filtered using a filter paper to separate the casein with water-soluble extract (WSE). The filtrate product in *freeze-dried* and stored at a temperature of -20°C, that process is to maintain the stability of pH.

2.3. Experimental animal

Animals used in this study were white rats (*Rattusnorvegicus*) *Wistar strain*, male, aged 8-12weeks, and weight 150-250 grams were 24 tails. Animals were adapted for seven days to adjust to the conditions in the laboratory.

2.4. Research design

This research was conducted an experimental study using a completely randomized design (CRD). Rats (*Rattusnorvegicus*) were divided into 6 groups, consist of the negative control, control casein (casein yogurt goat milk a dose of 600 mg/kg), a positive control (TCDD dose of 100 ng/kg), treatment 1 (casein yogurt goat milk 300 mg/kg and TCDD 100 ng/kg), treatment 2 (casein yogurt goat milk 600 mg/kg and TCDD 100 ng/kg) and treatment 3 (casein yogurt goat milk 900 mg/kg and TCDD 100 ng/kg). Casein volume goat milk yogurt and TCDD that goes into each rat (*Rattusnorvegicus*) 1 ml. Animals euthanation by cervical dislocation, then surgery by making an incision in the abdomen at the linea alba with open the skin and fascia. Gastric obtained by removing the abdominal contents first and washed with 0.9% physiological NaCl. The gastric cut into two parts. The first part wrapped with aluminum foil, dipped in liquid nitrogen for 10 seconds, then stored in a refrigerator at a temperature of 2°C as a material for examination MDA. The second part is inserted into an organ pot containing formaldehyde for histopathological examination of the stomach.

2.5. Data analysis

Measurement of MDA by TBA method using a gastric organ is cut into small pieces and crushed. Homogenized by centrifuge at speed of 1000rpm for 10 minutes. The supernatant result was measured with a spectrophotometer absorbance atλ = 532 nm. Data obtained from the results of treatment were analyzed by using Microsoft Office Excel and SPSS version 22 for Windows with one-way analysis of variance (ANOVA) and the advanced test Honestly Significant Difference (HSD) α = 5%.Histopathology of the stomach with Hematoxylin-Eosin staining method (HE) were observed the erosion of gastric mucosal cells by using a Nikon microscope H600L, DS Fi2 300 megapixel camera, 400x magnification.

3. Results and Discussion

This study aims to determine the preventive effects of goat milk yoghurt casein administration against gastric MDA levels of white rats (*Rattus norvegicus*) exposed to *2,3,7,8-tetrachlorinedibenzo-p-dioxin* (TCDD) showed significant differences from each treatment group which can be seen in **Table 1**. The test results of MDA with statistical analysis using one way ANOVA test followed by a test or Tukey's Honestly Significant difference at 95% confidence level. The average value of MDA stomach white rats (*Rattus novergicus*) treatment group are presented in **Table 1**.

Table 1. The average levels of Gastric MDA.

Group	The average levels of MDA (ng/ml)
K (–)	1644,50±89,09[a]
KK	1588,38±76,63[a]
K (+)	1928,38±116,10[c]
P (1)	1857,75±97,01[bc]
P (2)	1715,13±77,45[ab]
P (3)	1685,25±80,35[ab]

Information: Value of MDA with different notations indicate significant differences ($p < 0.05$) between the treatment groups.

Group K (-) were significantly different ($p < 0.05$) with group K (+) and group P (1), But not significantly different from group P (2) and P (3) together with a group notationKK, The value of the negative control group is the default value of MDA rat (*Rattus novergicus*) in a normal state in which the mice not given TCDD.Differences from group K (-) and KK, which is in the given preventive casein KK goat milk yogurt with a dose of 600 mg/kg, so hopefully, there is the effect of changes in levels of MDA of preventive given. Group KK significantly different from group K (+) and group P (1), but not significantly different from group P (2) and P (3) and same notation with the group K (-). MDA shown in the test results prove that the state of the body healthy mice still contained the content of free radicals which supported the statement Guyton & Hall (1996), that, concequently of the metabolic processes of biochemical systems (biological oxidation) in the body is able to produce free radicals as much as 2.5% of the total oxygen requirement or as much as 3.4 kg/24 hours.

In the group K (+) significantly different from group K (-), casein control group, the group P (2) and P (3), but not significantly different from group P (1). Indicate that exposure of TCDD will trigger a form Reactive Oxygen Species (ROS) that cause oxidative stress in the stomach that can increase levels of MDA [7].

In group P (1)significantly different from group K (-) and KK, while not significantly different from group K (+), group P (2) and P (3). Based on these data, showed that casein goat milk yoghurt acts as an antioxidant that prevents the increase in MDA levels in the stomach despite the preventive level of 300 mg/kg have not shown the average levels of MDA that approaching MDA levels in the negative control group (A).

Group P (2) not significantly different from group KK, Group C (-), and group P (1), but significantly different from group K (+), and have the same notation as groups P (3). In the group P (3) significantly different from group K (+), Not significantly different from group K (-) and group KK, and have the same notation as group P (2).Based on the results of the *Tukey* test average of 2 treatment groups with 3 treatment groups had the same notation as a negative control group. This shows that the group treated with casein therapy yoghurt 600 mg/kg (E) and 900 mg/kg (F) both have the same effect of decreasing levels of MDA in the gastric organ.

The treatment group therapy with casein yogurt goat milk 600 mg/kg (E) and 900 mg/kg (F) show decreased levels of MDA in gastric organ due yoghurt goat milk contains bioactive peptides that work as antioxidants through inhibition of lipid peroxidation by scavenger (trap) free radicals. This is consistent with the statement Korhonen & Pihlanto [8], that the bioactive peptides goat milk yogurt can reduce levels of MDA through the enzymatic inhibition of lipid peroxidation and nonenzymatic.Superoxide radicals captured by casein bioactive peptides and goat's milk yogurt will prevent the occurrence of the initial formation of lipid radicals that are unstable because of the loss of one hydrogen atom (H) of lipid molecules. Furthermore, it can inhibit the electron transfer molecular oxygen on peroxyl radicals and prevent free radical process propagated so that will not react with oxygen [9].

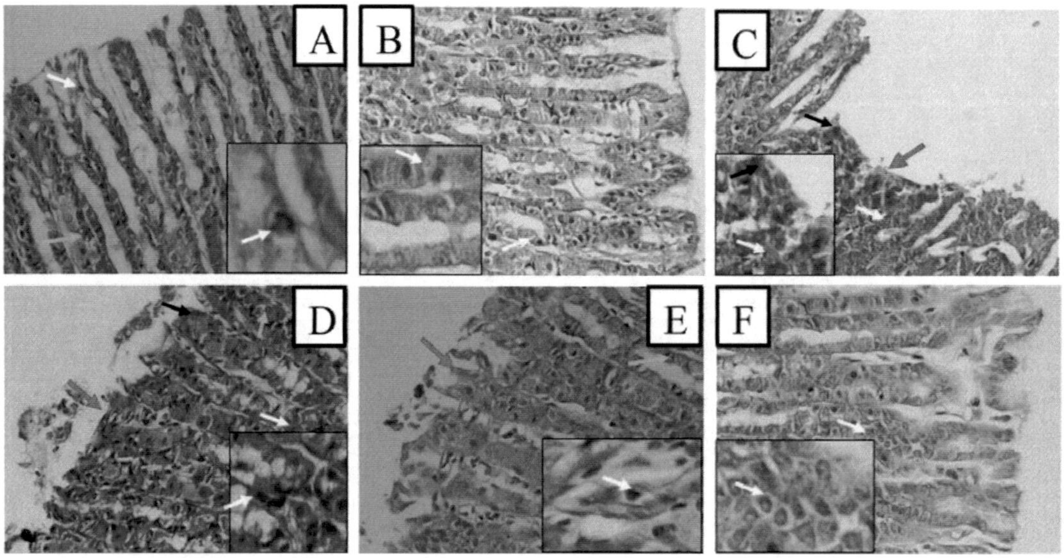

( ) gastric mucosal epithelerosion; ( ) neutrophil cell infiltration; ( ) picnotic necrosis; ( ) chief cell

Figure 1. Gastric histopathological profiles of rats in groups A to F.

Gastric histopathological results of each treatment group showed differences in gastric mucosal cell damage, necrosis pyknotic, neutrophil cell infiltration, and the number of chief cells. Figure of gastric negative control group (healthy mice) (A) shows at the mucosal columnar epithelium simplex seen with parietal cells and chief cells are still arranged and neatly lined with a cylindrical shape simplex. This is in accordance with the opinion from Puspitasari [10], those chief cells are basophilic because many contain mitochondria and granules for producing the enzyme pepsinogen, while the parietal cells are acidophilic because these cells produce HCL or stomach acid. The morphology of the parietal cells and large oval is normal. Also visible is the neutrophil cells in small amounts are still normal because neutrophils have a role to fight against foreign agents that are pathogenic.

Unlike the histopathologic picture in the positive control group (C) which indicates severe cell damage. Compounds TCDD can increase ROS levels found in the stomach. Through the intermediary of ROS, TCDD would cause lipid peroxidation of the cell membrane and will form lipid radicals. Lipid radicals formed binds with oxygen and forming a chain of free radicals, resulting in oxidative stress. TCDD induction of oxidative stress which causes an unbalance between free radicals and antioxidants is an important aetiological mechanism of many diseases [11,12]. Due to the oxidative stress can lead to damage to the gastric mucosa. On the histopathologic picture is seen on epithelial erosions accompanied by inflammatory cell infiltration (neutrophils) compared to the negative control. Erosion on the gastric mucosa as the effects of the free radicals that stimulate white blood cells to the gastric mucosal cells look much on the histopathological picture. White blood cells will produce H_2O_2 to kill some types of bacteria and fungi as well as for cell growth, but he did not attack specific targets so that he will also attack the fatty acids polyunsaturated cross cell membranes, organelles, or DNA, which can cause structural damage and cell function. The visible presence of inflammatory cell infiltration involving inflammatory cells such as neutrophils, lymphocytes, monocytes, and macrophages, but the inflammatory cells that dominate are neutrophils. Erosion on the gastric mucosa as the effects of the free radicals that stimulate white blood cells to the gastric mucosal cells look much on the histopathological picture. White blood cells will produce H_2O_2 to kill some types of bacteria and fungi as well as for cell growth, but he did not attack specific targets, so that he will also attack the fatty acids polyunsaturated cross cell membranes, organelles, or DNA, which can cause structural damage and cell function. The visible presence of inflammatory cell infiltration involving inflammatory cells such as neutrophils, lymphocytes, monocytes, and macrophages, but the inflammatory cells that dominate are neutrophils. Erosion on the gastric mucosa as the effects of the free radicals that stimulate white blood cells to the gastric mucosal cells look much on the histopathological picture. White blood cells will produce H_2O_2 to kill some types of bacteria and fungi as well as for cell growth, but he did not attack specific targets so that he will also attack the fatty acids polyunsaturated cross cell membranes, organelles, or DNA, which can cause structural damage and cell function. The visible presence of inflammatory cell infiltration involving inflammatory cells such as neutrophils, lymphocytes, monocytes, and macrophages, but the inflammatory cells that dominate are neutrophils. White blood cells will produce H_2O_2 to kill some types of bacteria and fungi as well as for cell growth, but he did not attack specific targets, so that he will also attack the fatty acids polyunsaturated cross cell membranes, organelles, or DNA, which can cause structural damage and cell function. The visible presence of inflammatory cell infiltration involving inflammatory cells such as neutrophils, lymphocytes, monocytes, and macrophages, but the inflammatory cells that dominate are neutrophils. White blood cells will produce H_2O_2 to kill some types of bacteria and fungi as well as for cell growth, but he did not attack specific targets, so that he will also attack the fatty acids polyunsaturated cross cell membranes, organelles, or DNA, which can cause structural damage and cell function. The visible presence of inflammatory cell infiltration involving inflammatory cells such as neutrophils, lymphocytes, monocytes, and macrophages, but the inflammatory cells that dominate are neutrophils.

Neutrophils are leukocyte cell types most in parts of white blood cell that is 50-70% among the other leukocyte cell. There are 2 kinds of neutrophils that neutrophils rod and (Stab) and neutrophil segments (polymorphonuclear). Differences stem from both the neutrophil is a young form of neutrophil segments are often referred to as neutrophils core having a horseshoe shape of a horseshoe. Along with the process of maturation, the form will essentially be segmented and segmented neutrophils. According Riswanto [13], the cell has a broad cytoplasm pale pink and purple colored fine granules. Neutrophils serves as the body's defense against foreign substances, especially against bacteria. Neutrophils in the blood circulation of approximately 10 hours and can live for 1-4 days, when in the extravascular tissue. According to Fournier [14], neutrophils are responsible for the process of the withdrawal of other inflammatory cells to localize the damage by releasing pro-inflammatory cytokines, in addition to neutrophils also play a role in mucosal repair. One hour after tissue injury, the neutrophils will be attracted and become the predominant cells in the injured tissue during the first 2 days and the numbers will increase on the second day. Neutrophils will phagocytosis damaged cells and kill bacteria, but it will issue the neutrophil proteases to break down damaged tissue, then the neutrophils will experience apoptosis (programmed cell death) and degradation by macrophages [15].

Damage to the gastric mucosal cells by TCDD also lead to necrosis. Necrosis is cell death induced pathological state and is irreversible [16]. Necrosis is visible is the nucleus of cell necrosis pyknotic which looked dark and wrinkled as DNA correlated into a solid mass. According to Hidayat et al. [17], the exposure to substances that can cause free radicals continuously will result in necrosis and cell damage.

Histopathologic features in the treatment group 1 (D) with therapy casein yogurt 300 mg/kg body weight did not differ much with the picture of the positive control group (C) is an only visible reduction in erosion of the epithelial cells of the gastric mucosa and is still visible necrosis pyknotic although not as much as the positive control group, In the treatment group 2 (E) with casein therapy of 600 mg/kg showed an improvement of epithelial erosion, absence of necrosis pyknotic, and accompanied by a reduction in inflammatory cells when compared to the treatment group 1. In the treatment group 3 (F) with therapeutic doses of casein yogurt 900 mg/kg, pyknotic necrosis and epithelial erosion had decreased significantly showed better results when compared to the treatment group 3 and shows the exact same results with gastric healthy mice.

Histopathologic features in the control group casein (B) with yoghurt casein therapy of 600 mg/kg, showed equally good results with the negative control group. The purpose of the casein control test to see the effects only given casein alone. It has been proven that goat milk yogurt casein act as anti-inflammatory and antioxidant. The content of bioactive peptides and it contains lactic acid bacteria play a role in eliminating microbial pathogens [18]. Goat milk casein which acts as antioxidants peroxyl radical change (the result of lipid peroxidation) become less reactive and break the chain reaction of free radicals [19].

4. Conclusion

The given of preventive therapy casein yoghurt goat milk at a dose of 600 mg/kg and 900 mg/kg has the same effect and the best dosage because it can prevent the increase in MDA levels of gastric rat (*Rattus norvegicus*) exposed to TCDD with decreased levels of MDA approach MDA group normal (K). And a dose of 900 mg kg is the best dose that is capable of preventing damage to the gastric fundus rat exposed to TCDD.

5. Acknowledgment

Researchers like to thank the Public Health Laboratory of Veterinary Medicine Brawijaya University, Brawijaya University Biosciences Laboratory, Laboratory of Physiology Medical Faculty of Brawijaya University, Brawijaya University Medical Anatomical Pathology Laboratory and Laboratory Testing Quality and Food Safety FTP Brawijaya University who have helped in this study.

References

[1] UNEP 2003 *Standardized toolkit for identification and quantification of dioxin and furan releases.* Geneva-Switzerland: Inter-Organization Programme for the Sound Management of Chemicals. Diambil 2 September 2004, darihttp://www.pops.int/documents/guidance/Toolkit_2003.pdf
[2] Price S A and Wilson L M 2006 *Patofisiologi. Ed 6* (Jakarta: EGC)
[3] Arkhaesi N 2008 *Kadar Malondialdehida (MDA) Serum SebagaiIndikator Prognosis Keluaran Pada Sepsis Neonatorum*[Tesis].Program Pascasarjana Magister Ilmu Biomedik dan Program Pendidikan Dokter Spesialis-I Ilmu Kesehatan Anak Universitas Diponegoro Semarang
[4] Luczaj W and Elzbieta S 2003 DNA Damage Caused by Lipid Peroxidation Products *Cellular and Molecular Biologi Letter* **8** 391-413
[5] Shah N, Caballero B, Trugo L C and Finlas P M 2003 Yoghurt: The Product and its Manufacture. *In Encyclopedia of Food Sciences and Nutrition Academic Press*: New York, U.S.A. pp 6252-6259
[6] Liu, Je-Ruei, *et al* 2005 Antioxidative Activities Of Kefir *Asian-Aust. J. Anim. Sci.* **18** (4) 567-573
[7] Doi H, Baba T, Tohyama C and Nohara K 2013 Functional Activation of Arylhydrocarbonrecepror (Ahr) in primary T-cell by 2,3,7,8 tetrachlorodibenzo-p-dioxin
[8] Korhonen H and Pihlanto A 2006 Bioactive Peptides: Production And Functionality *International dairy journal* **16** 945-960
[9] Kullisaar T, Songisepp E, Mikelsaar M, Zilmer K, Vihalemm T and Zilmer M 2003 Antioxidative Probiotic Fermented Goats Milk Decreases Oxidative-Stress-Mediated Atherogenicity in Human Subjects *British Journal of Nutrition* **90** 449-456
[10] Puspitasari D A 2008 *Gambaran Histopatologi Lambung Tikus Putih (Rattus novergicus) Akibat Pemberian Asam Asetil Salisilat* Institut Pertanian Bogor
[11] Ciftci, Osman., Murat, OlcayDisli and Timurkan N 2012 Protective effects of protocatechuic acid

on TCDD-induced oxidative and histopathological damage in the heart tissue of rats. *Toxicology and Industrial Health* **29** (9) 806-811 Department of Pharmaceutical Toxicology, University of Inonu, Turkey

[12] Yoshida R and Ogawa Y 2000 Oxidative Stress Induced by 2,3,7,8-TCDD: An Application of Oxidative Stress Markers to Cancer Risk Assessment of Dioxins *Indust. Health Journal* **38** 5-14

[13] Riswanto 2013 *Pemeriksaan Lab Hematologi* (Yogyakarta: Alfamedia and Kanal Medika)

[14] Fournier B M 2012 *The Role of Neutrophils During Intestinal Inflammation* Epithelial Pathology Research Unit, Department of Pathology and Laboratory Medicine, Emory University School of Medicine, Atlanta, Georgia:USA 5, 354-366

[15] Martin P and Leibovich S J 2005 Inflammatory cells during wound repair: the good, the bad, and the ugly *Trends in Cells Biology* **15** (11) 599-607

[16] Tambayong J 2000 *Patofisiologi untuk Keperawatan* (Jakarta: Buku Kedokteran EGC)

[17] Hidayat A, Christijanti W and Marianti A 2013 Pengaruh Vitamin E terhadap Kadar SGPT dan SGOT Tikus Putih Galur Wistar yang Dipapar Timbal *Unnes Journal Pangandan Argoindustri* **3** (4) 1412-1422

[18] Pessione E and Simona C 2016 Bioactive molecules released in food by Lactic Acid Bacteria: Encrypted Peptides and Biogenic Amines *Journal Frontiers in Microbiology*

[19] Jetawattana S 2005 Malondialdehida (MDA), A Lipid Oxidation Product : Free Radicals In Biology and Medicine *Journal Biology Mediciene* Department of Radiation Oncology Free Radical and Radiation Biology The University of Iowa for 77:222, Spring 2005 Iowa City, IA 52242-1181

Journal of Physics: Conference Series 1374 (2019) 012033 doi:10.1088/1742-6596/1374/1/012033

Anti-inflammatory Activity of Quercitrin on Hypoxia-induced EA.hy926

C N Ginting[1]*, I N E Lister[1], E Girsang[1], M S Mutia[1], Y E P Lubis[1], A Amalia[2], R Rizal[2] and W Widowati[3]

[1]Universitas Prima Indonesia, Jl. Belanga No. 1, Medan 20118, North Sumatera, Indonesia
[2]Aretha Medika Utama, Biomolecular and Biomedical Research Center, Jl. Babakan Jeruk II No 9, Bandung 40163, West Java, Indonesia
[3]Faculty of Medicine, Maranatha Christian University, Jl. Surya Sumantri No. 65, Bandung 40164, West Java Indonesia

*Corresponding author's email: chrismis@unprimdn.ac.id

Abstract. The evidence support the existence of oxygen deprivation involved in pregnancy disorder or preeclampsia. The intensive hypoxia also regulates production of placental pro-inflammatory cytokines. Quercitrin belongs to flavonoid group is known to have antioxidant and anti-inflammatory activity. This study aims to determine the potential of Quercitrin as anti-inflammatory in hypoxia-induced EA.hy926 as preeclampsia model. The cytotoxic assay of Quercitrin against EA.hy926 was conducted using MTS (3-4,5-dimethylthiazol-2-yl)-5-(3-carboxymethoxyphenyl)-2-(4-sulfophenyl)-2H-tetrazolium). The data were analyzed using SPSS ver 20.0 software. One-way analysis of variance (ANOVA) was conducted, followed Tukey HSD post-hoc test and $p<0.05$ was considered to be significant. The anti-inflammatory potential of Quercitrin (25 and 6.25 μg/ml) toward hypoxia-induced EA.hy926 was determined using ELISA to measure IL-10 and TNF-α. Quercitrin (3.13 – 25 μg/ml) were nontoxic to the EA.hy926 cells. Quercitrin (25 and 6.25 μg/ml) were capable to reduce TNF-α compared to positive control (2% O2 hypoxia-induced cells) but no significant increase in IL-10. Our study suggests that quercitrin possess anti-inflammatory properties through suppression of TNF-α in hypoxia-induced EA.hy926 cells.

1. Introduction

Preeclampsia is pregnancy disorder characterized by hypertension, proteinuria and increasing maternal systemic inflammatory response [1]. Preeclampsia affects 5% to 7% of pregnant women each year worldwide. Preeclampsia often ends with premature delivery of the fetus and also a major cause of fetal and perinatal morbidity and mortality [2].

TNF-α is one of the major proinflammatory cytokines invloved in the pathogenesis of chronic inflammatory diseases and is modulated by oxidative stress [3]. A couple of evidence support that tumor necrosis factor-α (TNF-α) has direct contact with maternal endothelial cell in vivo. There was increasing concentration of TNF-α in pregnant women with preeclampsia as compared to normal pregnant women [1]. It is pressumed that according to the recent studies, TNF-α is up-regulated during the preeclampsia's symptomps. TNF-α is produced by variety of cell, including endothelial cells, on appropiate stimulation.

Hypoxia has been suggested to be leading factor of preeclampsia. Under hypoxic condition, numerous genes are activated such as HIF-α and HIF-2α which is essentials for the inflammation process. Hypoxia conditions activates NF-kB which in turn increases the production of TNF-α [4]. Furthermore, inflammation was also occured when cells undergo oxidative stress.

Several studies have been demonstrated that phytochemical compounds possess rich pharmacological properties that play beneficial roles in inflammation-related diseases. Flavonoids are perceived to possess good effects on health. Flavonoids have been shown to reduce TNF-α in human and animal studies and cell culture models. Among flavonoids, quercitrin (3-rhamnosyl-quercetin) is mainly present in nature. Quercitrin is glycosylated forms of Quercetin that can be found commonly in vegetables. In vitro studies have shown quercitrin exerts down-regulation activity in inflammatory response [5].

Therefore, IL-10 plays an important anti-inflammatory role in the resolution of inflammation and has been shown to be upregulated by several flavonoids [6] including hesperidin, quercetin, and catechin. TNF-α and IL-10 are all regulated by the pro-inflammatory transcription factor nuclear factor kappa-light-chain-enhancer of activated B cells (NF-κB), and NF-κB has also been shown to be a target of several flavonoids including quercetin and cyanidin-3-glucoside (C3G) [6]. The aims of the present study was to evaluate the effects of Quercitrin toward TNF-α and IL-10 on hypoxia-induced EA.hy926.

2. Methods

2.1. Cell culture

EA.hy926 (ATCC ® CRL-2922™) cell line were cultured in Biomolecular and Biomedical Research Center, Aretha Medika Utama. The cell lines were grown in Dulbecco's Modified Eagle Medium DMEM high glucose (Biowest L0104) supplemented with 10% fetal bovine serum (FBS) (Biowest S181G), 1% antibiotic-antimycotic 100x (Biowest L0010), 1% Amphotericin B 100x (Biowest L0009). Cells were incubated at 37°C in a humidified atmosphere with 5% CO2. The cells were treated by 25 and 6.25 μg/ml of Quercitrin and incubated in 2% O2, 37°C for another 24 hours. The treatment of control, hypoxic (2% O_2), control DMSO, Hypoxic (25 μg/ml), and Hypoxic (6.25 μg/ml) repeated at 3 times for each group.

2.2. Cytotoxic Assay

EA.hy926 (ATCC ® CRL-2922™) cell line were cultured in DMEM high glucose (Biowest L0104) supplemented with 10% fetal bovine serum (FBS) (Biowest S181G), 1% antibiotic-antimycotic 100x (Biowest L0010), 1% Amphotericin B 100x (Biowest L0009). Cells were incubated at 37°C in a humidified atmosphere with 5% CO2. After the cells reach 80% confluency, 1x104 cells were seeded in each well of 96-well plate. After 24 hours incubation, the cells were treated with various concentration (0 - 100μg/ml at final concentration) of Quercitrin (Chengdu BioPurify BP1192) and incubated in 37°C, 5% CO2 for 24 hours. The viability of the cells were determined using MTS (3-4,5-dimethylthiazol-2-yl)-5-(3-carboxymethoxyphenyl)-2-(4-sulfophenyl)-2H-tetrazolium) (Abcam 197010). MTS was added to each well at a ration 1:10. The plate was incubated in 37°C, 5% CO2 for 3 hours. Absorbance was measured using spectrophotometer (Multiskan GO, Thermo) at 490 nm [7] [8] [9] [10].

2.3. Hypoxia-induced EA.hy926 (2% Oxygen)

Preeclampsia model was generated in vitro using hypoxia induced-EA.hy926. EA.hy926 was incubated in 2% O_2 to mimic preeclampsia condition. The cells were plated in 6-well plate (5 x 105 cells/well) and incubated for 24 h at 37°C, 5% CO2. On the next day, the cells were treated by 25 and 6.25 μg/ml of Quercitrin and incubated in 2% O2, 37°C for another 24 hours. After incubation, the conditioned medium was taken for ELISA assays and centrifuged at 1600rpm for 10 minutes. The supernatant was collected and stored at -80°C for quantification of IL-10 and TNF-α [11] [12].

2.4. Measurement of TNF-α

TNF-α levels in uninduced, hypoxia-induced and Quercitrin treated and hypoxia-induced EA.hy926 were determined using ELISA assay according to the manufacture's kit manual (BioLegend ELISA kit 421701). First, the plate was coated using capture antibody solution and incubated at 4°C overnight. The plate was washed four times using 300 µl of wash buffer, then incubated for one hour in orbital shaker. 50 µl of matrix C and 50 µl of assay buffer was added into each standard and sample well, respectively. Plates were shaken at room temperature and then washed for 4 times. Afterward, 100 µl of the detection antibody solution was added into each well and incubated at room temperature for 1 hr on the orbital shaker. The plate then was washed four times. Subsequently, 100 µl of diluted Avidin-HRP solution was added into each well, incubated at room temperature for 30 min in orbital shaker. The plate was washed again 5 times, then added with 100 µl of substrate solution, incubated for 10 min in the dark room. The reaction was stopped by adding 100 µl of stop solution briefly, and the absorbance was measured by Multiskan GO Microplate Reader at 450 nm [10] [13] [14].

2.5. Measurement of IL-10

IL-10 levels in uninduced, hypoxia-induced and Quercitrin treated and hypoxia-induced EA.hy926 were determined using ELISA assay according to the manufacture's kit manual (BioLegend ELISA kit). The plate was coated using capture antibody solution and incubated at 4°C overnight. The plate was washed four times using 300 µl of wash buffer, then incubated for one hour in shaker. Around 50 µl of matrix C and 50 µl of assay buffer was added into each standard and sample well, respectively. Plates were shaken at room temperature and then washed for 4 times. Afterward, 100 µl of the detection antibody solution was added into each well, incubated at room temperature for 1 hr on the orbital shaker. The plate then was washed four times. Subsequently, 100 µl of diluted Avidin-HRP solution was added into each well, incubated at room temperature for 30 min in orbital shaker. The plate was washed again 5 times, then added with 100 µl of substrate solution, incubated for 10 min in the dark room. The reaction was stopped by adding 100 µl of stop solution briefly, and the absorbance was measured by Multiskan GO Microplate Reader at 450 nm [10] [13] [14].

2.6. Statistical Analysis

The data were analyzed using SPSS ver 20.0 software. One-way analysis of variance (ANOVA) was conducted, followed Tukey HSD post-hoc test and $p<0.05$ was considered to be significant.

3. Results

3.1. Cell Viability

Cell viability was measured using MTS assay based on the transformation of yellow tetrazolium salt to a purple formazan product. Figure 1. showed EA.hy926 viability that has been treated by 0-25 µg/ml Quercitrin were over 80% compared to the control (untreated EA.hy926), indicated that Quercitrin was less toxic to the cells and able for the next assay. Meanwhile, Quercitrin at 50 and 100 µg/ml given EA.hy926's viability less than 80% indicated that the concentration given were quite toxic to the cells.

Journal of Physics: Conference Series **1374** (2019) 012033 doi:10.1088/1742-6596/1374/1/012033

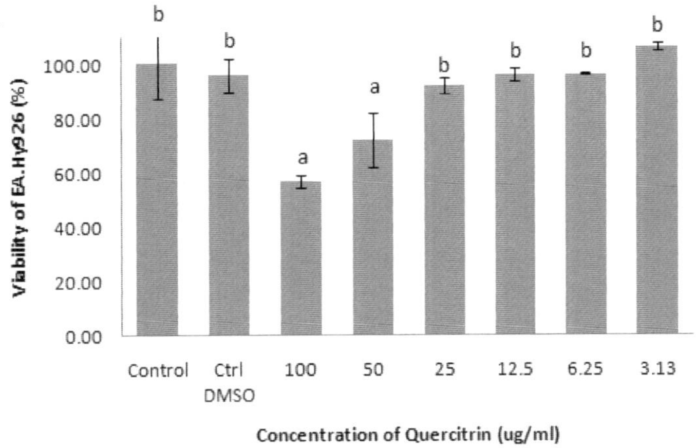

Figure 1. Cytotoxic of Quercitrin in various concentration (3.13-100 μg/ml) towards EA.hy926.

3.2. TNF-α level

A glance at the graphic bar provided reveals the level concentration of TNF-α on EA.hy926 cells which has been generated into preeclampsia model. Of the six Quercitrin's concentration screened at the cytotoxic assays, then two concentrations were chosen. The highest concentrations given significantly reduced hypoxia-induced TNF-α protein levels compared to control (Figure 2.).

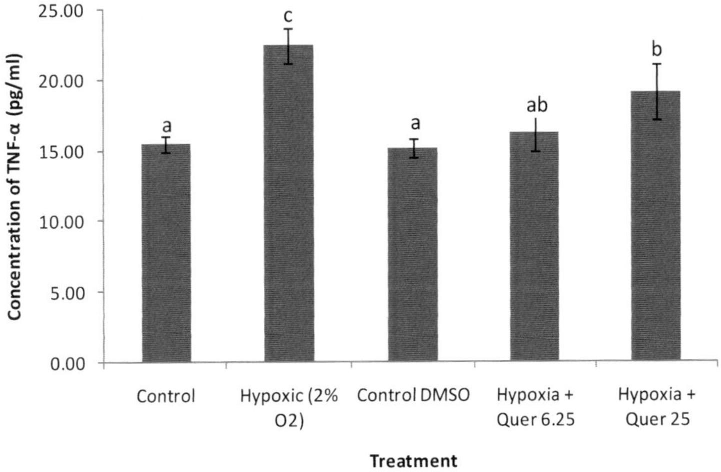

Figure 2. TNF-α levels on hypoxia-induced EA.hy926.

*The histograms are presented as mean ± standard deviation. The data were analyzed with ANOVA and continued with Tukey post hoc test. Different letters (a, b, c) indicate significant differences among treatment.Control represent uninduced cells. Hypoxia (2% O2) represent cell under hypoxia condition. Control DMSO represent uninduced cell treated by DMSO (1% at final concentration). Hypoxia + Quercitrin 6.25 or 25 μg/ml represent hypoxia-induced cell and treated by quercitrin.

3.3. IL-10 level

IL-10 is anti-inflammatory cytokine that possess several benefits to reduce inflammation throughout preeclampsia symptomps. The concentrations given of Quercitrin (25 and 6.25 μg/ml) to preeclampsia model proved that it may increase IL-10 levels as compared to the control hypoxia (Figure 3.) yet it was not significant.

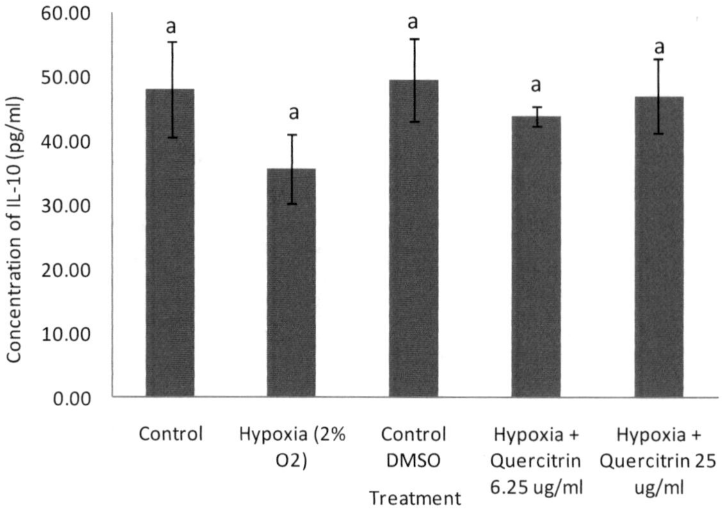

Figure 3. IL-10 levels on hypoxia-induced EA.hy926.

* The histograms are presented as mean ± standard deviation. The data were analyzed with ANOVA and continued with Tukey post hoc test. Different letters (a,b) indicate significant differences among treatment. Control represent uninduced cells. Hypoxia (2% O2) represent cell under hypoxia condition. Control DMSO represent uninduced cell treated by DMSO (1% at final concentration). Hypoxia + Quercitrin 6.25 or 25 µg/ml represent hypoxia-induced cell and treated by quercitrin.

4. Discussion

Preeclampsia is a systemic syndrome characterized by an antiangiogenic and inflammatory status [15]. In this study, EA.hy926 cell line was cultured under hypoxia condition (2% O2) to mimic placental hypoxia in preeclampsia and treated with Quercitrin. According to Zhou *et al.* (2013), persistence of low oxygen tension may cause development of preeclampsia [16]. Therefore, hypoxia and inflammation share an interdependent relationship [17].

Hypoxia condition (2% O2) caused an increase TNF-α and decrease IL-10 which was principally involved in inflammation process during preeclampsia [1]. Recent study has proved that TNF-α have been implicated in the pathogenesis of preeclampsia because of endhotelial effect [18]. While preeclampsia is associated with increased levels of proinflammatory cytokines, it is also associated with decreased placental production of the anti-inflammatory cytokine IL-10 [19] [20]. IL-10 is essentials during pregnancy as of its ability to inhibit secretion of TH1 inflammatory cytokines and thus provide an important counterbalance for controlled inflammation at the fetal-maternal interface [21].

In line with Hung *et al.*, (2004), tumor necrosis factor alfa (TNF-α) in EA.hy926 which cultured in hypoxia condition was significantly increased [1]. Albeit the explanation behind the increasing production of TNF-α is remain undebatable but the generation of reactive oxygen species (ROS) as a result of hypoxia condition may play an important role. According to Qutub & Popel (2008) and Azimi *et al.* (2017), under hypoxic conditions ROS are highly generated by mitochondrial complex III at the Q0 via electron transfer from ubisemiquinone to molecular oxygen. Increases in cellular ROS production can result in an increase in the expression of NF-kB, leading to the upregulation of factors involved in inflammation [22] [23]. Moreover, ROS contributes to intracellular signaling cascade leading to placental production of TNF-α. In addition, the balance between HIF-α and HIF-2α involved in the process of inflammation in the pathogenesis of preeclampsia [24].

Moreover, Quercitrin that belongs to polyphenols group decreased TNF-α. The mechanisms how Quercitrin exert anti-inflammatory effects is by inhibiting the endogenous production of the proinflammatory cytokine TNF-α. According to Yahfoufi *et al.* (2018), polyphenols undoubtedly contribute as anti-inflammatory by interupting the ROS-inflammation cycle. Polyphenols are known for its ability to scavenge a wide-range of free radicals and chelate metal ions [25] [26].

Quercitrin increased IL-10 cytokine that are linked to decrease inflammatory cytokines e.g. TNF-α. IL-10 is crucial due to its ability to improve characteristics of PE. IL-10 also significantly lower blood pressure in preeclampsia [20]. According to Leyva-Lopez *et al.* (2016), Quercetin down-regulated NF-kB expression that linked to up-regulation of IL-10. In this regard, the author proposed a hypothetical mode that quercitrin possess anti-inflammatory activity similar to Quercetin [27].

In other words, certain polyphenols such as Quercitrin exert their effects on the balance between pro- and anti- inflammatory cytokines production, they enhance IL-10 release while they inhibit the secretion of TNF-α [25] [28] [29].

5. Conclusion
Quercitrin possess anti-inflammatory activity in hypoxia-induced EA.hy926 showed by decreasing production of pro-inflammatory cytokines TNF-α but no significant increase in IL-10.

6. Acknowledgement
We are gratefully acknowledge the financial support of the Research Center and Service Community, Universitas Prima Indonesia, Medan, North Sumatera, Indonesia for research grant 2018. We gratefully acknowledge Biomolecular and Biomedical Research Center, Aretha Medika Utama, Bandung, West Java, Indonesia for research methodology and facilities support. We are thankful to Wahyu Setia Widodo, Satrio Haryo Benowo Wibowo, Hanna Sari Widya Kusuma, Ni Luh Wisma Ekayanti, Ika Adhani Sholihah, Dewani Tediana Yusepany, Dwi Surya Artie from Biomolecular and Biomedical Research Center, Aretha Medika Utama, Bandung, West Java, Indonesia for their valuable assistance

7. Conflict of Interest
All contributing authors declare no conflicts of interest.

References
[1] Hung T, Chamock-Jones D S, Skepper J N and Burton G J 2004 *American Journal of Pathology* **164** (3) 1049 – 1061
[2] Cornelius D C 2018 *Clin. Med. Insights Blood Disord.* **11** 1179545X17752325
[3] Nair M P, Mahajan S, Reynolds J L, Aalinkeel R, Nair H, Schwartz S A, and Kandaswami C 2006 *Clin. Vaccine Immunol.* **13** (3) 319 – 328
[4] Eltzschig H K and Carmeliet P 2014 *N. Engl. J. Med.* **364** (7) 656 – 665
[5] Comalada M, Camuesco D, Sierra S, Ballester I, Xaus J, Galvez J, and Zarzuelo A 2005 *European Journal of Immunology* **35** 584 – 592
[6] Gesso J L, Kerr J S, Zhang Q, Raheem S, Yalamanchili S K, O'Hagan D, Kay C D and O'Connell M 2015 *Mol. Nutr. Food Res.* **59** (6) 1143 – 1154
[7] Widowati W, Wijajya L, Setia T L W, and Yelliantty Y 2013a *J. Exp. Integr. Med.* **3** (3) 225 – 230
[8] Widowati W, Mozef T, Risdian C, and Yellianty Y 2013b *Oxid. Antioxid. Med. Sci.* **2** (2) 137 – 142
[9] Widowati W, Darsono L, Suherman J, Fauziah N, Maesaroh M, Erawijantari P E 2016a *Natural Product Sciences* **22** (3) 147-153
[10] Widowati W, Murti H, Jasaputra D K, Sumitro S B, Widodo M A, Fauziah N, Maesaroh M and Bachtiar I 2016b *Asian J. Cell Biol.* **11** (1) 1 – 12
[11] Luo R, Wang Y, Xu P, Cao G, Zhao Y, Shao X, Li Y, Chang C, Peng C and Wang Y 2016 Hypoxia-inducible miR-210 contributes to preeclampsia via targeting thrombospondin type I domain containing 7A *Sci. Rep.* **6** 19588
[12] Aouache R, Biquard L, Vaiman D and Miralles F 2018 Oxidative Stress in Preeclampsia and Placental Diseases *Int. J. Mol. Sci.* **19** (5) 1496
[13] Sandhiutami N M D, Moordiani N, Laksmitawati D R, Fauziah N, Maesaroh and Widowati W 2017 *Iran J. Basic Med. Sci.* **20** (1) 99 – 106

[14] Laksmitawati D R, Prasanti A P, Larasinta N, Syauta G A, Hilda R, Ramadaniati H U, Widyastuti A, Karami N, Afni M, Rihibiha D D and Kusuma H S W 2016 *Journal of Natural Remedies* **16** (2) 73 – 81
[15] Mihu D, Razvan C, Malutan A and Mihaela C 2015 *Taiwanese Journal of Obstetrics and Gynecology* **54** (2) 160 – 166
[16] Zhou J, Xiao D, Hu Y, Wang Z, Paradis A, Mata Greenwood E and Zhang L 2013 *Hypertension* **62** (3) 600 – 606
[17] Bartels K, Grenz A and Eltzschig H K 2013 *Proc. Natl. Acad. Sci. USA* **110** (46) 18351 – 18352
[18] Serin Y S, Ozcelik B, Bapbuoo M, Kylyc H, Okur D and Erez R 2002 *European Journal of Obstetrics & Gynecology and Reproductive Biology* **100** (2) 143 – 145
[19] Raghaupathy R 2013 *Medical Principles and Practice* **22** (1) 8 – 19
[20] Harmon A C, Cornelius D C, Amaral L M, Faulkner J L, Cunningham Jr M W, Wallace K and LaMarca B 2016 *Clinical Science* **130** 409 – 419
[21] Heo Y J, Joo Y B, Oh H J, Park M K, Heo Y M, Cho M L, Kwok S K, Ju J H, Park K S, Cho S G, Park S H, Kim H Y and Min J K 2010 *Immunol. Lett.* **127** (2) 150 – 156
[22] Qutub A A and Popel A S 2008 *Molecular and cellular biology* **28** (16) 5106 – 5119
[23] Azimi I, Petersen R M, Thompson E W, Roberts-Thomson S J and Monteith G R 2017 *Scientific Reports 7*
[24] Matsubara K 2018 *Hypertension Research In Pregnancy* **5** 46 – 51
[25] Yahfoufi N, Alsadi N, Jambi M and Matar C 2018 *Nutrients* **10** (1) 1 – 23
[26] Heim K E, Tagliaferro A R and Bobilya D J 2002 *J. Nutr. Biochem.* **13** 572 – 584
[27] Leyva-Lopez N, Gutierrez-Grijalva E P, Ambriz-Perez D L and Heredia J B 2016 *Int. J. Mol. Sci.* **17** (6) 921
[28] Comalada M, Ballester I, Bailon E, Sierra S, Xaus J, de Medina F and Zarzuelo A 2006 *Biochem. Pharmacol.* **72** (1) 1010 – 1021
[29] Crouvezier S, Powell B, Keir D and Yaqoob P 2001 *Cytokine* **13** (5) 280 – 286

Interaction of cadmium-cysteine binding and oxidation of protein causes of blood thrombosis

D Marisa[1], I Aflanie[2], N Muthmainah[3] and E Suhartono[4*]

[1]Department of Physiology, Faculty of Medicine, Lambung Mangkurat University, Banjarmasin, Indonesia
[2]Department of Forensic and Medico-Legal, Faculty of Medicine, Lambung Mangkurat University, Banjarmasin, Indonesia
[3]Department of Microbiology, Faculty of Medicine, Lambung Mangkurat University, Banjarmasin, Indonesia
[4]Department of Medical Chemistry/Biochemistry, Faculty of Medicine, Lambung Mangkurat University, Banjarmasin, Indonesia

*Corresponding author's e-mail: esuhartono@ulm.ac.id

Abstract. Thrombosis occurs if the balance between thrombogenic factors and protective mechanisms is disrupted. These disorders can be caused by endothelial disorders due to the presence of heavy metals such as cadmium (Cd). These amino acid residues can bind Cd metal covalently with the -S-S-, -C-S-, -SH, and phenyl groups, which are present in the amino acid residues. This situation causes changes in the structure of the protein, so that it becomes modified and forms a blood clot (thrombosis). However, the spectroscopic interaction of cadmium-cysteine as causes of thrombosis has not been widely studied. Therefore, this research needs to be done. The research conducted was an experimental study with posttest control with group design. This study used healthy blood obtained from PMI Banjar regency. This study used one control group and three treatment groups, namely P0 = control group, blood without Cd; P1 = blood group with Cd 0.003 mg/l; P2 = blood group with Cd 0.03 mg/l; P3 = blood group with Cd 0.3 mg/l; P4 = blood group with Cd 3 mg/l. Furthermore, each group was incubated for 45 minutes at 37°c. Then cloth weight, AOPP levels, and absorbance were measured at wavelengths 253 nm. The results showed that an increase in cadmium levels caused an increase in the rate of thrombosis. This increase is associated with an increase in AOPP levels. Cadmium can bind covalently with cysteine and other amino acids which results in the breakdown of peptide bonds. The study concluded that cd accelerated the rate of thrombosis due to increased protein oxidation (aopp). This oxidation is caused by binding cd to cysteine with kbind = 0.355.

1. Introduction

Thrombosis is the formation of blood clot (thrombus) in vein. Thrombosis occurs if the balance between thrombogenic factors and protective mechanisms is disrupted. These disorders can be caused by endothelial disorders due to the presence of heavy metals interacts with amino acid making up the constituent biomolecules of membranes.

Previous research said that Lead (Pb) heavy metal can interact with cysteine and tryptophan in rats induced by Pb with Spectroscopy enclosure. That research also concluded that Pb^{2+} - Pb interactions had higher affinity than Pb^{2+}–tryptophan interactions [1]. Zairin's research also stated that spectroscopy study in rats induced by Pb^{2+} can cause Pb^{2+}–tyrosine interaction in bone. This result in crystal structure becomes wider and bone calcium degraded that can lead to osteoporosis [2].

Metal Cd enters the body through food, and then absorbed and carried by blood through Ligand–blood protein binding [3]. That interaction is started with covalent binding to the functional group in amino acids of protein constituent, such as cysteine, tryptophan, and tyrosine. The amino acid residues can bind metal Cd covalently with the groups of –S-S–, –C-S–, –SH, as well as phenil, which are present in the amino acid residues [4]. This situation causes changes in the structure of the protein, so that it becomes modified and forms a blood clot (thrombosis). However, the spectroscopic interaction of cadmium-cysteine as causes of thrombosis has not been widely studied. Therefore, this research needs to be done.

2. Materials and Method

2.1. Study design

This research is an experimental in-vitro study for one week in May 2019.

2.2. Experimental section

This study used healthy blood obtained from PMI Banjar regency. This study used one control group and three treatment groups, namely P0 = control group, blood without Cd; P1 = blood group with Cd 0.003 mg/l; P2 = blood group with Cd 0.03 mg/l; P3 = blood group with Cd 0.3 mg/l; P4 = blood group with Cd 3 mg/l.

2.3. Measurement of blood thrombosis

Each 1 mL of vein blood was put in 4 tubes, that had been weighed before (Wo) and noted as P0, P1, P2, P3, and P4. Then, 1 mL of Cd was put in each tubes for its group. After that, they were incubated in 37 °C for 45 minutes. Next, the serum in each tubes was taken and kept for AOPP and cysteine measurement. The tubes consisted with cloth (W1) were weighed.

$$Thrombosis\ rate\ (\frac{mg}{minutes}) = \frac{(W1 - Wo)}{45}$$

2.4. Measurement of AOPP

The solution for test was made by mixing 200 μL of blood, 600 μL of phosphate buffer, and 100 μL of KI 1,16 M. Blank solution was made by mixing 800 μL buffer phosphate, 100 μL KI 1.16 M. The test and blank solutions were left for 2 minutes before added by 200 μL of acetate acid, measured its absorbance at =340 nm. AOPPs concentration was stated in $A = \varepsilon\, b\, C$ with $\varepsilon = 26\ mM^{-1}\ cm^{-1}$ and b = 1 cm [5].

2.5. Measurement of absorbance in Cd – cysteine interaction

Cysteine serum that was kept in each groups were measured its absorbance at λ=210-280 nm.

2.6. Metal-protein bonding constant determination

Interactions between metal (L) and protein (P) were based on reaction equations:

$$\text{Metal} + \text{Protein} \rightleftarrows \text{Metal:Protein.}$$

K_{bind} calculation used equation of the previous research [1-3]

3. Result and Discussion

Based on spectroscopy data of UV-Vis at wavelength 210-280 nm, resulted absorbance data such in figure 1. Figure 1 shows that the increase of Cd^{2+} can increase serum absorbance. This means Cd^{2+} binds covalently with amino acid that was represented in serum, so that the interaction between ligand and protein occurs. The interaction of Cd^{2+} with protein was due to the covalent binding between a) Cd^{2+} and

N-terminal in amide group (b) Cd^{2+} and N-terminal in imidazole group (c) Cd^{2+} and N-terminal in amin group [6].

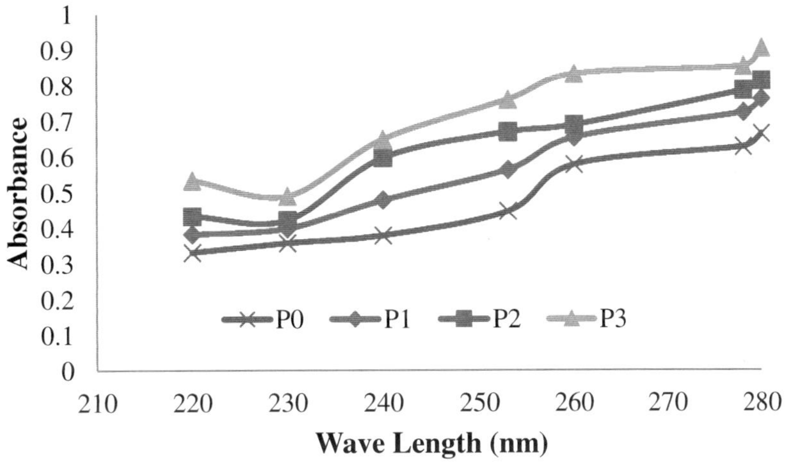

Figure 1. Absorbance Cd^{2+}- Protein at wavelength 210-280 nm.

Interaction of Cd^{2+}- protein can be determined from K_{bind} value which can be calculated with graphic (figure 2).

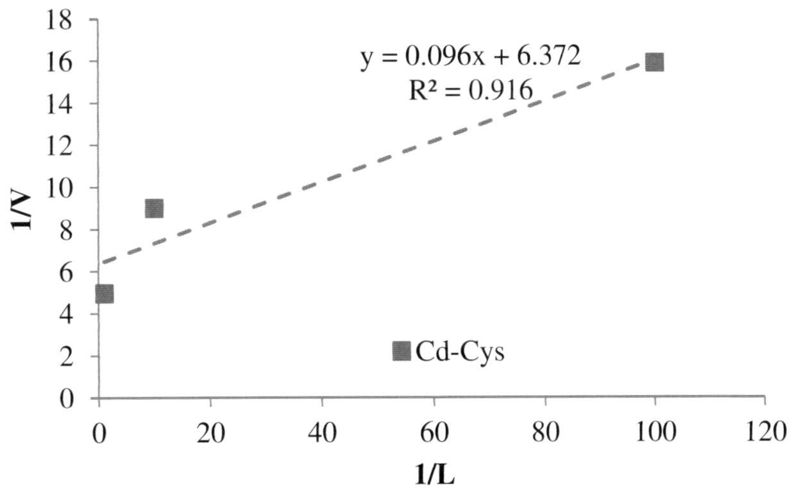

Figure 2. Constanta binding in the interaction of Cd^{2+}- Cys.

Based on figure 2, K_{bind} value of Cd^{2+}- Cys was resulted (K_{bind} = 5,467). This means Cd^{2+} has tendency to bind with cysteine that has binding of –S–S–. Next, the interaction results in protein oxidized and its level increases (figure 3).

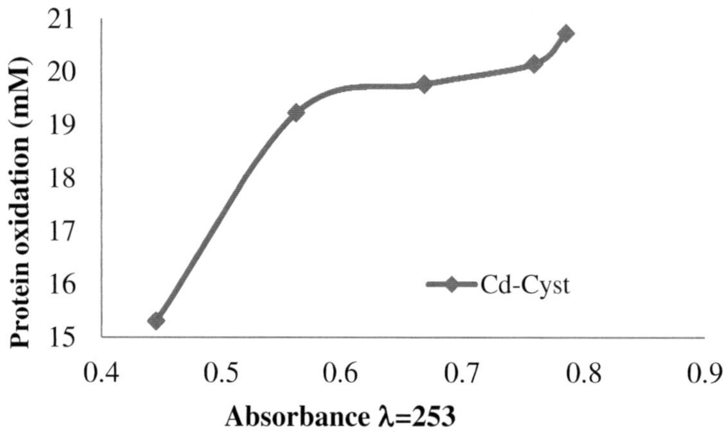

Figure 3. The relation between protein oxidation and Cd^{2+}-Cys interaction (λ=253).

Protein oxidation that increased results in thrombosis rate that also increased along with the increasing concentration of Cd^{2+} (figure 4).

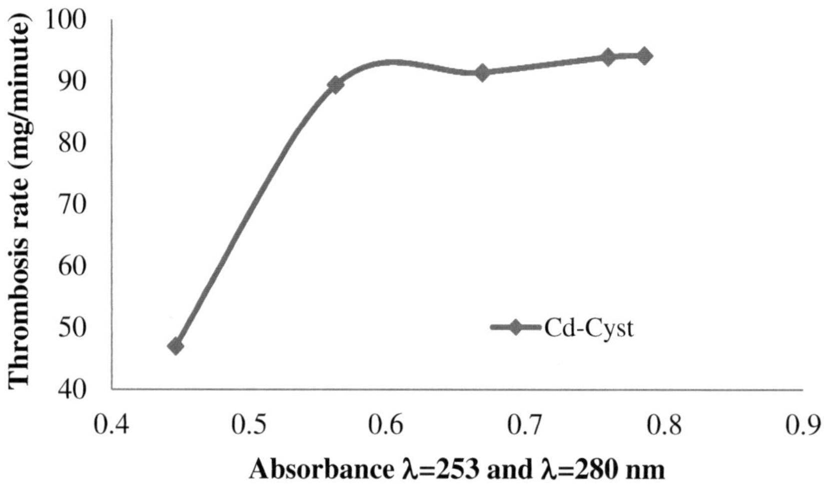

Figure 4. The relation between thrombosis rate and Cd^{2+}- Cys interaction (λ=253).

4. Conclusion

Therefore, in this study can be suggested for thrombotic mechanism with the existence of Cd^{2+}, which starts with covalent binding of –SH in cysteine. This binding causes modification of protein marked by the increase of protein oxidation level, so that resulted in increasing of thrombotic rate exponentially.

References

[1] Suhartono E, Thalib I, Aflanie I, Noor Z and Idroes R 2018 Study of interaction between cadmium and bovine serum albumin with UV-Vis spectroscopy approach *IOP conference series: Materials Science and Engineering* **350**

[2] Noor Z, Azharuddin A, Kania N, Aflanie I and Suhartono E 2018 Interaction betwen lead and bone protein to affect bone calcium level using UV-Vis spectroscopy *IOP conference series: Materials Science and Engineering* **350**

[3] Thalib I, Budianto W Y and Suhartono E 2017 Effect of cadmium exposure on increasing risk of diabetes melitus through the measurement of blood glucose level and liver glucokinase activity in rats *Berkala Kedokteran* **13** (2) 137-45

[4] Aflanie I, Thalib I and Suhartono E 2017 Study of interaction between lead and gastric mucosal protein of rats with forensic toxicology approach *Berkala Kedokteran* **13** (2) 119-24

[5] Edyson, Pardede A M E, Nugraha H G, Mashuri and Suhartono E 2019 In vivo antioxidant and UV photoprotective of extract pasak bumi (eurycoma longifolia jack.) *AIP Conference Proceedings* **2108** 020029

[6] Dieaconu M, Ioanid A, Iftimie S and Antohe S 2012 UV-absorption mechanisms of Ni-binding bovine serum albumin. *Dig. J. Nanomater. Biostruc.* **7** 1125-38

Changes of Biologic Marker in Neonatal Sepsis: Is it significance?

D I N Pratiwi[1,2,3*], P W Nurikhwan[4] and W R Devi[2]

[1]Medical Faculty, Lambung Mangkurat University, Banjarmasin, South Kalimantan, Indonesia
[2]Clinical Pathology Department and Medical Laboratory, Ulin Hospital, Banjarmasin, South Kalimantan, Indonesia
[3]The Education Program of Clinical Pathology Subspecialist for Infectious Disease, Medical Faculty, Gadjah Mada University, Yogyakarta
[4]Department of Medical Education, Faculty of Medicine, University of Lambung Mangkurat Banjarmasin, Indonesia

*Corresponding author's email: indahhariadi@gmail.com

Abstract. Introduction. Neonatal sepsis is the leading cause of death in newborns. The initial symptoms of sepsis are usually unclear, the clinician often judge these conditions based on risk factor and clinical examination. Blood culture is one of the parameters to be checked but the result takes time. It is also usually come with a positive result of 10 - 60% only. Another biologic marker such as hematologic one is used to evaluate the sepsis because it is considered faster enough. The aim of study is to evaluate the significant changes in biologic marker during neonatal sepsis with positive risk factors. **Methods.** This was a cross sectional study. All <72 hours old's neonates with positive risk factors, neonates who admitted in July to September 2015 in the Neonatal Intensive Care Unit (NICU), and clinically suspected of sepsis were included. Laboratory results of each case were recorded. The eight variables of biologic hematologic parameters of the neonates (hemoglobin, leukocytes, platelet counts, IT (immature to total neutrophil) ratio, IM (immature to mature neutrophil) ratio, polymorphonuclear, immature polymorphonuclear, and hematological scoring (HSS)); one variable of C-reactive protein, and one variable culture result were assessed. Each variable will be compared based on culture result using independent T-test of SPSS. **Results and Discussion.** Thirty-six samples were found which consist of twenty-nine negative and seven positive sample of culture results. All statistical variable's results weren't significant (p >0,05). These results showed there is no significant changes to these variables even if there are any risk factors within the neonates. Higher or lower value of these variable did also not prove whether the culture is positive. **Conclusion.** No significant changes in biologic marker during neonatal sepsis with positive risk factors.
Keywords: hematological parameters, Hematologic Scoring system, neonatal sepsis

1. Introduction

Sepsis is the main cause of death in newborns. In developing countries, the average mortality is between 11-68/1000 births, in Bangladesh an average of 42/1000 live births. Early diagnosis of neonatal sepsis is still very important. The incidence of sepsis in the Pediatric Intensive Care Unit (PICU) is 24%. While a recent French study conducted in 36 NICU PICUs, there was an incidence of sepsis of 3%, with a mortality rate of 30-60%. Of those with sepsis, approximately 49% suffer from bacteriology consisting of 58% with gram (+), and 42% with gram (-) bacteria. World Health Organization (WHO) estimates that 4 million neonates die each year, of which 35% including caused by neonatal sepsis [1, 2].

Neonatal septicemia is a typical clinical syndrome characterized by an infection accompanied by bacteremia in the first month of life. The initial symptoms of neonatal sepsis are usually not clear. Current diagnosis with microbiological culture also requires a long time, which is 3-14 days, with a positive result of 10-60% even if these microbiological parameters have high specificity [3]. Although diagnosis and therapy continue to develop, this condition remains a cause of high morbidity and mortality. The speed of diagnosis of neonatal sepsis is important because neonates who experience pain progression faster than in adults [1].

Another biologic marker such as hematologic one is used to evaluate the sepsis because it is considered faster enough. According author knowledge, study of hematologic biologic marker still few in Indonesia. The aim of study is to evaluate the significant changes in hematologic biologic marker during neonatal sepsis with positive risk factors. This study may become a grounded theory for future.

2. Material and Methods

This was a cross sectional study. All <72 hours old's neonates with positive risk factors, neonates who admitted in July to September 2015 in the Neonatal Intensive Care Unit (NICU), and clinically suspected of sepsis were included. Independent Variables were eight variables of biologic hematologic parameters of the neonates (hemoglobin, leukocytes, platelet counts, IT (immature to total neutrophil) ratio, IM (immature to mature neutrophil) ratio, polymorphonuclear, immature polymorphonuclear, and hematological scoring (HSS)); one variable of C-reactive protein and Dependent Variables neonates culture result. Inclusion criteria were mothers who have a risk of causing sepsis in infant and aged <72 hours with clinical symptoms of sepsis. Exclusion criteria were neonates aged >72 hours, no symptoms of sepsis. Subjects from affordable populations who met the inclusion and exclusion criteria were taken consecutively.

Two milliliters of neonatal venous blood taken was inserted in a tube of vacutainer anticoagulant EDTA (ethelenediamine tetra-acetic acid) for routine blood tests. Reagents for the automatic hematology analyzer and Cat Wright were provided. Used tool were Automated Hematology Analyzer Scan type BC-5800, micropipette, Object glass, Olympus Microscope

Each variable will be compared based on culture result using independent T-test of statistic software and advanced for logistic regression if the requirement was fulfilled.

3. Results and Discussion

Thirty-six samples were found which consist of twenty-nine negative and seven positive sample of culture results. Table below shown the characteristic and results.

Table 1. Risk Factor description.

No	Risk Factors	Culture Result (n = 36)	
		Negative (n= 29) (80%)	Positive (n= 7) (20%)
1	Mode of Delivery		
	Normal Vaginal	18	4
	Forceps or Vacuum	1	1
	Cesarean	10	2
2	Premature Rupture of the Membrane		
	No	28	7
	Yes	1	0
3	Hypertension in Pregnancy		
	No	27	6

Yes	2	1

Table 2. Statistical Results.

No	Parameter	Culture Result (n = 36)		P
		Positive (n = 7) Median (Min– Max)	Negative (n = 29) Median (Min – Max)	
1	Hemoglobin	16.4 (10.5 – 20.3)	15.9 (11.7 – 20.6)	0.873
2	Leukocytes	15.9 (11.1 – 33.70)	16.7 (8.4 – 261.3)	0.952
3	Platelet	287 (173 – 359)	228 (27 – 554)	0.436
4	IT (immature to total neutrophil) ratio	0.09 (0.04 – 0.24)	0.09 (0.03 – 9.2)	0.936
5	IM (immature to mature neutrophil) ratio	0.09 (0.01 – 0.3)	0.1 (0.01 – 11.5)	0.704
6	Polymorphonuclear,	12.14 (8.77 – 30.67)	12.12 (3.5 – 9724)	0.704
7	Immature polymorphonuclear.	1.02 (0.13 – 7.08)	1.03 (0.09 – 4.25)	0.984
8	Hematological scoring (HSS)	1 (0 – 5)	1 (0 – 6)	0.695
9	C-reactive protein (CRP)			0,645
	Positive	21	5	
	Negative	8	2	

All statistical variable's results weren't significant (p >0,05). These results showed there is no significant changes to these variables even if there are any risk factors within the neonates. Higher or lower value of these variable did also not prove whether the culture is positive. Logistic regression was unable to be done due to lack of requirement of statistical result of bivariate analysis.

Sepsis is a systemic inflammatory system response caused by infection in both the blood and tissue, This response is manifested with two or more of the following symptoms and signs: temperature> 38^{0}C or <36^{0}C, tachypnea, tachycardia, <30 torr, leukocytosis (> 12,000/mm3) or or leukopenia (<4,000/mm3) or neutrophil young cell count >10%.[4] Sepsis in children is generally caused by a bacterial infection consisting of 19% nosocomial infection, and bacteriemia in 49% of patients namely gram negative as much as 52% and gram positive 48%. The most common nosocomial infections are due to coagulase-negative staphylococcus, staphylococcus aureus and enterococcus [5]. This study found 36 samples consist of: 29 negative (80%) and 7 positive (20%) culture result with gram positive 2 (28.5%) and gram neg 5 (71.5%)

There are several risk factors that can increase the incidence of sepsis in children are: 1) Host factors, which consist of: malnutrition, immunodeficiency, chronic disease problems, trauma/burns, serious and critical illness. 2). Treatment factors: surgery, invasive procedures, invasive monitoring devices, antibiotics, immunosuppressive therapy, length of treatment and hospital environment. Sepsis alone was distinguished as early sepsis and advanced sepsis where in the early sepsis the mortality rate reached 5-50% [6].

In every infection, both primary and secondary due to nosocomial, the body will hold an acute phase immunological reaction that makes and releases various mediators that can cause changes in the blood. GCSF (Granulocyte Colony Stimulating Factor) mainly plays a role in increasing granulocyte production, especially neutrophils, increasing and releasing leukocytes, especially neutrophils in the bone marrow [7]. Other study said the number of leukocytes less than 5000/ul or more than 25,000/ul can predict neonatal sepsis but the value of the positive predictions of the abnormal leukocyte count is low. The total number of neutrophils is more sensitive than the total number of leukocytes in diagnosing neonatal sepsis [6]. The degree of neutropenia for prediction of neonatal sepsis is according to age, ie total neutrophils are less than 1800/ul at birth, 8100/ul at 12 hours, 7000/ul when 24 hours and less than

1800/ul after 72 hours [2]. The presence of neutropenia is a better indicator of neonatal sepsis than neutrophilia [2, 11]. The total number of immature neutrophils is more than 1100/ul, 1400/ul and 800/ul respectively when born, age 12 hours and above 60 hours is considered abnormal [2].

From the neutrophil index, the ratio of total immature/neutrophil neutrophils (I/T) is the single most sensitive and specific indicator for neonatal sepsis, with a sensitivity value of more than 96% and a specificity of more than 71% [2, 8, 12].

Thrombocytopenia occurs in 10-60% of cases of neonatal sepsis. Because it is not specific and the thrombocytopenia is slow, this value is not used alone especially for the diagnosis of early onset.[4,6]

Monroe et al provide criteria for neonatal sepsis, namely if found 2 or more of the following, the ratio of immature/neutrophil total neutrophils (I/T) >0.16, absolute neutrophil counts <7500/ul or >14,000/ul, number immature neutrophils >1400/mm [13].

In addition, complement that is activated both classically and alternatively can improve the phagocytic process seen in blood smear as vacuolization. Twelve subjects (33%) in this study with peripheral blood morphology obtained vacuolization neutrophil, normo-blast 3 subjects (8.3%) and 1 subject with toxic granulation (2.8%). TNF mediators produced by lymphocytes and monocytes that are activated due to infection will inhibit the neutrophil maturation process so that more neutrophil young cells form rods in the circulation. Activating Hageman factors due to endotoxin or due to vascular endothelial damage can eventually lead to DIC and thrombocytopenia. Infection can also cause red blood cell hemolysis which ultimately can cause anemia. Other mediators such as IL6 can cause hepatocytes to secrete acute phase reactants, namely CRP [7].

By looking at changes in blood, namely leukocyte shape changes, especially neutrophils (vacuolization, hyper granulation), leukocytosis or leukopenia, increasing I/T ratio > 0.2, increasing I/M ratio, immature neutrophils, red blood cell hemolysis process characterized by the presence of nucleated erythrocytes, thrombocytopenia which can all be calculated by scoring, better known as the Hematologic Score System, can detect suspicion of sepsis. Rodwell provides the hematological score for neonatal sepsis (HSS) as follows: The hematological scoring system according to Rodwell (Table 3) establishes a score of 1 for each significant hematologic finding, but there are exceptions to the number of PMNs, if no PMN is found in a portion of peripheral blood then the score is two [12].

Table 3. Hematological Scoring System [6, 12].

Criteria	**Abnormality**	**Score**
Total WBC	≤5.000/ul	1
	≥25.000 at birth	1
	≥30.000-12-24 h	
	≥21.000-Day 2 on ward	
Total PMN Count	No mature PMN seen	2
	Increased/decresed	1
Immature PMN Count	Increased	1
I/T PMN Ratio	Increased	1
Degenerative change in PMN	Toxic granules/cytoplasmic vacuoles	1
Platelet count	≤150,000/ul	1

Table 4. Interpretation of Hematological Scoring System [6, 12].

Score	**Interpretation**
≤ 2	Sepsis is unlikely
3 or 4	Sepsis is possible
≥ 5	Sepsis or infection is very likely

From the study of 287 neonates with hematological scores, it has 96% sensitivity, 78% specificity, positive predictive value 31% and negative predictive value 99% [14]. Other author reassessed this score with a larger number of samples reaching 1000 neonates with sensitivity 93%, specificity 82%, positive predictive value of 50% and negative predictive value of 98% [15]. If the score is only 0-2 then chances are not 99% sepsis [11]. Philip detected slow onset neonatal sepsis with the following 5 parameters, leukocyte count <5000 / ul, neutrophil ratio I / T 30.2, LED 315mm / hour, CRP (+), Haptoglobulin (+). The diagnosis of sepsis is made when two or more positive test results are found. With this approach the sensitivity is 88%, specificity is 80%, while if all five of these tests are negative (99%) is considered not sepsis [6, 16, 17]. Blood culture is the gold standard in neonatal sepsis, but in the literature, it is mentioned that patients with sepsis only 10-60% give a positive result, with a specificity of 100% [8, 9].

Complete blood count with type count is a very easy and fast laboratory examination for neonatal sepsis. The Manroe criteria state that if there are 2 abnormal values from 3 parameters (total neutrophils, immature granulocytes and I / T ratio) then the baby can be sepsis. C-reactive protein (CRP) is another parameter. CRP has several weaknesses, including liver maturity in infants hasn't mature; and the examination should be done serially, because the process of formation of CRP increases after 4-6 hours, 2x in 8 hours, peak 36 hours - 50 hours, half-life of 4 - 7 hours. While the strength of CRP are good for acute processes, relatively inexpensive, easy, can be obtained in a fast time, and does not require a lot of volume. Both combination of other rapid examination results such as CRP and LED, the accuracy of this blood eradication examination will also be increased, and there is no need to wait for the results of the culture long enough for the diagnosis of sepsis [7, 8, 10].

This study shown this parameter couldn't use as screening and diagnoses. There is no single parameter to diagnose neonatal sepsis, several and combined parameter (including clinical condition) need to assess. A new parameter is needed to develop for faster diagnosis of Neonatal Sepsis. Limitation of study was bigger sample size is needed to comprehend different result.

4. Conclusion

No significant changes in biologic marker during neonatal sepsis with positive risk factors.

References

[1] Laishram R S and Khuraijam R D 2013 Hematological and Biological Markers of Neonatal Sepsis *Iranian Journal of Pathology* **8** 137 – 146

[2] Sundareson R and Sheargean J N 2006 Current understanding and treatment of sepsis *Inf. Med. J.* **6** 261 – 68

[3] Buch A C, Srivastava, Kumar H and Jadhav P S 2011 Evaluation of Haematological profile in early diagnosis of clinically suspected cases of neonatal sepsis *International Journal of Basic and Applied Medical Sciences* **1** (1) 1 – 6

[4] Rohsiswatmo R 2009 Tatalaksana Sepsis Neonatorum. PBPK Fakultas Kedokteran Universitas Indonesia 106 – 116

[5] Hayun M, Alasiry E, Daud D *et al* 2015 The risk factor og early onset neonatal sepsis *American Journal of clinical and experimental medicine* 78 – 85

[6] Bintekholir K, Asadur Rahman M, Sutana T *et al* 2010 Role of Hematologic Scoring System in early diagnosis of neonatal Septicemia *Journal BSMMU* **3** (2) 62 – 67

[7] Anwer S K and Mustafa S 2000 Rapid Identification of neonatal sepsis

[8] Mayuga W A B and Isleta P E D 2005 Clinical correlation of neonatal and maternal hematological parameters as predictors of neonatal. Department of Pediatrics *PIDSP journal* **9** 36 - 43

[9] Puopolo K M and Escobar G J 2013 Early – onset sepsis : a predictive model based on maternal risk factor **25** 1 – 6

[10] Resch B, Hofer N and Muller W 2012 Challenges in the Diagnosis of sepsis of the neonate. *Reasearch for neonatal infectious diseases and epidemiology* **2012** 233 – 245

[11] Kries R C Sepsis in children. Dalam: Emergency Nursing World; 1-4 http:// enw. Org/

ReseachSeptic Kid. Htm
[12] Rodwell R L, Lesli A L and Tudehope D I 1988 Early Diagnosis of neonatal sepsis using a hematologic scoring system *J. Pediatr.* **112** 761-7
[13] Monroe B L, Rosenfeld C R, Weinberg A G and Browne R 1991 The differential leukocyte count in the assessment and outcome of early onset neonatal group B Streptococcus disease *J. Pediatr.* **91** 632-7
[14] Narasimha A and Harendra Kumar M L 2011 Signifcance of hematological scoring system (HSS) in early diagnosis of neonatal sepsis *Indian J. Hematol. Blood Transfus.* **27** 14-7
[15] Bhalodia M J, Hippargi S B and Patil M M 2017 Role of hematological scoring system in diagnosis of neonatal sepsis *J. Clin. Neonatol.* **6** 144-7
[16] Phillip A G S 1982 Detection of Neonatal Sepsis of Late Onset *JAMA* **247** 489-492
[17] Gotoff S P 1992 Sepsis in the Newborn. Dalam: Krugman S, Katz SL, Gerson AA, Wilfert CM, penyunting. Infectious disease of children; edisi ke-9. (St Louis: MosbyYear Book) 402-18

Effect of DHA Supplementation on The MDA and SOD Levels in Protein Malnourished Rats

D D Sanyoto[1], A Asnawati[2] and T Triawanti[3]

[1]Department of Anatomy, Faculty of Medicine, Lambung Mangkurat University, Banjarmasin
[2]Department of Physiology, Faculty of Medicine, Lambung Mangkurat University, Banjarmasin
[3] Department of Biochemistry of Medical Faculty of Lambung Mangkurat University, Banjarmasin

*Corresponding author's email: didikdwisanyoto@ulm.ac.id

Abstract. Protein malnutrition interferes with the synthesis of enzymes served as antioxidants while reducing antioxidants concentration in tissues, leading to oxidative stress. One of the most stable markers of oxidative stress is malondialdehyde (MDA). Docosahexaenoic acid (DHA) has an important role in inhibiting oxidative stress and protecting the tissue from peroxidative damage to lipids and proteins, thereby reducing oxidative stress in the tissue. This study aimed to analyze the effect of DHA on the MDA and superoxide dismutase (SOD) levels in protein malnourished rats. This study was carried out using the rat model Rattus norvegicus. After delivery, the rats were classified into two groups, namely, the malnutrition group (MG) and the non-malnutrition group (NMG). The MG was given low-protein diet since birth until 50 days old. The MG was further classified into two subgroups, namely the control malnutrition group with normal feed recovery (CMG), and the treated malnutrition group with normal feed recovery and DHA supplementation for 6 weeks (TMG). At the end of treatment, the blood MDA and SOD levels were assessed. The results showed that the MDA levels in the NMG, CMG, and TMG were 812.75 μM, 647.75 μM, and 624.00 μM respectively. The SOD levels were 19.40 μM, 19.20 μM, and 26.80 μM, respectively. The Kruskal Wallis test showed a significant difference in MDA levels between groups ($p = 0.001$). The posthoc test using the Mann Whitney test showed that the MDA level in TMG was significantly different from that in NMG, but not significantly different from CMG. The Anova test showed a significant difference in SOD levels between groups ($p = 0.000$). The Tukey posthoc test showed that the SOD level in TMG was significantly different from that in NMG and CMG. DHA has been shown to play an antioxidant role in malnutrition. DHA can increase the levels of SOD enzymes so that oxidative stress that occurs in malnutrition can be reduced. This is proven by the decrease in MDA levels as a lipid peroxidation product. It can be concluded that DHA supplementation can reduce the MDA levels and increase the SOD levels in protein malnourished rats.
Keywords: Protein malnutrition, DHA, MDA, SOD

1. Introduction

Indonesia is a developing country that still has a number of nutritional problems. Compared to other ASEAN countries, Indonesia is the most complete nutritional problem. The results of the 2018 basic health research (Riskesdas) reported that the prevalence of malnutrition are 19.6% (Riskesdas 2013) to 17.7% (Riskesdas 2018) [1].

Oxidative stress is the steady state level of oxidative damage in a cell, tissue or organs caused by an imbalance between the production of reactive oxygen species (ROS) and biological systems ability to detoxify the reactive intermediates or easily repair the resulting damage. All forms of life maintain a reducing environment within their cells. This reducing environment is preserved by enzymes that maintain the reduced state through a constant input of metabolic energy. Disturbance in this normal redox state can cause toxic effects through the production of peroxides and free radicals that damage all components of the cell [2].

Protein malnutrition interferes with the synthesis of enzymes served as antioxidants while reducing antioxidants concentration in tissues, leading to oxidative stress. ROS degrades polyunsaturated lipids, forming malondialdehyde (MDA). MDA is one of the most stable markers of oxidative stress. Increased levels of lipid peroxidation products in the serum are used as a marker for tissue damage [3].

Docosahexaenoic acid (DHA) has an important role in inhibiting oxidative stress and protecting the tissue from peroxidative damage to lipids and proteins, thereby reducing oxidative stress in the tissue both exvivo and in vivo [4].

This study aimed to analyze the effect of DHA on the MDA and superoxide dismutase (SOD) levels in protein malnourished rats.

2. Materials and Methods

2.1. Material

Docosahexaenoic acid (DHA), Rat Standard diet, Diet low protein AIN-76A Purified Rodent Diet contain 6.0% Casein (diets, USA).

2.2. Animal Study

This study was carried out using the rats *Rattus norvegicus* as the models. This work was approved by the Ethical Committee Faculty of Medicine Lambung Mangkurat University. This research used Posttest Control Group Design. After delivery, the rats were classified into two groups, namely, the malnutrition group (MG) and the non-malnutrition group (NMG). The MG was given low-protein diet since birth until 50 days old [5].

2.3. Administration of DHA

The MG was further classified into two subgroups, namely the control malnutrition group with normal feed recovery (CMG), and the treated malnutrition group with normal feed recovery and DHA supplementation for 6 weeks (TMG).

2.4. Measurement of MDA level and SOD level

After six weeks of feeding, the rats were sacrificed using cervical dislocation and blood was collected directly from the heart of the animals through 5 ml syringe into EDTA bottles. The blood collected into the plain bottles was allowed to clot, then centrifuged in 4000 x g and the supernatant was collected into another plain bottle for analysis MDA and SOD levels.

2.5. Statistical analysis.

The data obtained were analyzed using the Kruskall Wallis test and ANOVA with a confidence level of 95%

3. Results

The results showed that the MDA levels in the NMG, CMG, and TMG were 812.75 µM, 647.75 µM, and 624.00 µM respectively. The SOD levels were 19.40 µM, 19.20 µM, and 26.80 µM, respectively. The Kruskal Wallis test showed a significant difference in MDA levels between groups ($p = 0.001$). The posthoc test using the Mann Whitney test showed that the MDA level in TMG was significantly different from that in NMG, but not significantly different from CMG. The Anova test showed a significant difference in SOD levels between groups ($p = 0.000$). The Tukey posthoc test showed that the SOD level in TMG was significantly different from that in NMG and CMG.

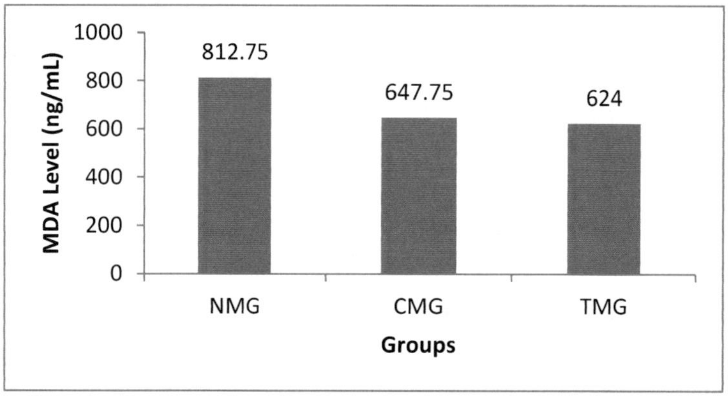

Figure 1. Mean of MDA level after supplementation DHA 1 mg/day. NMG= control (well-nourished), CMG = malnourished, TMG = malnourished + supplementation DHA (p=0.001).

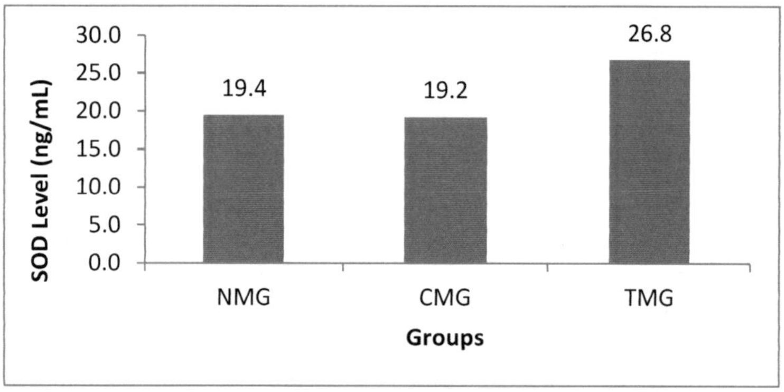

Figure 2. Mean of SOD level after supplementation DHA 1 mg/day. NMG= control (well-nourished), CMG = malnourished, TMG= malnourished + supplementation DHA (p=0.000).

4. Discussion

Experimental protein malnutrition in rats has been shown to reduce antioxidant cellular defence system and to enhance lipid peroxidation in plasma [6].

In the present study, DHA has been shown to play an antioxidant role in malnutrition. DHA can increase the levels of SOD enzymes so that oxidative stress that occurs in malnutrition can be reduced. This is proven by the decrease in MDA levels as a lipid peroxidation product.

Dietary deficiency of protein not only impairs the synthesis of plasma albumin and antioxidant enzymes but also reduces tissue concentrations of antioxidants, thereby resulting in a compromised antioxidant status. Malnourished children were found to have more oxidant damage products and less antioxidant levels. In children with PEM has been suggested to be caused by an imbalance between the production of these toxic free radicals and antioxidant potential. Experimental study, serum MDA in malnourished children is is a significant increase as compared to control [3].

Protein-deficient diet intake has been shown to influence the activity of drug metabolizing enzymes as well as antioxidant enzymes. Feeding of a protein-deficient diet to rats has been shown to increase lipid peroxidation and induce significant changes in activities of catalase, glutathione peroxidase and superoxide dismutase [7].

Docosahexaenoic acid (DHA) has an important role in inhibiting oxidative stress. DHA are described to have anti-inflammatory effects and to improve endothelial function. In experimental study is that DHA diminished ROS-induced DNA damage in human aortic endothelial cells. H_2O_2-induced γ-H2AX foci formation was decreased by DHA, which indicated that these n-3 PUFAs diminished DSBs, the most severe type of DNA damage [8].

In the present study, was show that intake of DHA causes decreased production of MDA, which may serve a redox signalling role, thereby activating one of the major cellular detoxification pathways. The nuclear factor-E2-related factor-2 (Nrf2) pathway. Nrf2 plays a key role in defense against oxidative stress throughout the body. Nrf2 plays an important role in regulating the expression of genes that encode proteins responsible for decreasing oxidative stress (MDA). It has previously been shown that Nrf2 is important for ischemic lesion repair [9]. Nrf2 including those involved in Phase II detoxification and antioxidant gene expression. Regulation of Nrf2 activity represents a critical step in initiating a cellular antioxidant response to reactive oxygen species [10].

Recent studies reported that daily administration of DHA increased antioxidant enzyme activities and decreased MDA production in the brain [11]. DHA supplementation reduced urinary F2-isoprotane levels, a marker for oxidative stress, as well as enhanced cellular antioxidant defense systems. Lipid peroxidation products, such as malondialdehyde, can be as deleterious as ROS and may even cause DNA damage. Therefore, DHA supplementation given in association with an antioxidant treatment to reduce the deleterious oxidative effects [12].

5. Conclusion

It can be concluded that DHA supplementation can reduce the MDA levels and increase the SOD levels in protein malnourished rats.

6. Acknowledgments

The authors are very much thankful to Dean of Medical Faculty Lambung Mangkurat University for providing funding and for the kind support.

References

[1] RI Ministry of Health, 2018
[2] Schafer F and Buettner G 2001 *Free. Radic. Biol. Med.* **30** (11) 1191-1212
[3] Khare M, Mohanty C, Das B K, Jyoti A, Mukhopadhyay B and Mishra S P 2014 *Intl J Pediatr.* 2014
[4] Kumar P, Pandey D K, Choudhary M, Singh A K and Tripathi S 2015 *Int J Res Pharm Sci* **5** (2) 43 –50
[5] Valadares C T, Fukuda M T H, Francolin-Silva A L, Hernandes A S and Almeida S D S 2010 *Nutritional Neuroscience* **13** (6) 274-82
[6] Ayala A, Muñoz M F and Argüelles S 2014 Oxidative Medicine and Cellular Longevity Hindawi Publishing Corporation
[7] Akinola F F, Oguntibeju O O and Alabi O O 2010 *Sci. Res. Essays* **5** 1145–1149
[8] Sakai C, Ishida M, Ohba H, Yamashita H, Uchida H, Yoshizumi M and Ishida T 2017 *PLOS ONE*
[9] Sahin N, Akdemir F, Orhan C, Aslan A, Agca C A, Gencoglu H, Ulas M, Tuzcu M, Viyaja J, Komorowsk J R and Sahin K 2012 *Nutritional Neuroscience* **15** 5

[10] Gao L, Wang J, Sekhar K R, Yin H, Yared N F, Schneider S N, Sasi S, Dalton T P, Anderson M E, Chan J Y, Morrow J D and Freeman M L 2007 *The Journal Of Biological Chemistry* **282** (4) 2529–2537

[11] Pan H C, Kaoc T K, Oud Y C, Yange D Y, Yenf Y J, Wang C C, Chuang Y H, Liao S L, Raung S L, Wu C W, Chiang A N and Chen C J 2009 *J Nutr Biochem* **20** 715–25

[12] Losano J D A, Angrimani D S R, Rui B R, Bicudo L C, Dalmazzo A, Silva B C S, Viana C H C, Mendes C M, Assumpção M E O A, Barnabe V H and Nichi M 2018 *Zygote* **26** (3) 199-206

The Role of Prophylactic Antibiotics on ANC Value and Fever Duration in Cancer Patients with Post-Chemotherapy Neutropenia Fever

E P Yunita[1,2]*, G J R Subagiyo[3] and S O Wardhani[4]

[1]Department of Pharmacy, Faculty of Medicine, Universitas Brawijaya, Malang, Indonesia
[2]Research Center of Smart Molecule of Natural Genetics Resource, Universitas Brawijaya, Malang, Indonesia
[3]Study Program of Pharmacy, Faculty of Medicine, Universitas Brawijaya, Malang, Indonesia
[4]Division of Hematology and Oncology, Department of Internal Medicine, Dr. Saiful Anwar General Hospital, Malang, Indonesia

*Corresponding author's email: emapristi@ub.ac.id

Abstract. Chemotherapy has immunosuppressive side effects, one of which is neutropenia. In neutropenia, fever often occurs as a manifestation of infection. Antibiotics are given in neutropenia condition to prevent complication of infection. The aim of this study was to investigate the role of different prophylactic antibiotics on the increase of Absolute Neutrophil Count (ANC), duration of fever, and degree of improvement in ANC values in cancer patients with fever neutropenia. This study was an observational with quantitative descriptive analyses. Data were taken from medical records for the period January 2015 to December 2016. The sample size that met the inclusion criteria was 45 patients including 15 using ceftriaxone-gentamicin, 8 using ceftriaxone, 16 using ceftazidime-ciprofloxacin, and 6 using ceftazidime. The results showed a significant difference in the increase ANC values for different antibiotics ($p = 0.001$). However, the administration of different antibiotics did not provide a significant difference in the mean duration of fever ($p = 0.341$). Administration of different antibiotics was known to give a degree of improvement in the same ANC value ($p = 0.711$). The conclusion of this study was that the administration of different antibiotics in post-chemotherapy neutropenia fever patients has a significant influence on the ANC value where the type of antibiotic that has the highest influence is a combination of ceftazidime-ciprofloxacin.
Keywords: ANC, Antibiotic, Cancer, Chemotherapy, Neutropenia Fever

1. Introduction

Neutropenia is one of the side effects of giving chemotherapy. After chemotherapy, hematology levels change every week, one of them is neutropenia. Neutropenia is a condition of the number of neutrophils in the blood less than 500/μL or less than 1000/μL with a tendency to decrease to 500/μL. Factors in the occurrence of neutropenia include disturbances in neutrophil formation, neutrophil distribution to tissues, increased consumption of neutrophils, and increased neutrophil damage to circulation.

Malignant cell infiltration and the myelosuppressive effects of chemotherapy are also one of the trigger factors for neutropenia. Through interference with Hexose Monophosphate Shunt activity, chemotherapy can reduce the ability of phagocytosis and bactericidal neutrophils [1].

Neutropenia fever is a fever that occurs in neutropenia with a body temperature of more than 38.3°C in one measurement or temperature increase > 38°C in two measurements in a 12 hour period or at least one hour of measurement on oral and axillary measurements [2]. Neutropenia fever is the most common complication in cancer patients who receive chemotherapy. This condition is a life-threatening condition with high mortality because it is susceptible to infection so that appropriate and efficient antibiotic therapy is needed to prevent the development of infection [3].

In Europe, with the right management guidelines for neutropenia fever produced a mortality rate due to neutropenia fever of 8.7% whereas in Indonesia, the reported mortality rate due to neutropenia fever is 12.5%-38.8%. This large difference in data is likely due to the fact that there are still no standard guidelines for diagnosis and management of neutropenia fever that are adapted to conditions and germ maps in Indonesia [1].

The management guidelines for neutropenia fever in Indonesia use the National Comprehensive Cancer Network (NCCN) guidelines, Clinical Practice Guideline for the Use of Antimicrobial Agent in Neutropenic Patients with Cancer, and Infectious Diseases Society of America (IDSA) guidelines. The guidelines recommend the use of ciprofloxacin and amoxicillin/clavulanic acid as empirical therapy in patients at low risk of infection. For second-line therapy, ciprofloxacin monotherapy or combination of ciprofloxacin and clindamycin can be chosen. Levofloxacin monotherapy can also be an alternative therapy of choice, but there is not enough data to support the efficacy of the use of levofloxacin. For patients with a high risk of infection, the recommended therapy is beta lactam (cefepime) or carbapenem (imipenem/cilastatin/tazobaktam) [4].

The aim of this study was to determine the role of different antibiotics in increasing ANC values, duration of fever, and the degree of improvement in ANC values in cancer patients with neutropenia fever after chemotherapy.

2. Materials and Methods

This study was non-experimental with descriptive design. Retrieval of data was retrospective using secondary data in the form of medical records by purposive sampling. The number of samples that met the inclusion and exclusion criteria were 45 patients. Inclusion criteria included (a) male and female patients diagnosed with cancer with neutropenia fever; (b) cancer patients who have fever at the nadir of neutropenia which was 5-14 days after chemotherapy; (c) cancer patients with neutropenia fever who receive antibiotic therapy and Leukogen® as G-CSF; and (d) cancer patients with neutropenia fever who receive antipyretic therapy. The study was conducted in the Medical Record Room of Saiful Anwar General Hospital in Malang for the period of January 2015 to December 2016. This study has been stated as ethical conduct based on Ethical Clearance No. 400/05/K.3/302/2017 by the Saiful Anwar General Hospital Health Research Ethics Commission.

In this study we will observe the effect of antibiotic use on increasing ANC values, duration of fever, and degree of improvement in ANC values. ANC value data obtained from the calculation of the multiplication of leukocyte and neutrophil values. Then, the data were processed and analyzed by Kruskall Wallis to determine the difference in increasing ANC values. One Way ANOVA test to determine the difference in duration of fever. The Chi-Square combines cells test to determine differences in antibiotic use against the degree of improvement in ANC values. Data were statistically significant if the p value was < 0.05.

3. Results and Discussion

3.1. Sample Distribution

From this study, there were 45 samples that met the inclusion and exclusion criteria. A total of 22 female patients (48.89%) and 23 male patients (51.11%). The age of patients varied, consisting of 20 children (44.44%) with ages 1-11 years, 3 early adolescents (6.67%) with ages 12-16 years, 5 late adolescents

(11.11%) with ages 17-25 years, 11 adults (24.44%) with ages 26-45 years, and 6 early elderly (13.33%) with ages 46-55 years. The frequency of the most types of cancer experienced by patients was AML (Acute Myeloid Leukemia) in 16 patients (35.56%). Meanwhile, the frequency of the fewest types of cancer was CML (Chronic Myeloid Leukemia), skin cancer, and bone cancer, respectively in 1 patient (2.22%).

3.2. Pattern of Use of Antibiotics

There are four types of antibiotics given to study subjects as a therapy for neutropenia fever, namely single therapy ceftriaxone and ceftazidime as well as combination therapy of ceftazidime-ciprofloxacin and a combination of cefriaxone-gentamicin. The number of patients receiving the ceftazidime-ciprofloxacin combination therapy regimen was 16 patients (35.56%), receiving a combination of ceftriaxone-gentamicin as many as 15 patient (33.33%), receiving single ceftriaxone therapy in 8 patients (17.78%), and received a single ceftazidime therapy in 6 patients (13.33%). The duration of antibiotic use ranges from 3-20 days.

Table 1. Characteristics of Study Subjects.

Patient Characteristics		Number of Patients	Percentage (%)
Gender	Male	23	51.11
	Female	22	48.89
Age	1-11 years old (children)	20	44.44
	12-16 years (early adolescents)	3	6.67
	17-25 years (late adolescents)	5	11.11
	26-45 years (adults)	11	24.44
	46-55 years (early elderly)	6	13.33
Duration of Neutropenia Fever	1-2 days	7	15.56
	3-4 days	31	68.88
	5-6 days	7	15.56
Antibiotics Used	ceftazidime-ciprofloxacin I.V	16	35.56
	ceftriaxone-gentamicin I.V	15	33.33
	ceftriaxone I.V	8	17.78
	ceftazidime I.V	6	13.33

Notes:
I.V = intravenous

3.3. The Effect of Different Types of Antibiotics on Increasing ANC Value

The Kruskall-Wallis test was carried out between different antibiotics for increasing ANC values. From Table 2 it was known that the administration of different types of antibiotics gives a significant difference in the increase in ANC values ($p = 0.001$). Then, post hoc testing was performed using the Mann Whitney test (Table 3) and it was found that the group that experienced a significant difference was a combination of ceftriaxone-gentamicin with a combination of ceftazidime-ciprofloxacin; single ceftriaxone with a combination of ceftazidime-ciprofloxacin; single ceftriaxone with single ceftazidime; and a combination of ceftazidime-ciprofloxacin with a single ceftazidime.

Based on the results of the Kruskall Wallis test it was found that the average increase in the highest ANC value was indicated by a combination of ceftazidime-ciprofloxacin with a value of 34.00. This shows that the highest increase in ANC value occurs in a combination of ceftazidime-ciprofloxacin. Previous study has shown that 98% of patients showed improvement by ceftazidime-ciprofloxacin combination in patients with neutropenia fever, while the other 2% failed because of intolerant patients with a combination of these antibiotics. The combination of ceftazidime-ciprofloxacin shows good efficacy and tolerance as initiation therapy in patients with neutropenia and immunosuppressive. The ceftazidime-ciprofloxacin combination regimen can replace the standard regimen namely a combination of antipseudomonal agents and aminoglycoside because the combination of ceftazidime-ciprofloxacin has no effect on nephrotoxicity and does not require monitoring of drug levels [6].

Table 2. Kruskall-Wallis Test Results for Increased ANC Value in Different Antibiotic Groups.

Antibiotics	n	Average ANC Value Increase	p
ceftazidime-ciprofloxacin I.V	16	34.00	0.001
ceftazidime I.V	6	21.08	
ceftriaxone-gentamicin I.V	15	18.37	
ceftriaxone I.V	8	11.13	
Total	45		

Table 3. Mann Whitney Test Results for ANC Value Increase.

No.	Antibiotic Group	n	p
1.	ceftriaxone-gentamicin	15	0.155
	ceftriaxone	8	
2.	ceftriaxone-gentamicin	15	0.001*
	ceftazidime-ciprofloxacin	16	
3.	ceftriaxone-gentamicin	15	0.668
	ceftazidime	6	
4.	ceftriaxone	8	0.001*
	ceftazidime-ciprofloxacin	16	
5.	ceftriaxone	8	0.039*
	ceftazidime	6	
6.	ceftazidime-ciprofloxacin	16	0.015*
	ceftazidime	6	

Notes:
* Statistically significant because of p value < 0.05

3.4. The Effect of Different Types of Antibiotics on the Duration of Fever

One way ANOVA test was carried out to determine the difference in mean duration of fever in administration of different antibiotics. Table 4 shows that there was no significant difference in duration of fever in the administration of different antibiotic groups (p = 0.341). In addition, it was also known that the average duration of fever in each group was 3.9 days for the combination of ceftriaxone-gentamycin; 3.3 days for ceftazidime-cyproloxacin combination; 3.2 days for single ceftriaxone; and 3 days for single ceftazidime. From the results of the study it was found that the highest average duration of fever was a combination of ceftriaxone-gentamicin which was equal to 3.9 days and the lowest average duration of fever was a single ceftazidime i.e. 3 days. This shows that giving different antibiotics does not provide a significant difference in the duration of fever experienced by patients. This condition was caused by all patients in this study getting antipyretic drugs during treatment with antibiotics. Therefore, giving different antibiotics does not have a significant effect on the duration of fever.

In this study the patients received antipyretics namely acetaminophen 3 x 500 mg if fever or metamizol 3 x 500 mg if fever. A study exhibits that acetamiophen and metamizol have similar efficacy as antipyretics and analgesics [8]. However, there are other studies which state that metamizol has a better effect as an antipyretic for children [9]. For this reason, antipyretic administration of neutropenia fever can be adjusted according to the age and condition of the patient to get better efficacy. However, statistically, the administration of different antibiotics does not have an effect on the difference in mean duration of fever.

Table 4. One Way ANOVA Test Results for Fever Duration in Different Antibiotic Groups.

		n	Average ± s.d	p
Antibiotics	ceftriaxone-gentamicin	15	3.9±1.33	0.341
	ceftazidime-ciprofloxacin	16	3.3±0.96	
	ceftriaxone	8	3.2±1.58	
	ceftazidime	6	3.0±0.89	

3.5. The Effect of Different Types of Antibiotics on the Status of ANC Value Improvement

The data of this study did not meet the requirements for Chi-Square testing because there were more than 20% of cells that had an expected count below 5. Therefore, the alternative test was chosen, namely the Chi-Square test with cell merging. The cell-merging Chi-Square test was conducted to determine the association between administration of different types of antibiotics with the status of ANC value improvement. Improvement status of ANC values was divided into two categories, namely improvements with ANC values > 1000 U/μL and improvements with ANC values < 1000 U/μL. Table 5 showed that there was no difference between the administration of different types of antibiotics with improvement in ANC values. Thus, it can be seen that whatever type of antibiotic given was able to provide results of improvement in the value of ANC > 1000 U/μL.

Table 5. Chi-Square Cell Merging Test Results for Degree of Improvement of ANC Value at Different Antibiotic Groups.

	Degree of ANC Value Improvement*				**p**
	> 1000 U/μL		**< 1000 U/μL**		
	n	**%**	**n**	**%**	
ceftriaxone-gentamicin + ceftriaxone	19	42.22	4	8.89	
ceftazidime-ciprofloxacin + ceftazidime	17	37.78	5	11.11	0.722
Total	36	80.00	9	20.00	

Notes:
* Grouping based on the severity of neutropenia namely low-moderate neutropenia if ANC values > 1000 U/μL and severe neutropenia if the ANC value is < 1000 U/μL [5].

From the results of this study, the value of p = 0.711 means that there was no significant association between the administration of different types of antibiotics with the improvement in the ANC value. From these results it can be concluded that whatever type of antibiotic was given tends to give an improvement in the value of ANC > 1000 U/μL. These results indicate that any type of antibiotic given tends to give an improvement in the value of ANC > 1000 U/μL. From crosstable data before cell merging can be seen in column ANC value > 1000 U/μL, the biggest percentage within antibiotics was indicated by ceftazidime-ciprofloxacin combination that is equal to 93.8%. This means that 93.8% of 16 patients who received a combination of ceftazidime-ciprofloxacin experienced an improvement in the ANC value > 1000 U/μL. Therefore, from the four groups of antibiotics the highest percentage giving an improvement in the value of ANC > 1000 U/μL was a combination of ceftazidime-ciprofloxacin.

This was in line with the results of the analysis of the association of giving different types of antibiotics to increasing ANC values. The results of the analysis also showed that the highest increase in ANC values occurred in the ceftazidime-ciprofloxacin group. Although the use of prophylactic antibiotics in cancer patients with neutropenia fever was less in accordance with IDSA and NCCN guidelines, however, the antibiotics used in this study have a high effectiveness on increasing the ANC value of patients.

The combination of cephalosporin and quinolone group antibiotics has the ability to inhibit gram negative and gram positive bacterial activity. This is because third generation cephalosporins such as ceftazidime have effective antibacterial activity against gram negative bacteria. Meanwhile, quinolone group antibiotics such as ciprofloxacin besides being able to inhibit the growth of gram negative bacteria also have good activity in fighting gram-positive bacteria [10]. In a cost effective manner, the administration of antibiotics in combination with ceftazidime in the treatment of neutropenia fever requires a lower cost than the administration of ceftriaxone antibiotics [11].

From the three analyzes, it was known that administration of antibiotics in patients with neutropenia fever has an influence on increasing ANC values. Previous study has shown that antibiotics have a role in accelerating the maturation of neutrophils in the blood so that the expected ANC value increases and the body's immunity gets better. Moreover, antibiotics have a role in removing bacteria and reducing translocation and bacteremia so that the risk of infection also decreases [12]. Therefore, antibiotic therapy should be given to cancer patients with neutropenia fever to reduce mortality.

From the results of this study, researchers recommend the use of a ceftazidime-ciprofloxacin combination as prophylactic antibiotics in patients with neutropenia fever. This was because this combination of antibiotics showed the best increase in ANC value. Third-generation cephalosporin antibiotics such as ceftazidime active against Streptococci, Haemophilus influenza, Moraxella catarrhalis, and other gram-negative bacteria whereas quinolone group antibiotics such as ciprofloxacin also have activity against gram negative bacteria but also have activity against gram-positive bacteria. Ciprofloxacin is also known to be the most sensitive quinolone antibiotic against Pseudomonas aeruginosa [13].

The National Agency of Drug and Food Control explains that quinolone group antibiotics can cause arthropathy in the joints so that they should not be used in children and adolescents. The use of fluoroquinolones can also increase the risk of tendonitis and tendon rupture in all age ranges. This antibiotic also has side effects on the child's central nervous system, which can disrupt a child's psychology, memory loss, and concentration. Therefore, although a combination of ceftazidime-ciprofloxacin was known to have effectiveness in increasing high ANC values, but for children patients other antibiotics should be chosen. Based on the Kruskall-Wallis test, antibiotics that have the highest mean increase in ANC values sequentially after a combination of ceftazidime-ciprofloxacin were single ceftazidime, a combination of ceftriaxone-gentamicin, and single ceftriaxone. For this reason, a single ceftazidime therapy or a ceftriaxone-gentamicin combination can be selected in pediatric patients. However, gentamicin antibiotics are known to have a ototoxic and nephrotoxic effect. Nephrotoxic effects are reversible but the ototoxic effects are irreversible [14]. Therefore, in cancer patients who have previously experienced impaired kidney function and hearing loss it is better to choose antibiotics other than gentamicin.

4. Conclusion

The conclusion of this study was that the use of different types of antibiotics in neutropenia fever has a significant influence on the difference in increasing ANC values, where the type of antibiotic that produces the highest increase in ANC value was a combination of ceftazidime-ciprofloxacin. However, the use of different types of antibiotics did not have a significant effect on the duration of fever. All groups of antibiotics in this study provide similar effectiveness to the degree of improvement in ANC values, namely increasing the ANC value to > 1000 U/μL. Further study is needed regarding the molecular mechanism of the effect of antibiotic administration on increasing ANC values.

References

[1] Rezeki S and Hadinegoro S 2002 *Demam pada Pasien Neutropenia Sari Pediatri* **3** (4) 235-241

[2] de Naurois J, Novitzky-Basso I, Gill M J, Marti F M, Cullen M H and Roila F 2010 Management of Febrile Neutropenia: ESMO Clinical Practice Guidelines *Annals of Oncology* **21** (5) v252-v256

[3] Glasmacher A, Lilienfeld-Toal M V, Schulte S, Hahn C, Schmidt-Wolf I G H and Prentice A 2005 An Evidence-Based Evaluation of Important Aspects of Empirical Antibiotic Therapy in Febrile Neutropenic Patients *Clinical Microbiology Infection* **11** (5) 17-23

[4] American Cancer Society 2010 *Prevention and Treatment of Cancer-Related Infection: National Comprehensive Cancer Network* (NCCN) and *Clinical Practice Guideline for the Use of Antimicrobial Agents in Neutropenic Patients with Cancer.* Update by the Infectious Diseases Society of America

[5] Newburger P E and Dale D C 2013 Evaluation and Management of Patient with Isolated Neutropenia *Seminars in Hematology* **50** (3) 198-206

[6] Ghazal H H, Ghazal C D and Tabbara I A 1997 Ceftazidime and Ciprofloxacin As Empiric Therapy in Febrile Neutropenic Patients Undergoing Hematopoietic Stem Cell Transplantation *Clinical Therapeutics* **19** (3) 520–526

[7] Pizzo P A, Hathorn J W, Hiemenz J, Browne M, Commers J, Cotton D *et al* 1986 A Randomized Trial Comparing Ceftazidime Alone with Combination Antibiotic Therapy in Cancer Patients With Fever and Neutropenia *The New England Journal of Medicine* **315** (9) 552–558

[8] Ohnesoge H, Bein B, Hans R, Francksen H, Mayer L, Scholz J *et al* 2009 Paracetamol versus Metamizol in Treatment of Postoperative Pain after Breast Surgery: A Randomized Controlled Trial *European Journal of Anaesthesiology* **26** (8) 648-653

[9] Oborilova A, Mayer J, Pospisil Z and Koristek Z 2002 Symptomatic Intravenous Antipyretic Therapy: Efficacy of Metamizol, Diclofenac, and Propacetamol *Journal of Pain and Symptom Management* **24** (6) 608–615

[10] Chamber H F 2007 *Antimycobacterial Drugs*, in: Katzung BG, Basic & Clinical Pharmacology, 10th Ed (New York: McGraw-Hill Companies) pp. 771-780

[11] Dietrich E S, Patz E, Frank U and Daschner F D 1999 Cost-Effectiveness of Ceftazidime or Imipenem/Cilastatin versus Ceftriaxone + Aminoglycoside in The Treatment of Febrile Episodes in Neutropenic Cancer Patients in Germany *Infection* **27** (1) 23-27

[12] Nguyen D N, Fuglsang E, Jiang P, Birck M M, Pan X, Kamal S B *et al* 2016 Oral Antibiotics Increase Blood Neutrophil Maturation and Reduce Bacteremia and Necrotizing Enterocolitis in The Immediate Postnatal Periode of Preterm Pigs *Innate Immunity* **22** (1) 51-62

[13] Oliphant C M and Green G M 2002 Quinolones: A Comprehensive Review. *American Academy of Family Physicians* **65** (3) 455-464

[14] Yulianti and Mahdiani S 2015 Gangguan Pendengaran Penderita Tuberkulosis Multidrug Resistant. *Oto Rhino Laryngologica Indonesiana* **45** (2) 83-89

Immunological Responses in Female Rats Model with Anemia Treated with Snake Fruit Seed Flour Extract

I K Ristanti[1,2], D G Tamtomo[1,4] and D Indarto[1,3,4*]

[1]Postgraduate of Nutrition Sciences, Universitas Sebelas Maret, Surakarta.
[2]Human Nutrition, Postgraduate of Nutrition Sciences, Universitas Sebelas Maret, Surakarta
[3]Department of Physiology, Faculty of Medicine, Universitas Sebelas Maret, Surakarta
[4]Biomedical Laboratory, Faculty of Medicine, Universitas Sebelas Maret, Surakarta
[5]Department of Anatomy, Faculty of Medicine, Universitas Sebelas Maret, Surakarta

[*]Corresponding author's email: dono@staff.edu.ac.id

Abstract. Natural substances become popular for alternative treatment of human diseases including anemia. A recent study has reported that administration of snake fruit seed flour (SSF) is able to increase hemoglobin level in rat model with anemia but it needs higher amount of SSF. In addition, SSF contains 3.71% protein/100g that potentially results in hypersensitivity. The aim of this study was to evaluate the side effects of SSF extract on leucocytes number and its differential account (granulocytes, lymphocytes, and monocytes) in rat model with anemia. We used 14 female rats model with anemia, aged 8-10 weeks, and had ± 200 g body weight (BW). After rats model with iron deficiency anemia (IDA) has been established, female rats were randomly divided into 2 groups: control (C) group only received a low iron diet, and treatment (T) group received a low iron diet and 1.75g/kg BW/day SSF extract for 14 days. Leucocytes number was measured using a veterinary haematology analyzer. Collected data were analyzed using the independent T-test with p value <0.05. Female rats in the C group had the mean leucocytes number higher than the T group, but not statistically different ($p=0.073$). From differential count, greater numbers of granulocytes and monocytes were observed in the C group, compared to the T group ($p=0.301$;$p=0.34$). In contrast to granulocyte and monocytes number, the mean lymphocytes number in the C group were statistically different than the T group ($p=0.046$). Overall, administration of SSF extract was able to decrease leucocytes number and differential count. Administration of 0.175g/100g BW/day SSF extract does not increase immunological responses in female rats model with anemia.
Keywords : Anemia, Leucocytes, Immunological responses, Snake fruit seed extract

1. Introduction

Iron deficiency is commonly found in some susceptible age groups such as children, female adolescents, and pregnant women [1,2]. Based on a global survey, the prevalence of anemia is 1.93 million people or 27% in 2013 and 59.5% anemia cases are due to iron deficiency [3,4]. In Indonesia, the prevalence of anemia belongs to moderate category. According to data of Riskedas (2013), the prevalence of anemia in children above 1 years old is 21.7% and 23.9% in women, higher than men

(18.4%) [5,6]. Meanwhile, the prevalence of anemia in pregnant women has increased from 37.1% in 2013 to 48.9% in 2018 [7].

Some studies have indicated that iron deficiency anemia (IDA) has conflicting results toward the immune system. In general, IDA can be due to increase of iron requirement, to decrease of iron intake or absorption or to increase of iron loss [8]. As a results, hemoglobin levels reduce below the normal range and it changes morphology of red blood cells [9]. A recent study conducted in 200 respondent (182 female and 18 male), age 32-52 years with IDA shows that total leucocytes number and its differential account are in normal range [10]. In contrast to this study, Samanta and Senapati (2018) reported that 30 respondent (19 female and 11 male), age 20-40 years with nutritional anemia have higher total leucocytes number than that of control group, but it is not statistically different [11]. In addition, hypoxia and ischemia are strongly associated with anemia, which lead to increase of white blood cells [12,13,14].

Iron-fortified foods and iron supplementation are the most recommended therapy for IDA [15, 16, 17,18]. Administration of iron tablet frequently has some side effects, such as abdominal pain, vomiting, and diarrhea, so that some anaemic patients do not complete their treatment. Now days, natural substances become popular for an alternative treatment of human diseases including anemia. A recent study has reported that administration of 18.6 g/kg BW/day SSF in rat model with anemia can restore hemoglobin levels after 4 weeks treatment [19]. It is not surprised that SSF administration can be used for anemia treatment because 100 g SSF contains 26.8 mg iron, 4.56 mg zinc, and 0.88 mg vitamin C. If this dose of SSF administration is converted to human with anemia, it will need high amount of SSF. In order to decrease the amount of SSF, we have extracted SSF by using a maceration method with ethanol solvent. Therefore, the aim of this study was to evaluate the side effects of SSF extract on total leucocytes number and its differential account in rat model with anemia.

2. Methods

An experimental laboratory with post test control group design was conducted in this study during 24 days in May 2019. Approval of this study protocol was obtained from The Ethics Committee of Medical Research, Faculty of Medicine, Univertas Sebelas Maret, Surakarta (No. 165/UNS27.6/KEPK/2018). Female Wistar rats, which used in this study were got from Laboratory Center for Food and Nutrition Studies, Gadjah Mada University, Yogyakarta.

2.1. Preparation of SSF Extract

Snake fruit seeds, which were waste of processed fruit production in three home industries at Turi district, Sleman Regency, Yogyakarta, were used as raw material for making SSF extract. We modified the existing method to extract SSF, using maceration with ethanol solvent [20]. SSF extract was prepared at Food Engineering Laboratory, Faculty of Agriculture, Universitas Sebelas Maret, Surakarta. Iron level in SSF extract was measured using Atomic Absorbance Spectrometry (AAS) in Research Laboratory, Faculty of Pharmacy, Universitas Muhammadiyah, Surakarta.

2.2. Rat Model with Iron Deficiency Anemia Care

Wistar rats were kept in clean plastic cage with a wire and controlled environmental conditions in the Animal Research Centre, Laboratory for Food and Nutrition Studies, Gadjah Mada University, Yogyakarta. The room was maintained at 27-29°C, 12 hours light and 12 hours dark, and 70-90% humidity [21]. Rats were given 10% BW of AIN-93M standard diet, and distilled water ad libitum. All foods and drink were provided by the Nutrition Laboratory. Development of rat model with IDA was adopted from Susanti et al (2017), giving by a low-iron diet (iron depletion from AIN-93M standar feed) for 10 days [22].

2.3. Study Design

A total of 14 female Wistar rats aged 8-10 weeks, had ± 200 g body weight (BW), and were healty condition were used in this study. Once rats model with IDA has been established, they were randomly divided into 2 groups: control (C) and (T) treatment groups. Anemic rats in the C group only received a low iron diet, whereas anemic rats in the T group received a low iron diet and 1.75g/kg BW/day SSF extract for 14 days. The administration of SSF extract proportionally correlated with daily iron requirement and was given to rats using a 3 ml gastric canulle every morning [19].

2.4. Data Analysis

Blood samples were taken from orbital vein for hematological studies and collected using vials with EDTA anticoagulant. Total leucocytes account was measured using a veterinary haematology analyzer at the Animal Hospital Prof. Soeparwi, Faculty of Veterinary Medicine, Gadjah Mada University, Yogyakarta. The independent t-test was used to analyze the differences of total leucocytes, granulocytes, lymphocytes, and monocytes between C and T groups with p value <0.05.

3. Results and Discussion

We developed IDA in 14 female rats by depletion of iron in the rats standard diet for 10 days. Table 1 showed basic characteristics of female rats with IDA. The average of BW in the C group (184.14±3.97 g) was significantly lower than the average of BW in the T group (193.29±3.35 g) with p= 0.001. Additionally, female rats in the C group had 9.65±0.57 g/dl haemoglobin level and lower than the average of haemoglobin level in the T group 9.88±0.49 g/dl, but not significantly different (p=0.444). IDA can affect the body weight of rats, that caused by reduced enzyme activity and oxygen supply to cells, so that decrease in cell energy and disturb body's metabolism [23].

Table 1. Characteristics of Female Rats Model with IDA.

Characteristics	C Group	T Group	p-value
BW (g)	184.14±3.97	193.29±3.35	0.001
Hemoglobin (g/dl)	9.65±0.57	9.88±0.49	0.444

In general, we found that female rats in the T group treated with SSF extract for 14 days had lower leucocytes, compared to female rats in the C group (Figure 1a-d). Higher mean total leucocytes account (13.78±4.5 x10^3/µl) was observed in female rats of C group, compared to female rats of T group (10.05±2.19 x10^3/µl) with p=0.073. From differential count, the average of granulocytes, lymphocytes, and monocytes in the C group was greater than the average of granulocytes, lymphocytes, and monocytes in the T group. However, a significant difference was only found in the average of lymphocytes (p=0.046).

Study by Singh (2010) reported in 20 patient with nutritional anemia, total number of leucocytes in the anemia group was higher than the control group, but not statistically different, and on differential count, there was an increase in the number of neutrophils, without changes in the number of lymphocytes [12]. This study indicate the same result with study by Samanta and Senapati (2018), reported that 30 respondent (19 female and 11 male), age group of 20–40 years with nutritional anemia, showed that total leucocytes number in anemia group is higher than control group, but statistically not significant [11]. However, these results are conflicting with the study by Burak et al (2016), a study conducted on 200 respondent (182 female and 18 male), age 32-52 years with IDA showed total leucocytes number and its differential number are in normal range [10]. Our study indicated the same result with Samanta and Senapati's, total leucocytes number in C group was higher than T group, but it was not statistically different.

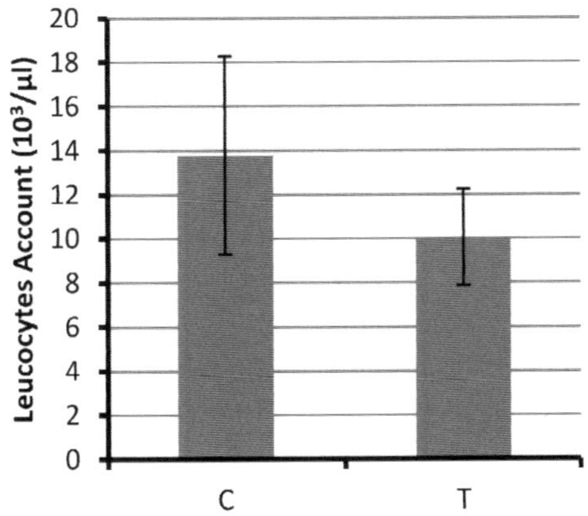

a

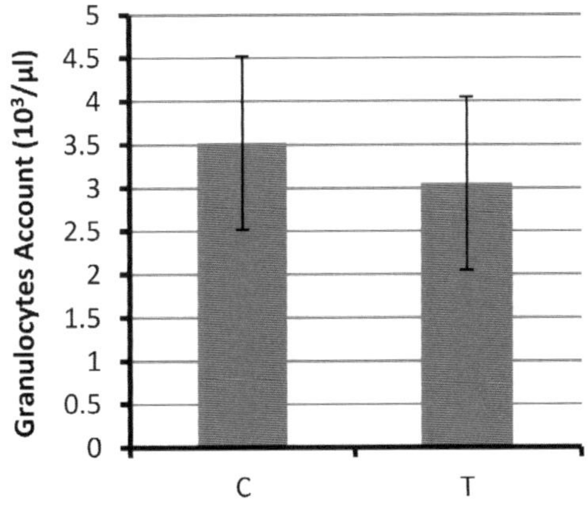

b

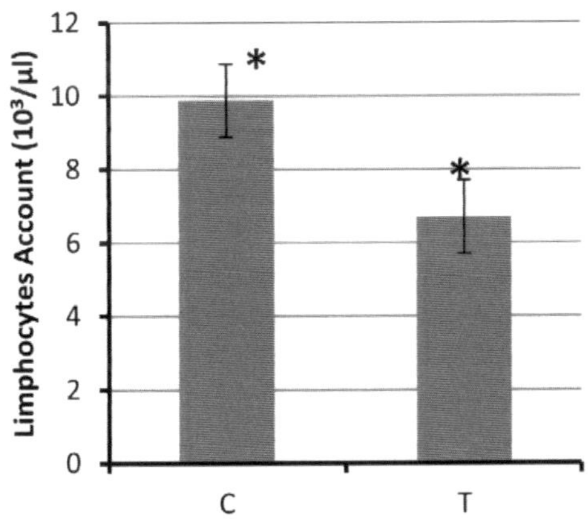

c

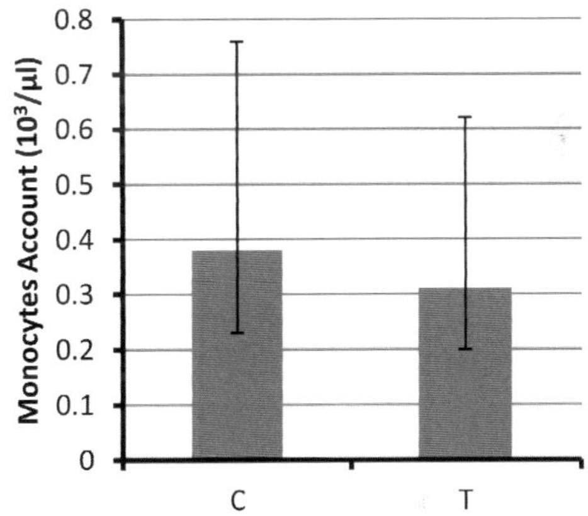

d

Figure 1. The effect of SSF extract administration on leucocytes number and cell types in female rats with IDA. Female rats with IDA in the C group (negative control) only received a low iron diet, whereas in the T group (treatment group) received a low iron diet and 1.75g/kg BW/day SSF extract for 14 days. The average of leucocytes in all groups (C group and T group) was presented as mean±SD. (*) indicated a significant difference of lymphocytes account between before and after treatment.

Anemia is associated with hypoxia and ischemia, an increase in total leukocytes (although still within the normal range) is associated with an increased risk of heart disease, and can be a sign of haematological stress syndrome [14,10]. Differential leucocytes that play a role as a marker of ischemic heart diseases is neutrophils to lymphocyte ratio (N/L ratio) [11].

Increased leucocytes number can occur because erythrocyte production is impaired due to a decrease in iron levels. The erythropoiesis process is inhibited because hypoferremia maintains the amount of iron for other metabolic needs during infection, thereby increasing glucocorticoid activity to increase the release of many nucleated leukocytes from the bone marrow. In addition, inflammatory cytokines directly suppress the erythropoiesis process, increase leucocyte production, and activate macrophages to improve erythrophagocytosis [12,11,24].

Granulocyte is a leukocyte group consisting of neutrophils, esinophils and basophils. Neutrophils will increase in response to inflammation. Our study showed granulocytes number of C group was higher than the T group, but statistically not significant. Iron deficiency will affect the regulatory function of T-Lymphocytes and immune cells. The N/L ratio is an inflammatory marker of IHD, myocardial infarction, solid tumour, and colon cancer [13,14,25,26]. The study by Gosai et al (2013), showed no significant difference of total leucocytes number in anemia and control group, but in the anemia group the N/L ratio was significantly higher than control group, this was due to the neutrophils numbers in anemia group significantly increase than control group [14]. Another study by Samanta and Senapati (2018), also reported an increase in the N/L ratio significantly in the anemia group compared to the control group [26]. However, in this study it has not been known whether the increase in leucocytes and lymphocytes in the control group will increase the risk of incidence IHD, MI and etc, because no neutrophils number is performed.

The role of monocytes in anemia cannot be determined with certainty. Inflammatory cytokines including TNF-α, IL-1 and Interferon-ɣ play a role in inhibiting the erythropoesis [27]. The study by Libgrets et al (2011), mice with overproduction of interferon-y caused mild to moderate normocytic anemia. Interferon-ɣ activates macrophages in the splenic red pulp causing a 50% reduction in erythrocytes. Increased interferon-ɣ is done by induction of viruses or intracellular pathogen infection, which causes increased production of monocytes and macrophages and suppresses the erythropoesis process [28].

4. Conclusion
Administration of 1.75g/kg BW/day SSF extract for 14 days not increases leucocytes number in female rats model with anemia. Lymphocyte number in female rats significantly higher in C group, while, total leucocyte, granulocytes, and monocytes amount in C group insignificantly different with T group. We require further investigation whether or not administration of SSF extract can result in allergic responses.

References
[1] WHO (World Health Organization) 2014 *WHO Global nutrition targets 2025: anaemia policy brief* (Geneva: World Health Organization)
[2] WHO (World Health Organization) 2015 *The global prevalence of anemia 2011* (Geneva: World Health Organization)
[3] Kassebaum N J, Jasrasaria R, Naghavi M, Wulf S K, Johns N, Lozano R, Regan M, Weatherall D, Chou D P, Eisele T P et al 2014 A systematic analysis of global anemia burden from 1990 to 2010 *Blood* **123** 615-24
[4] Kassebaum N J and GBD 2013 Anemia Collaborators 2016 The Global Burden of Anemia. *Hematol Oncol. Clin. North Am.* **30** 247–308
[5] WHO (World Health Organization) 2011 *Haemoglobin concentration for the diagnosis of anaemia and assesment of severity. vitamin and mineral nutrition information system* Geneva; World Health Organization
[6] Kemenkes RI 2013 *Riset Kesehatan Dasar 2013* (Jakarta : Badan Penelitian dan Pengembangan Kesehatan Kementrian Kesehatan Republik Indonesia)
[7] Kemenkes RI 2018 *Riset Kesehatan Dasar* 2018 (Jakarta : Badan Penelitian dan Pengembangan Kesehatan Kementrian Kesehatan Republik Indonesia)
[8] Clarke L and Dodds A J 2014 Iron deficiency anaemia: causes, symptoms and treatment *Medicine Today* **15** 36–42
[9] AlDallal S 2016 Iron deficiency anaemia : a short review *J. Cancer Res. Immunooncol.* **2** 106
[10] Burak U, Ilhan D, Hasan A and Deccane D 2016 The effect of iron deficiency anemia treatment on neutrophil to lymphocyte ratio and platelet to lymphocyte ratio *Global Journal of Hematology and Blood Transfusion* 2016 **3** 1–5
[11] Samanta P and Senapati L K 2018 Association of nutritional anemia with leukocyte and platelet counts in people of Odisha *Natl. J. Physiol. Pharm. Pharmacol.* **8** 526–29
[12] Singh K 2010 Leucocyte counts in anaemia *Indian J. Physiol. Pharmacol.* **54** 85–8

[13] Onabanjo O O, Jerling J C, Covic N, Van Graan A, Taljaard C and Mamabolo R L 2012 Association between iron status and white blood cell counts in African schoolchildren of the North-West Province, South Africa *J. Epidemiol. Glob. Health* **2** 103–10

[14] Gosai H, Akhani P, Kanchan A, Jasrotia R B, Palan B M and Harsoda J M 2013 Neutrophil to lymphocyte ratio and platelet count in anaemic female medical students *Int. J. Res. Med.* **2** 44–47

[15] Pritwani R and Mathur P 2015 Strategies to combat micronutrient deficiencies: a review *Int. J. Health Sci. Res.* 2015, **5** 362–73

[16] Taylor and Rampton 2015 Treatment of iron deficiency anemia: practical considerations *Pol. Arch. Med. Wewn.* **125** (6) 452–60

[17] Lin H M, Deng S G, Huang S B, Li Y J and Song R 2016 The effect of ferrous-chelating hairtail peptides on iron deficiency and intestinal flora in rats *J. Sci. Food Agric.* **96** 2839–44

[18] Gao F, Guo W , Zeng M , Feng Y and Feng G 2019 Effect of microalgae as iron supplements on iron-deficiency anemia in rats *Food Funct.* **10** 723–32

[19] Susanti T 2018 *Pengaruh dosis dan durasi pemberian tepung biji salak terhadap kadar hemoglobin dan indeks eritrosit pada tikus model anemia* (Surakarta: Universitas Sebelas Maret Surakarta)

[20] Nursucihta S, Thai'in H A, Putri D M, Utami D N and Ghani A P 2014 Antianemia Activity of *Parkia speciosa* Hassk Seed Ethanolic Extract *Trad. Med. J.* **19** 49–54

[21] Kusumawati D 2004 Bersahabatlah dengan hewan coba (Yogyakarta: Universitas Gajah Mada Press)

[22] Susanti, Dirgahayu P and Indarto D 2017 Development of rat model with iron deficiency anemia by modification of its standard food *2nd Public Health International Conference (PHICo 2017), Advances in Health Science Research* **9** 63–4

[23] Polin V, Coriat R, Perkins G, Dhooge M, Abitbol V, Leblanc S, Prat F and Chaussade S 2013 Iron deficiency: from diagnosis to treatment *Dig. Liver Dis.* **45** 803-9

[24] Ganz T 2017 Iron and infection *Int. J. Hematol.* **107** 7-15

[25] Park J J, Jang H J, Oh I Y, Yoon C H, Suh J W, Cho Y S, Youn T J, Cho G Y, Chae I H, Choi D J 2013 Prognostic value of neutrophil to lymphocyte ratio in patients presenting with ST-elevation myocardial infarction undergoing primary percutaneous coronary intervention *Am. J. Cardiol.* **111** 636–42

[26] Templeton A J, McNamara M G, Šeruga B, Vera-Badillo F E, Aneja P, Ocana A, Leibowitz-Amit R, Sonpavde G, Knox J J, Tran B et al 2014 Prognostic role of neutrophil-tolymphocyte ratio in solid tumors: a systematic review and meta-analysis *J. Natl. Cancer Inst.* **106** 1-11

[27] Nemeth and Ganz 2014 Anemia of inflammation *Hematol. Oncol. Clin. North Am.* **28** 671–681

[28] Libregts S F, Gutierrez L, de Bruin A M, Wensveen F M, Papadopoulos P, van I W, Özgür Z, Philipsen S and Nolte M A 2011 Chronic IFN-gamma production in mice induces anemia by reducing erythrocyte life span and inhibiting erythropoiesis through an IRF-1/PU.1 axis. *Blood* **118** 2578–88

Journal of Physics: Conference Series **1374** (2019) 012039 doi:10.1088/1742-6596/1374/1/012039

Effect of fat supplementation containing high palmitic acid enriched with lecithin on quail performances

I Syarif[1*], W Hermana[1] and Y Retnani[1]

[1]Department of Feed and Animal Science, Bogor Agricultural University, Indonesia

*Corresponding author's email: syarifirhamni@gmail.com

Abstract. The aim of this study was to observe the effect of fat supplementation containing high palmitic acid enriched with lecithin on the performances of quail starter-grower phase. This study used a completely randomized design with 330 female day old quails Coturnix Coturnix Japonica randomized and entered into 5 treatments consisted of commercial feed (T0), T0 + 2.5% fat local + 6% lecithin from added fat (T1), T0 + 2.5% fat commercial + 6% lecithin from added fat (T2), T0 + 5% fat local + 6% lecithin from added fat (T3), T0 + 5% fat commercial + 6% lecithin from added fat (T4). The results showed that fat supplementation containing high palmitic acid enriched with lecithin was not significantly different ($P<0.05$) on body weight, daily body weight gain at the age of 36-42 days, feed consumption, feed conversion ratio, and mortality but daily body weight gain was significantly different ($P<0.05$) at the age of 8-35 days. Local fat supplementation containing high palmitic acid produced a higher income over feed cost than the control treatment and fat commercial treatment. Fat supplementation containing high palmitic acid enriched with lecithin did not interfere with quail health status as a whole.

1. Introduction

The demand for livestock products as a source of protein for the human is increasing because people are increasingly aware of the importance of nutritional needs. One of the animal products that is popular in the society is quail products. The advantages of quail have been cheaper production costs, resistance to disease making it a superior laboratory animal, fast growth, rapid generation intervals, fast adult sex, and the smallest animals for high egg productivity [1]. Quail eggs with a small size have been 3-4 times more nutrient content than the nutrient content of chicken eggs. The high of amino acids, fatty acids, vitamin E, iron and zinc in quail eggs has been good sources of nutrients for human health [2]. The advantages of quail are very appropriate to be developed and used as a producer of eggs in order to meet the needs of eggs for human.

The needed of the human for quail eggs is done by increasing the quality of feed which is the largest production cost of quail farms. The feed is needed as a support for quail performances. One way to improve feed quality and reduce feed deficiency is supplementing feeds from abundant natural resources and not widely used. One of the potential natural resources is the byproduct of the agricultural industry.

The palm oil industry is a natural resource that has potential and not widely used in the livestock industry. The process of making palm oil produces byproducts in the form of fat containing high palmitic acid. This fat can increase metabolic energy and increase the use of other nutrients such as proteins and fat-soluble vitamins. Fat can also help reducing heat increments and increasing palatability when compared to high carbohydrate feeds [3]. Fats have insoluble properties in water so emulsifiers are needed to absorb fat in the body [4]. One emulsifier that can be used is lecithin. Supplementation of fat containing high palmitic acid enriched with lecithin is expected to improve quail performances. The purpose of the present study was to inspect the effect of fat supplementation containing high palmitic acid enriched with lecithin on the quail performances.

2. Materials and Methods

2.1. Time and Place

The study was conducted in December 2018 until May 2019, located at Kayumanis Quail Farm on Rawa Taman street number 180, Mekarwangi village, Tanahsereal Kodya district, Bogor. Proximate analysis of feed was carried out at the central laboratory of Bogor Agricultural University. Analysis of fatty acid profiles was carried out at the chemical integrated laboratory of Bogor Agricultural University.

2.2. Livestock

This study used 330 female day old quails (DOQ) Cortunix Cortunix Japonica and maintained until 42 days old. The initial body weight was weighed first at the beginning of maintenance. The animals were allocated into 5 treatments and 3 replications randomly with 22 quails each replication.

2.3. Lecithin

lecithin used in this study was soy lecithin. Soy lecithin was a byproduct of industrial biodiesel from soybean oil. Soy lecithin was obtained from Saraswati Indo Genetech Ltd. Commercial fat containing high palmitic acid mixed with lecithin as much as 6% meanwhile in local fat containing high palmitic acid, lecithin mixed in the commercial feed as much as 6%.

2.4. Fat containing high palmitic acid

Fat containing high palmitic acid was a byproduct of palm oil. Local fat containing high palmitic acid was obtained from Asianagro Agungjaya Ltd and was an authentic Indonesian natural resource. Commercial fat containing palmitic acid was obtained from another country. Fat commercially containing high palmitic acid mixed with 6% lecithin from fat supplemented meanwhile lecithin in local fat containing high palmitic acid mixed in commercial feed with the same amount as commercial fat containing high palmitic acid.

Table 1. Fatty acid composition of fat supplemented.

Fatty Acid		% w/w	
		Local Fat	Commercial Fat
Lauric Acid	C12:0	0.07	0.11
Myristic acid	C14:0	0.98	1.11
Pentadecanoic acid	C15:0	0.05	0.06
Palmitic acid	C16:0	58.73	61.73
Palmitoleic acid	C16:1	0.06	0.06
Heptadecanoic acid	C17:0	0.07	0.09
Stearic acid	C18:0	5.06	4.46
Elaidic acid	C18:1n9t	0.04	0.03
Oleic acid	C18:1n9c	23.64	16.41
Linoleic acid	C18:2n6c	4.84	5.47
Arachidic acid	C20:0	0.27	0.28
Linolenic acid	C18:3n3	0.12	0.38
Lignoseric acid	C24:0	0.04	0.04

2.5. Feeds

The feeds used commercial feed from Sinta Feedmill Ltd. The feed was given in the form of mash. The feed ingredients used in BR2 and SP2 were corn, bran, corn gluten, pollard, meat and bone flour, soybean meal, oil, calcium phosphate, calcium carbonate, sodium chloride, amino acids, vitamins, trace minerals, and antioxidants. The feed used for 1 to 35 days old was BR2 feed. Age 36 days to complete, quail was given SP2 feed. The nutrient content of the feed can be seen in table 2.

Table 2. Nutrient composition of feed.

Nutrients	BR2	SP2
Water content (%)	10.23	9.20
Ash (%)	4.69	13.94
Crude fat (%)	6.32	5.83
Crude protein (%)	17.41	17.37
Crude fiber (%)	1.96	2.33
Gross Energy (kal/gram)	4270	3886

2.6. Cages, Tools, and Materials

This study used 15 flocks of the cage and the sizes were 20 cm x 30 cm x 20 cm for the starter-grower phase (8-35 days) and 50 cm x 60 cm x 40 cm for layers phase (36-42 days). The equipment used in this study were cages, newspapers, cardboard boxes, 25-watt incandescent lamps, plastic, food and drink containers, scales, thermometers, calculators, and stationary. This material used in this study were 330 female DOQ, soy lecithin, local fat containing high palmitic acid, commercial fat containing high palmitic acid, and quail feed.

2.7. Procedures

2.7.1. Sample Testing. Analysis of fatty acid profiles of fat samples was measured by using AOAC (2012) method in the chemical integrated laboratory of Bogor Agricultural University using gas chromatography. The proximate analysis of feed was measured by using AOAC (2005) method in the central laboratory of Bogor Agricultural University.

2.7.2. Cage Preparation and Maintenance. The cage was prepared before DOQ arrived. The inside and outside of the cage, a place for feed, drinking water, and lights as heating, were cleaned and prepared. The study was conducted for 42 days. Activities carried out during maintenance were cleaning cages, cleaning drinking water places, and cleaning the environment around the cage.

Giving incandescent lights with 25 watts of power for each flock was used as lighting and heating for livestock. 25-watt incandescent lamp was used until 28 days in each flock and was used as lighting and heating. The age of 29 days until completion only used one 25 watt incandescent lamp as lighting.

Quail was given feed treatment for the adaptation process since the age of 1 day. Feed adaptation was carried out for 7 days with a ratio of 25:75% on days 1 and 2, 50:50% on days 3 and 4, 75:25% on days 5 and 6, and 100% on day 7. In health management, quails were given regular vaccinations to prevent an illness from coming. The first vaccine given to quails at 6 days was the ND IB vaccine and the second vaccine given to quails at 28 days was the ND AI Kill vaccine.

Quail was given feed and drinking water as ad libitum. Quail 1 to 24 days old were given drinking water using 1-liter gallon container. The manual gallon was replaced and washed every morning and afternoon. 25 days old quail to completion was given drinking water using 3-liter gallon container.

2.7.3. Temperature Measurement. The measurement of the temperature of the cage was carried out in the morning (6:00 a.m to 7:00 a.m), during the day (12:00 p.m to 1:00 p.m), and in the afternoon (04:00 p.m to 05:00 p.m). The daily temperature value was obtained by calculating the average temperature in the morning, noon, and afternoon. The night temperature was considered same as in the morning temperature.

2.7.4. Experimental Design and Data Analysis. The experimental design in this study used a completely randomized design because the quails had relatively uniform initial body weights, same cage, and same equipment. The data were analyzed by analysis of variance and continued with Duncan test to see differences between treatments (Steel and Torie 1993). Statistical analysis used SAS University 2019. The treatments were:

T0 = Commercial feed (CE)
T1 = Commercial feed (CE) + 2.5% local fat containing high palmitic acid (LF) + 6% lecithin from added fat (L)
T2 = Commercial feed (CE) + 2.5% commercial fat containing high palmitic acid (CF)
T3 = Commercial feed (CE) + 5% local fat containing high palmitic acid (LF) + 6% lecithin from added fat (L)
T4 = Commercial feed (CE) + 5% commercial fat containing high palmitic acid (CF)

2.8. Measured Variables

2.8.1. Body Weight (g). Body weight was measured by weighing quails in each flock and divided by the number of quails in the flock.

2.8.2. Daily weight gain (g day $^{-1}$). Body weight gain was measured by weighing body weight per week and reduced body weight at the last week.

2.8.3. Feed consumption (g day^{-1}). Feed consumption was calculated from the average amount of daily feed and divided by the number of quails that live.

2.8.4. Feed conversion ratio (g/g). The feed conversion ratio was calculated from the average amount of feed consumption and divided by the body weight produced.

2.8.5. Mortality (%). Mortality (death) was obtained from the number of quails that die during the maintenance process and divided by the number of quails that survive.

2.8.6. Income Over Feed Cost (IOFC). IOFC was calculated by knowing the body weight produced on the amount of feed consumption and the price of quail. IOFC was obtained from a reduction in the total price of quail with the total cost of feed during the maintenance.

3. Results and Discussion

3.1. General Conditions of Cage Environment

Table 3. Environmental Temperature.

Week-	(06.00-07.00 a.m)	(12.00 a.m-01.00 p.m)	(04.00-05.00 p.m)
	$\bar{X} \pm SD$	$\bar{X} \pm SD$	$\bar{X} \pm SD$
1	30.75±0.50	33.75±1.50	34.50±1.00
2	29.86±1.21	31.71±1.70	32.43±2.30
3	30.29±0.95	33.86±0.90	33.14±0.69
4	28.86±1.57	32.00±1.53	30.43±1.40
5	27.14±1.07	29.86±1.21	29.14±0.90
6	26.83±0.75	28.43±1.27	28.57±1.51
Average	28.87±1.80	31.44±2.34	31.13±2.46

The intensive farm requires good management. One factor that needs to be considered in the intensive farm is temperature. The temperature will affect the level of feed consumption and consumption of drinking water. Randall and Bolla [5] reported that the ideal temperature for the first week DOQ was 35°C and decreased 3.5°C per week. The different temperatures at week 1 to week 6 was due to different weather conditions each week. In table 3 shows that quails are in the normal temperature which ranges from 28-35°C.

3.2. Quail Performances

Table 4. Performances of quail 8-42 days old.

Variables	T0	T1	T2	T3	T4
Body weight g (8 days)	21.12±0.82	22.13±1.54	23.02±0.62	21.70±1.13	23.37±1.16
Body weight g (35 days)	122.95±8.79	121.30±3.92	124.08±3.63	111.54±1.62	114.55±5.05
Body weight g (42 days)	151.98±8.83	145.14±5.23	147.28±4.33	139.83±5.60	143.67±8.12
Daily weight gain g day^{-1} (8-35 days)	3.64±0.29^{a}	3.54±0.09^{ab}	3.61±0.15^{a}	3.21±0.09^{b}	3.26±0.15^{b}
Daily weight gain g day^{-1} (36-42 days)	4.15±0.43	3.41±0.25	3.31±0.62	4.04±1.01	4.16±0.71
Daily weight gain g day^{-1} (8-42 days)	3.74±0.24	3.51±0.11	3.55±0.12	3.38±0.13	3.44±0.20
Feed consumption g day^{-1} (8-35 days)	12.20±0.41	12.01±0.61	12.01±0.52	11.37±0.26	11.48±0.47
Feed consumption g day^{-1} (36-42 days)	19.29±0.86	18.10±0.31	19.32±0.85	17.72±0.75	18.27±1.33
Feed consumption g day^{-1} (8-42 days)	12.84±0.40	12.36±0.57	12.64±0.44	11.80±0.41	12.03±0.53
Feed conversion ratio g/g (8-35 days)	3.36±0.17	3.39±0.13	3.33±0.22	3.55±0.18	3.53±0.13
Feed conversion ratio g/g (36-42 days)	4.69±0.65	5.08±0.53	5.97±1.17	4.58±1.19	4.45±0.58
Feed conversion ratio g/g (8-42 days)	3.44±0.21	3.52±0.06	3.56±0.03	3.50±0.05	3.50±0.06
Mortality % (8-42 days)	0.20±0.17	0.20±0.17	0.30±0.00	0.40±0.46	0.30±0.00

*Treatments: T0:Commercial feed (CE), T1: CE+LF2.5%(L6%), T2: CE+CF2.5%(L6%), T3: CE+LF5% (L6%) T4: CE+CF5%(L6%)

Fat supplementation containing high palmitic acid enriched with lecithin was not significantly different (P<0.05) on body weight, body weight gain at age 36-42 days, feed consumption and feed conversion but it was significantly different (P<0.05) on daily body weight gain at age 8-35 days. Fat supplementation did not damage overall quail performances, this is also in accordance with the recent studies [6,7] reported that supplementation fat for poultry feed formulated did not damage overall quail health status. The fat containing high palmitic acid which used is a byproduct of crude palm oil production. The fat was chosen because of its low price and high saturated fatty acid, and then it can reduce oxidation and peroxidation in adipose tissue rather than fat containing high polyunsaturated fatty acid [8]. Lecithin used in this study was 6% from added fat. According to Heugten and Odle [9] and Roy *et al.* [6] reported that lecithin levels more than 1% of added fat can improve fat digestion efficiency so that it can be used as an energy source and increased body weight. 6% of lecithin levels from the fat added in this study are not enough to help absorb fat perfectly. Fat absorption depends on factors such as carbon chain length, number of bonds, the presence or absence of ester bonds (triglycerides or free fatty acids), special regulation of saturated and unsaturated fatty acids on glycerol, free fatty acid composition, feed composition, type and the amount of triglycerides added in feed, intestinal flora, gender and age [10-12].

Body weights up to 42 days were not significantly different (P<0.05) but the average quail body weight at 42 days met the standard body weights for quail which will enter the layer phase. This result is in agreement with Filho *et al.* [13] reported that quail which have a body weight below 140 g will have lower egg weight and lower egg mass in the layer phase. The results of this study are also in agreement with several studies [6,14-16] which reported that supplementation of fat and emulsifiers was not significantly affecting body weight in the starter phase. This may be caused by the digestive system in the quail starter-grower phase is not perfect so that the absorption of fat supplemented is not perfect. That is also in agreement with several studies [3,14,17-27] reported that insufficient secretion of bile salts and the absence of lipase enzyme activity in the starter phase resulted in incomplete absorption of fat. Bile salts also stimulate the production of pancreatic lipase enzymes, which are affected mainly by the hormone Cholecystokinin (CCK) which is located in the intestinal mucosal crypt which is sensitive to the presence of fat.

Daily weight gain was significantly different (P<0.05) at 8-35 days but not significantly different (P<0.05) at 36-42 days. The daily weight gain in treatment T1 and T2 were not significantly different (P<0.05) with the control treatment (T0) but were significantly different (P<0.05) with T3 and T4 treatments. 6% of lecithin levels can help digestion of fat so that it can be used as an energy source for the quail starter-grower phase but the 6% of lecithin levels from the fat added are not enough to help digestion of fat perfectly in the quail starter-grower phase. This is in agreement with Lammasak *et al.* [3] which reported that there was a limit on the capacity of lecithin to form micelles with fatty acids and glycerol.

Feed consumption was not significantly different (P<0.05) because the quail starter-grower phase has an imperfective digestive organ so it is not perfect for digesting the fat supplemented, this is in agreement with recent studies [28-30] which reported that fat has a limited digestibility in young quails because the physiological functions needed for efficient fat digestion in young quails are immature and continue to develop for several weeks after hatching. The results of this study are in agreement with recent studies [31-33] which reported that an increase in fat content did not affect feed consumption.

The feed conversion ratio was used to determine feed efficiency consumed by quails. The feed conversion ratio was not significantly different (P<0.05). The results of the study were in agreement with recent studies [25-26,34] which reported that feed conversion which was not significantly different from increasing fat levels in the starter phase may be caused by the absence of lipase activity in the starter phase which causes lower fat absorption so that energy absorption was low. Quail feed conversion at the age of 42 days has met the standard of feed conversion in agreement with Widyatmoko *et al.* [35] which reported that the average feed conversion in quail was 3.77 with a range of 3.47-4.07.

Table 5. Income Over Feed Cost 8-42 days old.

Variables	T0	T1	T2	T3	T4
Feed consumption (kg)	0.45	0.43	0.44	0.41	0.42
Feed price (Rp kg)	6550.00	6676.43	6861.25	6802.85	7172.5
Feed price total (Rp)	2943.17	2888.07	3034.91	2809.78	3019.82
Quail price (Rp)	12000	12000	12000	12000	12000
IOFC	9056.83	9111.93	8965.09	9190.22	8980.18

*Treatments: T0:Commercial feed (CE), T1: CE+LF2.5%(L6%), T2: CE+CF2.5%(L6%), T3: CE+LF5% (L6%), T4: CE+CF5%(L6%)

The grower quail phase can also be sold to farmers as quails who are ready to lay. The highest IOFC available in T1 and T3 treatment because this is related to T1 and T3 treatments having a lower feed consumption than other treatment. The treatment of T3 and T1 has the advantage of a greater economic aspect than the other treatments as a whole in term of body weight was not significantly different ($P<0.05$) from other treatments.

4. Conclusion

Fat supplementation containing high palmitic acid enriched with lecithin did not interfere quail health status as a whole in term of mortality quails was not significantly different ($p<0.05$) with control treatment (T0). Local fat was not significantly affected ($P<0.05$) to body weight, daily weight gain, feed consumption and feed conversion ratio but the IOFC value was higher than the control treatment and fat commercial treatment. 6% of lecithin levels from the added fat in this study were not enough to help absorb fat perfectly. Further studies are therefore necessary to determine the effects of fat supplementation enriched with lecithin more than 6% from the added fat.

References

[1] Vali N 2008 The japanese quail: a review *Int. J. Poult. Sci.* **7** (9) 925-31
[2] Tunsaringkarn T, Tungjaroenchai W and Siriwong W 2013 Nutrient benefits of quail (coturnix coturnix japonica) eggs *Int. J. Sci. Res.* **3** (5) 1–8
[3] Lammasak K, Kijparkorn S and Angkanaporn K 2018 Porcine bile powder supplementation of a high fat broiler diet in relation to growth performance and nutrient digestion *Crop. Pasture. Sci.* A-H
[4] Siyal F A *et al.* 2017 Effect of soy lecithin on growth performance, nutrient digestibility and hepatic antioxidant parameters of broiler chickens *Int. J. Pharmacol.* **13** (4) 396-402
[5] Randall and Bolla 2008 *Primefacts 602 Second Edition: Raising Japanese Quail* (New South Wales: Department of Primary Industries)
[6] Roy A, Haldar S, Mondal S and Ghosh T K 2010 Effects of supplemental exogenous emulsifier on performance, nutrient metabolism, and serum lipid profile in broiler chickens *Vet. Med. Int.* 1-9
[7] Donaldson J, Pillay K, Madziva M T and Erlwanger K H 2014 The effect of different high-fat diets on erythrocyte osmotic fragility, growth performance and serum lipid concentrations inmale, japanese quail (coturnix coturnix japonica) *J. Anim. Physiol. Anim. Nutr.* 1-9
[8] Lemieux H, Bulteau A L, Friguet B, Tardif J C and Pierre U B 2011 Dietary fatty acids and oxidative stress in the heart mitochondria *Mitochondrion.* **11** 97–103
[9] Heugten E V and Odle J 2000 Evaluation of lysolecithin as an emulsifier for weaning pigs *ANS Report Number 248*, North Carolina University Department of Animal Science
[10] Renner R and Hill FW1961 Utilization of fatty acids by the chicken *J. Nutr.* **74** (3) 259-264
[11] Leeson S and Summers JD 2001 *Nutrition of the chicken* 4th ed (Ontario: University Books)
[12] Nascif C C C, Gomes P C, Albino L F T and Rostagno H S 2004 Determinação dos valores

energéticos de alguns óleos e gorduras para pintos de corte machos e fêmeas aos 21 dias de idade *Rev. Bras. Hematol. Hemoter.* **33** (2) 375-385

[13] Filho J A V, Garcia E A, Molino A B, Santos T A, Paz I C L A and Baldo G A A Productivity of japanese quails in relation to body weight at the end of the rearing phase *Maringá.* **38** (2)213-217

[14] Neto G A C, Pezzato A C, Sartori J R, Mori C, Cruz V C, Fascina F B, Pinheiro D F, Madeira L A and Gonçalvez J C 2011 Emulsifier in broiler diets containing different fat sources *Rev. Bras. Cienc. Avic.* **13** (2) 119-125

[15] Luc M, Ludo S, Marc R, Arno A, Saskia L and Van der Aa A *2013 The effect of different emulsifiers on fat and energy digestibility in broilers* Proceedings of the 19th European Symposium on Poultry Nutrition, August 26-29, 2013, Potsdam, Germany, pp: 1-4

[16] Abbas M T, Arif M, Saeed M, Reyad-ul-ferdous M, Hasan M A, Arain M A and Rehman A 2016 article emulsifier effect on fat utilization in broiler chicken *Asian J. Anim. Vet. Adv.* **11** (3) 158-67

[17] Serafin J A and Nesheim M C 1970 Influence of dietary heat-labile factors in soybean meal upon bile acid pools and turnover in the chick *J. Nutr.* **100** (7) 786–796

[18] Jackson B T, Smallwood R A, Piasecki G J, Brown A S, Rauschecker H F and Lester R 1971 Fetal bile salt metabolism i the metabolism of sodium cholate-14c in the fetaldog *J. Clin. Invest.* **50** (6) 1286–94

[19] Hakansson J 1974 Factors affecting the digestibility of fats and fatty acids in chicks and hens *Swed. J. Agric. Res.* **4** 33–47

[20] Lester R, Smallwood R A, Little J M, Brown A S, Piasecki G J and Jackson B T 1977 Fetal bile salt metabolism *J. Clin. Invest.* **59** 1009-1016

[21] Sell J L, Krogdahl A and Hanyu N 1986 Influence of age on utilization of supplemental fats by young turkeys *Poult. Sci. J.* **65** (3) 546–554

[22] Krogdahl A and Sell J L 1989 Influence of age on lipase, amylase, and protease activities in pancreatic tissue and intestinal contents of young turkeys *Poult. Sci. J.* **68** (11) 1561–1568

[23] Noy Y and Sklan D 1995 Digestion and absorption in the young chick *Poult. Sci. J.* **74** (2) 366-73

[24] Al-Marzooqi W and Leeson S 1999 Evaluation of dietary supplements of lipase, detergent, and crude porcine pancreas on fat utilization by young broiler chicks *Poult. Sci. J.* **78** (11) 1561-66

[25] Gaiotto J B, Menten J F M, Racanicci A M C and Iafigliola M C 2001 Soybean oil, acidulated soapstock, beef tallow and mixtures of fat sources in broilers diets *Rev. Bras. Cienc. Avic.* **2** (3) 219-227

[26] Wongsuthavas S, Yuangklang C, Wittayakun S, Vasupen K, Mitchaothai J, Srenanual P and Beynen A C 2007 Dietary soybean oil, but not krabok oil, diminishes abdominal fat deposition in broiler chickens *Int. J. Poult. Sci.* **6** (11) 792-795

[27] Tan H S, Zulkifli I, Farjam A S, Goh Y M, Croes E, Partha S K and Tee A K 2016 Effect of exogenous emulsifier on growth performance, fat digestibility, apparent metabolisable energy in broiler chickens *JOBIMB.* **4** (1) 7-10

[28] Katongole J B D and March B E 1980 Fat utilization in relation to intestinal fatty acid binding protein and bile salts in chicks of different ages and different genetic sources *Poult. Sci. J.* **59** (4) 819–827

[29] Chen H Y and Chiang H S 2005 Effect of dietary polyunsaturated/saturated fatty acid ratio on heat production and growth performance of chicks under different ambient temperature *Anim. Feed Sci. Technol.* **120** (3-4) 299-308

[30] Nagargoje S B, Dhumal M V, Nikam M G and Khose K K 2016 Effect of crude soy lecithin with or without lipase on performance and carcass traits, meat keeping quality and economics of broiler chicken *Int. J. Livest. Res.* **6** (12) 46-54

[31] Crespo N and Esteve-Garcia E 2001 Dietary fatty acid profile modifies abdominal fat deposition

in broiler chickens *Poult. Sci. J.* **80** (1) 71-78

[32] Shahryar H A, Salamatdoust-Nobar R, Lak A and Lotfi A 2011 Effect of dietary supplemented canola oil and poultry fat on the performance and carcass characterizes of broiler chickens *Curr. Res. J. Biol. Sci.* **3** (4) 388-392

[33] Nobakht A, Ariyana A and Mazlum F 2012 Effect of different levels of canola oil with vitamin E on performance and carcass traits of broilers *Int. Res. J. Applied Basic Sci.* **3** (5) 1059-1064

[34] Gracia M I, Latorre M A, Garcia M, Lazaro R and Mateos G G 2003 Heat processing of barley and enzyme supplementation of diets for broilers *Poult. Sci. J.* **82** 1281-1291

[35] Widyatmoko H, Zuprizal and Wihandoyo 2013 Effect of utilization of corn dried distiller grains with solubles in the diets on performance of male coturnix *Buletin Peternakan.* **37** (2) 120-124

Dihydroeponemycin inhibits ubiquitin proteasome system of *Plasmodium falciparum in silico*

K Mardhiyyah[1,2,7*], M H Arif[3], L E Fitri[4,7], S Winarsih[5], D Chandradikusuma[3,7] and T Nurseta[6]

[1]Doctoral Program in Medical Sciences, Faculty of Medicine Universitas Brawijaya, Indonesia
[2]Department of Biochemistry and Biomolecular, Faculty of Medicine Universitas Brawijaya, Indonesia
[3]Department of Interna Medicine, Faculty of Medicine Universitas Brawijaya, Indonesia
[4]Department of Parasitology, Faculty of Medicine Universitas Brawijaya, Indonesia
[5]Department of Pharmacy, Faculty of Medicine Universitas Brawijaya, Indonesia
[6]Department of Obstetry and Ginecology, Faculty of Medicine Universitas Brawijaya, Saiful Anwar General Hospital Malang, Indonesia
[7]Malaria Research Group, Faculty of Medicine Universitas Brawijaya, Indonesia

*Corresponding author's email: kanamardhiyyah@ub.ac.id

Abstract. Drug resistance in malaria treatment becomes major problems all over the world. Finding a novel potent antimalarial candidate is mandatory. Previous study proved that secondary metabolite of *Streptomyces hygroscopicus* Hygroscopicus containing eponemycin analog significantly inhibits Ubiquitin Proteasome System (UPS) of Plasmodium in vivo. The aims of this study are to know toxicity profile of dihydroeponemycin and its potency as Ubiquitin Proteasome System inhibitor of *Plasmodium falciparum*. The methods of this study were ADMET SAR Toxicity (ACD/LABS), Protein Modelling (SWISS), and Molecular Docking (AUTODOCK VINA PYRX). The result of Absorption, Distribution, Metabolism, Excretion and Toxicity profiles (ADMET) of dihydroeponemycin was absorbed 83.5% via intestinal route. Dihydroeponemycin belonged to Category 4 of toxicity. E3 is a potential drug target because of its different structure to that of the human. The C-score (confidence score of the prediction) of the active side E3 was 0.06 and E2 was 0.45. Dihydroeponemycin had the ability to bind to E3 and E2 with an affinity score binding of -3.8 and -4.1. The more negative the affinity binding value the easier the compound binds to the target protein. Based on this in silico study, dihydroeponemycin has potential effect as antimalaria by inhibiting Ubiquitin Proteasome System of *Plasmodium falciparum*.

1. Introduction

Malaria is a disease caused by a parasite. People with malaria often experience fever, chills, and flu-like illness. They may develop severe complications and die if untreated. Million cases of malaria occurred worldwide and people died, mostly children in the African Region. Many cases of malaria are diagnosed in the United States each year. The vast majority of cases in the United States are in travelers and immigrants returning from countries where malaria transmission occurs, many from sub-Saharan Africa and South Asia [1].

Plasmodium falciparum is the most prevalent malaria parasite in the WHO African Region, accounting for 99.7% of estimated malaria cases in 2017, as well as in the WHO regions of South-East Asia (62.8%), the Eastern Mediterranean (69%) and the Western Pacific (71.9%). *P. vivax* is the predominant parasite in the WHO Region of the Americas, representing 74.1% of malaria cases [1].

Plasmodium falciparum strains have acquired resistance to Artemisinin (ART)-based combination therapies across much of Southeast Asia. ART creates widespread protein and lipid damage inside intraerythrocytic parasites, necessitating macromolecule degradation. The proteasome is the main engine of Plasmodium protein degradation. Proteasome inhibition and ART have shown synergy in ART-resistant parasites. Ubiquitin modification is associated with altered parasite susceptibility to multiple antimalarials. Targeting the ubiquitin-proteasome system (UPS), is an attractive avenue to combat drug resistance [2]. Previous study proved that secondary metabolite of *Streptomyces hygroscopicus* Hygroscopicus containing eponemycin significantly inhibits Ubiquitin Proteasome System (UPS) of *Plasmodium* in vivo [3].

2. Methods
The methods of this study were ADMET SAR Toxicity (ACD/LABS), Protein Modelling (SWISS), and Molecular Docking (AUTODOCK VINA PYRX).

3. Results and Discussion
The result of Absorption, Distribution, Metabolism, Excretion and Toxicity profiles (ADMET) of dihydroeponemycin was absorbed 83.5% via intestinal route and belonged to Category 4 of toxicity. Smaller toxicity category means the compound has higher toxicity. The Lipinski-type properties of dihydroeponemycin showed that it can be ligand to bind with E2 and E3. This is based on Lipinski's rule of five [4].

Table 1. Lipinski-type properties of dihydroeponemycin based on pubchem database.

Properties	
Molecular weight	400.5 g/mol
No. of Hydrogen Bond Donors	4
No. of Hydrogen Bond Acceptors	6
TPSA	128 A^2
No of Rotatable Bonds	14

Several possible drug targets within *Plasmodium* UPS were 26S Proteasome, PfhsIU/pfhsIV proteasome complex, E3 ligases, E3 target substrates, and DUBs. E3 is a potential drug target because of its different structure to that of the human. The E1 and E2 are proteins that conserved in eukaryotic. E3 is important to maturation of schizont, so that the development of Plasmodium was restricted [5][6].

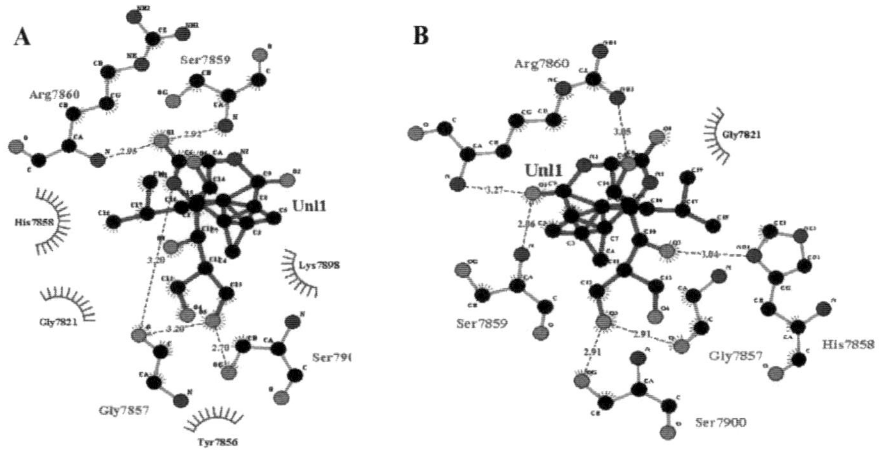

Figure 1. Molecular docking. A. Interaction of E2 and dihydroeponemycin. B. Interaction of E3 and dihydroeponemycin.

Table 2. Binding affinity score of dihydroeponemycin with E2 compared with other proteasome inhibitor.

Macromolecule	**Ligand**	Binding Affinity (kcal/mol)
	Epoxomicin	-4.9
	MG123	-5.3
E2	Lactacyin	-5.5
	Dihydroeponemycin	-4.9
	Chloroquine	-4.7

Table 3. Binding affinity score of dihydroeponemycin with E2 compared with other proteasome inhibitor.

Macromolecule	**Ligand**	Binding Affinity (kcal/mol)
	Epoxomicin	-5.1
	MG123	-5.4
E3	Lactacyin	-5
	Dihydroeponemycin	-4.8
	Chloroquine	-4.7

The C-score (confidence score of the prediction) of the active side E3 was 0.06 and E2 was 0.45 (COACH ZHANG LAB). Dihydroeponemycin had the ability to bind to E3 with an affinity score binding of -4.8. Dihydroeponemycin also had the ability to bind to E2 with an affinity score binding of -4.9The more negative the affinity binding value the easier the compound binds to the target protein.

Through the complex life cycle in the hosts, Plasmodium invades and replicates in totally different cells thus making the study of the biology of the parasite and the identification of targets for drug development affecting all stages very difficult. It was shown that host molecules, have a role in the progression and regulation of the parasite cell cycle. When the parasite is exposed to host molecule there is an increase in transcription levels of genes encoding for proteins related to the Ubiquitin Proteasome (UPS) System. This system is essential for the survival of the parasite. The Plasmodium UPS shows low similarity to the ubiquitin proteasome system in Humans; the identification of unique targets to be used for therapeutic molecules development increases the importance of UPS studies in malaria challenging. The study of the mechanism of action of the UPS and the identification of potential targets for new drugs development are promising alternative strategies to fight the drug-resistance problem in malaria parasites [7].

4. Conclusion

Based on this *in silico* study, dihydroeponemycin has potential effect as antimalaria by inhibiting Ubiquitin Proteasome System of *Plasmodium falciparum*.

5. Conflicts of interest

The authors declared that there was no potential conflict of interest relevant to this article.

6. Acknowledgment

The authors thank the Malaria Research Group, Faculty of Medicine, Universitas Brawijaya, for providing the support for the study.

References

[1] World Health Organization 2018 global health issues 2018
[2] Ng C L, Fidock D A, and Bogyo M 2017 protein degradation systems as antimalarial therapeutic targets *Trends Parasitol* **9** 731-43
[3] Rivo Y B N, Alkarimah A, Ramadhani N N, Cahyono A W, Laksmi D A, Winarsih S, Fitri L E 2013 metabolite extract of *Streptomyces hygroscopicus* Hygroscopicus inhibit the growth of *Plasmodium berghei* through inhibition of ubiquitin-proteasome system *Tropical Biomedicine* **2** 1–10.
[4] Leslie Z B, Chelsea M H, Tudor I O, and Tudor I 2016 the rule of 5 and drugability. *OpreaAdv Drug Deliv Rev* **101** 89–98
[5] Krishnan K M, and Williamson K C 2018 the proteasome as a target to combat malaria: hits and misses *Transl Res* **198** 40-47
[6] Laura A. Kirkman, Wenhu Zhan, Joseph Visone, Alexis Dziedziech, Pradeep K. Singh, Hao Fan, et al., 2018 antimalarial proteasome inhibitor reveals collateral sensitivity from intersubunit interactions and fitness cost of resistance *PNAS* **29** E6863-70
[7] Pereira P H S, Curra C, Garcia C R S 2018 ubiquitin proteasome system as a potential drug target for malaria *Curr Top Med Chem* **5** 315-20

In Silico: γ-Oryzanol as Anti-inflammatory During Folliculogenesis in Rattus Novergicus Exposed to Pyrethroid Aerosol

L Lisnawati[1*], M I E Santoso[2], S Poeranto[3] and A T Endharti[4]

[1]Doctoral Program of Medical Science, Faculty of Medicine, University Brawijaya, Malang, East Java, Indonesia.
[2]Departement of Health Research Ethics Commission, Faculty of Medicine, University Brawijaya, Malang, East Java, Indonesia.
[3]Departement of specialists and clinical microbiology, Faculty of Medicine, University Brawijaya, Malang, East Java, Indonesia.
[4]Departement of Doctoral Program Medical Science, Faculty of Medicine, University Brawijaya, Malang, East Java, Indonesia.

*Corresponding author's email: aura8277@yahoo.co.id

Abstract. The use of insect repellent is a disease preventive effort due to mosquito vectors as well as sources of pollutants that disrupt human health, especially reproductive health. This study aimed to determine the effect of γ-Oryzanol on the expression of Foxo3a, TNF-α, GDF-9 in the folliculogenesis of mice (Rattus Novergicus) exposed to pyrethroid aerosol mosquitoes. The research use bioinformatics technique with in silico approach. The result of study showing that γ-oryzanol has the potential as an anti-inflammatory and NO scavenger, the influence of the active compound γ-oryzanol on Foxo3 expression obtained pathway analysis with the results of the prediction of the target HITPICK able to bind to APP, AKT1 and Mtor expressions through modification of post-translational Foxo1, Effect of γ-oryzanol on Gdf9 expression, through bonding with APP, AKT1, AKT3, Pik3cg and modifying Smad 4 post-translation, The effect of the active compound γ-oryzanol on TNF-α expression via APP binds with AKT1, AKT3, AKT2 then activates IkbkB and bindings with Casp3, Fadd, Tradd, Tnfrsf1a, Tnfrsf1b, Traf3, Birc2, Ikbkg. Based on average part length analysis AKT1 is the fastest interact with protein target. The conclusion, γ-oryzanol has predicted to be used for the treatment of insecticide exposure, because it can act as an anti-inflamatory, and lipid perosidase inhibitor. This condition influences normally folliculogenesis with stable expression of Foxo3a, GDF 9 and TNF-α.
Keywords: γ-Oryzanol, Folliculogenesis, Pyrethroid

1. Introduction

Indonesia as a tropical climate has the potential to become a hotbed for developing mosquitoes. The results of a survey by Sunaryo (2016) that most people use household insecticides (86.33%) with the highest intensity of use once a day (85.4%) for more than 5 years (74.51%) [1]. The types of use of mosquito repellent, spray or aerosol (12.2%) are widely used [2].

The insect repellent uses synthetic parathyroids [3,4]. Pyrethroids as xenoestrogens and act as hormone disruptors or known as Endocrine Disruptor Chemical (EDC) [5].

The use of pyrethroid active ingredients in aerosol mosquito repellents causes adverse effects on health by causing oxidative stress. This is indicated by a decrease in the level of Extraceluller Superdioxide (EC-SOD) to fight ROS [5].

Through the PI3K-Akt approach Exposure to parathyroid aerosol mosquitoes during cellular folliculogenesis can result in cell damage through increased ROS levels and indicates the occurrence of oxidative stress. PI3K pathway it can be seen how the expression of Foxo3a, GDF-9 and TNF-α so that in this condition there is a need for antioxidant and anti-inflammatory sources to repair damaged cells during the folliculogenesis process [6,7,8].

This condition needs to be followed by changes in healthy lifestyles including through healthy food supplements that contain antioxidants and anti-inflammatory. Brown rice bran has higher antioxidant activity than other bran variants, which is 68.66 ± 3.682 ppm and it is known that the bioactive composition of red rice is as follows; -oryzanol 24,201 ± 945 ppm, β-carotene 15,007 ± 482 ppm, tocopherol 3,706 ± 460 ppm [9]. This study use γ-Oryzanol an inflammatory and antioxidant source to suppressing complication insecticide exposes by parathyroid in folliculogenesis process.

2. Methods

2.1. Target selection

2.1.1. Pass Server. Pass server is a structural approach the server will compare the inputted compounds with compounds that have been shown to have certain activities. The higher the Pa value, the more similar the structure and function of the compound.

2.1.2. Way2drug. The analysis uses the way2drug web server to predict the bioinformatics potential of the γ-oryzanol compound.

2.1.3. SMILE. The target protein for γ-oryzanol is obtained by entering the PUBCHEM SMILES server, (http://pubchem.ncbi.nlm.nih.gov/ The server provides a list of predicted target proteins for γ-oryzanol; these are compared, and target proteins with the most potential as a ligand γ-oryzanol (PUBCHEM ID 5282164).

2.2. Intermolecullar interaction

2.2.1. *STRING.* The interaction of protein using STRING from γ-oryzanol interacts with the target protein GDF9, FOXO3, TNF through the intermediary of several proteins.

2.3. Average analysis transport part length

Average transport path length analysis is an analysis to find the fastest and most effective path from a network pathway.

3. Results and Discussion

3.1. Pass server

Table 1. γ-Oryzanol analysis.

Bioactivity	Probability Active of Score (Pa)
Antiinflamatory	0.66
Immunosuppressant	0.538
Antioxidant	0.456
Lipid peroxidase inhibitor	0.457
Free radical scavenger	0.196
Apoptosis agonist	0.844
Caspase 3 stimulant	0.832

Table 1 shows that the Pass Server analysis, the γ-oryzanol is predicted to be used for the treatment of insexide exposure, because it can act as an antiinflamatory, and lipid perosidase inhibitor.

Insecticides can trigger oxidative stress. This oxidative stress can induce lipid peroxidase. Lipid peroxidase makes vasodilation change due to increased Selectin, ICAM, VCAM resulting in displacement of proinflammatory molecules such as cytokines, macrophages, LDL and immune cells. This molecule then undergoes oxidation and inflammation. As a result, endothelial cells become damaged and inflammation occurs, the appearance of plaque and other disorders because the barrier is dysfunctional.

The final product of lipid peroxidase is MDA (Reactive Aldehyde / Malodialdehyde) and 4-HNE. Both of these molecules can cause DNA adducts. DNA adduct is a DNA segment attached to a chemical, which can induce mutations. This mutation can cause the emergence of certain diseases such as cancer. While γ-oryzanol has a role that is predicted to neutralize oxidative stress because it can act as an antioxidant and lipid peroxidase inhibitor

3.2. Way2drug

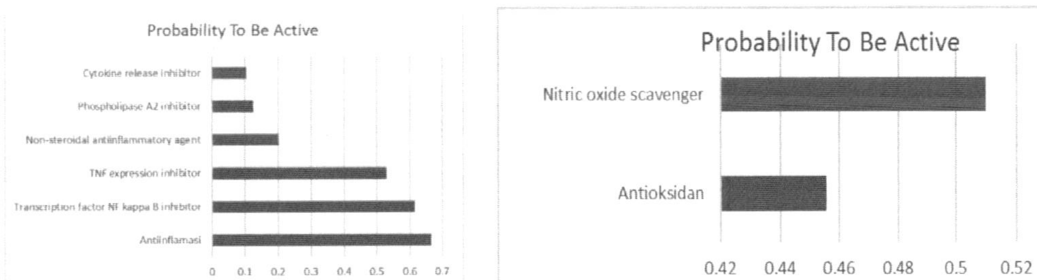

Figure 1. Probability active γ-Oryzanol.

Figure 1. show that The bioactive potential of γ-oryzanol through the in silico approach has the potential as an anti-inflammatory and NO scavenger.

3.3. SMILE

Objective: To find out the interactions of Cypemethrine, Transfluthrin, and Prallethrine on TNF, FOXO3, and GDF9 proteins
Sample: Gamma oryzanol (PUBCHEM ID 5282164)
SMILE: CC (CCC = C (C) C) C1CCC2 (C1 (CCC34C2CCC5C3 (C4) CCC (C5 (C) C) OC (= O) C = CC6 = CC (= C (C = C6) O) OC) C) C

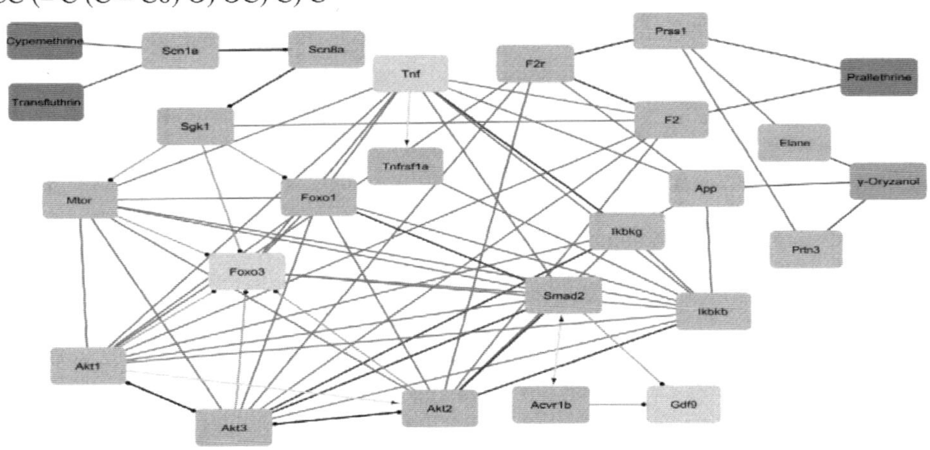

Figure 2. Intermediate protein target of γ-Oryzanol.

Table 2. Prediction signaling target.

Compound	Signaling target	Result
Cypemethrine and Transfluthrin	Signaling Foxo3	Cypemethrin dan Transfluthrin are predicted to target SCN1A protein. Protein SCN1Aprotein canbind with SCN8A dan SGK1,I to influence FOXO3 expression
	Signaling GDF9	Cypemethrin and Transfluthrin are predicted to target SCN1A protein. The SCN1A protein can bind with SCN8A and SGK1, to then influence FOXO1 expression. Foxo 1 can bind with SMAD2 and then SMAD2 can affect GDF9 expression either directly or through ACVR1B activation
	Signaling TNF	Cypemethrin and Transfluthrin are predicted to target SCN1A protein. The SCN1A protein can bind with SCN8A and SGK1, to then affect MTOR expressions. MTOR is predicted to have interactions with TNF even though the interactions that occur do not yet know the type of interaction
Pralethrin	Signaling Foxo3	Prallethrine is predicted to target PRSS1 and F2 proteins. F2 itself has interactions with SGK1 even though the type of interaction is not yet known. SGK1 itself is predicted to affect the expression of FOXO3
	Signaling GDF9	Prallethrine is predicted to target PRSS1 and F2 proteins. F2 itself has interactions with SGK1 even though the type of interaction is not yet known. SGK1 itself is predicted to affect the expression of FOXO1. Foxo 1 can bind with SMAD2 and then SMAD2 can affect GDF9 expression either directly or through ACVR1B activation
	Signaling TNF	Prallethrine is predicted to target PRSS1 and F2 proteins. F2 itself has interactions with TNF even though the type of interaction is not yet known
γ-oryzanol	Signaling Foxo3	γ-oryzanol is predicted to target Elane, APP, and Prtn3 proteins. Elane and PRTN3 proteins can interact with PRSS1. PRSS1 interacts with F2R and F2. F2 itself has interactions with SGK1 even though the type of interaction is not yet known. SGK1 is predicted to affect the expression of FOXO3
	Signaling GDF9	γ-oryzanol is predicted to target Elane, APP, and Prtn3 proteins. The APP interacts with IKBKG and IKBKB to then interact with SMAD2 to go to GDF
	Signaling TNF	γ-oryzanol is predicted to target Elane, APP, and Prtn3 proteins. APP proteins are predicted to be able to interact with TNF, although we cannot yet know the types of interactions that occur.

3.4. STRING

Table 3. Prediction of bioactive target protein and biology process.

No	Compound	Target	Interaction	Biology Process	Precission (%)	TC Similiarity
1	γ-oryzanol	APP	Direct-activation	Foxo signaling pathway	26.4	0.35
2	Pralethrin	PRSSI	Direct-binding	Foxo signaling pathway	15.7	0.25
		F2		TNF signaling pathway	15.4	0.25
3	Transflutrin	CYP2B6	Direct	Foxo signaling pathway	43.5	0.49
		CYP3A4			24.6	0.49
		SCN1A			53.3	0.49

3.5. Average analysis transport part length

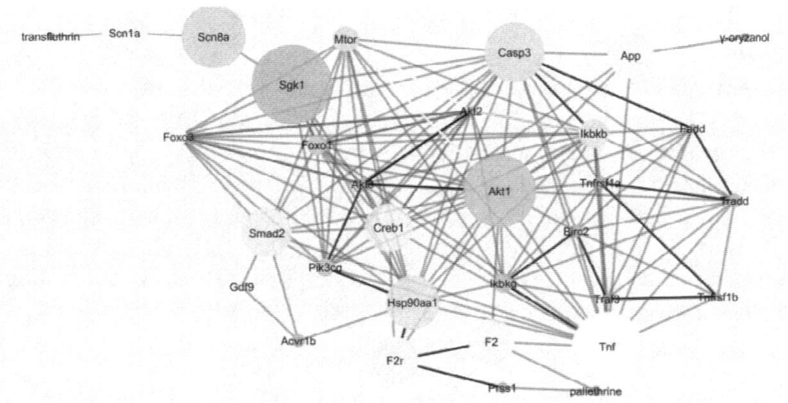

Figure 3. Average Analysis Transport Part Length γ-oryzanol.

Figure 3 shows that in is average transport path length AKT1 is predicted to be the fastest pathway intermediary protein to interact with the target protein (Foxo3a, GDF-9 and TNF-α).

4. Conclusion

The γ-oryzanol compound has a role that is predicted to neutralize oxidative stress from perithroid aerosol, because it can act as an inflamatory and lipid peroxidase inhibitor. This condition influences normally folliculogenesis with stable expression of Foxo3a, GDF 9 and TNF-α. Based on the analysis of betweenes centrality protein AKT1 is a mediator protein that is widely passed to interact with the target protein. The larger the size and the darker the color of the visualization of the protein, the more dominant the role of the protein in reacting.

References

[1] Sunaryo, Puji A and Dyah W 2015 Gambaran pemakaian insektisida rumah tangga di daerah endemis DBD kabupaten Grobogan. *BALABA.* **11(1)** 9-14

[2] Hazel 2013 *Proporsi penggunaan obat nyamuk perkotaan dan pedesaan, Riskesda. Badan Penelitian dan Pengembangan Kesehatan* Jakarta Departemen Kesehatan RI.

[3] Djojosumarto P 2008 *Teknik aplikasi pestisida edisi revisi* Jakarta PT Agromedia Pustaka 200-211

[4] Environmental Protection Agency (EPA) 2016 Pyrethrins and pyretroid. www.epa.gov, (downloaded on May 2018)

[5] American Mosquito Control Association (AMCA) 2014 *Mosquito control.* www.amca@mosquito.org, (downloaded on May 2017)

[6] Sanchez A M I, Candau R B, Csibi A, Pagano A F, Raibon A and Bernardi H 2012 The role of AMP-activated protein kinase in the coordination of skeletal muscle turnover and energy homeostasis. *Am J Physiol Cell Physiol.* **303**(5) C475-85

[7] Sasseville M I, Ritter L J, Nguyen T M, Liu F, Mottershead D G, Russell D L and Gilchrist R B 2010 Growth differentiation factor 9 signaling aequires ERK1/2 activity in mouse granulosa and cumulus cell's. *J.Cell Sci.* **15**(123) 3166-76

[8] Mumpuni D P 2013 Analisis kadar tokoferol gamma oryzanol dan beta karoten serta aktivitas antioksidan minyak bekatul kasar artikel penelitian; Program studi ilmu Gizi Fakultas Kedokteran Universitas Diponegoro

[9] Ashton M, Barnard J, Casset F, Charlton M, Downs G, Gorse D, Holliday J D, Lahana R and Willett P 2003 Identification of diverse database subsets using property-based and fragment-

based mollecular descriptions. *Quantitative structure-activity relationships.* **21**(6) 598-604

[10] Filimonov D A, Lagunin A A, Gloriozova T A, Rudik A V, Druzhilovskii D S, Pogodin PV and Poroikov V V 2014 Prediction of the biological activity spectra of organic compounds using the PASS online web resource. *Chemistry of Heterocyclic Compounds.* **50**(3) 444-457

[11] Giray B, Gurbay A and Hincal F 2001 Cypermethrin-induced oxidative stress in rat brain and liver is prevented by vitamin E or allopurinol. *Toxicol Lett.***118**(3) 139-46

[12] John A Ambrose, Rajat S and Barua 2004 The pathophysiology of cigarette smoking and cardiovascular disease. *Journal of the American College of Cardiology.* **43**(10) 1731-1737

[13] Szklarczyk D, Morris J H, Cook H, Kuhn M, Wyder S, Simonovic M, Santos A, Doncheva N T, Roth A, Bork P, Jensen L J, Von Mering C 2017 The STRING database in 2017: Quality-Controlled Protein-Protein Association Networks, Made Broadly Accessible. *Nucleic Acids Res*, **45**:D362-68

The Effect of High Fat High Fructose Diet (Modification of AIN-93M) on Nuclear Factor Kappa Beta Expression in the Liver Tissue of Male Sprague Dawley Rats

M M S Engel[1], I Kusumastuty[1], K W Anita[2] and D Handayani[1,3*]

[1] School of Nutrition, Faculty of Medicine, Brawijaya University, Malang, Indonesia
[2]Department of Anatomical Pathology, Faculty of Medicine, Brawijaya University, Malang, Indonesia
[3]Research Centre for Smart Molecules of Natural Genetic Resources, Brawijaya University, Malang, 65145, Indonesia

*Corresponding author's e-mail: handayani_dian@ub.ac.id

Abstract. Metabolic inflammation (low-grade inflammation) remains an etiopathogenic key factor in the development of metabolic syndrome. Nuclear Factor Kappa Beta (NF-κβ) is a transcription regulator of genes having a role in immunity, the inflammatory response which can be associated with obesity-related pathological conditions like nonalcoholic fatty liver (NAFLD). Various stimuli, such as metabolic stress (hyperglycemia, ROS, fat metabolism) and proinflammatory cytokines (TNF-α, IL-6, IL-1β) could activate NF-κβ. This study was aimed to investigate the underlying molecular mechanisms of NAFLD in rats fed a modified AIN-93M HFHF (High Fat High Fructose) diet. The design of this study was experimental post-test only controlled group design. Thirty male Sprague Dawley rats were distributed into 2 treatment groups by a completely randomized design (CRD) technique. The sacrifice was performed after 17 weeks of treatment. NF-κβ expression was assessed by an immunohistochemical method (IRS score). The results showed there were significant differences in feed intake and energy intake between groups P1 and P2 ($p = 0{,}000$, $p = 0{,}000$). The average NF-κβ expression in the P2 group was significantly higher ($p = 0.000$) compared to the control group (P1). The correlation test between dietary intake and NF-κβ expression proved that there was a positive correlation between energy, carbohydrate and fat intake on NF-κβ expression (p = 0.001, 0.000, 0.046). However, there was a negative relationship between protein intake and NF-κβ expression ($p = 0.000$). This study concluded the modified AIN-93M HFHF diet increased NF-κβ expression in the liver tissue of male Sprague Dawley rats.

1. Introduction

Obesity is a complex and chronic condition, originating from several physiological interactions, specific biological, individual, and environmental factors affecting appetite and the process of energy metabolism [1, 2]. Obesity occurs due to the enlargement and increases fat/adipose tissue that accumulates excessively, that could lead into health problems and lower life expectancy [1, 3]. The prevalence of obesity has increased over the past three decades. Globally, there are 650 million people affected by obesity, while Indonesia is among the 10 highest prevalence of obesity in the world. National prevalence of obesity in adults (> 18 years) is 15.4% [3-5]. The accumulation of adipose tissue would induce metabolic inflammation that triggers insulin resistance, type 2 Diabetes Mellitus, dyslipidemia, atherosclerosis, and nonalcoholic fatty liver (NAFLD) [6, 7]. Therefore, obesity becomes one of the risk factors of NAFLD [8].

Obesity can be induced to experimental animals with high fat diets with various percentages, ranging from 20% to 60% [8-13]. In some studies, high-fat and fructose (HFHF) diets consisting of high-fat diets (HFD) and fructose, can also trigger oxidative stress in tissue and plasma, metabolic syndrome (obesity), and obesity-related-disease, as non-alcoholic fatty liver [14-16]. The key factors in the development of metabolic diseases including obesity are chronic low-grade inflammation triggered by nutrients and metabolic signals, also known as metabolic inflammation [7].

Nuclear Factor Kappa Beta (NF-κβ) is a transcription factor that regulates inflammation, specific and nonspecific immune system, wound healing response, and cell homeostasis, activated through 2 signal cascade: classical (canonical) and alternative (non-canonical) pathways [17]. Activation of NF-κβ could be triggered by various stimuli, such as metabolic stress (hyperglycemia, ROS, fat metabolism), radiation (ultraviolet) and proinflammatory cytokines (TNF-α, IL-1β, IL-6). Active NF-κβ will translocate into the nucleus and transcribe various specific genes that produce cytokines such as TNF-α, IL-6, IL-1β, IL-10. This cytokines lead to insulin resistance and liver damage [14, 18-21]. Consequently, NF-κB is the key that can connect metabolism, inflammation & insulin work to each other [21].

Previous studies related to the expression of NF-κβ in experimental animals with metabolic syndrome (obesity and type 2 Diabetes Mellitus) used HFD as an induction feeding [22-24]. According to Cai, identification of NF-κβ expression related to insulin resistance could be carried out in the central nervous system (hypothalamus) and peripheral metabolic tissues such as the adipocytes and liver, due to its function in nutrient metabolism, where NF-κβ has an important role [22, 25, 26]. Globally, the effect of AIN-93M modification as high fat high fructose (HFHF) diet on NF-κβ expression in liver tissue of experimental animals have not been widely studied. Therefore, this research was aimed to broaden knowledge and improve science in the field of molecular nutrition.

2. Material and Methods

2.1. Study Design

The design of this study was true experimental in vivo *Post Test Only Controlled Group Design*. The subjects of this study were male *Sprague Dawley* rats randomized into two groups, namely P1 receiving AIN-93M standard diet (control) and P2 receiving HFHF (High Fat High Fructose) AIN-93M modification diet. This research has been approved by ethical research committee from the Medical Faculty, Universitas Brawijaya Malang, Indonesia, Number 368/EC/KEPK/10/2017.

2.1.1. Criteria and Sample Size of Research Subject. The inclusion criteria of samples were male Sprague Dawley rat, age ± 3 months old, complete body organs and clear eyes, body weight of 200-250 grams. Meanwhile the exclusion critea were rats with disabilities and experience anatomic abnormalities. Formula for number of samples [Federrer Formula: *(*t-n) *(*n-1) ≥ 15, t: number of treatment groups, n: number of samples in each group]. The total sample was 36 rats.

2.2. Treatment

Twelve week old male Sprague Dawley rats were randomized into 2 groups, namely the AIN-93M standard diet group or P1, (nutritional value of 42.87%, fat 25.81%, protein 31.32% with energy density (DE) of 4.21 kcal.gram) and the HFHF with AIN-93M modification diet group or P2 (with nutritional value of carbohydrates, fat, and protein are 29.9 %, 51.64%, and 21.81%, respectively, with DE 5.08 kcal / gram and Liquid fructose with 30% concentration was added for P2 group for rat drinking). All groups were given food and drink ad libitum for 17 weeks of treatment. At the end of the study period, surgeries were performed and Nuclear Factor Kappa Beta (NF-κβ) expression in liver of subjects were assessed. The study was conducted in the Nutrition and Anafarma Laboratory, Malang Health Politeknik, Health Ministry, Malang, BioSains Laboratory, Biochemistry-Biomolecular Laboratory,

Pathology-Anatomy Laboratory, Faculty of Medicine, Brawijaya University Malang and Pathology-Veterinary Laboratory, Faculty of Veterinary Medicine, Airlangga University Surabaya.

2.3. Collecting Data

1) Data on food intake was obtained by weighing the remaining feed given, done every 2 days. The average would act as daily intake data.
2) Data on fluid intake was obtained by measuring the remaining beverage to the initial volume of daily supply, which is 250 ml.
3) Body weight data was obtained by weighing each subjects every week.
4) The NF-κβ expression data was obtained by semiquantitative IRS (Immuno-reactive Score) method. NF-κβ expression was identified by brownish color in rats liver cells (cytoplasm and nucleus), using a light microscope / slide scanner in 5 different fields of view (LP) with 100x and 400x magnifications [27, 28]
5) Histopathological examination and identification of NF-κβ expression were carried out by pathologist from the Department of Veterinary Pathology, Faculty of Veterinary Medicine, Airlangga University.

2.4. Data Analysis

Normal distributed and homogeneous data was analyzed by Independent T-test and Pearson statistical test while non-parametric tests would be used if the terms were not fulfilled. The analysis was processed by IBM SPSS Statistics 16 for Windows with a confidence level of 95%. The p value was significant if it was less than 5% ($p < 0{,}05$).

3. Results

3.1. Subjects Characteristic

The characteristic of the rats in this study was presented in **Table 1.**

Table 1. Characteristics of Experimental Rats.

	P1	P2	p-*value**
Type	*Sprague Dawley*	*Sprague Dawley*	-
Number (n)	14	16	-
Sex	Male	Male	-
Age	± 3 months	± 3 months	-
Initial Body Weight**	246.11 ± 19.10	245.58 ± 22.05	0.945

P1 = AIN-93M standard diet, P2 = High Fat High Fructose Diet (HFHF) AIN-93M modification, *p-value = independent T-test results ($p < 0.05$), **Mean ± SD

The number of samples survived until the end of the study were 30 rats. Data on the initial body weight of the treatment group were normally distributed and homogeneous. Based on the results of the comparative independent T-test, it was found that there was no significant difference between the initial body weight of group P1 (AIN-93M standard diet) and P2 (AIN-93M modified HFHF diet) ($p = 0.945$).

3.2. Diet Characteristic

Nutrient composition on feeding was presented in **Table 2.**

Table 2. Nutritional composition per 100 grams diet.

	Treatment group	
	P1	**P2**
Energy (kcal)	421.00	508.00
Carbohydrate (gram; %)	45.12; 42.87	37.33; 29.39
Protein (gram; %)	32.96; 31.32	27.70; 21.81

Fat (gram; %)	12.07; 25.81	29.15; 51.64
ED (kkal/gram)	**4.21**	**5.08**

P1 = AIN-93M standard diet, P2 = High Fat High Fructose Diet (HFHF) AIN-93M modification, ED: Energy Density

3.3. Food and Nutrient Intake

Food and nutrient intake of P1 and P2 during the 17-week treatment period was shown in **Figure 1.**

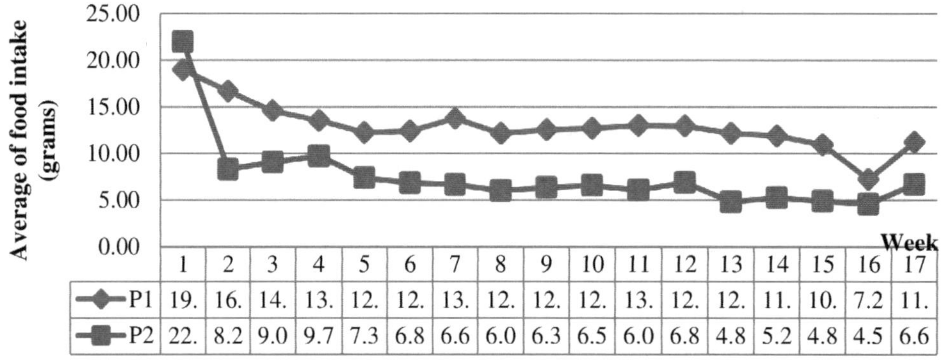

Figure 1. Average Food Intake per Week (gram).

The trend of total energy intake in each week were shown in **Figure 2** below.

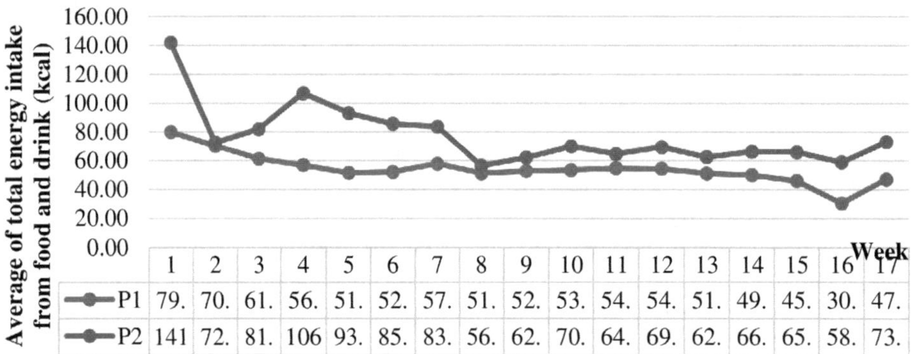

Figure 2. Average Total Energy Intake per Week (mL).

The analysis of intake data, namely the average weight of feed intake (AIN-93M standard diet and AIN-93M modified HFHF diet), total energy, and nutrients (from food and beverages) during the 17-week treatment period are presented in **Table 3.**

Table 3. The Average of Food Intake, Total Energy, and Nutrient per Day.

Parameter	Treatment Group		p-*value*
	P1***	P2***	
Food (g)	12.02 ± 1.62	6.31 ± 1.12	0.000**
Fructose 30% (mL)	[redacted]	29.73 ± 3.88	[redacted]
Energy (kcal)	50.60 ± 6.85	67.75 ± 6.38	0.000*
Carbohydrate (g); %	5.42 ± 0.73; 42.84%	11.27 ± 1.13; 66.53%	0.000**
Protein (g); %	3.96 ± 0.53; 31.30%	1.74 ± 0.31; 10.27%	0.000**
Fat (g); %	1.45 ± 0.19; 25,79%	1.84 ± 0.32; 24.44%	0.001*

P1 = AIN-93M standard diet, P2 = High Fat High Fructose Diet (HFHF) AIN-93M modification, *p-value = value of independent T-test (p <0.05), **p-value = value of Mann Whitney test (p <0.05), ***Mean ± SD

3.4. Body Weight of Experimental Rats

The description on body weight changes of the rats during the study periode was shown in **Table 4.**

Table 4. The Weight and Its changes in Experimental Rats (gram).

Parameter	Treatment Group		p-value*	p-value**	p-value***
	P1****	P2****			
Initial Body Weight	246.11 ± 19.10	245.58 ± 22.05	0.945	0.027	0.000
Final Body Weight	261.93 ± 29.29	276.92 ± 35.65	0.223		
Body Weight Changes	15.82 ± 2.37	31.33 ± 2.63	0.103	[redacted]	

P1 = AIN-93M standard diet, P2 = High Fat High Fructose Diet (HFHF) AIN-93M modification, *p-value = value of independent T-test (p <0.05), **p-value = value of paired T-test P1, ***p-value =value of paired T-test P2, * ***Mean ± SD

Based on **Table 4**, the independent T-test revealed no significant differences between initial and final body weight in both treatment groups (p-value initial weight = 0.945, final p-value BB = 0.223). If final body weight and initial body weight were compared, there was a weight increase during the 17 weeks of the study. The weight changes in rats in P2 group (31.33 ± 2.63) were greater than those in group P1 rats (15.82 ± 2.37).

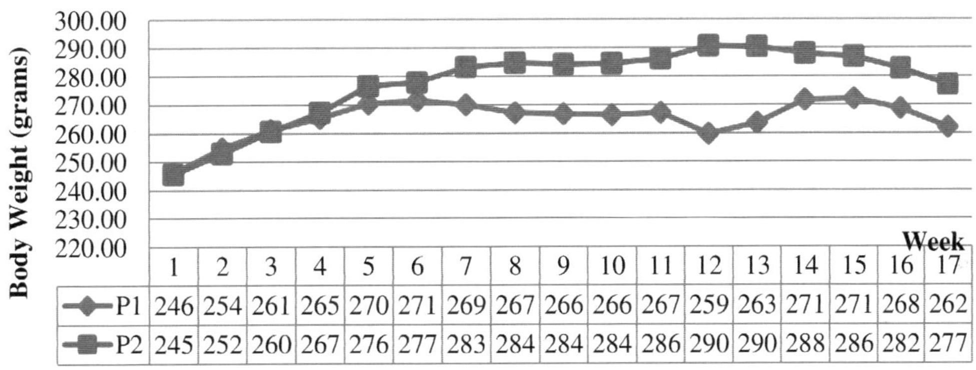

Figure 3. Average Body Weight per Week (grams).

The paired T-test revealed a significant difference between initial body weight and final weight at the end of the study period in each treatment group (p-value P1 = 0.027, meanwhile p-value P2 = 0.000). However, independen t test found no significant difference in body weight changes between P1 and P2 (p-value = 0.103). The fluctuation of the body weight from the first week to the end of the study (17th week) was shown in **Figure 3.**

3.5. NF-κβ Expression in the Liver Tissue of Experimental Rats

The images of liver tissue scans of P1 and P2 can be seen in **Figure 4** below.

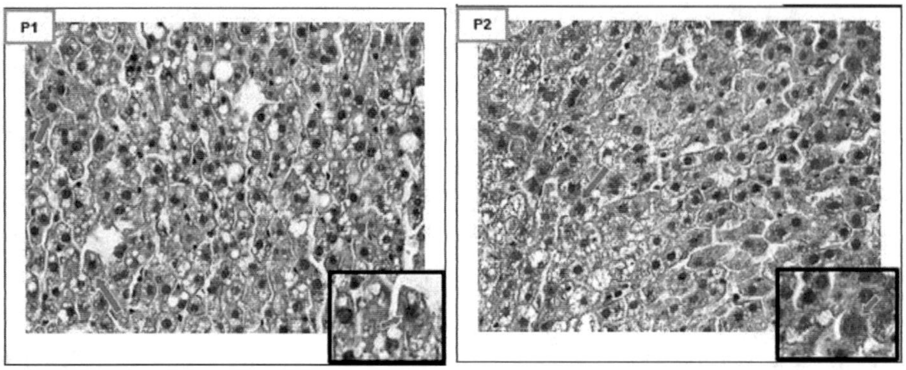

Figure 4. Expression of Nuclear Factor-κβ (NF-κβ) in the Liver, with Immunohistochemical Staining, Magnification of 400x.

Description: P1 = a group of rats given a standard AIN-93M diet, P2 = a group of rats given a modified AIN-93M diet. The arrow showed the NF-κβ expression identified in the field of view. In P1, NF-κβ expression shows that the total IRS value is 1 (negative category), while P2 group shows NF-κβ expression with IRS 4 value (medium category). Arrows: NF-κβ expression on hepatocytes (nucleus and cytoplasm)

The Shapiro-Wilk test showed normally distributed NF-κβ expression ($p = 0.069$). The average of NF-κβ expression in liver in P1 and P2 groups is presented in **Figure 5** below.

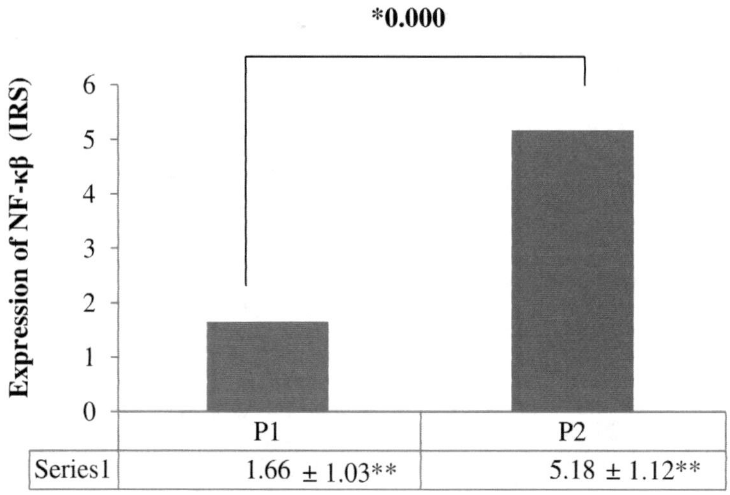

Figure 5. The Average of NF-κβ Expression (IRS) in Rat Liver.

Description: P1 = AIN-93M standard diet, P2 = High Fat High Fructose Diet (HFHF) AIN-93M modification, *p-value = value of independent T-test ($p < 0.05$), **Mean ± SD

Data showing the correlation between average dietary intake and NF-κβ expression (IRS) in liver tissue of rats are listed in **Table 5.**

Table 5. Correlation between average dietary intake and NF-κβ expression (IRS) in liver tissue of male *Sprague Dawley* rats.

	p-value **NF-κβ expression (IRS)**	**r*****
Energy (kkal)	0.001*	0.564
Carbohydrate (g)	0.000**	0.652
Protein (g)	0.000**	-0.820

Fat (g)	0.046*	0.367

*p-value = value of Pearson test ($p < 0.05$), **p-value = Spearman test ($p < 0.05$), ***p-value = the strength of correlation test ($n = 30$)

The data in Table 5 were obtained through Pearson and Spearman correlation test showing that there was a significant correlation between the average energy, carbohydrate, protein, and fat intake on the expression of NF-κβ (IRS) (p-value < 0.05). The strength of the correlation was moderate, strong, very strong, and weak correlation, respectively. The value of r on energy, carbohydrate, and fat were positive which means that the direction of the three groups is in the same direction. The greater the average intake of energy, carbohydrates, or fats, the greater the NF-κβ expression (IRS) in rats. Meanwhile, in the protein group, the r value was negative, therefore the greater the average protein intake, the lesser the expression of NF-κβ (IRS) in rats.

4. Discussion

4.1. Average Food Intake of Experimental Rats

There were two types of feeding given for experimental animals (male *Sprague Dawey* rats) in this study, the standard AIN-93M diet in the form of pellets for the control group (P1) and the AF-93M modified HFHF (High Fat High Fructose) diet in the form of liquid and pellets for group P2. Food and beverages for experimental animals were given in ad libitum. Every 100 grams of the AIN-93M standard diet in this study contained 421 kcal of energy, 45.12 grams of carbohydrates (42.87%), 32.96 grams of protein (31.32%), and 12.07 grams of fat (25.81%). Reeves et al. in his study stated that every 100 grams of the AIN-93M diet contained 360.1 kcal of energy, 72.73 grams of carbohydrates (75.9%), 12.58 grams of protein (14.1%), and 4 grams of fat (10.0%). Meanwhile, the AIN-93M standard diet in the study of Santos et al. contained 371.9 kcal of energy, 72.05 grams of carbohydrates (77.5%), 11.90 grams of protein (12.8%), and 4 grams of fat (9.7%) [29, 30].

The energy content of this study was higher compared to studies conducted by Reeves et al. and Santos et al., with a range of differences of 49.1-60.9 kcal [29, 30]. Protein and fat nutrients in this study were also up to 2 times higher than the percentage of two other studies. However, the carbohydrate content in this study was lower with a difference of 33.03% with Reeves et al. and 34.63% with Santos et al. HFHF (High Fat High Fructose) with AIN-93M modification is a modified diet that combines High Fat Diet (HFD) and High Fructose (HF) feeding. The energy and nutrient content in 100 grams of HFD used by this study were 508 kcal of energy, 37.33 grams of carbohydrates (29.39%), 27.70 grams of protein (21.81%), and 29.15 grams of fat (51.64%). Madsen et al. used HFD with an energy content of 441 kcal, 51.4% carbohydrates, 16.8% protein, and 31.8% fat per 100 grams of diet [31]. Handayani et al. stated that the energy from the used-HFD was 4.49 kcal/g. (each 100 gram contained 449 kcal of energy), 32% carbohydrate, 16% protein, 50% fat, and 2% fiber [32]. The value of energy, protein, and fat from HFD in this study were higher than the other two studies.

However, unsimilarly, research by So et al. used 100 grams of HFD containeing 516 kcal of energy, 20% carbohydrate, 20% protein, and 60% fat. Castillo et al. also supported the study that in 100 grams of HFD contained 560 kcal of energy, 24% carbohydrate, 18% protein, and 58% fat [33, 34]. The energy and fat content of this study was lower, compared to the latters with a difference of 8-52 kcal in energy and 6.36% -8.36% in fat content. The High Fructose liquid of 30% concentration was given ad libitum as a drink for P2 group, with a content of 120 kcal of energy and 30 grams of carbohydrate per 100. This concentration was similar with the study conducted by Lee et al., Lambertz et al., Yoo et al. using 30% fructose solution in their research [35-37]. There are also several studies with lower or higher concentrations of fructose solution than this study, such as the studies of Vilà et al. and Fariba et al. with a concentration of 10% and Vasiljević et al. with a concentration of 60% [38-40].

The HFHF diet in this study had a high proportion of fat and fructose, 51.64% and 30%, respectively. Lee et al. using the HFHF diet for one of the treatment groups, used HFD containing 60% fat percentage and 20% fructose solution concentration. Zarghani et al. in a study related to a diet that induces NAFLD utilizing a high fat high fructose (HFHFr) diet containing 53% fat and 20% fructose solution [35, 41].

The fat percentage in this study was lower than fat percentage in the studied conducted by Lee et al. and Zarghani et al., but already classified as a high-fat diet (High Fat Diet). The concentration of fructose solution in this study was higher than the two.

Based on the results of the Mann Whitney statistical test, it was found that the average of daily food intake of group P1 (AIN-93M standard diet) and P2 (AIN-93M modified HFHF diet) differed significantly ($p < 0.05$). The results in Table 5.3 showed that the average feed intake in group P1 was higher than group P2, with a value of 12.02 ± 1.62 grams in group P1 and 6.31 ± 1.12 gram in group P2. This was in accordance with the results of the study of Zarghani et al. and Yoo et al. that feed intake in the control group was greater than the HFHF treatment group [36, 41]. According to Curfs et al. in the Handbook of Laboratory Animal Science, the estimated average of daily food intake in adult mice is 25-30 grams per day [42]. When compared with this number, the average of daily food intake of rats in both group was below the average.

The level of feed consumption in experimental animals was influenced by body weight, individual animals, feed types, and environmental factors. In addition there are also palability factors for feed, taste, texture, size and consistency of feed [43]. Some factors that may cause lower intake of HFD diet (atherogenic) were rat physiological factors and feed factors. Rat physiological factors, related to eating behavior, may occur due to saturation while getting a HFHF diet. Rat feed factor was related to the form of feed, texture and state of organoleptics [44]. Moreover, there were also several environmental factors such as the state of the cage and lighting that can affect the feed intake of mice [45, 46].

Based on the texture, the modified AIN-93M HFHF feed was softer and greasy than the AIN-93M standard feed due to the addition of lard (cow fat) to the modified AIN-93M HFHF. Lard was added to increase the level of saturated fat (SFA) in the feed. Lard plays a significant role in SFA intake, because every 100 grams of lard has 39 grams of SFA [47, 48]. This soft and oily feed texture could be related to low intake because rats' preference of solid and hard feed texture [49].

The average intake of 30% fructose in P2 group rats was 29.73 ± 3.88 mL daily ~ energy 31.02-40.33 kcal per day. The intake in the P2 group had met the standards (rat daily water intake was about 15-30 mL) [43]. Another study showed that drinking intake in the treatment group given a *standard chow* diet and fructose solution was 30% higher than the control group, with a value of 38.6 ± 7.6 mL per day [50]. This shows that giving fructose can affect the average volume of drink intake in rats.

The total daily energy intake was obtained by summing the energy from feed and drinks. The results of the Independent T-test stated that the energy intake of the P1 and P2 groups differed significantly ($p < 0.05$). Based on the results, the average daily energy intake of P2 group (67.75 ± 6.38 kcal) was higher than that in group P1 (50.60 ± 6.85 kcal), although the feed intake of the P2 group was actually lower than group P1. This occurs because the energy density of the P2 group is greater (5.08 kcal / gram) and was also added by the energy intake from 30% fructose liquid, while the energy density in group P1 feed was 4.21 kcal / gram without any addition from the drink. The results of this study were similar with Yoo's study which is stated that calorie intake of the treatment group (30Frc + 45Fat) was greater than the control group (RD) [36]. Decreasing feed intake could originate from subjects learning to compensate for high energy density diets by consuming less amounts of feed [51].

4.2. Changes in Experimental Rat Weight

The development of body weight of experimental animals was observed during the study period (17 weeks) and body weight was measured every 7 days. The average body weight description in experimental animals can be seen in Figure 3 where the increase in body weight in group P1 continued to occur from week 1 to 7, then fluctuates along the rest of study period. While in P2 group, body weight tended to increase starting from week 1 to week 13, then began to experience weight loss in the 14th week. The average initial body weight of group P1 rats was 246.11 ± 19.10 grams while the P2 group was 245.58 ± 22.05 grams. Based on the data in Table 4, there was an increase in body weight in both groups, 15.82 ± 2.37 grams in the P1 (6.42%) and 31.33 ± 2.63 groups in the P2 (12.76%). Although there were differences in weight gain between groups P1 and P2, the values did not differ significantly.

On the contrary, Zarghani's study in Wistar rats with a composition of dietary HFHF concentration of 20% fructose and 53% fat content of total energy and the composition of a standard diet containing 14.5% fat from thermal energys showed that body weight in the HFHF diet group was significantly higher than other groups [41]. Another study conducted by Yoo also showed the results of the group given the HFHF diet significantly had higher body weight than the control group [36]. Although the increase in body weight in groups P1 and P2 in this study did not differ significantly, the pattern of differences between treatment groups had been seen.

Meanwhile, the paired T-test stated that there was a significant difference in body weight between the initial and final body weight at the end of the study period in each treatment group. So that it can be said that the weight of both group had increased significantly. The results of this study were in line with of Lee et al. stating that weight in the group given the HFHF diet increased significantly [35]. Obesity in experimental animals can be assessed based on several criteria, such as (1) increase in body weight or Lee Index and/or (2) increase in body fat levels [52]. According to some studies, experimental animals were said to be moderately obese if there was an 10-25% increase in body weight, and severe obese if the increase in body weight exceeded 40% of the initial body weight [53-55]. Based on the cut-off, the P2 group was not classified as obese. Lee's index is an approach to measure BMI (Body Mass Index) of the rats using the body weight and body length of mice [56]. The previous part of this study stated that the average Lee Index calculation in P2 group (287.78 ± 10.344 g / cm) was greater than group P1 (285.06 ± 10.15 g / cm) but not significantly different [57]. It was also written that the P2 group rats were more obese compared to group P1 rats.

Hariri & Thibault wrote that obesity occurs when the stored energy is greater than the energy released, so that the reserved energy in body fat increases, especially in adipose tissue. The state of obesity can involve one or both of the increase in the number of adipocytes (hyperplasia) or their size (hypertrophy) [52]. The previous part of this study, discussed the effect of the modified AIN-93M HFHF diet on body fat index [BFI) of male Sprague Dawley rats [58]. The results of this study found that BFI in P2 group (3.69 ± 0.77%) was smaller than group P1 (3.93 ± 1.18%) and did not differ significantly. Rats in P2 group cannot be categorized as obese because they have a low BFI value. Based on these parameters, it can be seen that the P2 rats could not been classified as obese. It can be concluded that the modified AIN-93M HFHF diet in this study had not been proven to cause obesity in male Sprague Dawley rats.

4.3. NF-κβ Expression in Experimental Rat Liver

Obesity is one of the risk factors for nonalcoholic fatty liver disease (NAFLD). NAFLD can be identified by the presence of fat accumulation (triglycerides) in hepatocytes [15]. The NAFLD pathogenesis is known as the 'two-hit' hypothesis, (1) fat metabolism and insulin resistance could induce simple fat accumulation and fatty liver and (2) associated with oxidative stress, inflammation and certain other factors. Oxidative stress in NAFLD can cause accumulation of triglycerides and free fatty acids in the liver, and mitochondrial dysfunction. Fatty acid oxidation can trigger oxidative stress in the liver, decreasing of antioxidants, and mitochondrial dysfunction, as a result, proinflammatory cytokines such as TNF-α, IL-6, IL-1β were released [21, 23].

Nuclear beta kappa factor (NF-κβ) is a transcription factor that becomes a central mediator of inflammation and can be associated with pathological conditions related to obesity [9]. NF-κβ is also one of the key regulators in the inflammatory process that can stimulate the synthesis of TNF-α, IL-6, and adipokine, which in turn, increase the level of inflammation [59]. Based on immunohistochemical examination, it was found that the average NF-κβ (IRS) expression in the P2 group (5.18 ± 1.12) was greater than group P1 (1.66 ± 1.03). The difference in IRS of NF-κβ expression in both groups was significantly high, ± 3.53. The results of the independent T-test showed a significant difference between the NF-κβ expression values (IRS) between P1 group rats and P2 group rats (p-value = 0.000).

The results of this study were in line with the research which was aimed to determine the role of various types of diets on hepatic inflammation and expression of lipogenic factors mRNA. The study also included male Sprague Dawley rats as subjects and had 4 treatment groups, the control group (chow diet standard), HF (high fructose 20%), HFAT (high fat) and HFHF (high fructose high fat). After a 16-week treatment period, it was found that NF-κβ expression in the HF, HFHF and HFAT groups were significantly higher than the control group ($p < 0.05$), with the highest expression in the HFHF group [60]. Another study showed that the liver tissue from the HF (high fructose 30%) group induced for 8 weeks was significantly having higher NF-κβ levels than the control group [61].

The previous part of this study, examined the effect of the AIN-93M modified HFHF diet on fatty liver in male Sprague Dawley mice. The study found significant difference in the normal group of AIN-93M modification compared with P2 group, p value = 0.000 ($\alpha = 0.05$). In the group of rats given a normal diet of 93M AIN modification, the average percentage of fatty liver cells was 19.64%, whereas in the group of rats fed a modified AIN-93M HFHF diet, the average number of fatty liver cells was higher (62.19%) [62]. Research by He stated that HFAT and HFHF treatment groups showed higher NAFLD scores than the control group and HF, and the percentage of NASH was 80% in the HFAT group and 90% in the HFHF group [60]. Another study by Jarukamjorn also stated that the group given HFFD (high fat fructose diet) had more steatosis than the control group [15].

5. Limitations of Research

The high energy density and texture of HFHF feed which was soft and oily due to the high fat proportion can reduce the appetite of mice, leading to lower feed intake in HFHF group. Consequently, the final weight was not significantly different between the two groups. In addition, the bottle drinking containing fructose solution was leaked several times, and certainly affects the volume of drinking intake of mice and causing the rat's hair wetness and moist condition of the husk.

6. Conclusion

1) The average daily feed intake of the P1 group was higher than the P2 group. But the average energy intake in P2 group was higher than in group P1. The results of statistical tests found significant differences between daily feed intake and energy intake of group P1 mice (AIN-93M standard diet) with mice P2 (AIN-93M modified HFHF diet), p = 0.000 in both.
2) The average body weight of rats in P2 group was higher than in group P1. But the results of the statistical test showed that there was no significant difference between the average final weight of the end of the group P1 and the P2 group with the p value = 0.223.
3) The average NF-κβ expression in P2 group was higher than in group P1. The results showed a significant difference between the average values of NF-κβ expression (IRS) between P1 group rats and P2 group rats (p = 0.000). The results of the correlation test between dietary intake and NF-κβ expression in rat liver tissue showed that there was a positive correlation between energy, carbohydrate and fat intake with NF-κβ expression (p = 0.001, 0.000, 0.046). On the other hand, there was a negative correlation between protein intake and NF-κβ expression in rat liver tissue (p = 0.000).

References

[1] Sudoyo A W, Setiyohadi B, Alwi I, Simadibrata M and Setiati S 2009 Buku Ajar Ilmu Penyakit Dalam Jilid II edisi V *Jakarta: Interna Publishing* **310** 1973-82
[2] Judith E. Brown JI, Bea Krinke, Ellen Lechtenberg, Maureen Murtaugh, Carolyn Sharbaugh, M.S., R.D., Patricia Splett, Jamie Stang, Nancy Wooldridge 2011. Nutrition Through the Life Cycle, 4th Edition. USA: *Wadsworth Cengage Learning*.
[3] WHO. 2017. Obesity and Overweight. available from: http://www.who.int/mediacentre/ factsheets/ fs311/en/].
[4] Kemenkes-RI. 2013. Riset Kesehatan Dasar Nasional
[5] Ng M, Fleming T, Robinson M, et al. 2014. Global, regional, and national prevalence of overweight and obesity in children and adults during 1980-2013: a systematic analysis for the Global Burden of Disease Study 2013. *The Lancet*, **384** (9945): 766-81.

[6] Ferrante AW, Jr. 2007. Obesity-induced inflammation: a metabolic dialogue in the language of inflammation. *Journal of internal medicine*, **262** (4): 408-14.
[7] Fabbrini E, Sullivan S, and Klein S. 2010. Obesity and nonalcoholic fatty liver disease: biochemical, metabolic, and clinical implications. *Hepatology (Baltimore, Md)*, **51** (2) 679-89
[8] Alpert MA, Lavie CJ, Agrawal H, Aggarwal KB, and Kumar SA. 2014. Obesity and heart failure: epidemiology, pathophysiology, clinical manifestations, and management. *Translational research : the journal of laboratory and clinical medicine*, **164** (4): 345-56.
[9] Carlsen H, Haugen F, Zadelaar S, et al. 2009. Diet-induced obesity increases NF-kappaB signaling in reporter mice. *Genes Nutr*, **4** (3) 215-22
[10] Noeman SA, Hamooda HE, and Baalash AA. 2011. Biochemical study of oxidative stress markers in the liver, kidney and heart of high fat diet induced obesity in rats. *Diabetology & metabolic syndrome*, **3** (1): 17.
[11] Li XH, McGrath KC, Nammi S, Heather AK, and Roufogalis BD. 2012. Attenuation of liver pro-inflammatory responses by Zingiber officinale via inhibition of NF-kappa B activation in high-fat diet-fed rats. *Basic & clinical pharmacology & toxicology*, **110** (3): 238-44.
[12] Williams LM. 2012. Hypothalamic dysfunction in obesity. *The Proceedings of the Nutrition Society*, **71** (4) 521-33
[13] Wang L, Lin Q, Yang T, et al. 2017. Oryzanol Modifies High Fat Diet-Induced Obesity, Liver Gene Expression Profile, and Inflammation Response in Mice. *Journal of agricultural and food chemistry*, **65**(38): 8374-85.
[14] Seki E, and Schwabe RF. 2015. Hepatic inflammation and fibrosis: functional links and key pathways. *Hepatology (Baltimore, Md)*, **61**(3): 1066-79.
[15] Jarukamjorn K, Jearapong N, Pimson C, and Chatuphonprasert W. 2016. A High-Fat, High-Fructose Diet Induces Antioxidant Imbalance and Increases the Risk and Progression of Nonalcoholic Fatty Liver Disease in Mice. *Scientifica (Cairo)*, **2016** 5029414
[16] Lozano I, Van der Werf R, Bietiger W, et al. 2016. High-fructose and high-fat diet-induced disorders in rats: impact on diabetes risk, hepatic and vascular complications. *Nutrition & metabolism*, **13** 15
[17] Elsharkawy AM, and Mann DA. 2007. Nuclear factor-kappaB and the hepatic inflammation-fibrosis-cancer axis. *Hepatology (Baltimore, Md)*, **46** (2): 590-7.
[18] Rutledge AC, and Adeli K. 2007. Fructose and the metabolic syndrome: pathophysiology and molecular mechanisms. *Nutrition reviews*, **65** (6 Pt 2): S13-23.
[19] Jou J, Choi SS, and Diehl AM. 2008. Mechanisms of disease progression in nonalcoholic fatty liver disease. *Seminars in liver disease*, **28** (4): 370-9.
[20] O. Beenhouwer D. 2010. Molecular Basis of Diseases of Immunity. In: Molecular Pathology, 2nd Edition. USA: *Elsevier Academic Press*.
[21] Tornatore L, Thotakura AK, Bennett J, Moretti M, and Franzoso G. 2012. The nuclear factor kappa B signaling pathway: integrating metabolism with inflammation. *Trends in cell biology*, **22** (11): 557-66.
[22] Sun B, and Karin M. 2008. NF-kappaB signaling, liver disease and hepatoprotective agents. *Oncogene*, **27** (48): 6228-44.
[23] Lin Z, Cai F, Lin N, Ye J, Zheng Q, and Ding G. 2014. Effects of glutamine on oxidative stress and nuclear factor-κB expression in the livers of rats with nonalcoholic fatty liver disease. *Exp Ther Med*, **7** (2): 365-70.
[24] Nugraheni F. 2014. Pengaruh Tepung Daun Katuk terhadap Ekspresi Nuclear Factor Kappa Beta pada Jaringan Hepar Tikus Putih Jantan yang Diberi Diet Aterogenik. Tugas Akhir. Universitas Brawijaya Malang.
[25] Cai D. 2009. NFκB-mediated metabolic inflammation in peripheral tissues versus central nervous system. *Cell Cycle*, **8** (16): 2542-8.
[26] Hall J, and Guyton AC. 2014. Guyton dan Hall Buku Ajar Fisiologi Kedokteran, Edisi 12, Ermita Ilyas (penerjemah). Singapore: *Elsevier (Singapore) Pte Ltd.*

[27] Nowak M, A. Madej J, and Dziegiel P. 2007. Intensity of COX2 expression in cells of soft tissue fibrosacrcomas in dogs as related to grade of tumour malignancy. *Bulletin of the Veterinary Institute in Pulawy*, **51**: 275-9.
[28] Kaemmerer D, Peter L, Lupp A, et al. 2012. Comparing of IRS and Her2 as immunohistochemical scoring schemes in gastroenteropancreatic neuroendocrine tumors. *International journal of clinical and experimental pathology*, **5**(3): 187-94.
[29] Reeves PG, Nielsen FH, and Fahey GC, Jr. 1993. AIN-93 purified diets for laboratory rodents: final report of the American Institute of Nutrition ad hoc writing committee on the reformulation of the AIN-76A rodent diet. *The Journal of nutrition*, **123**(11): 1939-51.
[30] Farias Santos J, Suruagy Amaral M, Lima Oliveira S, et al. 2015. Dietary intake of ain-93 standard diet induces Fatty liver with altered hepatic fatty acid profile in Wistar rats. *Nutricion hospitalaria*, **31**(5): 2140-6.
[31] Madsen AN, Hansen G, Paulsen SJ, et al. 2010. Long-term characterization of the diet-induced obese and diet-resistant rat model: a polygenetic rat model mimicking the human obesity syndrome. *The Journal of endocrinology*, **206**(3): 287-96.
[32] Handayani D, J. Meyer B, Chen J, et al. 2012. The Comparison of the Effect of Oat and Shiitake Mushroom Powder to Prevent Body Weight Gain in Rats Fed High Fat Diet. *Food and Nutrition Sciences*, **3**: 1009-19.
[33] So M, Gaidhu MP, Maghdoori B, and Ceddia RB. 2011. Analysis of time-dependent adaptations in whole-body energy balance in obesity induced by high-fat diet in rats. *Lipids in health and disease*, **10**: 99.
[34] Ble-Castillo JL, Aparicio-Trapala MA, Juarez-Rojop IE, et al. 2012. Differential effects of high-carbohydrate and high-fat diet composition on metabolic control and insulin resistance in normal rats. *International journal of environmental research and public health*, **9**(5): 1663-76.
[35] Lee JS, Jun DW, Kim EK, Jeon HJ, Nam HH, and Saeed WK. 2015. Histologic and Metabolic Derangement in High-Fat, High-Fructose, and Combination Diet Animal Models. *The Scientific World Journal*, **2015**: 9.
[36] Yoo S, Ahn H, and Park YK. 2016. High Dietary Fructose Intake on Cardiovascular Disease Related Parameters in Growing Rats. *Nutrients*, **9**(1): 11.
[37] Lambertz J, Berger T, Mak TW, van Helden J, and Weiskirchen R. 2017. Lipocalin-2 in Fructose-Induced Fatty Liver Disease. *Front Physiol*, **8**: 964-.
[38] Vila L, Rebollo A, Adalsteisson GS, et al. 2011. Reduction of liver fructokinase expression and improved hepatic inflammation and metabolism in liquid fructose-fed rats after atorvastatin treatment. *Toxicology and applied pharmacology*, **251** (1): 32-40.
[39] Fariba B, Hassan S, Shahin G, and Yavar R. 2013. Effects of dietary fructose, sucrose and lactose in induction of nonalcoholic fatty liver in rat. *World Applied Sciences Journal*, **22** (3): 368-73.
[40] Vasiljevic A, Bursac B, Djordjevic A, et al. 2014. Hepatic inflammation induced by high-fructose diet is associated with altered 11betaHSD1 expression in the liver of Wistar rats. *European journal of nutrition*, **53**(6): 1393-402.
[41] Shojaei Zarghani S, Soraya H, Zarei L, and Alizadeh M. 2016. Comparison of Three Different Diet-Induced Non Alcoholic Fatty Liver Disease Protocols in Rats: A Pilot Study. *Pharmaceutical Sciences*, **22**: 9-15.
[42] Curfs J, Chwalibog A, Savenije BS, and Ritskes-Hoitinga M. 2011. Nutrient requirements, experimental design, and feeding schedules in animal experimentation. In: Essential principles and practices.
[43] Permana Z. 2010. Konsumsi, Kecernaan dan Performa Tikus Putih (Rattus norvegicus) yang Disuplementasi Bio-mineral Cairan Rumen dalam Ransum. Tugas Akhir. Program Studi Ilmu Nutrisi dan Makanan Ternak, Fakultas Peternakan, Institut Pertanian Bogor.
[44] Kusumastuty I. 2014. Sari Buah Markisa Ungu Mencegah Peningkatan MDA Serum Tikus dengan Diet Aterogenik. *Indonesian Journal of Human Nutrition*, **1**(1): 50-6.

[45] Pendergast JS, Branecky KL, Yang W, Ellacott KL, Niswender KD, and Yamazaki S. 2013. High-fat diet acutely affects circadian organisation and eating behavior. *The European journal of neuroscience*, **37**(8): 1350-6.
[46] Syarkiah S, Fitri LE, and Pudjirahaju A. 2013. Pengaruh Pemberian Minyak Buah Merah (Pandanus conoideus oil) Terhadap Pembentukan Foam Cellpada Aorta Tikus Galur Wistar (Rattus norvegicus) Dengan Diet Aterogenik. *Jurnal Kedokteran Brawijaya*, **24**(1): 1-8.
[47] Zevenbergen H, de Bree A, Zeelenberg M, Laitinen K, van Duijn G, and Floter E. 2009. Foods with a high fat quality are essential for healthy diets. *Annals of nutrition & metabolism*, **54** Suppl 1: 15-24.
[48] Heymsfield SB, and Wadden TA. 2017. Mechanisms, Pathophysiology, and Management of Obesity. *New England Journal of Medicine*, **376**(3): 254-66.
[49] Sholichah Z. 2007. Mengenal Jenis Tikus. Banjarnegara: *BALABA*.
[50] Francisqueti FV, Santos KC, Ferron AJ, et al. 2016. Fructose: Toxic effect on cardiorenal risk factors and redox state. *SAGE open medicine*, **4**: 2050312116684294.
[51] Stubbs RJ, and Whybrow S. 2004. Energy density, diet composition and palatability: influences on overall food energy intake in humans. *Physiology & behavior*, **81**(5): 755-64.
[52] Hariri N, and Thibault L. 2010. High-fat diet-induced obesity in animal models. *Nutrition research reviews*, **23**(2): 270-99.
[53] Harrold JA, Williams G, and Widdowson PS. 2000. Early leptin response to a palatable diet predicts dietary obesity in rats: key role of melanocortin-4 receptors in the ventromedial hypothalamic nucleus. *Journal of neurochemistry*, **74**(3): 1224-8.
[54] Levin BE, and Dunn-Meynell AA. 2002. Reduced central leptin sensitivity in rats with diet-induced obesity. *American journal of physiology Regulatory, integrative and comparative physiology*, **283**(4): R941-8.
[55] Woods SC, Seeley RJ, Rushing PA, D'Alessio D, and Tso P. 2003. A controlled high-fat diet induces an obese syndrome in rats. *The Journal of nutrition*, **133**(4): 1081-7.
[56] de Lartigue G, Barbier de la Serre C, Espero E, Lee J, and Raybould HE. 2011. Diet-induced obesity leads to the development of leptin resistance in vagal afferent neurons. *American journal of physiology Endocrinology and metabolism*, **301**(1): E187-95.
[57] Fabiana BA. 2018. Pengaruh Pemberian Diet High Fat High Fructose [HFHF) Modifikasi AIN-93M Terhadap Antropometri Tikus [Sprague Dawley) Jantan. Tugas Akhir. Program Studi Ilmu Gizi, Fakultas Kedokteran Universitas Brawijaya.
[58] Indirasari AP. 2018. Pengaruh Pemberian Diet High Fat High Fructose (HFHF) Modifikasi Standar AIN-93M Terhadap Body Fat Index (BFI) Pada Tikus Sprague Dawley Jantan. Tugas Akhir. Program Studi Ilmu Gizi, Fakultas Kedokteran Universitas Brawijaya.
[59] Basaranoglu M, Basaranoglu G, Sabuncu T, and Senturk H. 2013. Fructose as a key player in the development of fatty liver disease. *World journal of gastroenterology*, **19**(8): 1166-72.
[60] Chongxin He XG, Jiangao Fan, Zhengjie Xu. 2017. Different dietary contribution to hepatic inflammatory and lipogenic factor mRNA expression. *Int J Clin Exp Med*, **10**(1): 325-33.
[61] Iskender H, Dokumacıoğlu E, Saral S, Yenice G, and Sevim Ç. 2018. NF-κB, TNF-α and IL-6 Levels in Liver and Kidney of High-Fructose-Fed Rats. *International Journal of Research in Pharmaceutical and Biomedical Sciences*, **18**: 1-7.
[62] Nurmalitasari S. 2018. Pengaruh High Fat High Fructose (HFHF) Diet terhadap Perlemakan Hati Tikus Sprague Dawley Jantan. Tugas Akhir. Program Studi Ilmu Gizi, Fakultas Kedokteran Universitas Brawijaya.

The Improvement of INS Gene Sequences on Rat Model of Type 1 Diabetes in Goat Milk Casein Protein Treatment, but not in Glibenclamide

M Mulyati[1], E Y Mu'jizah[2] and F Fatchiyah[2,3*]

[1]Faculty of Biology, Gadjah Mada University, Indonesia
[2]Biology Department, Faculty of Mathematics and Natural Sciences, Brawijaya University, Indonesia
[3]Research Center of Smart Molecule of Natural Genetics Resource, Brawijaya University, Indonesia

*Corresponding author's email: fatchiya@ub.ac.id

Abstract. This study was aimed to determine the effect of CSN1S2 protein from Etawah Crossbred goat milk for repairing DNA sequence of INS gene in T1DM rat model. We divided the experimental rats into control group, diabetes group, control with CSN1S2 protein treatment group, diabetes with CSN1S2 protein treatment group, and diabetes with glibenclamide treatment group. The dose of CSN1S2 protein and glibenclamide used was 800 mg/kg BW and 5 mg/kg BW, respectively. We isolated the DNA from rat pancreas tissue and amplified it with a specific primer of INS gene. Our study investigated that there were transition, transversion, and deletion mutation in diabetic rat. The glibenclamide administration had changed 6 point mutations in INS of the diabetic rat to normal, but it created the other 9 new mutations. The CSN1S2 protein treatment showed mutation repairing in INS of diabetes group. Consequently, CSN1S2 have potential effect as anti-diabetes and is safe to consume than glibenclamide.
Keywords: CSN1S2, insulin gene, T1DM

1. Introduction

Type 1 diabetes (T1DM) is an autoimmune disease leading to the deficiency of pancreatic beta cells when insulin secretion in the islets of Langerhans decrease [1]. Rat insulin is synthesized by two genes, preproinsulin-1 (Ins-1) and preproinsulin-2 (Ins-2) [2]. Preproinsulin-1 (Ins-1) is rodent-specific retrogene and ortholog to human gene. The preproinsulin-1 and preproinsulin-2 encode proinsulin peptide which expressed in pancreas. Insulin action occurs because of the interaction between insulin and its receptors on the target cell membrane. It activates the downstream proteins, such as Insulin Receptor Substrate-1 (IRS-1), Insulin Receptor Substrate-2 (IRS-2) and other proteins [3-4].

The triggers for T1DM can vary. The environmental and dietary factor may cause someone with the genetic susceptibility to develop this disease. Bovine serum albumin and bovine insulin in cow's milk are the implicated dietary proteins in diabetes development. Bovine insulin activates T-cells in someone with diabetes predisposition. It leads to the destruction of beta-cell through the direct attack from T-killer cells [5]. One of the drugs used in the management of diabetes is glibenclamide. Glibenclamide also known as glyburide is derivative of sufonylurea with sequester characteristic that have a role for hypoglycemia stimulation as farmakogenomic agent [6]. This drug acts by inhibiting ATP-sensitive potassium channel (K-ATP), then the cells membrane will depolarize and open the calcium channel in β cells pancreas. As a result, the intracellular calcium level will increase and stimulate insulin secretion from granula secretory by exocytosis [7-8]. However, glibenclamide can cause hypoglycaemia as a side effect [9].

The alternative that can be implemented to deal with diabetes management is the consumption of healthy nutrition. Goat milk can act as a potential health-promoting food due to its nutritional properties and its lower allergenicity when compared to cow milk [10]. Etawah crossbred goat from Indonesia has a specific protein in its milk, namely α-s2 casein (CSN1S2) protein. The CSN1S2 administration in rheumatoid arthritis rat can induce the STAT3 gene improvement through frameshift mutation [11]. We predicted that the CSN1S2 protein might affect the repairing mechanism of INS gene in type 1 diabetic rat as well. Therefore, this study aimed to analyse the potential effects of CSN1S2 protein from Etawah crossbred goat milk as antidiabetes through the INS gene sequence improvement.

2. Materials and Methods

2.1. Animal model

We used the 2-3 months old adult Wistar rats (*Rattus norvegicus*) for the experimental animal. Their weight was about 150-200 gram. We purchased the rats from The Integrated Research and Testing Laboratory, Gadjah Mada University. These rats underwent acclimatization for one week. We divided the rats into five groups, namely control (C), control with 800 mg/kg BW of CSN1S2 protein treatment (CM), diabetes (DM), diabetes with 800 mg/kg BW of CSN1S2 teratment (DMM), and diabetes 5 mg/kg BW of glibenclamide treatment (DMG). We induced the rats by 60 mg/kg BW of streptozotocin for T1DM rat model establisment. These animals treatment was evaluated and approved by ethic committee of Brawijaya University (Ethical certificate number: 417 KEP UB).

2.2. Preparation of CSN1S2 protein from Etawah Crossbred goat milk

We isolated the CSN1S2 protein from Etawah Crossbred goat milk. A 250 mL of goat milk was boiled on 40℃ of temperature and added 5 mL of glacial acetic acid. The separation of protein precipitation was done using the mesh of nylon membrane. The protein concentration measurement was performed by NanoDrop spectrophotometer. The CSN1S2 protein was stored at -20℃ [12].

2.3. DNA isolation and amplification

We isolated the DNA from rat pancreas tissue. The tissue was weighed 0,3 gram, then homogenized. The next procedures referred to Fatchiyah *et. al.*, 2011 with few modifications in the extraction buffer [13-14]. The DNA was quantified by using NanoDrop spectrophotometer. The quality of the DNA was evaluated using agarose gel 1% with chemidoc gel imaging. The amplification mix solution composition was referred to Fatchiyah *et. al.*, 2011. We designed the primer based on NCBI sequence. The primer pair for INS was NC_019129.3: rINS-Foward 34-5' AAT GTT CCA ACA TGG CCC TG 3' and rINS-Reverse 333-5' TGC AGC ACT GAT CCA CAA TG 3'. The PCR program for this primer was one cycle for the hot start at 95℃ for 1 minute, then 35 cycles for denaturation at 95℃ for 1 minute, annealing at 51℃ for 1 minute, and extension at 72℃ for 30 seconds. The post extension was at 72℃ for and 10 minutes. The amplification results were tested by agarose gel 1,5%. They could be stored at -20℃ for long-term storage.

2.4. DNA sequencing and analysis

A 5 μl of purified PCR products was mixed with 5 μl of primer. We homogenized the mixture before sequenced it by using ABI 3730xl DNA Sequencer (Koeln, Germany) [15]. The DNA sequences data were aligned and analyzed using multiple alignment program by BioEdit version 7.0 [16].

3. Results and Discussion

The INS DNA sequences from our samples was aligned with the sequence from NCBI database with gene ID: NM_019129. The aligned sequences was started from nucleotide 65 to 463. There were some frameshift mutations in nucleotides of all treatments. Transition mutations had been occurred in CM group from $_{102}$C to $_{102}$T, $_{106}$T to $_{106}$C and deletion sequence in nucleotides 100, 101, and 114. The sequence of DM group showed transversion mutations in $_{99}$C to $_{99}$G, $_{112}$C to $_{112}$G, and $_{115}$A to $_{115}$C, transition mutation in nucleotide $_{102}$C to $_{102}$T and also deletion sequences in nucleotide 97, 100, 101, 114 and 124. The CSN1S2 protein treatment with the dose of 800 mg/kg BW in diabetic rat was able to improve the DNA sequence of INS. In addition, the DMM group had similar nucleotide sequence compared to CM group, but it had transversion mutation in $_{96}$G to $_{96}$C. Glibenclamide has been known as the common drug for diabetes mellitus patient management. However, this drug treatment had changed many of INS DNA sequence in diabetic rats. The DMG group underwent DNA sequence alteration, namely deletion in nucleotides 101, 114 and 115, transversion mutations in $_{99}$C to $_{99}$G, $_{108}$G to $_{108}$T, $_{109}$G to $_{109}$T, $_{111}$G to $_{111}$T, $_{121}$G to $_{121}$T, and, $_{125}$A to $_{125}$C, and $_{127}$G to $_{127}$C, and transition mutation $_{103}$C to $_{103}$T and $_{128}$C to $_{128}$T (Figure 1).

Mutations caused by streptozotozin can be repaired through the base excision repair (BER). The administration of CSN1S2 protein as nutrition for rats with T1DM might induce the DNA polymerase β, then the BER mechanism in INS gene would be stimulated. This repairing mechanism happened through the elimination of 5'deoxyribose phosphate (5'dRP) compound and catalyzation of DNA synthesis in single-nucleotide-gap-filling by lyase activity which stimulated by DNA polymerase β [17-18]. The glibenclamide treatment to animals with DM as a synthetic drug may cause the condensation and DNA fragmentation in cells and lead to specific apoptosis in β cells pancreas [19-20]. In other study, mutation in insulin gene occurred in normal animal treated with CSN1S2 protein. It can be prevented by consuming the whole milk which contain vitamins, several kinds of protein, mineral, and conjugated linoleic acid [21]. Therefore, the consumption of Etawah Crossbred goat milk for a normal condition with appropriate concentration also give a good impact.

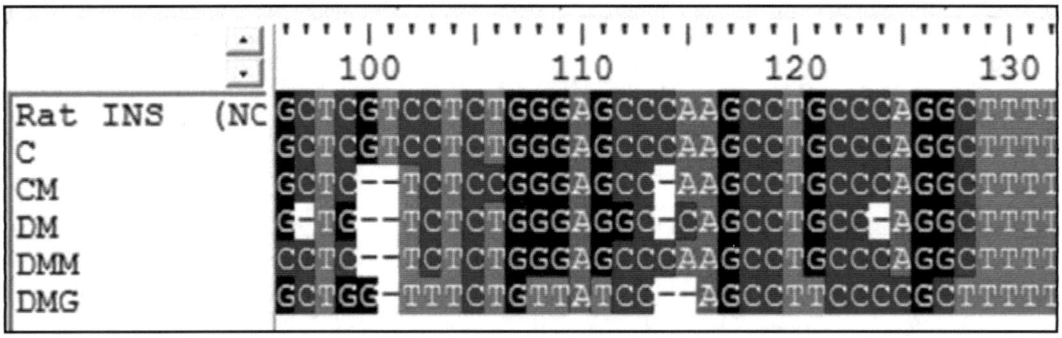

Figure 1. The DNA sequences of INS in all rat groups.

4. Conclusion

According to this research, we concluded that CSN1S2 protein from Etawah Crossbred goat milk can affect the repairing mechanism of gene damage. This protein plays role as anti-diabetes for repairing the damage in insulin gene of T1DM rat model.

References

[1] King A J 2012 The use of animal models in diabetes research *Br. J. Pharmacol.* **166** 877–894
[2] Shiao M, Liao B, Long M and Yu H 2008 Adaptive Evolution of the Insulin Two-Gene System in Mouse *Genetics* **178** 1683–1691
[3] Chai R, Chen S, Ding J and Grayburn PA 2009 Efficient, glucose responsive and islet-specific transgene expression by a modified rat insulin promoter *Gene Ther.* **16** 1202–1209
[4] Hay C W and Docherty K 2006 Comparative Analysis of Insulin Gene Promoters Implications for Diabetes Research *Diabetes* **55** 3201–3213
[5] Pizzorno J E, Murray M T and Joiner-Bey H 2016 Diabetes mellitus *The Clinician's Handbook of Natural Medicine* 249–286

[6] Gardner D G 2004 *Basic and Clinical Endocrinology*, 7th ed (San Fransisco: The McGraw-Hill Companies)

[7] Rai A, Eapen C and Prasanth V G 2012 Interaction of Herbs and Glibenclamide : A Review *ISRN Pharmacology* **2012** 1-5

[8] Proks P, Reimann F, Green N, Gribble F and Ashcroft F 2002 Sulfonylurea Stimulation of Insulin Secretion *Diabetes* **51** S368-S376

[9] Pandarekandy S T, Sreejesh P G, Thampi B S H and Sreekumaran E 2017 Hypoglycaemic Effect of Glibenclamide: A Critical Study on the Basis of Creatinine and Lipid Peroxidation Status of Streptozotocin-induced Diabetic Rat *Indian J. Pharm. Sci.* **79** 768-777

[10] Turck D 2013 Cow's Milk and Goat's Milk Szajewska H, Shamir R (eds): Evidence-Based Research in Pediatric Nutrition *World Rev. Nutr. Diet.* Basel, Karger **108** 56–62

[11] Rohmah R N, Hardiyanti F, Fatchiyah F 2015 Inhibition on JAK-STAT3 Signaling Transduction Cascade Is Taken by Bioactive Peptide Alpha-S2 Casein Protein from Goat Ethawah Breed Milk *Acta Inform. Med.* **23** 233-238

[12] Fatchiyah F, Setiawan B, Suharjono S and Noor Z 2015 The anti-osteoporosis effects of CSN1S2 protein of goat milk and yoghurt on a complete Freund's adjuvant-induced rheumatoid arthritis model in rats *Biomarkers Genomic Med.* **7** 139–146

[13] Fatchiyah F, Arumingtyas L E, Widyarti S and Rahayu S 2011 *Biologi molekular: Prinsip dasar analisis* (Jakarta: Erlangga)

[14] Schnepp B C, Jensen R L, Clark K R and Johnson P R 2009 Infectious molecular clones of adeno-associated virus isolated directly from human tissues *J. Virol.* **83** 1456–64

[15] Pareek C S, Smoczynski R and Tretyn A 2011 Sequencing technologies and genome sequencing *J. Appl. Genet.* **52** 413–435

[16] Hall T 2004 BioEdit Help Contents

[17] Sukhanova M V, Khodyreva S N, Lebedeva N A, Prasad R, Wilson S H and Lavrik O I 2005 Human base excision repair enzymes apurinic / apyrimidinic endonuclease1 (APE1), DNA polymerase b and poly (ADP-ribose) polymerase 1 : interplay between strand-displacement DNA synthesis and proofreading exonuclease activity *Nucleic Acids Res.* **33** 1222–9

[18] Çağlayan M and Wilson S H 2016 Oxidant and environmental toxicant-induced effects compromise DNA ligation during base excision DNA repair Melike *DNA Repair* **35** 85–89

[19] Fink S L and Cookson B T 2005 Apoptosis, Pyroptosis, and Necrosis : Mechanistic Description of Dead and Dying Eukaryotic Cells *Infect. Immun.* **73** 1907–1916

[20] Hambrock A, de Oliveira Franz C B, Hiller S, Osswald H 2006 Glibenclamide-Induced Apoptosis Is Specifically Enhanced by Expression of the Sulfonylurea Receptor Isoform SUR1 but Not by Expression of SUR2B or the Mutant SUR1 (M1289T) *J. Pharmacol. Exp. Ther.* **316** 1031–1037

[21] Miller G D, Judith J K and McBean L D 2000 *Handbook of Dairy Foods and Nutrition, Second Edition*. (Florida: CRC Press LLC)

Systemic lupus erythematosus: PKCA is an inhibition pathway for mTOR by the active ingredient of green tea

H Mawarti[1]*, J Nugraha[2], D A Purwanto[3] and J Soeroso[4]

[1]Doctoral Program of Medical Science, Faculty of Medicine, Airlangga University, Indonesia
[2]Department of Clinical Patology, Faculty of Medicine, Airlangga University, Indonesia
[3]Department of Pharmaceutical Chemistry, Faculty of Pharmacy, Airlangga University, Indonesia
[4]Department of Internal Medicine, Faculty of Medicine, Airlangga University, Indonesia

*Corresponding author's email: herinmawarti@fik.unipdu.ac.id

Abstract. This study aims to evaluate the interaction of green tea active compounds with proteins related to mTOR signals. The in silico study uses SEA protein target software, DB strings, and AUTODOCK PYRX 9.5. There are twenty target proteins that can interact with the active compounds of green tea. Of the twenty proteins, only six proteins are connected to the mTOR pathway. Of the six proteins, one that is a regulator of mTOR inhibitors is PKCA. Epigallocatechin has the strongest interaction with PKCA 4ARA (-8 kcal / mol). Cianidanol has the strongest interaction with PKCA 3IW4 (-9.3 kcal / mol). To analyze the involvement of the autophagy, a docking between ULK1 and AMPK was conducted, and there was an interaction between ULK1 and AMPK (bond energy of −1446.11 kcal). For the interaction between mTOR and ULK1, the bond energy is −624.5 kcal. For active green tea compounds, the bonding energy is more positive than the mTOR bond with ULK1. It was concluded that the green tea active ingredient as an inhibitor control against mTOR through PKCA and ULK1-AMPK (autophagy pathway).

Keywords: autophagy, green tea, SLE, mTOR, inhibitor

1. Introduction

Systemic lupus erythematosus is a multisystem autoimmune-inflammatory disease, characterized by T and B cell dysfunction and antinuclear antibody production. In 100,000 population, 20-150 people with SLE are found [1]. Factors that cause SLE come from genetic factors and environmental exposure [2]. The spectrum of clinical manifestations is very broad, in the form of mild skin lesions to serious organ damage. Although the etiology of this disease is still unclear, several mechanisms are thought to be involved in the pathogenesis of this disease, including the Toll-like receptor pathway, apoptotic defects, and abnormal activation of the interferon pathway [3].

The mammalian target of rapamycin (mTOR) is a protein complex as a member of phosphoinositide 3-kinase (PI3K) -related protein kinase family [4]. mTOR integrates signal growth factor and nutrition to support cell survival, growth and proliferation [5, 6]. The mTOR mRNA levels in SLE patients were not significantly different compared to controls [7]. MTOR activation contributes to the pathogenesis of SLE. The contribution of mTOR activation is found in various SLE cell cases, including T cells, B cells, mesenchymal stem cells, and hepatocytes [8].

Autophagy is a catabolic process mediated by lysosomes to eliminate long lived-misfolded proteins and damaged organelles. Autophagy plays a role in maintaining cellular homeostasis and survival under stressful conditions [9, 10]. This process takes place in basal conditions, but increases in stress response and disease [11]. Autophagy dysregulation was found in SLE, in the form of an increase in basal autophagy levels in T cells that have the potential to increase autophagy cell death. In addition, the lack of autophagy response due to induction by certain stimuli causes a decrease in cell survival and an increase in apoptosis which triggers an increase in the release of autoantigen [12].

The chemical composition of green tea varies according to weather, season, horticultural practice, and leaf position. The main composition of the active ingredient is polyphenols. The main polyphenols in green tea are flavonoids. The four main green tea flavonoids include epicatechin (EC), epigallocatechin (EGC), epicatechin gallate (ECG), and epigallocatechin gallate (EGCG). Epigallocatechin gallate is the most significant component [13]. The use of green tea in autoimmune diseases is still very little. EGCG can suppress RNA and autoantigen proteins in normal cells in the salivary glands and skin [14]. In addition, EGCG is also a competitive inhibitor of active ATP against mTOR [15]. To the knowledge of the researchers, there are still rare studies that analyze the mechanism of green tea active compounds against the mTOR pathway in the autophagy target. Therefore this study aims to investigate the mechanism of green tea active compounds against the mTOR pathway for autophagy as an effort to inhibit the development of SLE.

2. Material and Methods

2.1. Protein target prediction analysis

The green tea compound content was obtained from Dr. Duke's database (https://phytochem.nal.usda.gov/phytochem/search/list). One of the main compounds of green tea is known as EGCG, so further exploration is needed. To find out the target of the EGCG compound protein from green tea, analysis was carried out using the SEA Target webserver (http://sea.bkslab.org/). The parameter used in SEA Target is the Max TC value that reflects the accuracy of predictions if it approaches a score of 1.

2.2. Network analysis

To determine the interaction of the relationship between the target of EGCG protein and mTOR pathway, a network analysis was performed using STRING webserver (https://string-db.org). The STRING webserver is able to show the interaction of several proteins taken from several study pathways namely Biocarta, BioCyc, GO, KEGG, and Reactome (Szklarczyk et al. 2015). Some proteins obtained from the target SEA and have a value of less than 1 are considered potential and play a role in finding new pathways, such as PRKCA, PRKCE, PRKCH also included as input at STRING.

2.3. Molecular docking

To determine the relationship between PRKCA and bioactive compounds in green tea, molecular docking analysis was performed using PyRx v0.9.5. 3D samples of PRKCA proteins (ID 4ARA and 31W4) were obtained from the GDP database (https://www.rcsb.org), while the green tea bioactive compounds were theoflavin (ID 1147), epigallocathecin (ID 65064), and cianidanol (ID 73160) obtained from the PubChem database (https://pubchem.ncbi.nlm.nih.gov). When docking, validation using two 3D structures with different controls to show that the potential possessed by bioactive Green Tea has a target MTOR protein through PRKCA. Furthermore, the docking protein is visualized using PyMol software.

3. Results

First, we analyze the content of green tea compounds with Dr. Duke Phytochemical. The analysis shows that green tea contains EGCG. Then the target protein analysis of EGCG was carried out with the SEA Target Protein badger device. From the target analysis the accuracy values are obtained as in Table 1.

Table 1. Target protein using SEA Target Protein.

Target Name	Description	Max TC
FUTP4	Alpha-(1,3)-fucosyltransferase 4	1
PGD	6-phosphogluconate dehydrogenase, decarboxylating	1
ST3GAL3	CMP-N-acetylneuraminate-beta-1,4-galactoside alpha-2,3-sialyltransferase	1
FUT7	Alpha-(1,3)-fucosyltransferase 7	1
BCL2	Apoptosis regulator Bcl-2	1
MMP14	Matrix metalloproteinase-14	1
MAPT	Microtubule-associated protein tau	1
STAT1	Signal transducer and activator of transcription 1-alpha/beta	1
DNMT1	DNA (cytosine-5)-methyltransferase 1	1
TERT	Telomerase reverse transcriptase	1
MMP2	72 kDa type IV collagenase	1
MET	Hepatocyte growth factor receptor	1
DYRK1A	Dual specificity tyrosine-phosphorylation-regulated kinase 1A	1
MAPK14	Mitogen-activated protein kinase 14	1
KCNH2	Potassium voltage-gated channel subfamily H member 2	1
PGF	Placenta growth factor	0.32
PRKCA	Camp dependent protein kinase catalytic subunit alpha	0.31
PRKCH	Protein kinase c eta type	0.31
PRKCE	protein kinase c epsilon type	0.31
SERPINE1	plasminogen activator inhibitor 1	0.48

Table 2. Docking With PKCA 4ARA.

Reseptor	**Ligand**	**Binding Affinity (kkal/mol)**
PKCA 4ARA	Theoflavin 11477	-7.4
PKCA 4ARA	Epifalocatechin 65064	-8
PKCA 4ARA	Cianidanol 73160	-7.3
PKCA 4ARA	Control inhibitor 4ara 1R)-9-[(3S,4S)-1,3-dimethylpiperidin-4-yl]-8-(2-fluorophenyl)-1-methyl-3,5-dihydro[1,2,4]triazino[3,4-c][1,4]benzoxazin-2(1H)-one	-9.8

Table 3. Docking With PKCA 3IW4.

Reseptor	**Ligand**	**Binding Affinity (kkal/mol)**
PKCA 3IW4	Theoflavin 11477	-7.4
PKCA 3IW4	Epifalocatechin 65064	-9.0

PKCA 3IW4	Cianidanol 73160	-9.3
PKCA 3IW4	Control inhibitor 3iw4 -(1H-indol-3-yl)-4-[2-(4-methylpiperazin-1-yl)quinazolin-4-yl]-1H-pyrrole-2,5-dione	-11.3

All proteins presented in table 1 are then analyzed by STRING DB to find out their interactions with mTOR. Some proteins that have interactions with mTOR include PRKCA, PRKCE, TERT, BCL2, MAPK14, and PRKCH, as presented in Figure 1.

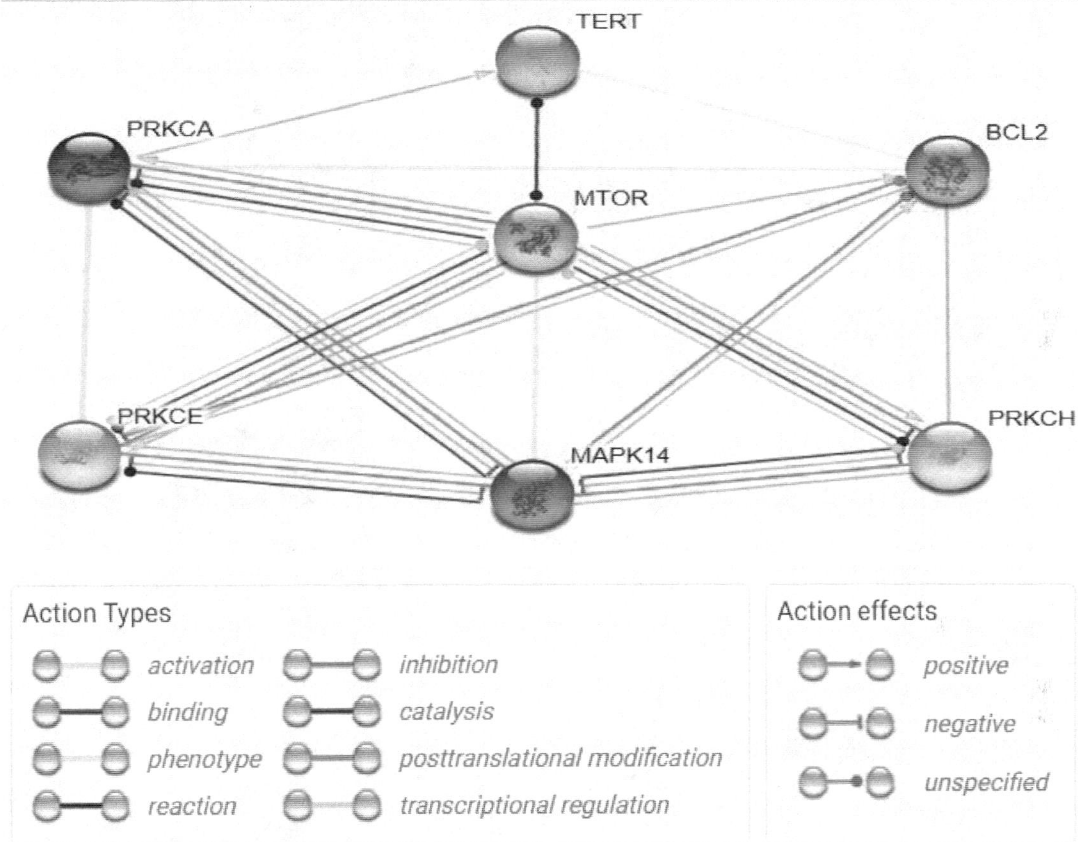

Figure 1. Prediction of Interaction of Compounds with Interest Proteins.

PKCA 4ARA is docking with EGCG and the comparative compounds include theaflavin, cianidanol, and control inhibitor compounds, namely 1R) -9- [(3S, 4S) -1,3-dimethylpiperidin-4-yl] -8- (2-fluorophenyl) -1-methyl-3,5-dihydro [1,2,4] triazino [3,4] [1,4] benzo-xazin-2 (1H) -one. Bond affinity can be seen in Table 2.

PKCA 3IW4 is docking with EGCG and the comparative compounds include theoflavin, cianidanol, and control inhibitor compounds, namely (1H-indol-3-yl) -4- [2- (4-methylpiperazin-1-yl) quinazolin-4-yl] -1H-pyrrole-2,5-dione. Bond affinity can be seen in Table 3.

Furthermore, to analyze the involvement of the autophagy pathway, a docking between ULK1 and AMPK was carried out, there was an interaction between ULK1 and AMPK (bond energy of 41,446.11 kcal. For interactions between mTOR and ULK1, bonding energy was −624.5 kcal. (+)-Epigallocatechin has docking energy with mTOR of −242.8 kcal. (+)-catechin has docking energy with mTOR of −245.8 kcal. (-)-Epicatechin gallate has docking energy with mTOR of −279.7 kcal. -

catechin has docking energy with mTOR of −246.9 kcal. (-)-gallocatechin has docking energy with mTOR of −253.1 kcal. Epicatechin it has docking energy with mTOR of −247.3 kcal. Epigallocatechin gallate has docking energy with mTOR of 272.2 kcal (as shown in Table 4).

Table 4. Docking With mTOR.

Docking	Energy Docking	Figure
1. ULK1 with AMPK	41.446,11kkal	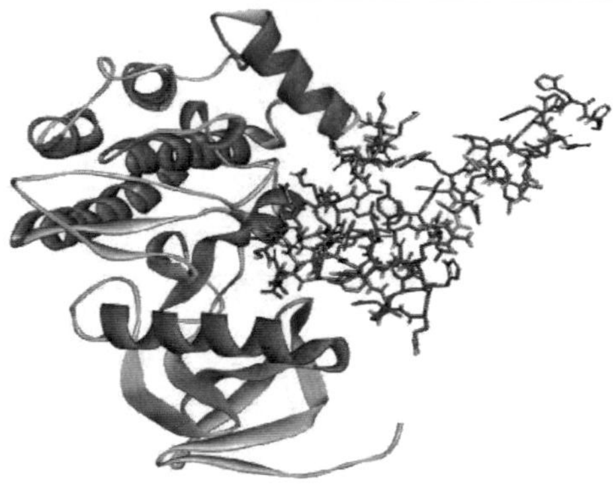
2. ULK1 with mTOR	-624,5 kkal	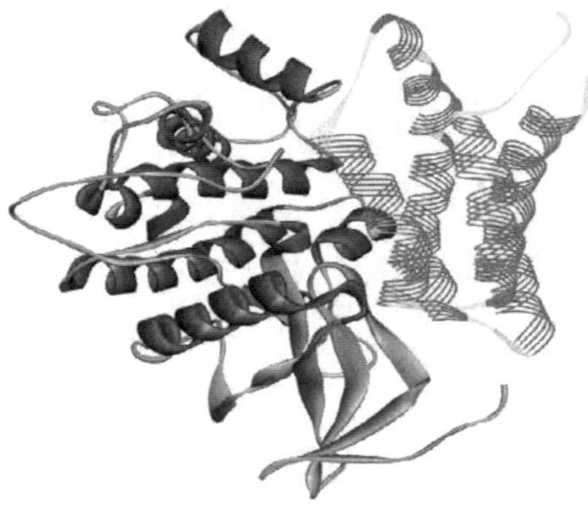
3. Epigallocatechin_CID_104 25234 with bmTOR complexes	-242,8 kkal)	

4. catechin_CID_9064 with mTOR complexes	-245,0 kkal/ mol	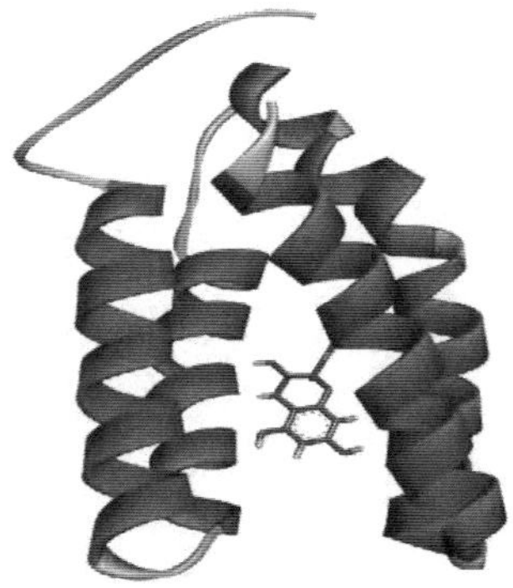
5. Epicatechin gallate_CID_107905 with mTOR complexes	-279,7 kkal/ mol	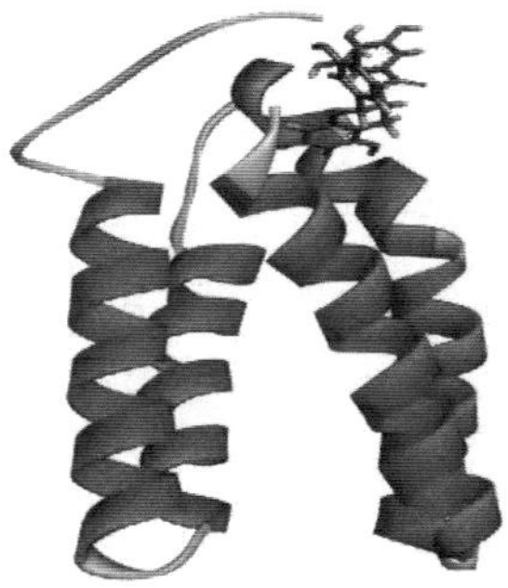
6. -catechinCID_73160 with mTOR complexes	-246,9 kkal/ mol	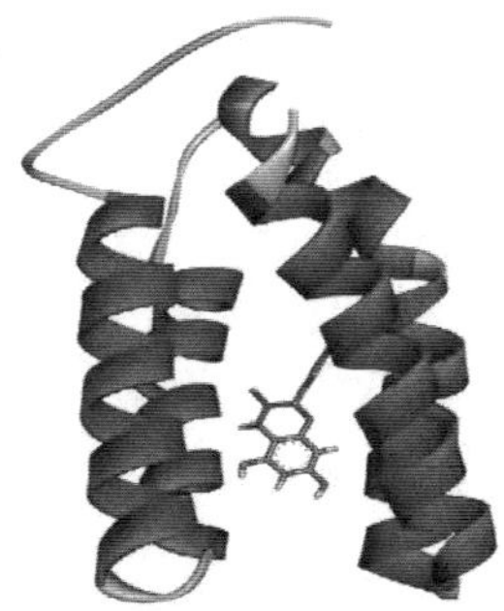
7. - gallocatechinCID_9882981 with mTOR complexes	-253,1 kkal/ mol	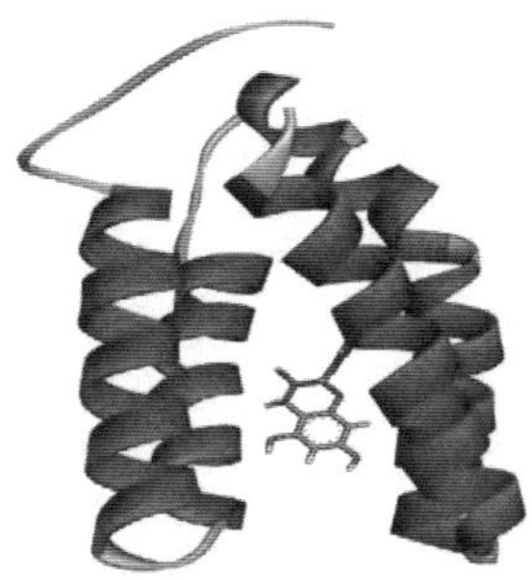

8.	EpicatechinCID_72276 with mTOR complexes	-247,3 kkal/ mol	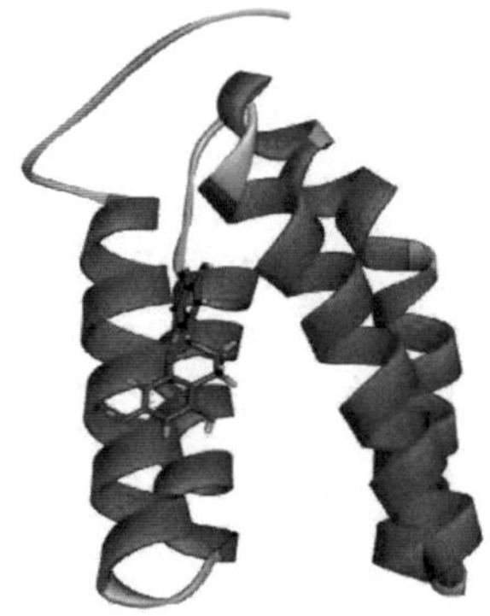
9.	(-)-epigallocatechin gallate (EGCG) with mTOR complexes	-272,7kkal	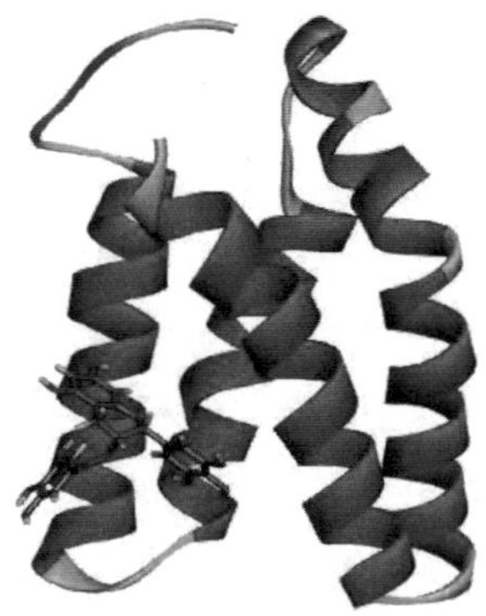

4. Discussion

mTOR is a master regulator of cellular metabolism and plays an important role in autophagy. When there is energy deficiency, mTOR activity is low, autophagy is upregulated to carry out nutrient recycling. When nutrients and energy are available, mTOR is active and downregulated. In SLE mTOR levels were found to be comparable to controls. This indicates that in SLE the formation of phagolysosomes is obtained but there is a degradation blockade [7]. In this study it was proven that EGCG as the active ingredient of green tea has the ability with proteins in the mTOR pathway, namely PKCA. Regarding PKCA 4ARA it was found that EGCG was easier to interact than theaflavin and cianidinol. Meanwhile, against PKCA 3IW4 it was found that EGCG was easier to interact with theaflavin. This indicates that the effect of EGCG as one of the active ingredients of green tea is at least through the PKCA 4ARA pathway. PKC is one of the regulators that is involved in pathway transduction that is related to various cell functions [16]. This study is consistent with previous findings that EGCG can inhibit PKC activity [17, 18].

Against the autophagy pathway, this study proved that the interaction between ULK1 and AMPK has more negative docking energy compared to the docking energy between ULK1 and mTOR. This indicates that in basal conditions there is an easier interaction between ULK1 and AMPK than ULK1 with mTOR. Interestingly, of the various green tea active compounds, the docking energy formed in the interaction with mTOR was more positive than that between mTOR and ULK1. This indicates that there is an easier interaction between mTOR and ULK1 compared to the interaction of green tea active ingredients with mTOR. This study extends previous findings that EGCG is able to modulate and inhibit mTOR [19, 20].

We conclude that the green tea active ingredient which is an inhibitor control against mTOR through PKCA and ULK1-AMPK pathway. Thus, in the pathomechanism of SLE, EGCG can be a candidate in SLE treatment through the PKCA-mTOR pathway and ULK1-AMPK (autophagy).

References

[1] Tsokos G C 2011 Systemic lupus erythematosus *N. Engl. J. Med.* **365** 2110–2121

[2] Leffers H C B, Lange T, Collins C, Ulff-Moller D J and Jacobsen S 2019 The study of interactions between genome and exposome in the development of systemic lupus erythematosus *Autoimmun Rev.* **18** 382-392

[3] Hsu S and Dickinson D 2006 A new approach to managing oral manifestations of Sjogren's syndrome and skin manifestations of lupus *J. Biochem Mol. Biol.* **39** 229-239

[4] Laplante M and Sabatini D M 2012 mTOR signaling in growth control and disease *Cell* **149** (2) 274–293

[5] Guertin D A and Sabatini D M 2007 Defining the role of mTOR in cancer *Cancer Cell* **12** (1) 9–22

[6] Duzgun Z, Eroglu Z and Avci C B 2016 Role of mTOR in glioblastoma *Gene* **575** 187-190

[7] Wu Z Z, Zhang J J, Gao C C, Zhao M, Liu S Y, Gao G M and Zheng Z H 2017 Expression of autophagy related genes mTOR, Becline-1, LC3 and p62 in the peripheral blood mononuclear cells of systemic lupus erythematosus *Am. J. Clin. Exp. Immunol.* **6** (1) 1-8

[8] Scheinecker C, Bonelli M and Smolen J S 2010 Pathogenetic aspects of systemic lupus erythematosus with an emphasis on regulatory T cells *J. Autoimmun.* **35** (3) 269–275

[9] Klionsky D J 2007 Autophagy: from phenomenology to molecular understanding in less than a decade *Nat. Rev. Mol. Cell Biol.* **8** 931–937

[10] Mizushima N and Klionsky D J 2007 Protein turnover via autophagy: implications for metabolism *Annu. Rev. Nutr.* **27** 19–40

[11] Pan Q, Gao C, Chen Y, Feng Y, Liu W J and Liu H F 2015 Update on the role of autophagy insystemic lupus erythematosus: a novel therapeutic target *Biomed. Pharmacother.* **71** 190–193

[12] Alessandri C, Barbati C, Vacirca D, Piscopo P, Confaloni A, Sanchez M *et al* 2012 Tlymphocytes from patients with systemic lupus erythematosus are resistant to in-duction of autophagy *FASEB J.* **26** (11) 4722–4732

[13] Pastore R L and Frarellone P 2006 Potential health benefits of green tea (Camellia sinensis): a narrative review *Explore* **2** (6) 531-539

[14] Hsu S and Dickinson D 2006 A new approach to managing oral manifestations of Sjogren's syndrome and skin manifestations of lupus *J. Biochem. Mol. Biol.* **39** 229-239

[15] Van Aller G S, Carson J D, Tang W, Peng H, Zhao L, Copelanda R A, Tummino P J and Luo L 2011 Epigallocatechin gallate (EGCG), a major component of green tea, is a dualphosphoinositide-3-kinase/mTOR inhibitor *Biochem. & Biophys. Res. Comm.* **406** 194-199

[16] Hall J L, Matter C M, Wang X and Gibbons G H 2000 Hyperglycemia inhibits vascular smooth musclecell apoptosis through a protein kinase C-dependent pathway *Circ. Res.* **87** 574–80

[17] Singh B N, Shankar S and Srivastav R K 2011 Green tea catechin, epigallocatechin-3-gallate (EGCG): Mechanisms, perspectives and clinical applications *Biochem. Pharmacol.* **82** 1807-182

[18] Das J, Ramani R and Suraju M O 2016 Polyphenol compounds and PKC signaling *Biochem. Biophys. Acta* **1860** 2107-2121

[19] Glenn S. Van Aller G S, Carson J D, Tang W, Peng H, Zhao L, Copeland R A, Tummino P J and Luo L 2011 Epigallocatechin gallate (EGCG), a major component of green tea, is a dualphosphoinositide-3-kinase/mTOR inhibitor *Biochem. Biphys. Res. Comm.* **406** 194-199

[20] Sato S, Mukai Y, Hamaya M, Sun Y and Kurasaki M 2013 Long-term effect of green tea extract during lactation on AMPK expressionin rat offspring exposed to fetal malnutrition *Nutrition* **29** 1152-1158

FSH Receptor Gene Polymorphism in Indonesian Women with Polycystic Ovarian Syndrome (PCOS)

R Panghiyangani[1]*, M Kurniati[2], P Soeharso[3], A Andrijono[4], D A Suryandari[3] and B Wiweko[4]

[1]Department of Biomedical Sciences, Faculty of Medicine-Lambung Mangkurat University-South Borneo
[2]Faculty of Medicine, Malahayati University- Lampung
[3]Department of Biology, Faculty of Medicine, University of Indonesia-Jakarta
[4]Department of Obstetrics and Gynecology, Faculty of Medicine, University of Indonesia-Jakarta

*Corresponding author's email: roselina.darma@gmail.com

Abstract. Polycystic ovary syndrome (PCOS) is a reproductive problem that often occurs among women in reproductive age, but until now, the etiopathogenesis of PCOS is still unclear. One of the alleles for candidate genes associated with the PCOS incident is the FSH receptor (FSHR) gene. This research was aimed at analyzing the genotypes within Asn680Ser FSHR with PCOS pathogenesis. The total research subject was 142 subjects. DNA isolation was used for performing the RFLP analysis toward Asn680Ser FSHR polymorphism. There was no mean value difference of the genotype distribution within Asn680Ser FSHR between the PCOS and non-PCOS groups ($p>0.05$). There was no difference of the basal FSHR levels based on the Asn680Ser FSHR genotyping variation between the PCOS and the non-PCOS groups ($p>0.05$). The Asn680Ser FSHR genotype does not determine the individual susceptibility to suffering from PCOS. There was no difference in basal FSH levels between PCOS and non-PCOS groups based on the Asn680Ser FSHR genotype variation. However, there are differences in the frequency distribution pattern of the FSHR genotype at position 680 in the PCOS group compared to the previous research findings in several countries. **Keywords**: FSHR, PCOS, Polymorphism

1. Introduction

Polycystic ovary syndrome (PCOS) is a reproductive problem that frequently happens among 5-10% of women in reproductive age to be fertile, however, the main cause has not been found yet up until now [1][2]. PCOS can be considered as a heterogeneous disorder with elevated androgen levels and various levels of reproductive and metabolic problems determined by the interaction of various genetic factors and environment [3]. The general illustrations for PCOS patients are including oligomenorrhoea or amenorrhea, syndrome of ovarian dysfunction characterized by the existence of anovulation, and polycystic ovarian morphology, hyperandrogenism such as hirsutism, acne, alopecia, and often followed by obesity [4]. Genome-wide association studies (GWAS) has identified some variances of genetic significantly related to PCOS. Up until now, a large number of genetic studies have identified more than 200 genes functionally related to PCOS. One of the genes related to PCOS is Follicle stimulating hormone (FSH) receptor gene [5].

FSH plays a role in inducing the proliferation of granulosa cells, the recruitment of secondary follicle, the selection of dominant follicle, and regulating the activity of aromatase enzyme in granulosa cells for converting androgen into estrogen (aromatization process). In running the function, FSH interacts with the receptor namely FSHR. FSH and LH will interact with the targeted cell and they will affect the function through a cell surface receptors in the form of integral proteins in the cellular membrane. In women, the affinity of FSH binding toward the receptor in the surface of follicle granulosa cells will determine the capacity of a follicular cell to proliferate and synthesize the estrogen [6].

The mutation in FSHR gene can cause a problem in the activity of the FSH receptor. One of the most investigated things is rs6166 FSHR SNP (Asn680Ser). From the previous studies, it was reported that polymorphism of the FSHR gene is Ser680 mutant allele associated with the lower response of receptor and it is significantly connected with the high level of basal FSH; besides, it changes the response towards the ovarian stimulation by using exogenous FSH for in-vitro fertilization [7] [8]. In most of PCOS patients, there is an increase in excessive secretion of LH that can result in the increase of androgen synthesis in ovarian. It has been found out an increased LH/FSH ratio (>3) [9]. The study in PCOS patients in *RSCM Jakarta* shows that 79% of the PCOS patients have an increased LH to FSH ratio. Both PCOS patients with and without insulin resistance experience an increased LH to FSH ratio of 81.5% and 90.9% respectively [10]

The first study conducted in Singapore at Thr307Ala and Ser680Asn, the variants of FSHR gene mutation related to the susceptibility to PCOS in 124 PCOS patients found that there was no mutation in the PCOS patient group. The study proceeded to analyze the polymorphism at the same point with 236 individuals classified in a healthy group as the control and the finding showed that the distribution was not different from the variants of allele and protein isoform in PCOS patients who had been investigated previously. The research finding showed that mutation in the coding area of the FSHR gene was not the factor of PCOS in women in Singapore [11]. In the study of polymorphism of FSHR Ser680Asn and Ala307Thr in Korea that involved 235 PCOS patients and 128 healthy individuals as the control subject, and all samples classified into reproductive age found that the polymorphism was significantly related to PCOS. The FSHR polymorphism is also associated with the concentration of endogenous serum FSH and Prolactin. However, based on the data, there was no hormone from the endocrine system related to other PCOS and no clinical pregnancy levels in PCOS patients recorded in that study [12].

In the report written by Du, et al., (2010), only a study in Chinese women revealed the significance between polymorphism of FSHR and PCOS. This study involved 60 PCOS patients and 92 healthy individuals as control. It stated that in one side, FSHR gene SNPs determines the susceptibility to PCOS and in another side; it also affects the sensitivity of the receptor of exogenous FSH during the therapy of induced ovulation [13]. Based on previous research reported that the Asn680Ser polymorphism of the FSH receptor gene is known to reduce FSH receptor activity with different results on race and ethnicity, this study aims to analyze the genotypes within Asn680Ser FSHR with PCOS pathogenesis in Indonesian Women.

2. Research Method

2.1. Subjects

The research subject was women at 18-40 years old who suffered from PCOS and non-PCOS women collected from the patients who underwent the IVF assisted fertilization program with a total of 142 people (Approval letter No. 106/H2.F1/ETIK/2013). The isolated DNA was collected from 2 subject groups namely women with PCOS who underwent the IVF assisted fertilization program diagnosed based on the Rotterdam 2003 criteria and had a willingness to be the research subject. The women with PCOS who underwent IVF assisted fertilization program and the infertility cause was diagnosed such as the male factor issues, fallopian tube, idiopathic infertility (unexplained), and had a willingness to be the research subject.

2.2. Fragment Amplification of Targeted DNA
The Asn680Ser rs 6166 polymorphism of the FSHR gene was identified by using PCR-RFLP technique based on the site of BsrI restriction enzyme (Promega) in the area of exon 10. The amplification of DNA fragment was performed by using PCR method with BioMix Red kit (BiolineTM) (IDT). Pairs of primer for the amplification of FSHR gene in exon 10 at position 680 (Asn680Ser), was left primer /forward (F): 5'GAG CAA GTG TGG CTG CTA TG 3', and right primer/reverse (R): 5'TGT AGA AGC ACT GTC AGC TC 3'. The amplification was conducted for 30 cycles, with initial denaturation at 95°C for 6 minutes, continued with the denaturation cycle at 95°C for 1 minute, annealing at 49°C for 1 minute, and elongation at 72°C for 1.5 minutes. At the end of the cycle, it was continued with elongation at 72°C for 5 minutes [14][15][16]. By using a pair of prime above, it would result in a PCR product with a size of 258bp. An amplicon was separated using electrophoresis technique in 2% agarose gel at a voltage of 80 volts for 1 hour.

2.3. Restriction Fragment Length Polymorphism (RFLP)
The restriction fragment was performed by using BsrI restriction enzyme (New England Biolabs). The digestion mixture was made using a composition consisting of a 10μl DNA fragment from the amplification result (DNA concentration ≈ 10μg) and 2μl buffer 3(1x), and 10 units of BsrI restriction enzyme, then ddH20 was added until the volume reached 30μL. The mixture was incubated at 65°C for 3 hours. The result of RFLP was detected using electrophoresis in 3% agarose gel in 1X TAE Buffer solution at a voltage of 80 volts for 1 hour. The visualization of DNA bands was observed using illuminator UV (UV long life™ Filter Spectroline® and it was captured using a digital camera. The result of RFLP toward FSHR gene of Asn680Ser using BsrI restriction enzyme was in the form of single DNA band that was not cut by BsrI restriction enzyme of 258pb in size with homozygous Asn/Asn genotype (wild type); it was formed two DNA bands of 136pb and 122pb respectively with homozygous Ser/Ser genotype (mutant type) and three DNA bands of 258pb, 136pb, and 122pb respectively with Asn/Ser genotype (heterozygote).

2.4. Sequencing Analysis
The sequencing of DNA was conducted using the Sanger method to determine the polymorphic site and the change of nucleotide from some samples by using the result of amplicon that had polymorphic sites collected from the result of PCR-RFLP. The sequencing of DNA was conducted in the 1st Base, Singapore. The analysis of sequencing result was done using *Finch TV program* Version 1.4.0 and then the process of BLAST in the NCBI site to conduct alignments between the DNA sequences of Asn680Ser RefSeq NG_008146.1 FSHR gene and the DNA sequences from the sample taken from a sequencing result.

2.5. Statistical Analysis
The statistical analysis conducted to determine the relevancy between the Asn680Ser polymorphism of the FSHR gene and the susceptibility of an individual towards PCOS was the Pearson Chi-square test (X^2 test) with a p-value of ≤ 0.05. Statistical analysis namely one way ANOVA test ($p \leq 0.05$) was done to prove the difference of the variation in the genotype of FSHR in exon 10 at position 680 and the basal FSH levels in both PCOS and non PCOS women [17].

3. Results and Discussion
3.1. The Characteristic of research subject
The total research subject was 142 people consisting of 66 PCOS women (46.6%) and 76 non-PCOS women (53.5%) (Table 1).

Table 1. The Characteristic of the Research Subjects in PCOS and Non-PCOS groups.

Paramater	Group		*p-value*
	PCOS	Non PCOS	
Age (years)	32.21 ± 0.499 (n=66)	32.80 ± 0.516 (n=76)	0.416[b]
BMI (Kg/m^2)	24.425 ± 0.585 (n=53)	22.84 ± 0.494 (n=59)	0.034[b]
FSH (mIU/mL)	5.715 ± 0.299 (n=48)	5.186 ± 1.033 (n=33)	0.233[b]
LH (mIU/mL)	5.626 ± 1.444 (n=56)	6.234 ± 0.976 (n=65)	0.002[b]
Prl (ng/mL)	12.27 (3.26-148.40) (n=50)	15.98 (6.32-179.80) (n=25)	0.018[a]
P (ng/mL)	2.133 ± 1.084 (n=22)	0.740 ± 0.163 (n=44)	0.083[b]
E2 (pg/mL)	32.40 (5.0-198.6) (n=41)	41.49 (2.07-300) (n=33)	0.132[a]
AMH (ng/mL)	5.705 ± 1.092 (n=49)	2.988 ± 0.793 (n=43)	0.000[b]

The data is presented in mean value ± SEM (*Standard Error Mean*)
[a] Non-parametric Mann Whitney test, significant at a p-value of <0.05
[b] Independent t-test, significant at a p-value of <0.05; n = total subject

3.2. Genotyping Analysis

The PCR technique was conducted to amplify the targeted DNA for FSHR in exon 10 at position 680 containing polymorphic sites and recognized by BsrI restriction enzyme that has a restriction site (5'AGTGGN3'). The polymorphism of FSHR gene at position 680 (Asn680Ser) was detected by finding the amino acid code namely Serine (A**G**T) (mutant type), which replaced the amino acid code namely Asparagine (A**A**T) (wild type); there was a change from nucleotide base namely Adenine into Guanine (A→G). The DNA amplicon was cut by BsrI enzyme to know the susceptibility of DNA toward restriction enzyme. If the result of the PCR was cut, the sample experienced mutation at that point. The result of DNA amplicon cutting would be electrophoresed on 3% agarose gel afterward (Figure 1.)

It was obtained three RFLP illustrations namely homozygote Ser/Ser (mutant type) with a restriction site for BsrI so that the PCR amplicon was divided into two bands of 136pb and 122pb respectively, Asn/Asn homozygote (wild type) that was not cut so that it has one band of 258pb, heterozygote Asn/Ser that has RFLP illustration with three DNA bands of 258pb, 136pb, and 122pb respectively. To clarify the polymorphism of the FSHR gene at position 680, the sequencing was done towards 10 samples that were randomly chosen.

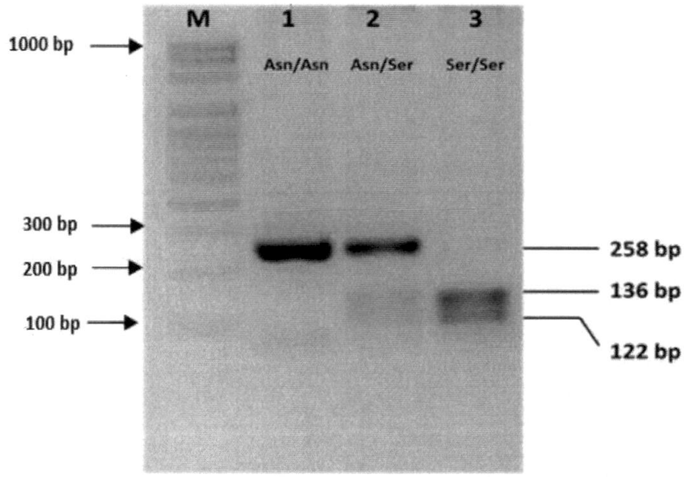

Figure 1. The result of the PCR-RFLP for Asn680Ser FSHR gene was cut using BsrI Enzyme M= DNA 100bp marker, well 1 = sample 1; homozygous Asn/Asn genotype, well 2 = sample 2; heterozygous Asn/Ser genotype, well 3 =sample 3; homozygous Ser/Ser genotype.

3.3. Sequencing Analysis

The result of sequencing showed that there was a change from nucleotide base namely Adenine into Guanine (A→G) at position 196710 that caused the change in amino acid code namely Asparagine (A**A**T) (wild type) into Serine (A**G**T) (mutant type) (see Figure 2.)

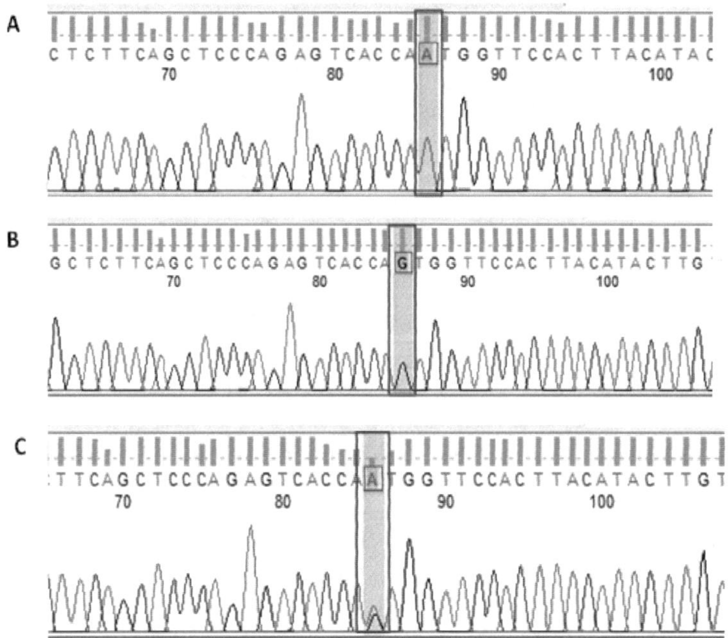

Figure 2. The result of sequencing uses the primer of FSHR at position 680. The result of sequencing shows the change in base namely the wild type adenine (A), into Guanine (G). A. The illustration of a sequencing result with a wild type sample (Asn/Asn). B. The sequencing result with a mutant type sample (Ser/Ser) C. The sequencing result with a sample that has a heterozygote type (Asn/Ser).

3.4. The Frequency Distribution of Genotype and Allotype within Asn680Ser FSHR gene in exon 10
In PCOS group, around 36 people (54%) had Asn/Asn genotype, 24 people (37%) had Asn/Ser genotype, and 6 people (9%) had Ser/Ser genotype. In non-PCOS group, around 40 people (52.63%) had Asn/Asn genotype, 25 people (32.90%) had Asn/Ser genotype, and 11 people (14.47%) had Ser/Ser genotype.

In the PCOS group, the frequency of Asn allele was 0.73 (73%) and Ser allele was 0.27 (27%). In the non-PCOS group, the frequency of Asn allele was 0.69 (69%) and Ser allele was 0.31 (31%). It is the comparison between the distribution of genotype and allotype within Asn680Ser FSHR gene in both PCOS and non-PCOS groups. The distribution of Asn680Ser FSHR genotype in PCOS group was in accordance with Hardy–Weinberg law ($p>0.05$), while in the non-PCOS group, it was not in accordance with Hardy–Weinberg law ($p<0.05$). The Asn allele that is the normal type (wild type) has a more dominant proportion towards Ser allele as a mutant allele.

Table 2. The frequency distribution of genotype and allotype within Asn680Ser FSHR in exon 10 in both PCOS group and non-PCOS group.

Group	Genotype			Alleles	
	Asn/Asn	Asn/Ser	Ser/Ser	Asn	Ser
PCOS (66)	54% (36)	37% (24)	9% (6)	0.73	0.27
Non-PCOS (76)	52.63% (40)	32.90 (25)	14.47 (11)	0.69	0.31

The result of the Person-Chi-Square test showed that the distribution of Asn680Ser FSHR genotype in PCOS group was not extremely different from the non-PCOS group (n= 142; df=2; p-value >0.05) and the distribution of FSHR allotype (Asn680Ser) in PCOS group was not extremely different from the non-PCOS group (n=284; df=1; p-value>0.05). This research finding showed that the allotypes within Asn680Ser FSHR gene in exon 10 did not affect the susceptibility for the occurrence of polycystic ovarian syndrome.

The genotype frequency within Asn680Ser FSHR in this research showed that the individual with homozygous Asn/Asn genotype (wild type) had a higher frequency than an individual with heterozygote Asn/Ser and an individual with homozygote Ser/Ser (mutant type) in both PCOS group and non-PCOS group. The result of the Person-Chi-Square test showed that the genotype distribution within Asn680Ser FSHR in PCOS group was not extremely different from the non-PCOS group ($p>0.05$) and the allotype distribution within Asn680Ser FSHR in PCOS group was not different from the non-PCOS group ($p>0.05$).

The distribution of Asn680Ser FSHR genotype in PCOS group in this research was different if it was compared to the similar research finding conducted by the previous researchers. In this research, the Asn/Asn genotype was more dominant than other genotype variants such as Asn/Ser and Ser/Ser. The previous studies related to the polymorphism of FSHR gene have been reported with various results in different races. Sudo, et al., [18] investigated the influence of the FSHR gene polymorphism variant at codon 307 (Alanine → Threonine) and at codon 680 (Asparagine → Serine) toward the ovarian function in 522 women in Japan. The frequency of Asn680Ser FSHR genotype in PCOS group was 16.7% Asn/Asn, 66.7% Asn/Ser, and 16.7% Ser/Ser. In the PCOS group, the percentage of Thr307-Asn680/Ala307-Ser680 variant was significantly bigger than a woman with a normal ovulation cycle. Conway, et al., [19] investigated the prevalence of FSHR isoforms in 93 PCOS women in England. The frequency of genotype was 24.7% Asn/Asn, 51.6% Asn/Ser, and 23.7% Ser/Ser. There was no significant difference in genotype distribution [20]. The research about Asn680Ser FSHR polymorphism in 518 PCOS women in a Caucasian population obtained a genotype frequency as follows: 24.8% Asn//Asn, 50.1% Asn/Ser, and 25.1% Ser/Ser [21]. Bon, et al reported the frequency of Asn680Ser

FSHR genotype in 235 PCOS women in Korea namely 58.72% Asn/Asn, 38.72% Asn/Ser, and 2.56% Ser/Ser [19]. Wahyono, et al investigated the FSHR polymorphism genotype in 91 women at reproductive age and normogonadotropic. The finding showed that among the Indonesian women population, the frequency of FSHR genotype at position 680 was 48.4% Asn/Asn, 41.7% Asn/Ser, and 9.9 % Ser/Ser [15].

Figure 3 shows the comparison of the frequency of Asn680Ser FSHR genotype in PCOS women in various race populations. It is very clear the there is a difference in the frequency distribution pattern of FSHR genotype at position 680. Caucasian population (Europe), represented by England, Italy, and Turkey, has a higher genotype frequency in the form of heterozygote Asn/Ser, while for Mongoloid population (Asian) represented by Japan, China, Korea, Thailand, and Indonesia, the higher genotype frequency is homozygote Asn/Asn (wild type), except for Japan. The frequency of Asn680Ser FSHR genotype in Japanese people population is almost similar to the Caucasian race namely heterozygote Asn/Ser that is higher. The distribution pattern of the European population is 1:2:1 for homozygote:heterozygote:homozygote ratio. Apparently, there is an association between the distribution of Asn680Ser FSHR genotype and the geographical variations and races. To prove those things, further research with more number of samples is needed. Besides, the research should be multicenter.

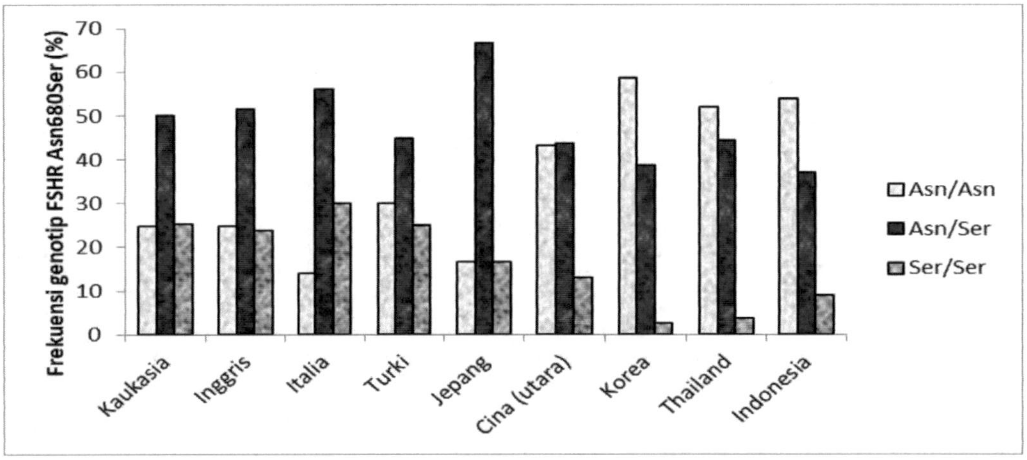

Figure 3. The comparison of the frequency of Asn680Serine FSHR genotype in PCOS women in a Caucasian population, England, Italy, Turkey, Japan, China (north), Korea, Thailand, and Indonesia.

The analysis of FSHR gene polymorphism at codon 680 in PCOS women has been conducted by the previous researchers in 50 PCOS patients compared to the normal women. The result of genotype frequency in PCOS group was 32% Asn/Asn, 44% Asn/Ser, and 24% Ser/Ser while in normal group was 34% Asn/Asn, 50% Asn/Ser, and 16% Ser/Ser [22]. However, the result is different from the illustration of Asn680Ser FSHR genotype frequency found in this research. In the previous studies, it was obtained a higher frequency of heterozygous Asn/Ser genotype (44%) than two genotype variations namely Asn/Asn (32%) and Ser/Ser (24%). In this research, it is shown that the higher frequency of homozygous genotype namely Asn/Asn (54%) than the other genotype variations namely Asn/Ser (37%) and Ser/Ser (9%).

Based on the explanation above, the distribution of FSHR genotype at position 680 between the PCOS group and non-PCOS group did show a significant difference. This explains that there is no relationship between the polymorphism of Asn680Ser FSHR gene and the pathogenesis of PCOS in PCOS women in Indonesia. It is in line with the research finding in Thailand about the polymorphism of FSHR in women who suffer from chronic anovulation [23]. The study conducted by Valkenburg, et al [21] in Caucasian women stated that the FSHR variant was not associated with the risk of being affected by PCOS; yet the FSHR variant was strongly associated with the severity level of PCOS characteristic illustration such as gonadotropin levels and hyperandrogenism. The result of a meta-analysis study conducted by Dao, et al [24] reported that there was no relationship between two FSHR polymorphisms namely Thr307Ala and Asn680Ser and the susceptibility toward PCOS. The FSHR polymorphism was not associated with the increased risk of PCOS.

The result of a study on Asn680Ser FSHR polymorphism associated with the pathogenesis of PCOS in PCOS women in Korea successfully proved that Asn680Ser FSHR was identified as the potential factor for the change of PCOS pathophysiology. In that study, it was shown that the Ser680Asn FSHR polymorphism caused infertility and folliculogenesis that deviates in PCOS patients [19]. The pathogenesis of PCOS can be caused by the other involved genes by the regulation of folliculogenesis, the regulation of synthesis, and androgen action, the regulation of synthesis, and insulin action. It is associated with the hyperandrogenemia, the insulin resistance or the increased level of LH associated with the anovulation situation as the cause of infertility in PCOS women. Another factor that plays a role in determining the severity of PCOS is a gene involved in lipid metabolism and chronic inflammation [25][26][21].

The different ovarian response and the success of ovarian stimulation towards the FSH stimulation depend on the FSHR genotype. The FSHR genotype variation at codon 680 from each individual that affects the condition has been proven by some previous researchers [27]. Other researchers found that the splicing alternative that could affect the ligand-binding part in the extracellular domain from the FSHR receptor caused the reduced cAMP activation in the FSHR variants compared to the normal one. The research was successful in identifying that the FSHR alternative splicing variant (deletion in exon 2) was associated with the reduced follicle growth. It is supposed that the different individual response toward the ovarian stimulation in the in-vitro fertilization program [26]. One of the problems of PCOS patients is the different response towards the ovarian stimulation and the low oocyte quality that can affect the pregnancy success rate, which until now, it cannot be explained well. Further knowledge about the etiology that can help in determining the parameter that determines the response of an individual towards the ovarian stimulation in an assisted reproductive program in treating infertility in PCOS [12][27][28].

3.5. The relationship between Asn680Ser FSHR genotype and basal FSH levels in both PCOS and non-PCOS groups

A statistical analysis namely ANOVA test to prove the difference between the FSHR genotype variation at position 680 in exon 10 toward the basal FSH levels in each PCOS group and non-PCOS group. The result of the one-way ANOVA test in PCOS group is shown in Table 3.

Table 3. The relationship between the Asn680Ser FSHR genotype variation and the basal FSH levels in PCOS group.

Genotype	n	The Average of basal FSH levels	p
Asn/Asn	27	5.774 ± 1.471	
Asn/Ser	17	5.308 ± 1.477	0.512
Ser/Ser	4	6.125 ± 2.458	

Based on the Asn680Ser FSHR genotype variation namely Asn/Asn, Asn/Ser, and Ser/Ser, there was no significant difference in the average level of basal FSH levels in PCOS group ($p > 0.05$). The result of the one-way ANOVA test in the non-PCOS group is shown in Table 4.

Table 4. The relationship between the Asn680Ser FSHR genotype variation and the basal FSH levels in the non-PCOS group.

Genotype	n	The average of basal FSH levels	p
Asn/Asn	20	6.159 ± 1.801	
Asn/Ser	11	6.701 ± 2.298	0.709
Ser/Ser	2	5.800 ± 0.424	

Based on the Asn680Ser FSHR genotype variation namely Asn/Asn, Asn/Ser, and Ser/Ser, there was no significant difference in the average of basal FSH levels in the non-PCOS group ($p > 0.05$).

The average of basal FSH levels was 5.715±0.299 mIU/mL in the PCOS group and 5.186±1.033mIU/mL in the non-PCOS group. It is seen that the average of basal hormone in PCOS group tends to be higher than the non-PCOS group, however, the difference is not statistically significant ($p>0.05$). The basal FSH levels in both PCOS group and non-PCOS group are still within normal limits of around 5-9IU/L [30]. The average basal FSH levels based on the genotype variation in PCOS group show a difference, but based on the statistical analysis, the difference is not significant ($p >0.05$).

Table 4 shows a tendency that Ser/Ser genotype group has a higher average basal FSH level of 6.125 ± 2.458 mU/mL than the Asn/Asn genotype group and Asn/Ser have namely 5.774 ± 1.471mIU/mL and 5.308 ± 1.477mIU/mL respectively. In the non-PCOS group, the highest average of basal FSH levels was in Asn/Ser genotype that was 6.701±2.298 mIU/mL, while the Asn/Asn genotype was 6.159± 1.801mIU/mL and Ser/Ser genotype was 5.800±0.424 mIU/mL. It is shown the there is a difference but it is not statistically significant ($p >0.05$).

The previous research findings stated that the individual as the carrier of Ser680 allele had a higher basal FSH level than the individual the carrier of Asn680 allele had [30][18][31]. This research could not prove that there was a significant difference in basal FSH levels in both allele groups. Nevertheless, it tended to have a difference in basal FSH levels based on FSHR genotype variation in both PCOS group and non-PCOS group.

The inconsistency of result obtained from some studies that investigated the relationship between polymorphism and PCOS has been reported by some researchers. It might be because of the methodological reason such as Population stratification (genetic heterogeneity, the history of the population), the definition of illness/conflicting characteristics, problems in selecting the subjects or an incidental finding. One of the main problems with the study about candidate gene was the sample size that was frequently insufficient. The inconsistent result might occur because of the characteristic for polymorphism such as incomplete penetration, genetic heterogeneity, and interaction between genes or interaction between gene and environment. Other weaknesses of using candidate gene/marker to analyze the association are including the fact that only a small part of the genome that is being studied, and this is performed independently from each interaction that may be involved [12]. Another thing that should be considered is the responsibility factor toward the association of PCOS with some disturbances, which is important for clinical diagnosis in PCOS patients.

4. Conclusion

Based on this research finding, it can be concluded that: (1) there is a frequency distribution pattern of FSHR genotype at position 680 in PCOS group that is different from the previous research findings, (2) Asn680Ser FSHR genotype does not determine the susceptibility of an individual to suffer from PCOS, and (3) there is no difference of basal FSH levels in women with PCOS compared to non PCOS patients based on the Asn680Ser FSHR genotype.

5. Acknowledgment

The researcher would express gratitude to *Yasmin* Fertility Clinic, *RSCM Kencana*, Jakarta and the Integrated Laboratory of Faculty of Medicine, the University of Indonesia for supporting this research.

References

[1] Hadibroto B 2005 Sindrom ovarium polikistik *Maj. Kedok. Nusant. FK USU* **38** 333-7
[2] Alvero R and Schlaff W D 2007 *Reproductive endrocrinology and infertility: the requisites in obstretrics and gynecology* (USA: Elsevier Inc) p.65-76
[3] Dasgupta S and Reddy B 2008 Present status of understanding on the genetic etiology of polycystic ovary syndrome *J. Postgrad. Med.* **54** 115
[4] Xita N, Georgiou I and Tsatsoulis A 2002 The genetic basis of polycystic ovary syndrome *Eur. J. Endocrinol* **147** 715-25
[5] Joop S E L 2019 Follicle Stimulating Hormone Receptor (FSHR) Polymorphism and Polycystic Ovary Syndrome (PCOS) *Frontier in Endocrinology* **10** 23
[6] Fritz M A and Speroff L 2011 *Clinical gynecologic endocrinology and infertility: chronic anovulation and the polycystic ovary syndrome 8th ed* (North Caroline: Lippincott Williams and Wilkins) p.495-531
[7] Simoni M, Gromoll J and Nieschlag E 1997 The follicle-stimulating hormone reseptor: biochemistry, molecular biology, physiology and pathophysiology *Endocr. Rev.* **18** 739-73
[8] Mayorga M P, Gromoll J, Behre H M, Gassnerr C, Nieschlag E and Simoni M 2000 Ovarian respon to follicle-stimulating hormone (FSH) stimulating depends on the FSH receptor genotype *J. Clin. Endocrinol. and Metab.* **85** 3365-9
[9] Simoni M, Gromoll J and Nieschlag E 1997 The follicle-stimulating hormone reseptor: biochemistry, molecular biology, physiology and pathophysiology *Endocr. Rev.* **18** 739-73
[10] Wiweko B and Mulya R 2008 Profil resistensi insulin pada pasien sindrom ovarium polikistik (SOPK) di RS Dr. Cipto Mangunkusumo Jakarta *J. Obstet. Gynecol. Indonesia* **32** 93-8
[11] Tong Y, Liao W X, Roy A C and Ng S C 2001 Absence of mutations in the coding regions of follicle-stimulating hormone receptor gene in Singapore Chinese women with premature ovarian failure and polycystic ovary syndrome *Horm. Metab. Res.* **33** 221–6
[12] Wu X Q, Xu S M, Liu J F, Bi X Y, Wu Y X and Liu J 2014 Association between FSHR polymorphisms and polycystic ovary syndrome among Chinese women in north China *J. Assist. Reprod. Genet.* **31** 371–7
[13] Du J, Zhang W, Guo L, Zhang Z, Shi H, Wang J *et al* 2010 Two FSHR variants, haplotypes and meta-analysis in Chinese women with premature ovarian failure and polycystic ovary syndrome *Mol. Genet. Metab.* **100** 292–5
[14] Suryandari D A 2005 *Polimorfisme gen reseptor FSHR: pengaruhnya pada sensitivitas ovarium wanita yang mengikuti program reprosduksi berbantuan terhadap stimulasi FSH eksogen [disertasi]* (Jakarta: Program Doktor Ilmu Biomedik-Pascasarjana FKUI)
[15] Wahyono D J 2005 *Perbandingan Efektivitas metode PCR_SSCP dan PCR-RFLP untuk deteksi polimorfisme kodon 680 ekson 10 gen FSHR [tesis]* (Jakarta: Program Pasca Sarjana Ilmu Biomedik FKUI FKUI)
[16] Valkenburg O, Uitterlinden A G and Piersma D 2009 Genetic polymorphisms of GnRH and gonadotrophic hormone receptors affect the phenotype of polycystic ovary syndrome *Hum. Reprod.* **24** 2014–22
[17] Dahlan M S 2003 *Statistik untuk Kedokteran dan Kesehatan: Deskriptif, Bivariat, dan Multivariat Dilengkapi Aplikasi dengan menggunakan SPSS* (Jakarta: Penerbit Salemba Medika) p: 61-86; 167-79
[18] Sudo S, Kudo M, Wada S, Sato O, Hsueh A W J and Fujimoto S 2002 Genetic and functional analyses of polymorphism in the human follicle stimulating hormone receptor gene *Mol. Hum. Reprod.* **8** 893-9
[19] Bon H G, Jung M P and Kwang H B 2010 Genetic variations of follicle stimulating hormone receptor are associated with polycystic ovary syndrome *Inter. J. of Mol. Med.* **26** 107-12
[20] Conway G, Conway E, Hoopner W, Gromoll J and Simoni M 1999 Mutation screening and isoform prevalence of the follicle-stimulation hormone receptor gene in women with

premature ovarium failure, resistant ovary syndrome and polycystic ovary syndrome *Endocrinol.* **51** 97-9

[21] Valkenburg O, Uitterlinden A G and Piersma D 2009 Genetic polymorphisms of GnRH and gonadotrophic hormone receptors affect the phenotype of polycystic ovary syndrome *Hum. Reprod.* **24** 2014–22

[22] Ciptaning L W 2005 *Analisis polimorfisme gen FSH reseptor dan kadar FSH basal pada penderita sindrom ovarium polikistik di Indonesia [tesis]* (Jakarta: Program Pascasarjana Ilmu Biomedik FKUI)

[23] Singhasena W, Pantasri T, Piromlertamorn W, Samchimchom S and Vutyavanich T 2014 Follicle-stimulating hormone receptor gene polymorphism in chronic anovulatory women, with or without polycystic ovary syndrome: a cross-sectional study *Reprod. Biol. and Endocrinol.* **12** 86

[24] Dao J C, Rui D, Ji Y C, Jin X Z, Jia X Z and Dong Q Y 2014 Two follicle-stimulating hormone receptor polymorphisms and polycystic ovary syndrome risk: a meta-analysis *Eur. J. of Obst. Gyn.* (EJOG) **182** 27-32

[25] Gersak K and Ferk P 2007 Genetics of polycystic ovary syndrome *Gynaecol. Perinatol.* **16** 53–7

[26] Gerasimova T, Thanasoula M N, Zattas D, Seli E, Sakkas D and Lalioti M D 2010 Identification and in vitro characterization of follicle stimulating hormone (FSH) receptor variants associated with abnormal ovarian response to FSH *J. Clin. Endocrinol. Metab.* **95** 529–36

[27] Mayorga M P, Gromoll J, Behre H M, Gassnerr C, Nieschlag E and Simoni M 2000 Ovarian respon to follicle-stimulating hormone (FSH) stimulating depends on the FSH receptor genotype *J. Clin. Endocrinol. and Metab.* **85** 3365-9

[28] Trimeche S, Paris F, Brum I S and Sultan C 2003 Genetic and molecular aspects of polycystic ovary syndrome *Therapeutic Med. Endocrinol.* **5** 134-45

[29] Haggstrom M 2014 Reference ranges for estradiol, progesterone, luteinizing hormone and follicle-stimulating hormone during the menstrual cycle_ *Wikiversity J. of Med.* **1**

[30] Laven J S E, Mulders A G M, Suryandari D A, Gromoll J, Nieschlag E, Fauser B C J M *et al* 2003 Follicle stimulating hormone reseptor polymorphisms in women with normogonadotropic anovulatory infertility *Fertil Steril* **80** 986-91

[31] Simoni M, Nieschlag E, Gromoll J 2002 Isoforms and single nucleotide polymorphisms of the FSH receptor gene: implications for human reproduction *Hum. Reprod. Update* **8** 413-21

Genomic Profile of rIR and rIRR in Type 2 Diabetes Mellitus Rat Model toward Effect of Goat Milk CSN1S2 Protein

S H Tambunan[1], H N Meidinna[1,2], R N Rohmah[2] and F Fatchiyah[1,2*]

[1]Biology Department, Faculty of Mathematics and Natural Sciences, Brawijaya University, Indonesia
[2]Research Center of Smart Molecule of Natural Genetics Resource, Brawijaya University, Indonesia

*Corresponding author's email: fatchiya@ub.ac.id

Abstract. Diabetes mellitus is metabolic disorder with hypergycemia condition. Diabetes meilitus is commonly divided into two types, type 1 diabetes mellitus (T1DM) and type 2 diabetes mellitus (T2DM). The research focus was to determine the effect of CSN1S2 protein from Etawah Crossbred goat milk on DNA sequence of rIR exon 17 and rIRR exon 2 in T2DM rat model. The experimental rats were divided into four main groups, namely control group, diabetes group, control with CSN1S2 protein treatment group, and diabetes with CSN1S2 protein treatment group. We used 375, 750 and 1500 mg/kg BW of goat milk CSN1S2 protein for the treatment. The DNA was extracted from the rat liver, then amplified with a spesific primer of rIR and rIRR genes. The results of the amplification will be sequenced by using ABI 3730xl DNA Sequencer. The CSN1S2 protein treatment in rIR gene exon 17 did not affect a specific DNA sequence alteration. The diabetic rats treated with 375, 750 and 1500 mg/kg BW of CSN1S2 protein had the nucleotide changed to normal ($_{3950}A \rightarrow _{3950}G$). Therefore, the CSN1S2 protein may have the function in triggering the DNA repair system.
Keywords: CSN1S2, IR, IRR, T2DM

1. Introduction

Diabetes mellitus is a group of metabolic disorders characterized by increased blood sugar levels that result from insulin deficiency, insulin resistance, or a combination of both. Chronic effects of diabetes can damage several organs, including hearts, eyes, kidneys, nerves, and blood vessels [1]. Diabetes has been reported attacking 10 million people in Indonesia in 2015 and will be elevated in 2030 as many as 21,3 million people. The disease is caused by a lifestyle, ethnic and heredity history [2, 3].

Type 1 diabetes mellitus (T1DM) and Type 2 diabetes mellitus (T2DM) are two of four diabetes categories classified by the America Diabetes Association (ADA). They were previously known as Insulin dependent or juvenile-onset diabetes and non insulin dependent or adult-onset diabetes [4]. T2DM is caused by insulin resistance combined with relative insulin deficiency [5]. Insulin resistance comes from insulin receptor damage and consequently disturbs the regulation of glycogen synthesis and sugar homeostasis in blood [6]. Insulin receptor family (IRF) consists of three members: insulin receptor (IR), insulin-like growth factor receptor (IGFR) and insulin receptor-related receptor (IRR). These receptors can be found in the liver tissue [7]. Insulin receptor has been the main role for activation of tyrosine kinase and leads to glycogen synthesis [8].

Research of food materials targetting the improvement of insulin receptor of T2DM patient becomes important. One of the food materials that can be explored is milk. Milk is the source of macro and micro nutrition, including the bioactive protein for body health [9]. Cow milk is the most common milk consumed by people. The other milk source comes from the goat. This goat milk is rarely to consume. Etawah crossbred goat from Indonesia is a crossing product of etawah goat from India and domestic goat from Indonesia [10]. This goat has a specific protein, namely CSN1S2 with 36 kDa of weight [11]. This protein has the advantage as anti-inflammatory, immunomodulatory, and anti-oxidant [12–14].

The influence of CSN1S2 protein from etawah crossbred goat milk for repairing the abnormality of insulin receptor gene in T2DM is not yet observed. Therefore, this research analyzed the CSN1S2 protein potential effects on the insulin receptor (rIR and rIRR) genes of the Wistar rat (*Rattus novegicus*) with T2DM.

2. Materials and Methods

2.1. Animal model

The 2-3 months old adult Wistar rat (*Rattus norvegicus*) with weight range of 150 gram were collected from The Integrated Research and Testing Laboratory, Gadjah Mada University. These rats were acclimatized for one week. The experimental animals were clustered into four main groups: control (C), control with CSN1S2 protein from Etawah Crossbred goat milk treatment (CM), diabetes (DM), diabetes with CSN1S2 protein from Etawah Crossbred goat milk treatment. The doses of the protein in CM and DM group consisted of 375, 750 and 1500 mg/kg BW of CSN1S2 protein. The CSN1S2 protein administration was carried out for 28 days. These animals treatment was evaluated and recognized by the ethic committee of Brawijaya University (Ethical certificate number: 417 KEP UB).

2.2. The Streptozotocin-induced type 2 diabetes mellitus in rat

The rats in diabetes (DM) group were fed with a high-fat diet for 2 months. The rats with cholesterol levels exceeding 200 mg/dl were injected with 25 mg/kg of streptozotocin (STZ) [15]. The cholesterol level and glucose blood level of the animal were measured with cholesterol stick and glucometer stick, respectively.

2.3. Preparation of CSN1S2 protein from etawah crossbred goat milk

The CSN1S2 protein was isolated from 250 mL of Etawah Crossbred goat milk. The milk was boiled on 40℃ and added by 5 mL of glacial acetic acid. Nylon membrane was used for separating the protein precipitation through the mesh of the nylon. Protein concentration was measured by UV-vis NanoDrop spectrophotometer. The protein was stored at -20℃ [16].

2.4. DNA extraction

The rat liver tissues were collected and weighed 0,3 gr. The tissues were homogenized by mortar, pestle and extraction buffer. The next procedures referred to Fatchiyah *et. al.*, 2011 with fews modification in buffer of extraction [17,18]. The isolated DNA was quantified by UV-vis NanoDrop spectrophotometer. It was also examined using agarose gel 1% with chemidoc gel imaging.

2.5. DNA amplification

The composition of amplification mix solution referred to Fatchiyah *et. al.*, 2011. The primer was designed from NCBI database sequence. The primer pair for rIR in exon 17 (ID:NC_005111.4) was: rIR-Forward 124954-124974- 5' GAA AAA ATT GCG GCG GGA GG 3' and rIR-Reverse 125340-125321-5' TGG CCC TAG GCA GAA CAA AG 3'. The PCR program was hot start at 94℃ for 3 minutes (1 cycle), 32 cycles consisted of denaturation at 95℃ for 30 seconds, annealing at 54℃ for 45 seconds, and extension at 72℃ for 30 seconds. The post extension was at 72℃ for seven minutes. The primer for rIRR in exon 2 amplification (ID NC_005101) was rIRR-*forward* 3510-3539-5' CCA GTC TTG ACA TCC GCT CG 3' and rIRR-*reverse* 4058-4039-5' ACA GTG GCT GGA GGT CCA AC 3'. The rIRR gene had the PCR program of hot start at 94℃ for 3 minutes (1 cycle), 32 cycles consisted of denaturation at 95℃ for 30 seconds, annealing at 54℃ for 30 seconds, and extension at 72℃ for 30 seconds. The post extension was at 72 ℃ for seven minutes (1 cycle). Then, the amplification results were evaluated by agarose gel 1,5% and stored at -20℃.

2.6. DNA sequencing
The 5 µl of purified DNA amplification product was mixed with 5 µl primer in 1.5 ml microtube. The solution was transferred to ABI 3730xl DNA Sequencer (Koeln, Germany) for DNA sequencing [19]. The sequences data were aligned using multiple alignment program by BioEdit version 7.0 [20].

3. Results and Discussion
The rIR amplification product had the size of 386 bp, while the rIRR gene had base the DNA size of 687 bp. The results of the rIR gene sequencing in each treatment were aligned with the rIR gene sequence from NCBI database (ID: NC_005111.4). The rIR exon 17 alignment was started from the nucleotide sequences number 124820 to 125135. We revealed that all rat groups had similar rIR DNA sequences (Figure 1a). There was no sequence alteration in them. The DM group and all CSN1S2-treated DM group did not experienced insulin resistance in the insulin receptor. Insulin resistance in the rIR gene is indicated by mutations in gene sequences. A factor that might affect the absence of rIR sequence mutations in the diabetic rat is there is no experimental animal model establishment based on the changes in genetic material. The genetically modified T2DM animal model is created mostly using the knockout system. This program can efficiently regulate the metabolic damage in experimental animals so that they experience glucose intolerance [21-22].

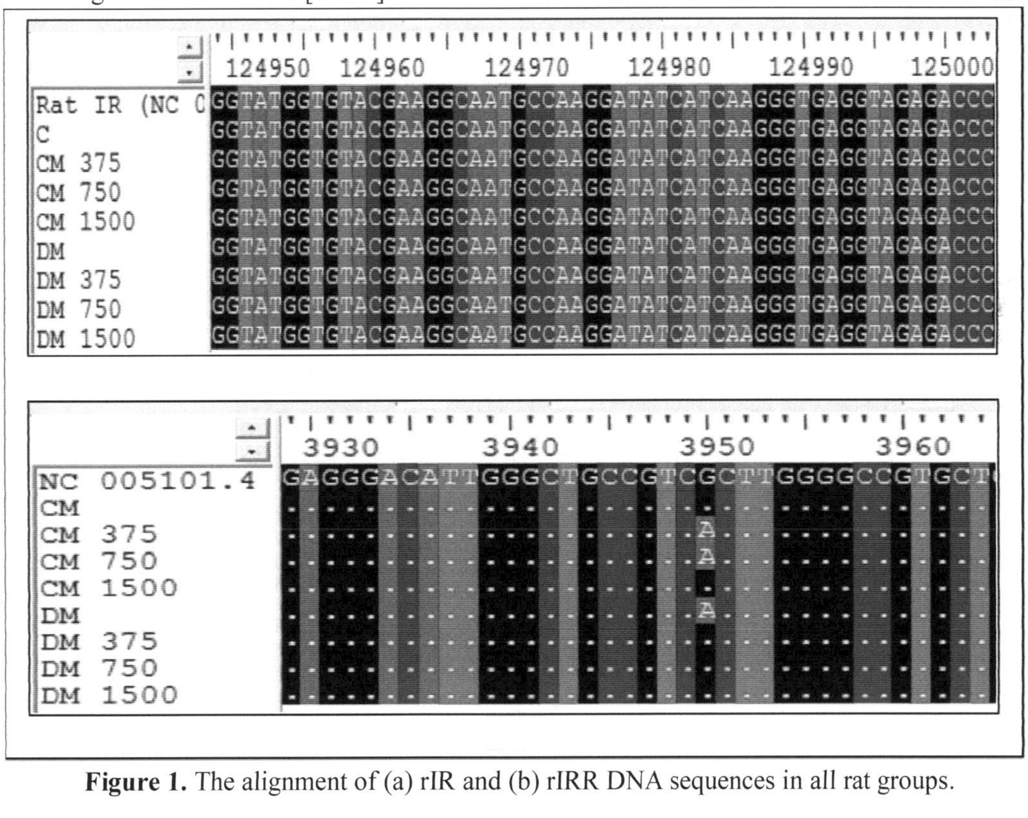

Figure 1. The alignment of (a) rIR and (b) rIRR DNA sequences in all rat groups.

The rIRR exon 2 sequences of all rat groups were aligned started from the nucleotide number 3710 to 4180. The rIRR gene (ID: NC_005101) from NCBI database was used in the alignment. The high cholesterol condition and STZ administration with the dose of 25 mg/kg BW were able to induce T2DM in rat. This is indicated by the mutation in rIRR exon 2 sequence of DM group (Figure 1b). The mutation occurred was point mutations which changed the nucleotide from $_{3950}$G to $_{3950}$A. The CM 375 and CM 750 group had the same mutation as in the diabetic rat. However, the 375, 750, and 1500 mg/kg BW of CSN1S2 protein administration in the diabetic rat resulted in the alteration of rIRR sequence to normal. In addition, the control rat receiving 1500 mg/kg BW of CSN1S2 protein had no mutation in its rIRR sequence. This result indicated that the 1500 mg/kg BW of CSN1S2 protein is the safe dose for normal and diabetic rats.

The hypercholesterolemia and hyperglycemia condition in diabetic rat will trigger the accumulation of triglycerides and LDL in cells. Fatty acids and triglycerides cause mitochondrial dysfunction, so free radicals form in the cells. The free radicals can inactivate the action of insulin receptors by inhibiting the phosphorylation process of tyrosine kinase on the IRR [23-24]. Interestingly, the CSN1S2 protein was able to trigger the improvement of rIRR nucleotide sequences in all CSN1S2-treated diabetic rat. It might play a role in counteracting free radicals or as antioxidants in the rIRR exon 2 gene. It was able to suppress the inhibitory activity of fatty acids and triglycerides as well. This inhibition of free radicals will trigger the repair of damaged nucleotide sequences through DNA repair. Base excision repair (BER) is an efficient program in repairing DNA damage through mechanisms of oxidation, deamination and alkylation [25]. BER occurs in women with T2DM is regulated by the activation of enzymes, such as Apyrimidine endonuclease (APEX1) to initiate DNA repair through cutting phosphate sugar, X-ray repair cross-complementing protein 1 (XRCC1) and DNA ligase 3 (LIG3) to insert the correct complementary base the sequence [26].

4. Conclusion

The CSN1S2 protein of Etawah Crossbred goat milk at a dose of 375 mg/kg BW, 750 mg/kg BW, and 1500 mg/kg BW treatment did not cause mutations in the rIR gene sequence of T2DM rat model. This protein was able to induce the rIRR sequences improvement in T2DM rats.

References

[1] Stein S A, Maloney K A and Pollin T I 2014 Genetic Counseling for Diabetes Mellitus *Curr Genet Med Rep* **2** 56–67
[2] International Diabetes Federation 2015 Executive Summary of Diabetes Atlas 7th Edition *IDF Diabetes Atlas*
[3] Wild S, Roglic G, Green A, Sicree R and King H 2004 Global Prevalence of Diabetes: Estimates for the year 2000 and projection for 2030 *Diabetes Care* **27** 1047–1053
[4] American Diabetes Asosiation 2014 Diagnosis and classification of diabetes mellitus *Diabetes Care* **37** S81–S90
[5] Forbes J M and Cooper M E 2013 Mechanisms of diabetic complications *Physiol. Rev.* **93** 137–88
[6] Seino S, Seino M and Bell G I 1990 Human insulin-receptor gene: partial sequence and amplification of exons by polymerase chain reaction *Diabetes* **39** 123–128
[7] Hale L J and Coward R J M 2013 Insulin signalling to the kidney in health and disease *Clin. Sci. (Lond)* **124** 351–70
[8] Fernandez A M and Torres-Alemán I 2012 The many faces of insulin-like peptide signalling in the brain *Nat. Rev. Neurosci.* **13** 225–239
[9] Pereira P C 2014 Milk nutritional composition and its role in human health *Nutrition* **30** 619–627
[10] Moeljanto W T B, Rini D and Wiryanta 2002 *Khasiat & manfaat susu Kambing: Susu terbaik dari hewan ruminansia* (Jakarta: ArgoMedia)
[11] Budiarti I K, Padaga M and Fatchiyah 2013 Nutritional Composition and Protein Profile of Goat Yogurt PE with Double Culture between *Streptococcus thermophilus* and *Lactobacilus* Species *Cukurova Med. J.* **38** 681–686
[12] Chotimah C, Ciptadi G, Setiawan B and Fatchiyah F 2015 CSN1S2 protein of goat milk inhibits the decrease of viability and increases the proliferation of MC3T3E1 pre-osteoblast cell in methyl glyoxal exposure *Asian Pacific J. Trop. Dis.* **5** 219–223

[13] Rohmah R N, Widjajanto E and Fatchiyah F 2015 Protective effect of CSN1S2 protein of goat milk on ileum microstructure and inflammation in rat-CFA-induced rheumatoid arthritis *Asian Pacific J. Trop. Dis.* **5** 564–568

[14] Bia R R, Virgirinia R P, Setiawan B, Soewondo A and Fatchiyah F 2015 Goat milk CSN1S2 is able to decrease the severity scoring, TNF-a, and RAGE expression in complete Freund's adjuvant-induced rheumatoid arthritis model of rats *Biomarkers Genomic Med.* **7** 64–71

[15] Wang L, Duan G, Lu Y, Pang S, Huang X, Jiang Q and Dang N 2013 The Effect of Simvastatin on Glucose Homeostasis in Streptozotocin Induced Type 2 Diabetic Rats *J. Diabetes Res.* **2013** 1–5

[16] Fatchiyah F, Setiawan B, Suharjono S and Noor Z 2015 The anti-osteoporosis effects of CSN1S2 protein of goat milk and yoghurt on a complete Freund's adjuvant-induced rheumatoid arthritis model in rats *Biomarkers Genomic Med.* **7** 139–146

[17] Fatchiyah F, Arumingtyas L E, Widyarti S, Rahayu S 2011 *Biologi molekular: Prinsip dasar analisis* (Jakarta: Erlangga)

[18] Schnepp B C, Jensen R L, Clark K R and Johnson P R 2009 Infectious molecular clones of adeno-associated virus isolated directly from human tissues *J. Virol.* **83** 1456–64

[19] Pareek C S, Smoczynski R and Tretyn A 2011 Sequencing technologies and genome sequencing *J. Appl. Genet.* **52** 413–435

[20] Hall T 2004 BioEdit Help Contents

[21] Kadowaki T 2000 Insights into insulin resistance and type 2 diabetes from knockout mouse models *J. Clin. Invest.* **106** 459–65

[22] Betram C E and Hanson 2001 Animal models and programming of the metabolic syndrome *British Medical Bulletin* **60** 103-121

[23] Miccoli R, Bianchi C, Penno G and Del Parto S 2008 Insulin resistance and lipid disorders *Future lipidol.* **3** 651-664

[24] Saini V 2010 Molecular Mechanisms of insulin resistance in type 2 diabetes mellitus *World J Diabetes* **1** 68-75

[25] Chiruvella K K, Liang Z and Wilson T E 2013 Repair of double-strand breaks by end joining *Cold Spring Harb. Perspect. Biol.* **5** a012757

[26] Grindel A, Guggenberger B, Eichberger L, Poppelmeyeer C, Gschaider M, Tosevska A, Mare G, Briskey D, Brath H and Wagner K 2016 Oxidative stress, DNA damage and DNA repair in female patients with diabetes mellitus type 2 *PLoS ONE* **11** e0162082

Construction and Optimization of Prethrombin-2 Human Genes in *E. coli* for the Production of Active Thrombin

S Silaban[1]*, S Gaffar[2], M Simorangkir[1], I P Maksum[2] and T Subroto[2]

[1]Department of Chemistry, Universitas Negeri Medan, Medan 20221, Indonesia
[2]Department of Chemistry, Universitas Padjadjaran, Sumedang 45363, Indonesia

*Corresponding author's email: saronomsilaban@unimed.ac.id

Abstract. Prothrombin-2 is a thrombin precursor, which plays an essential role in the conversion of fibrinogen to fibrin during the blood clotting process. In this study, we designed and constructed the human prethrombin-2 (hPT2) gene. hPT2 gene fused with a tag at the n terminal position, which contains the sequence intein followed by chitin binding domain (CBD) that useful for the purification process. The digestion process of hPT2 tag induced by changing its pH/ temperature. The hPT2 codon that used in this study designed according to the preference codon of *E. coli*. hPT2 gene was created using the OPTIMIZER software with the addition of two side restriction at the 5' and 3', *Bam*HI and *Xho*I respectively then were cloned by using the pMAT vector in *E. coli*. Moreover, hPT2 fragment ligated into an expression vector, pTWIN1, for *E. coli*. The result from the characterization hPT2 gene in pTWIN1 that using DNA sequencing method shown that the designed hPT2 gene cloned. Thus, the cloned hPT2 gene used as a precursor for PT2 expression in the *E. coli* host.
Keywords. prethrombin-2, thrombin, codon preferences, *E. coli*

1. Introduction

Covering the wound surgery method used a stitches technique, the gold standard applied. Although it is the gold standard, this technique caused several problems, such as prolonged time wound healing time and surgery process time, additional trauma (during sew and retraction threads), increasing the inflammatory, and the possibility of the complication relating to stitches of infection [1].

Fibrin glue (LF) replacement stitching techniques can glue and close the wound. LF as bioadhesive material, composed of thrombin, fibrinogen, calcium, and XIII factors. The content is designed to resemble the final stages of coagulation to form a fibrin clot. LF used as a hemostatic material that stops bleeding from the incision gap, matrix for wound healing and tissue adhesive [2], currently, thrombin in the commercial LF usually made from frozen and fresh beef plasm [3], commercial LF instead of stitching gives a lot of advantages, such as the surgery is more convenient, faster and can avoid losses due to suture [4].

The extensive use of *E. coli* as a host in the production of recombinant proteins is due to its nature which can overgrow with a short life cycle, information and genomes' characters completely so that easily manipulated, production costs are relatively cheap, high-level expression of the target protein, fast, and the technology is established [5]. However, in addition to the advantages mentioned above, the host also has a weakness, such as the phenomenon of codon refraction [6], and the potential to produce complex and inactive aggregates protein that it knew as inclusion bodies [7].

The first strategy needs to be done to solve this weakness that mentioned above is by optimizing codon of the target gene to codon preference of the host [8]. This strategy aims to address the lack of protein expression of the target gene. This optimization process is done by changing the codon encoding a particular amino acid derived from another source into codons with high frequency in the expression host. The expression level of genes composed codon optimization results more elevated than the genes without optimization [9]. The results of gene expression increased at least threefold with codon optimization [10]. The second strategy was used to the synthetic gene technology based on the ability to change codon refraction from the target gene into the host's match with codon preference recombinant [11], Another advantage of this technology is the effectiveness and efficiency than the its isolation process, and to avoid the diseases transmission and allergic reactions [12].

2. Materials and Methods

2.1. Strains, vectors, media, chemicals

E. coli TOP10F' is a host strain for plasmid cloning and rejuvenation. pMAT is a commercial cloning vector. Strains were grown in Luria Bertani medium (LB) with composition (tryptone 1%, yeast extract 0.5%, and 1% sodium chloride) supplemented with tetracycline antibiotics (100 ug / mL) and ampicillin (100 ug / mL). For solid media, LB media components added with 2% agar. All restriction enzymes and T4 DNA ligase obtained commercially from Fermentas (Canada). PTWIN1 expression vector obtained commercially from New England Biolabs, NEB. hPT2 synthetic gene (CBD-intein Ssp DnaB-hPT2) synthesized by GeneArt AG (Germany).

2.2. Design and codon optimization of hPT2

The hPT2 synthetic gene was designed based on the amino acid sequence in GenBank (Accession number: NM_000506.3). Codon preference of *E. coli* that contained in the Codon Usage Database (http://www.kazusa.or.jp/codon/). Codon optimization is performed using software Optimizer (http://gnomes.urv.es/OPTIMIZER) and Graphical Codon Usage Analyzer (http://gcua.schoedl.de/).

2.3. Fusion Construction of hPT2 and vector of pTWIN1

Fusion expression protein of hPT2 in *E. coli* and purification using IMPACT-TWIN system, the synthetic design of hPT2 genes equipped with *Bam*HI and *Xho*I restriction site at the intern. To combine the hPT2 synthetic and the pTWIN1 expression vector, pMAT-hPT2 cut using restriction enzymes *Bam*HI and *Xho*I. In parallel, cutting pTWIN1 with the same enzymes. Then, combined pTWIN1 by using T4 DNA ligase to produce pTWIN1-hPT2 plasmid.

2.4. Transformation of E. coli TOP10F'

Transformation of pTWIN1-hPT2 used competent cells of *E. coli* TOP10F' procured by heat shock [13]. *E. coli* transformed colonies selected through media contained the tetracycline and ampicillin antibiotic. pTWIN1-hPT2 recombinant plasmid isolated from *E. coli* TOP10F' transformed colonies using the QIAgen Spin Plasmid Miniprep Test Kit according to Qiagen's protocol. The recombinant plasmid purification results, analyzed by 1% agarose gel electrophoresis. Then, plasmids can be used for restriction analysis and determined the nucleotide sequence using DNA sequencing methods. Sequencing results aligned using Seqman on the Bioedit program.

3. Results and Discussion

3.1. Codon optimization of hPT2

Synthetic designing of hPT2 gene made through software. hPT2 synthetic designed based on the amino acid sequence in GenBank (Accession number: NM_000506.3) and used codon preferences of *E. coli* that contained in the Codon Usage Database. Based on data in GenBank, hPT2 gene consists of 308 amino acids.

The sequence analysis results of hPT2 codons in *E. coli* signed there was an unmatch codon preference of *E. coli*. Some human hPT2 codons encoding amino acids which have less than 50% similarity to the codon preference of *E. coli*, among others (1) S: art, tcg, tcc, tca, (2) E: gag, (3) T: act, aca, (4) G: gga, ggg, (5) R: agg, aga, cga, cgg, (6) P: cct, ccc, (7) K: aag, (8) L: ctc, ctt, ttg, (9) V: gtc, (10) Q: caa. While hPT2 codon that have the relatively more than 50% to nearly 100% consists of (1) A: gcc, gca, gct, (2) Y: tac, (3) F: ttc, (4) N: aat, (5) D: gac, (6) C: tgt, (7) I: atc, ata, (8) H: cac, (9) V: gtt, (10) T: acg, dan (11) G: ggt. hPT2 codons which have not yet reached 100% relative similarity to *E. coli* optimized up to 100%.

Synthetic gen utilizing can simplify and accelerate desired genes acquisition because unlimited at the natural biological source [8]. Also, Gen Bank's data can be accessed easily as the basis of determining the sequence of genes to be synthesized. Codon optimization is done because the target genes have potential differences of preference codon from the host genome. Although, some of the target genes have sufficient similarity with the host so the codon optimization is not necessary. In this study, hPT2 gene that expressed in the *E. coli* host, hPT2 codon choice has low preference codon than *E. coli* [14]. The previous study reported that the hPT2 gene expressed in *E. coli* would produce an inclusion body [15-18]. Therefore, codon optimization of the original organism against preference codon of the host is needed [19-20].

3.2. Recombinant plasmid construction

pTWIN1 has a gene encoding the chitin binding domain, which can bind chitin to the purification matrix [21]. In this research, hPT2 constructed into fusion form in the N-terminal position (CBD-intein Ssp DnaB-hPT2). At the end both inserted cutting sides of *Bam*HI (**ggatcc**) at the 5' end and *Xho*I (**ctcgag**) at the 3' end that the hPT2 gene combined with pTWIN1. The synthetic design of hPT2 gene for CBD-intein-hPT2 as follows:

ctcgagaccgcgaccagcgaatatcagacctttttaacccgcgcacctttggcagcggcgaagcggattgcggcctgcgcccgctgtttgaaaaa
aaaagcctggaagataaaaccgaacgcgaactgctggaaagctatattgatggccgcattgtggaaggcagcgatgcggaaattggcatgagccc
gtggcaggtgatgctgtttcgcaaaagcccgcaggaactgctgtgcggcgcgagcctgattagcgatcgctgggtgctgaccgcggcgcattgcc
tgctgtatccgccgtgggataaaaactttaccgaaaacgatctgctggtgcgcattggcaaacatagccgcacccgctatgaacgcaacattgaaaa
aattagcatgctggaaaaaatttatattcatccgcgctataactggcgcgaaaacctggatcgcgatattgcgctgatgaaactgaaaaaaccggtgg
cgtttagcgattatattcatccggtgtgcctgccggatcgcgaaaccgcggcgagcctgctgcaggcgggctataaaggccgcgtgaccggctgg
ggcaacctgaaagaaacctggaccgcgaacgtgggcaaaggccagccgagcgtgctgcaggtggtgaacctgccgattgtggaacgcccggtg
tgcaaagatagcacccgcattcgcattaccgataacatgttttgcgcgggctataaaccggatgaaggcaaacgcggcgatgcgtgcgaaggcgat
agcggcggcccgtttgtgatgaaaagcccgtttaacaaccgctggtatcagatgggcattgtgagctggggcgaaggctgcgatcgcgatggcaa
atatggcttttatacccatgtgtttcgcctgaaaaaatggattcagaaagtgattgatcagtttggcgaataa**ggatcc**

Inserted of hPT2 synthetic gene in pTWIN1 was done on the intein Ssp DnaB. To enable the process of protein expression, at the 3'end hPT2 added a stop codon (taa) before the *Bam*HI fusion restriction so that Mxe GyrA did not express because it would increase the size of the resulting protein. The start codon added because it included in the CBD-intein SspDnaB in pTWIN1.

3.3. Cloning of hPT2 genes

To obtained hPT2 fragment, the pMAT-hPT2 plasmid cut by using BamHI and XhoI based on the previous synthetic genes design. Cutting result of pMAT-hPT2 characterized using 1% agarose gel electrophoresis. The analysis showed that the hPT2 fragment successfully cut from the pMAT vector. Two bands with molecular weight approximately 2374 bp and 939 bp indicated as pMA-T and hPT2, respectively (Figure 1). Using the same enzyme, pTWIN1 was cut and then isolated for hPT2 ligation process. Figure 2 shown that pTWIN1 expression vector successfully cut. There were two bands with a molecular weight of 6536 bp as pTWIN1 vector, and bands with the molecular weight are 839 bp as Multi Cloning Site (MCS) fragment, which released from pTWIN1 vector (Figure 2).

hPT2 fragments ligated to the pTWIN1 expression vector using T4 DNA ligase (Fermentas), Next, the ligation results of isolated and purified by DNA isolation kit (Roche). Refining result characterized by 1% agarose gel electrophoresis. As seen in Figure 3, agarose gel electrophoresis results showed that hPT2 fragments (939 bp) and pTWIN1 vector (6536 bp) was successfully isolated and purified from the agarose gel.

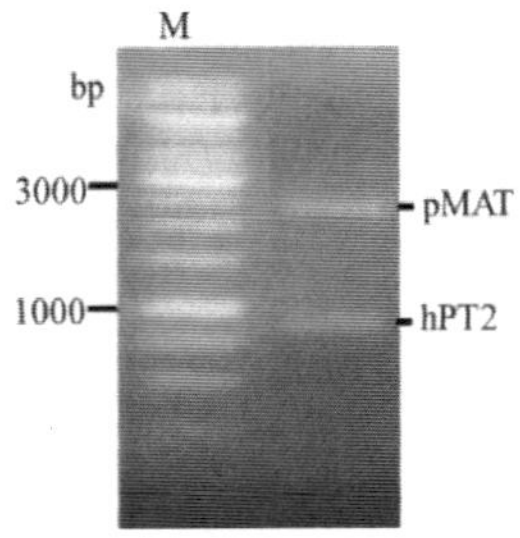

Figure 1. Electrophoresis analyses of results of pMAT-hPT2 (3313 bp) plasmid cut with *Bam*HI and *Xho*I enzymes. M: 1 kb DNA Marker, pMAT with a molecular weight of 2374 bp, hPT2 with a molecular weight of 939 bp.

3.4. Gene cloning characterization

Confirming the successful ligation result, and to ensure similarity the nucleotide sequence of hPT2 gene cloned results with the designed results, as many as 10 μL hPT2-pTWIN1 plasmid with a concentration is 100 ng / mL of the nucleotide sequences determined by DNA sequencer by MacroGene (Korea). Comparison of the nucleotide sequence pt2 hPT2 optimization results with cloned presented in Figure 4.

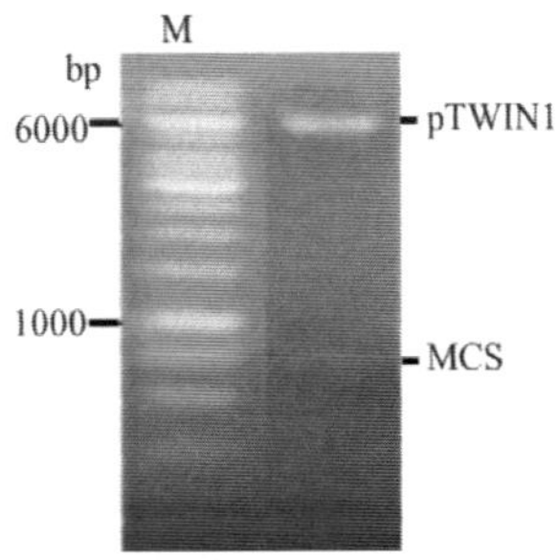

Figure 2. Electrophoresis analyses of results of pTWIN1 (7375 bp) plasmid cut with BamHI and XhoI enzymes. M: 1 kb DNA Marker, pTWIN1 with a molecular weight of 6536 bp, MCS with a molecular weight of 839 bp.

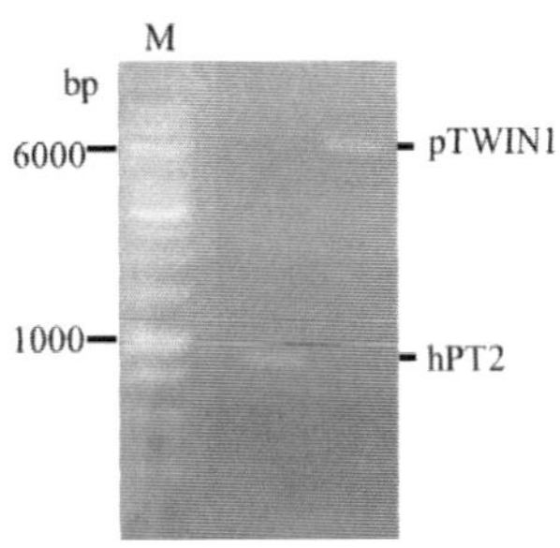

Figure 3. Extraction analyses of pTWIN1 and hPT2 by using agarose gel. M: 1 kb DNA Marker, pTWIN1 with a molecular weight is 6536 bp, hPT2 with a molecular weight is 939 bp.

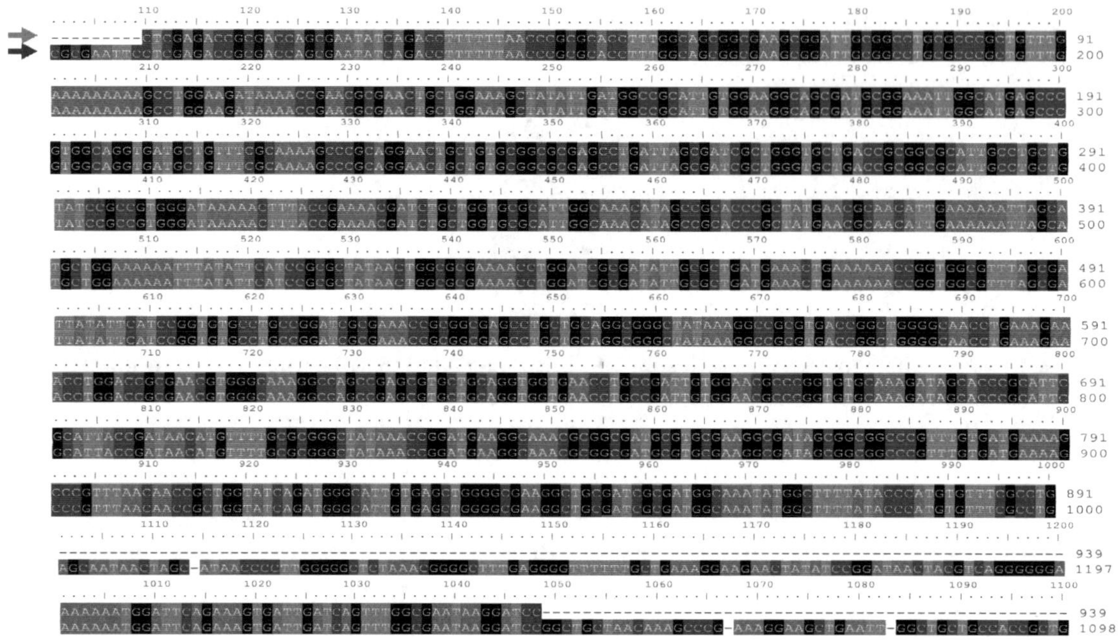

Figure 4. Sequencing results. Comparison of the DNA sequences of the designed hPT2 gene (→) with the results of cloning (→).

4. Conclusions

hPT2 gene codon optimization according to codon preference of *E. coli* can minimize the effects of codon refraction, which the impact on the expression of recombinant proteins. Use of the synthetic gene is more efficient and effective than isolation process from a natural source.

5. Acknowledgments

This work is supported by grant "Penelitian Terapan Unggulan Perguruan Tinggi (PTUPT with a contract **No. 60/UN33.8/PL-DRPM/2019**) Year 2019" from Direktorat Jenderal Penguatan Riset and Pengembangan, Ministry of Technology Research and Higher Education, Republic of Indonesia.

References

[1] Enus S, Natadisastra G, Shahib M N and Sulaeman R 2011 Peran lem fibrin otologus pada penempelan tandur konjungtiva bulbi mata kelinci terhadap ekspresi gen fibronektin dan integrin *Makalah Kedokteran Bandung* **43** 183-188

[2] Spotnitz W D and Prabhu R 2005 Fibrin sealant tissue adhesive--review and update *Journal of Long-Term Effects of Medical Implants* **15** 245

[3] Kaufman H E, Insler M S, Ibrahim-Elzembly H A and Kaufman S C 2003 Human fibrin tissue adhesives for suturless lamellar keratoplasty and scleral patch adhesion: a pilot study *Ophthalmology* **110** 2168-2172

[4] Koranji G, Seregard S and Kopp E D 2004 Cut and paste: a no suture, small incision approach to pterygium surgery British Journal of Ophthalmology **88** 911-914

[5] Cabrita L D, Weiwen D and Stephen P B 2006 A family of *Escherichia coli* expression vectors for laboratory scale and high throughput soluble protein production *BMC biotechnology* **6** 12

[6] Sorensen S P and Mortensen K K 2005 Advanced genetic strategies for recombinant protein expression in *Escherichia coli Journal of Biotechnology* **115** 113-128

[7] Freydell E J, Ottens M, Eppink M, van Dedam G and van der Wielen L 2007 Efficient solubilization of inclusion bodies *Journal of Biotechnology* **2** 678-684

[8] Gustafsson C, Govindrajan S and Minshull J 2004 Codon bias and heterologous protein expression *Trends in Biotechnology* **22** 346-353
[9] Wang X, Li X, Zhang Z, Shen X and Zhong F 2010 Codon optimization enhances secretory expression of *Pseudomonas aeruginosa* exotoxin A in *E. coli Protein Expression and Purification* **72** 101-106
[10] Zhou Z, Schnake P, Xiao L and Lal A A 2004 Enhanced expression of a recombinant malaria candidate vaccine in *Escherichia coli* by codon optimization *Protein Expression and Purification* **34** 87-94
[11] Gräslund S, Nordlund P, Weigelt J, Hallberg B M, Bray J, Gileadi O, Ming J *et al* 2008 Protein production and purification *Nature Methods* **5** 135-146
[12] Hughes R A, Miklos A E and Ellington A D 2011 Gene synthesis: Methods and applications *Methods in Enzimology* **498** 277-309
[13] Sambrook J, Fritsch E F and Maniatis T 1989 Molecular cloning: a laboratory manual, 2nd Ed (New York: Cold Spring Harbor Laboratory Press)
[14] Welch M, Villalobos A, Gustafsson C and Minshull J 2009 You're one in a googol: optimizing genes for protein expression *Journal of the Royal Society Interface* **6** S467-S476
[15] Soejima K, Mimura N, Yonemura H, Nakatake H, Imamura T and Nozaki C 2001 An efficient refolding method for the preparation of recombinant human prethrombin-2 and characterization of the recombinant-derived α-thrombin *The Journal of Biochemistry* **130** 269-277
[16] Silaban S, Maksum I P, Ghaffar S, Hasan K, Enus S, Subroto T and Soemitro S 2015 Codon optimization and chaperone assisted solubilization of recombinant human prethrombin-2 expressed in *Escherichia coli Microbiology Indonesia* **8** 177-182
[17] Silaban S, Gaffar S, Simorangkir M, Maksum I P and Subroto T 2018 Effect of IPTG concentration on recombinant human prethrombin-2 expression in *Escherichia coli* BL21(DE3) ArcticExpress *IOP Conf. Ser.: Earth Environ. Sci.* **217** 012039
[18] Rizkia P R, Silaban S, Hasan K, Kamara D S, Subroto T, Soemitro S and Maksum I P 2015 Effect of Isopropyl-β-D-thiogalactopyranoside concentration on prethrombin-2 recombinan gene expression in *Escherichia coli* ER2566 *Procedia Chemistry* **17** 118-124
[19] Silaban S, Simorangkir M, Maksum I P, Subroto T and Hasan K 2018 Pengaruh suhu pada ekspresi pretrombin-2 manusia rekombinan di *Escherichia coli* ER2566 *Jurnal Sains Indonesia* **42** 25-30
[20] Silaban S, Maksum I P, Hasan K, Enus S, Subroto T and Soemitro S 2017 Purification of recombinant human pretrombin-2 in Escherichia coli for thrombin production as fibrin glue components *Jurnal Pendidikan Kimia* **9** 265-272
[21] Chong S, Montello G E, Zhang A, Cantor E J, Liao W, Xu M Q and Benner J 1998 Utilizing the C-terminal cleavage activity of a protein splicing element to purify recombinant proteins in a single chromatographic step *Nucleic Acids Research* **26** 5109-5115

Protein-based virtual screening: SIRT-1 as modulator maturation of circulating endothelial progenitor cells

T A Wihastuti[1*], W Nurwidyaningtyas[2], C T Tjahjono[3] and T Heriansyah[4]

[1]Department of Biomedical Nursing Science, Faculty of Medicine, Brawijaya University, Malang
[2]STIKES Kendedes, Malang, Indonesia
[3]Department of Cardiology, Faculty of Medicine, Brawijaya University, Malang, Indonesia
[4]Department of Cardiology, Faculty of Medicine, Syah Kuala University, Banda Aceh, Indonesia

*Corresponding author's e-mail: titinwihastuti@gmail.com

Abstract. Circulating endothelial progenitor cells (cEPCs) are having involvement in the replacing vascular endothelial cells damaged or detachment from the basement membrane. cEPC needs to fulfill both quantity and quality requirements in order to play this important role. After efflux from bone marrow to circulation, niche EPC necessary of maturation to reach the potential for re-endothelialization. Risk factors exposure for cardiovascular disease not only affect the quantity and quality of cEPC, but it is also to be involved in downregulation of SIRT1. SIRT1 downregulation is mostly associated with the mechanism of senescence. Objective. to predict the direction of SIRT-1 interaction with cEPC maturation that is validated with cEPC marker surface using the STRING DB method of protein interaction. Result. SIRT1 interacts with two proteins, inhibiting VEGFA and activating and inhibiting P53. P53 inhibits the expression of p16ink4a, a protein involved in cellular senessence and P53 through AKT involved in CD 309 and ITGA2B expression. Whereas there was a direct interaction of the expression of CD 309 through the VEGFA line; CD117; TIE-2; CD 144; CD62a. Conclusion. SIRT-1 is an intracellular protein that is involved in cellular anti-aging processes but is not directly involved in the expression of EPC marker surfaces.
Keywords: surface marker, cEPC, SIRT1, maturation

1. Introduction

Circulating EPC is one of haematopoietic stem cells can be isolated either from the umbilical cord or peripheral blood [1,2,3,4]. The amount in circulation is almost associated with various conditions, including exposure to risk factors for cardiovascular disease (CVD). CEPC existence in the circulation occurs due to increased plasma levels of stromal derived factor-1 (SDF-1) which was released platelets due to injury to vascular endothelium, causing efflux of EPC from the bone marrow [5].

cEPC in circulation expresses the marker CD133 or c. Kit Ligand (CD 117) are called EPC Niche. Niche EPC is a type of hematopoietic stem cells that can differentiate into various types of single nucleated cells, one of which is EPC [6], niche EPC is still unable to perform transendothelial differentiation to replace detachment endothelial cells from the basement membrane [1]. EPC niche needs to pass maturation to the late EPC phase which is characterized by changes in marker surface to pre-existing endothelial residents [1,7,8]. Decreased number of cEPC or loss of potential cEPC in endothelial repair due to oxidative stress or hyperglycemia will have an impact on decreasing intracellular EPC NO synthesis. In addition, EPC functional changes were also caused by premature senescence initiation which was associated with failure of cEPC maturation due to SIRT-1 depletion [11]. SIRT-1 or Sirtuin 1 is one of the 7 target protein deacetylation enzymes that have antiaging effects in various biological processes starting from genomic maintenance, metabolic regulation, tumor suppressor and many other roles [14].

Massive oxidative stress or increase in plasma glucose will cause increased acetylation of Fox O1, which is one of the deacetylation targets of SIRT-1 which is a regulator negative of angiogenesis [10] thought to be part of premature senescence modulation in cEPC. Premature senescence in cEPC is characterized by an increase in p16 cell cycle inhibitor expression and is not related to the telomere shortening mechanism [9]. Another statement calls the expression of SIRT-1 blockade also has an impact on decreasing heterochromatin expression of protein 1α (HP1α) which is an indicator of EPC differentiation into OECs [11].

To describe the function of intracellular proteins which in this case is SIRT-1 towards cEPC maturity, an understanding of specific interactions is needed as functional partnership information which is the core of cellular processing and their systematic characterization helps to provide a context in molecular system biology [13]. However, partitioning of different pathways or complexes can be somewhat arbitrary and may not do justice to the prevalence of crosstalk and dynamic variation in the interaction landscape [13]. This study discusses the predictions of interaction between SIRT-1 and EPC maturation based on the expression of each surface marker from the niche to the late EPC.

2. Method

The STRING DB method of interaction used for analysis of the direction of interaction between SIRT-1 and various markers of maturation of endothelial progenitor cells. The STRING database (http: //string-db.org) aims to provide a critical assessment and integration of protein interactions, including direct (physical) as well as indirect (functional) associations. The basic interaction unit in STRING is the functional association, i.e. a specific and productive functional relationship between two proteins, likely contributing to a common biological purpose [13] The associations in STRING include direct (physical) interactions, as well as indirect (functional) interactions, as long as they are specific and biologically meaningful [12].

The prediction of the direction of SIRT interaction will be directed at several different cEPC markers including:

a. CD 133 (prominin I)
b. CD 117 (cKitL)
c. CD34 , Asahara et al, 1997
d. CD 309/ KDR/ VEGFR-2
e. TEK/ TIE-2
f. CD31/PECAM/
g. CD144 / CDH5/ VE-cadherin
h. CD41 /vWf
i. CD45
j. E-selectin / CD62E/ SELE

3. Result

Figure 1 shows that directly SIRT-1 interacts with two proteins that inhibit the activation and inhibition VEGFA and P53. P53 inhibits the expression of p16ink4a / CDKN2a, a protein involved in cellular senescence and P53 through AKT involved in the expression CD 309 / KDR and ITGA2B. While through the VEGFA pathway, there is a direct interaction of the expression CD 309 / KDR; CD117 / KIT; TIE-2 / TEK; HBA1; CD 144 / CDH5; CD62a / SELE.

SIRT-1 has a non-specific direction with Akt1, but with p53 showing many types of interactions that can be identified include post-translational modification, activation, transcriptional regulation, conversely p53 shows the interaction of inhibition of p53 activity.

CDKN2 as a downstream effect of p53 appears that the interaction model of p53 inhibits CDKN2A expression but on the other hand CDKN2A expression is reported to activate p53. Between CDKN2A and p53 there is an interaction binding or in post translational modification with an unspecified type of effect. There is an indirect interaction between SIRT-1 and CDKN2A or p16 through p53 as in the previous publication. SIRT-1 does not directly interact with the entire cEPC marker in all phases of maturation. And SIRT-1 regulates VEGFA transcription (inhibition) and VEGFA expression directly inhibits SELE, CDH5 and TEK expression but there are publications that report VEGFA inducing CD34 and TEK expression.

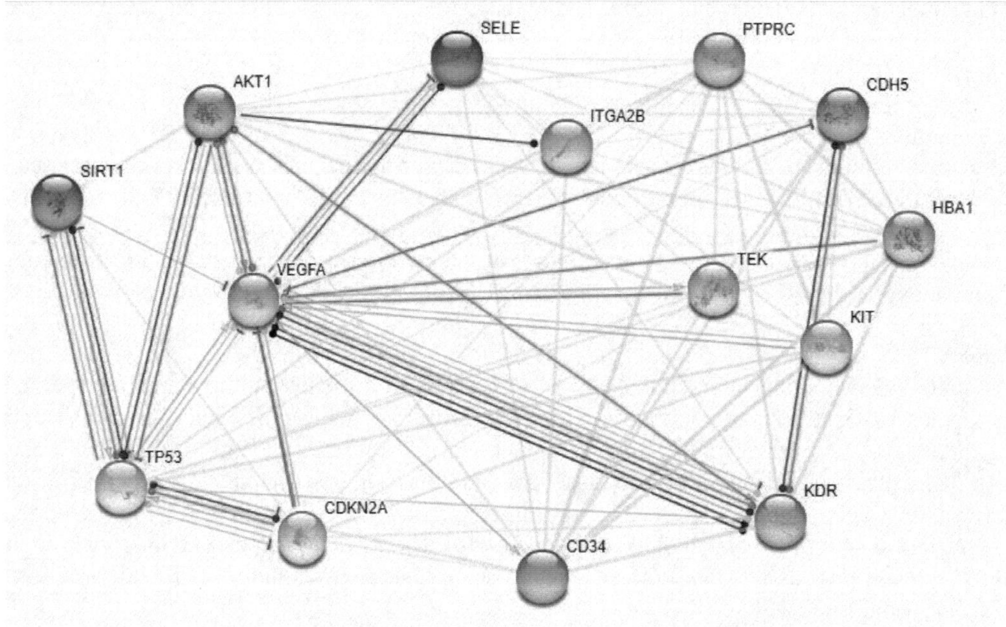

Figure 1. SIRT1 interaction model with various cEPC marker surfaces.

4. Discussion

Protein interactions (PPIs) handle a wide range of biological processes, including cell-to-cell interactions and metabolic and developmental control [15]. In silico is one of three forms of method that are widely used to identify protein-protein interaction.

This method uses computer simulations to understand the total context of the interaction potential, and it is better to develop a range of possible interactions between proteins. The results of this study showed no direct interaction between SIRT1 and CD antigen expression used as a marker of cEPC maturation.

Cluster differentiation (CD) antigen is surface molecules expressed on many types of cells and designate a spliced variant of the extracellular domain [16]. The CD antigen is protocol used for identification and investigation of molecules providing targets for immunophenotyping of cells. CD antigens can act in a lot of ways, like receptors or ligands in terms of physiology. As a signal, antigens usually initiated CD, altering the behavior of the cell. Some CD proteins do not play a role in cell signaling, but have other functions, such as cell adhesion [9]. The results of this study mention that there is a direct interaction between SIRT1 and p53 but it is not an interaction with cEPC marker surface expression.

The involvement of SIRT1 depletion in premature senescence was identified through p53 acetylation activity which would later modulate promyelocytic leukemia protein (PML) - (a protein that mediates premature senescence at cellular level). The inhibition of SIRT1 expression by miR-127 causes senescence in endothelial cells, including exposure to conditions for hyperglycemia, which also have the same effect. In other words the decrease in deacetylation activity and fungi by SIRT1 will have an impact on increasing acetylation and stabilization of p53, which then mediates various senescence cellular [14].

SIRT1 is also maintenance of haematopoetic stem-cell (HSC) homeostasis so that the decrease in SIRT1 expression will increase DNA damage, age-related molecular accumulation and aging phenotype in HSC and also on endothelial progenitor cells (EPC).

Failure to repair damaged DNA causes instability of genomics and contributes to premature cellular senescence and increased aging. Under these conditions, SIRT1 induces these improvements through deacetylation of Ku70 proteins, including in cases of atherosclerosis in vascular smooth muscle cells (VSMC). SIRT1 initiates activation of Nijmegen breakage syndrome-1 repair protein (NBS1) and not p53 [14].

Another possibility is that cEPC maturation failure is associated with a decrease in the amount of cellular ATP due to increased production of mitocondria-derived reactive oxygen species (mROS) which reacts with NO to form ONOO- which inhibits PGC-1α protein synthesis by mitochondria [18]. Another controversy states that premature senescence due to a decrease in SIRT1 is not related to activation of p53 / p21 but rather the stress induced premature senescence mechanism along with the increase in p16INK4a [19].

So this study provides information that surface molecule expression is not related to intra cellular signaling by SIRT1, p53, p16 or PGC1α.

5. Conclusion

SIRT1 is an intracellular protein that is involved in the anti-aging process at cellular level. But SIRT1 is not directly involved in the expression of marker surfaces from EPC. The clusters of differentiation are a molecule in the form of a ligand or receptor that is widely used to identify cell characters based on surface markers with an immunophenotyping approach and this molecule does not always play a role in cell signaling.

References

[1] Tagawa S, Nakanishi C, MasayukiMori, Yoshimuta T, Yoshida S, Shimojima M, Yokawa J, Kawashiri M, Yamagishi M, and Hayashi K 2015 Determination of early and late endothelial progenitor cells in peripheral circulation and their clinical association with coronary artery disease, Hindawi Publishing Corporation International Journal of Vascular Medicine Volume **2015**, Article ID 674213, 7 pages

[2] Zhao J, Mitrofan CG, Appleby SL, Morrell NW, and Lever AML, Disrupted Endothelial Cell Layer and Exposed Extracellular Matrix Proteins Promote Capture of Late Outgrowth Endothelial Progenitor Cells, Stem Cells International Volume **2016**, Article ID 1406304, 13 pages

[3] Zhang M, Rehman J, Malik AB, Endothelial progenitor cell and vascular repair, Curr Opin Hematol. 2014 May ; **21**(3):224-228

[4] Liu W, Ren L, Wang T, Navarro N, Tang L, The Involving Roles of Intrahepatic and Extrahepatic Stem/Progenitor Cells (SPCs) to Liver Regeneration, International Journal of Biological Sciences 2016; **12**(8): 954-963

[5] Pecchi VG, Sara Valdés , Véronique Pons , Paula Honorato, Laurent O. Martinez, Liliana Lamperti, Claudio Aguayo, Claudia Radojkovic; Apolipoprotein A-I enhances proliferation of human endothelial progenitor cells and promotes angiogenesis through the cell surface ATP synthase, Microvascular Research **98** (2015) 9–15

[6] Hirstov M, Erl W, Weber PC. Endothelial Progenitor Cells Mobilization, Differentiation, and

Homing, Arterioscler Thromb Vasc Biol. 2003 **23** 1185-1189

[7] Malinovskaya NA, Komleva YK, Salmin VV, Morgun AV, Shuvaev AN, Panina YA, Boitsova EB and Salmina AB (2016) Endothelial Progenitor Cells Physiology and Metabolic Plasticity in Brain Angiogenesis and Blood-Brain Barrier Modeling. Front. Physiol. **7** 599

[8] Hur J, Yoon CH, Kim HS, Choi J, Kang H, Hwang KK, Oh B, Lee M, Park YB (2004) Characterization of Two Types of Endothelial Progenitor Cells and Their Different Contributions to Neovasculogenesis, Arterioscler Thromb Vasc Biol. 2004;**24**:288-293. available at http://www.atvbaha.org

[9] CD antigens / Cluster of Differentiation. https://www.sinobiological.com/cd-antigens-cluster-of-differentiation.html.

[10] Tarantino C, Paolella G, Cozzuto L, Minopoli G, Pastore L, Parisi S, Russo T. miRNA 34a, 100, and 137 modulate differentiation of mouse embryonic stem cells. FASEB J. 2010; **24** 3255–3263

[11] Berezin AE. Epigenetically Modified Endothelial Progenitor Cells in Heart Failure. J Clin Epigenet. 2016, **2**:2

[12] Sklarczyk D, Morris J, Cook H, Kuhn M, Wyder S, Simonovic M, Santos A, Doncheva T. Roth A, Bork P, Jensen LJ, Mering CV, The STRING database in 2017: quality-controlled protein–protein association networks, made broadly accessible, Nucleic Acids Research, 2017, Vol. **45**, Database issue

[13] Szklarczyk D, Franceschini A, Wyder S, Foslund K, heller D, Huerta JC, Simonovic M, Roth A, Santos A, Tsafao KP, Kuhn M, Borl P, Jensen LJ, Mering C, STRING v10: protein-protein interaction network, integrated over the tree of life, Nucleic Acids Research, 2015, Vol. **43**, Database issue D447–D452

[14] Ghosh S, Zhou Z, SIRTain regulators of premature senescence and accelerated aging, Protein Cell 2015 **6** (5) 322–333

[15] Srinivasa V, Srinivas K, Sunand GN, Protein-protein interaction detection : method and Analysis, International Journal of Proteomics Volume **2014**, Article ID 147648, 12 pages

[16] Engel P, Boumcell L, Balderas R, Bensussan A, Gattei V, Horejsi V, Jin B, Malavasi F, Mortari F, Schwartz R, Stockinger H, Zelm MC, Zola H, Clark G, CD Nomenclature 2015: Human Leukocyte Differentiation Antigen Workshops as a Driving Force in Immunology, J Immunol 2015; **195**:4555-4563

[17] Maeng YS , Young Kwon J, Kweon Kim E, Kwon YG, (2015). Heterochromatin Protein 1 Alpha (HP1α: CBX5) is a Key Regulator in Differentiation of Endothelial Progenitor Cells to Endothelial Cells: HP1α is an EPC Differentiation Inducing Factor. STEM CELLS. **33**

[18] Gomes AP, Price NL, Ling AJ, Moslehi JJ, Montgomery MK, Rajman L, White JP, Teodoro JS, Wrann CD, Hubbard BP, Mercken EM, Palmeira CM, de Cabo R, Rolo AP, Turner N, Bell EL, Sinclair DA 2013 Declining NAD(+) induces a pseudohypoxic state disrupting nuclear-mitochondrial communication during aging. *Cell*. **155**(7):1624-38.

[19] Vassallo PF, Simoncini S, Ligi I, Chateau AL, Bachelier R, Robert S, Morere J, Fernandez S, Guillet B, Marcelli M, Tellier E, Pascal A, Simeoni U, Anfosso F, Magdinier F, George FD and Sabatier F, (2014) Accelerated senescence of cord blood endothelial progenitor cells in premature neonates is driven by SIRT1 decreased expression, *Blood* **123** 2116-2126

Identification and molecular docking analysis alkaloids *Polyalthia longifolia* leaves from Indonesia and the Philippines as anti-inflammatory

T Z Anita[1], C Mahdi[1], M A G Beltran[2] and A Aulanni'am[3*]

[1]Chemistry Department, Brawijaya University, Malang, 65151, Indonesia
[2]College of Veterinary Medicine, Tarlac Agricultural University, Camiling, Tarlac, 2306, Philippines
[3]Faculty of Veterinary Medicine, Brawijaya University, Malang, 65151, Indonesia

*Corresponding author's email: aulani@ub.ac.id

Abstract. *Polyalthia longifolia* (family: Annonaceae) is widely planted to effectively reduces noise pollution. This plant spread in many countries including Indonesia and the Philippines. Alkaloids are the main active compounds other than terpenoids in *P.longifolia* and it has the potential to be anti-inflammatory. Each leaves considered have different active compounds because of the different geographic factor from each country. To confirm this, we investigated the differences alkaloid compounds from two different sources of plants and predicted their anti-inflammatory potential. Shade-dried leaves from Indonesia and the Philippines were extracted by ethanol 70%. Two extracts were analyzed with LC-MS to ensure alkaloid compounds. Ensured alkaloid compounds further take on molecular docking. The compounds were drawn with ChemDraw then convert to .pdb with Open Babel. The protein COX-2 obtained from .pdb then prepared with PyMol. The docking process held by PyRx and the interaction was visualized by LigPlot+. LC-MS analysis identified 5 alkaloids contained from the ethanol extract of *P.longifolia* leaves from Indonesia and the Philippines. O-methylbulbocapnine-N-oxide was found only in ethanol extract leaves from Indonesia, while N- methylnandingerine-β-N-oxide was only found in ethanol extract leaves from the Philippines. All compounds have the potential as an anti-inflammation. Liriodenine as the most potent compound with binding energy -10.9kcal/mol. O-methylbulbocapnine-N-oxide has lower binding energy than N- methylnandingerine-β-N-oxide. In conclusion, there are differences between the alkaloid compounds and anti-inflammatory potential of the ethanol extract of *P.longifolia* leaves from Indonesia and the Philippines. Moreover, Indonesia's ethanol extract leaves showed more potential than Philippines's.

1. Introduction

Polyalthia longifolia is a plant commonly used to reduce noise pollution [1]. *P.longifloia* can be found in tropical and sub-tropical countries [2] such as Indonesia and the Philippines. This plant has been used as a traditional medicine to treat fever, skin disease, hypertension, diabetes. P.longifolia has pharmacological and biological activities such as anti-oxidant, anti-bacterial, anti-fungal, anti-cancer and anti-inflammatory [3].

Previous studies have reported that this plant contains flavonoids, alkaloids, sesquiterpenes, diterpenes, saponins, quercetin, bulbocapnine [2]. Geographical differences will cause differences in the content of chemical compounds in plants [4]. *P.longifolia* has the main secondary metabolites of alkaloids and diterpenoids. Various derivatives of the identified alkaloid compounds can be used as an anti-inflammatory [5]. For example, alkaloids are anti-inflammatory for IBD [6].

The anti-inflammatory mechanism can inhibit the enzyme cyclooxygenase (COX) [7]. The cyclooxygenase enzyme has two isoforms, namely COX-1 and COX-2. COX-2 is inducible and responsible for inflammatory [8]. Selective inhibition of COX-2 will reduce inflammation with a low risk of gastrointestinal [9]

In silico methods such as molecular docking are used to predict the ability of active compounds (ligands) to cause a biological effect computationally [7]. This study aims to determine the differences in the active compounds of *P.longifolia* plants from the regions of Indonesia and the Philippines. In addition, molecular docking is used to determine the ability of active compounds as anti-inflammatory compounds.

2. Materials and Methods

2.1. Plant Material
P.longifolia leaves ware collected from two different countries. *P.longifolia* leaves from Indonesia ware collected from Malang, East Java. While *P.longifolia* leaves from the Philippines were collected from Camiling, Tarlac. The Plants identified in plant taxonomy laboratories, Brawijaya University.

2.2. Hardware and Software
The computer used had the specifications of the Inter® Core ™ i3-4005U CPU @ 1.70GHz 1.70 GHz, Random Access Memory (RAM) 6.00 gigabytes. The software used is ChemDraw Ultra 12, PyMol, PyRx 0.8 and LigPlot + v.2.

2.3. Extraction
P.longifolia leaves from Indonesia and the Philippines were shade-dried and powdered. The powder was macerated with 70% ethanol for 72h. The filtrate was evaporated to get the extract. The extracts obtained were stored at 4°C until further analysis.

2.4. Alkaloid Analysis with LC-MS
LC-MS (TSQ Quantum Access MAX Triple, Thermo-Scientific) with C18 column (1.7 μm 100Å 50x2.1 mm, Kinetex) was adjusted according to the desired conditions (7.50 min separation process and the flow rate of 300 μl/min). The mobile phase was in the form of two solvents (eluent A = water and eluent B = acetonitrile). The gradient elution was carried out as follows, 0-0.60 minutes 90%: 10% (A: B), 5.00-5.50 minutes 25%: 75% (A: B) and 6.00-7.50 minutes 90%: 10% (A: B).

2.5. Preparation of Ligand Structures
The ligands used for molecular docking were alkaloid compounds present in the ethanol extract of leaves of *P. longifolia* from LC-MS. The structure of the ligands was drawn using ChemDraw Ultra 12 and Open Babel is used to convert to .pdb. Alkaloid compounds included in the study were Polylogine, noroliveroline, liriodenine, oliveroline-β-N-oxide, N-methylnandingerine-β-N-Oxide and O-methylbulbocapnine-N-oxide (Results from LC-MS).

2.6. Protein Preparation
Preparation of COX-2 protein begins by selecting an active form protein that binds native ligand (GDP code: 3LN1). Water molecules, ligands and B, C and D chains were removed from proteins using PyMol.

2.7. Molecular docking
Molecular docking proteins with ligands was carried out using AutoDock Vina in PyRx. Docking was carried out on a grid with a center of 31,2552; -23,4834; -16,1872 and Dimensions 17,4839; 17,3785;

21,3620 (Angstrom). Docking results in the form of bond strength interactions between ligands and receptors. Analysis of interactions with amino acids residues was used by the LigPlot + v.2 program.

3. Result and Discussion

3.1. Alkaloids by LC-MS

Alkaloid compounds were identified using LC-MS with 7 targets molecular weight of compounds which included polylongine, polyfothine, liriodenine, noroliveroline, oliveroline β-N-oxide N-methylnandingerine-β-N-oxide and O-methylbulbocapnine-N-oxide. Identification of compounds from the target molecular weight using m/z [M+H]+. Five alkaloid compounds identified on *P.longifolia* leaves from Indonesia and the Philippines. The alkaloid compounds detected are shown in table 1. The structure of the alkaloid compounds is shown in figure 1. There are differences in the compounds found, namely N-methylnandingerine-β-N-oxide in Philippine extracts and O-methylbulbocapnine-N-oxide in Indonesian extracts. Differences in compounds were due to different geographies between Indonesia and the Philippines. This condition would affect environmental factors such as temperature, altitude and sun duration [4].

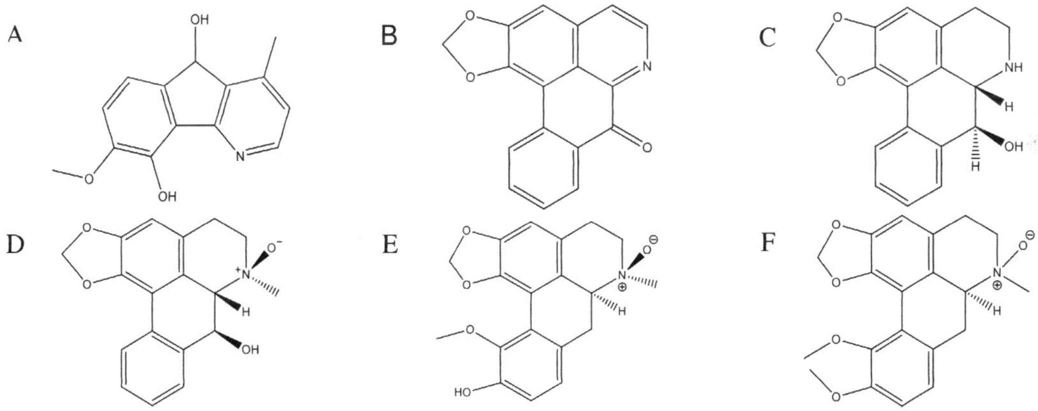

Figure 1. Structure of alkaloid compounds. A) Polylongine, B) Liriodenine, C) Noroliveroline, D) Oliveroline β-N-oxide, E) N-methylnandingerine-β-N-oxide, F) O-methylbulbocapnine-N-oxide.

Table 1. Alkaloid compounds identified in the ethanol extract of *P.longifolia* leaves from Indonesia and the Philippines.

Compounds	MW (g/mol)	From	RT(min)
Polylongine	243.261	Indonesia	0.68
		Filipina	0.71
Liriodenine	275.263	Indonesia	3.09
		Filipina	3.09
Noroliveroline	281.31	Indonesia	2.75
		Filipina	2.75
Oliveroline-β-N-Oxide	311.337	Indonesia	2.99
		Filipina	2.99
N-methylnandingerine β-N-oxide	341.363	Indonesia	-
		Filipina	4.87
O-methylbulbocapnine-N-oxide	355.39	Indonesia	4.92
		Filipina	-

The polarity of alkaloid compounds was indicated by the retention time (RT) resulted. Table 1 showed the polarity of the alkaloid compounds found in *P.longifolia* leaves. Polylongine had the highest polarity then noroliveroline, oliveroline-β-N-oxide, liriodenine, N-methylnandingerine-β-N-oxide and O-methylbulbocapnine-N-oxide. The greater retention time indicates that the compound was retained by non-polar column. While compounds with the smallest retention time such as polylongine were not held up in column because they were polar in contrast to the column. Liriodenine in previous studies had retention time (RT) 3.51 [10]. Retention time differences were possible due to differences in flow speed and type of column used. Liriodenine was a compound that had been shown antiradical activity [11] and antioxidants [12].

3.2. Docking Results

Alkaloid compounds from *P.longifolia* leaves are thought to have anti-inflammatory effects. The cyclooxygenase-2 enzyme is an inducible enzyme caused by inflammation[8]. Inhibition of COX-2 is clinically effective as an anti-inflammatory agent that does not adversely affect the gastrointestinal tract. Molecular docking was done to determine the potential of alkaloid compounds as an anti-inflammatory by looking at the interaction of compounds with the active site of COX-2. Molecular docking was done with the PyRx program to assess the potential of compounds to be anti-inflammatory.

The 3D structures of alkaloid compounds detected from the ethanol extract of *P.longifolia* leaves were created using ChemDraw (Figure 2).

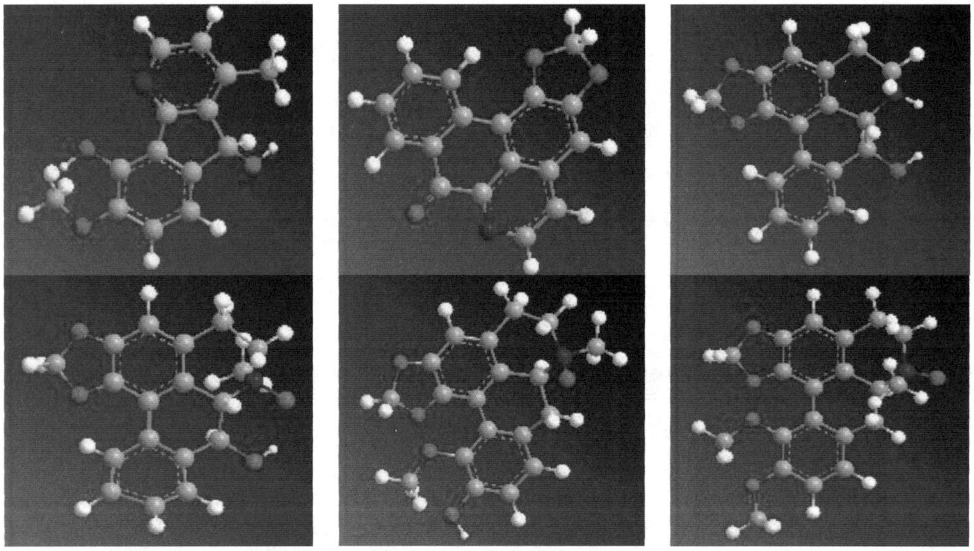

Figure 2. The 3D structure of Alkaloids. A) Polylongine, B) Liriodenine, C) Noroliveroline, D) Oliveroline β-N-oxide, E) N-methylnandingerine-β-N-oxide, F) O-methylbulbocapnine-N-oxide.

The 3D structure of COX-2 proteins that binds to the original inhibitor is obtained at the RCSB GDP online (Figure 3A). Water molecules, other ligands and chain B, C and D removed by PyMol software (Figure 3B).

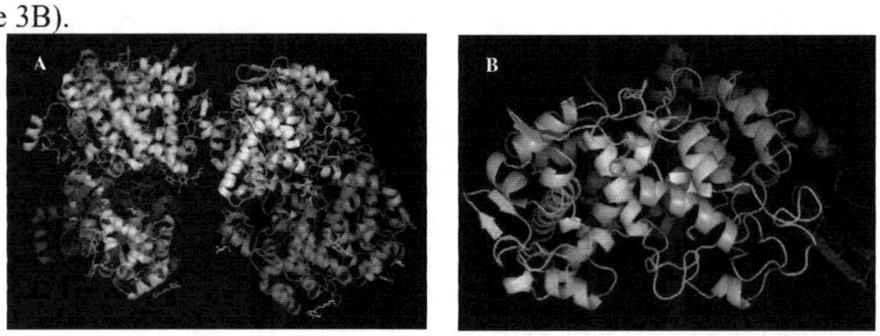

Figure 3. A) The 3D of COX-2 protein, B) The 3D chain A structure of COX-2 protein.

Molecular docking results and interactions with residual amino acids were shown in Table 2. Table 2 showed the potential of each compound as anti-inflammatory seen from the binding energy and amino acid interactions. The total binding energy of the compounds that were docking had a value of RMSD (Root Mean Square Deviation) <2. RMSD was the difference between predictive and experimental observations, the smaller the RMSD value obtained, the predicted position of the compound was closer to native [7]. Therefore, it can be concluded that this molecular docking was valid for predicting COX-2 inhibition.

Table 2. The results of molecular docking between alkaloid compounds in *P.longifolia* leaves with COX-2 enzyme.

Compounds/Ligands	Binding energy (kcal/mol)	Hydrogen bonds distance (Å)	Amino acid residues from hydrogen binding	Amino acid residues from hydrophobic bonds
Polylongine	-8.7	3.29 2.83	Tyr341	Leu517, **Ala513***, Ser516, Val335, **Phe504***, **Leu338***, **Ser339***, **Val509***
Liriodenin	-10.9			**Val335***, **Ala513***, Ser516, **Leu338***, **Val509***, **Phe504***, **Ser339***, **Tyr341***
Noroliveroline	-10.7	3.27 3.12	**Ala513*** Ser516	Val335, Leu338, **Phe504***, **Ser339***, **Val509***, **Tyr341***, Leu517, Arg106
Oliveroline-β-N-oxide	-9.3			**Phe504***, **Val509***, **Met508***, **Leu338***, Tyr371, Trp373, Ser516, Leu517, **Val335***, **Ala513***, **Tyr341***, **Ser339***
N-Methylnandingerine-β-N-Oxide	-7.4	3.05	Ser516	Leu517, **Val335***, **Ala513***, **Gly512***, **Leu338***, **Ser339***, **Phe504***, **Val509***, **Tyr341***, Leu345
O-Methylbulbocapnine-N-oxide	-8.2			Tyr371, **Trp373***, **Phe504***, **Leu338***, **Tyr341***, Ser516, **Ala513***, **Val509***, Arg106, **Ser339***, **Val335***, Leu517, Leu345, Val102

*= same amino acid interactions with reference inhibitor

Binding energy was a conformational stability parameter between receptors (COX-2) and ligands (compounds). All alkaloid compounds that had been docking had low binding energy which ranges from -7.4 to -10.9 kcal/mol. Low binding energy causes the protein-ligand complex (compound) to become stable so that the ability to inhibit COX-2 was getting better [13]. This showed that these compounds had the potential as an anti-inflammatory. Based on the binding energy, it can be classified as anti-inflammatory compounds as follows: liriodenin> noroliveroline> oliveroline-β-N-oxide> polylongine> O-methibulbocapnine-N-Oxide> N-methylnandingerine-β-N-oxide.

The LigPlot+ program was used to visualize protein-ligand interactions. The results obtained illustrate the interaction of amino acid residues with ligands, which were mediated by hydrogen bonds with hydrophobic interactions. The interaction of ligands with amino acid residues of the O-methylbulbocapnine-β-N-oxide compound and N-methylnandingerine-N-oxide was shown in Figure 4.

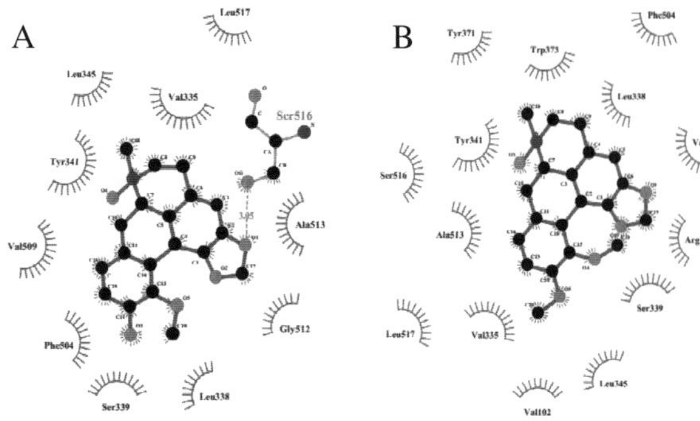

Figure 4. Visualize the results of LigPlot+. A) N-methylnandingerine-β-N-oxide, B) O-methylbulbocapnine-N-oxide.

In addition to binding energy, a comparison of amino acid residue interactions from reference inhibitors was carried out [14], [15]. This was done to see that the alkaloid compounds were on the same side (active side) as celecoxib. Redocking from celecoxib as a reference inhibitor obtained of hydrophobic interaction on the amino acid residues of Met508, Ala513, Gly512, Val509, Trp373, Leu370, Try341, Val335, His75, Phe504, Ile503 and Ala502 and hydrogen bonds acceptor with Ser339, Gln178, Leu338 and Arg499. Based on amino acids binding to the compounds in Table 2, it was found that all alkaloid compounds interacted on the active site of the receptor (COX-2). Hence, it could be concluded that the alkaloid compounds of ethanol extract of *P.longifolia* leaves had the potential as an anti-inflammatory by selectively inhibiting COX-2. Based on the potential of anti-inflammatory compounds O-methylbulbocapnine-N-Oxide was better than N-methylnandingerine-β-N-oxide. Thus, extract ethanol of *P.longifolia* leaves from Indonesia was better than the Philippines.

4. Conclusion

In conclusion, *P.longifolia* leaves from Indonesia and the Philippines contained different alkaloid compounds. N-methylnandingerine-β-N-oxide was only found in extracts from the Philippines, whereas O-methylbulbocapnine-N-oxide was only found in extracts originating from Indonesia. All alkaloid compounds found had anti-inflammatory abilities by inhibiting COX-2 in silico. Molecular docking results showed that compound from Indonesia had better anti-inflammatory potential than the Philippines.

References

[1] Thenmozhi M and Sivaraj R 2010 Phytochemical Analysis and Antimicrobial Activity of Polyalthia longifoia *Int. J. Pharma Bio Sci.* **1** (3) 1–7

[2] Sampath M and Vasanthi M 2013 'Isolation, Structure Elucidation of Flavonoids From Polyalthia longifolia (Sonn.) Thaiwaites and Evaluation of Antibacterial, Antioxidant and Anticancer Potential', *Int. J. Pharm. Pharm. Sci.*, **5** (1) 336–341

[3] C. Lavanya, B. G. Rao, and D. Ramadevi 2018 Phytochemical and Pharmacological Studies on Polyalthia longifolia', *Int. J. Pharm. Sci. Res.* **3** (4) 1–7

[4] W. Liu, Y. Dongxue, N. Li, H. Xiaogai, W. Dongmei, D. Li, and J. Liu, 'Influence of Environmental Factors on the Active Substance Production and Antioxidant Activity in Potentilla fruticosa L . and Its Quality Assessment', *Nature*, no. July, pp. 1–18, 2016.

[5] A. L. Souto, J. F. Tavares, M. Sobral, M. De Fátima, F. Melo, P. F. De Athayde-filho, J. Maria, and B. Filho, 'Anti-Inflammatory Activity of Alkaloids: An Update from 2000 to 2010', *Molecules*, **16**, pp. 8515–8534, 2011.

[6] J. Peng, T. Zheng, X. Li, Y. Liang, L. Wang, Y. Huang, and L. Wang 2019 Plant-Derived Alkaloids : The Promising Disease-Modifying Agents for Inflammatory Bowel Disease *Front. Pharmacol.* **10** 1–15

[7] D. D. Agistia, H. Purnomo, M. Tegar, and A. E. Nugroho, 'Interaction Between Active Compounds From Aegle marmelos Correa as Anti Inflammation Agent With COX-1 and COX-2 Receptor', *Tradit. Med. J.* **18** (2) 80–87, 2013.

[8] G. C. Krisnamurti and F. Fatchiyah, 'Interaction of acetaminophen and caffeine towards cyclooxygenase-2 (COX-2) in inhibition of prostaglandin (PGH 2) synthesis Interaction of acetaminophen and caffeine towards cyclooxygenase-2 (COX-2) in inhibition of prostaglandin', in *IOP Conf. Series: Journal of Physics: Conf. Series*, 2019 **2**, pp. 1–6.

[9] V. Shofia, A. Aulanni'am, and S. Prasetyawan, 'In Silico Study of Apigenin as COX-2 Inhibitor and In vivo Studies of Apigenin contained in Methanolic Extract of of Atactodea striata for decreasing Malondialdehyde Levels On Inflammatory Bowel Disease Rats', *Int. J. ChemTech Res.* **8** (6) 640–644, 2015.

[10] I. de la C. Chaco′ and A. R. G. Lez-Esquinca, 'Liriodenine Alkaloid in Annona Diversifolia During Early Development', *Nat. Prod. Res.*, **21** (1) 42–49, 2012.

[11] C. N. Loumpangou, L. Justin, C. Bonazaba, A. Murphy, E. Ndinga, A. D. Empfani, Y. Mahmout, and J. Moutsabote, 'Dosage and evaluation of the antiradical activity of alkaloids of the root barks of Oulotricha Le Thomas : isolation of Liriodenine', *Int. J. Herb. Med.* **6** (6) 96–100, 2018.

[12] A. K. Obaid, S. Salbiah, S. Abdul, Y. M. Bakri, M. A. Nafiah, K. Awang, S. A. Aowda, N. M. Hassan, Y. Z. H. Hashim, and F. Abdullah, 'Antioxidative and cytotoxic activities of crude and isolated compounds of P . Lateriflora (bl .) King', *J. Pharm. Sci. Res.* **10** (11) 2718–2721, 2018.

[13] N. P. L. Laksmiani, N. L. P. V. Paramita, and I. M. A. . Wirasuta, 'In Vitro and In silico Antioxidant Activity of Purified Fractions From Purple Sweet Potato Ethanolic Extract', *Int. J. Pharm. Pharm. Sci.*, **8** (8) 177–181, 2016.

[14] A. Yanuar, I. Pratiwi, and R. R. Syahdi, 'In silico Activity Analysis of Saponins and 2 , 5-Piperazinedione from Marine Organism against Murine Double Minute-2 Inhibitor and Procaspase-3 Activator', *J. Young Pharm.* **10** (2) 16–19, 2018.

[15] S. R. Lestari, B. Lukiati, S. N. Arifah, A. R. Nurul, and A. Gofur, 'Medicinal Uses of Single Garlic in Hyperlipidemia by Fatty Acid Synthase Enzyme Inhibitory : Molecular Docking', in *IOP Conference Series: Earth and Environmental Science PAPER* 2019 **276** 1–8

Swallow nest extract can be able to prevent nephropathy diabetic of the white rats streptozotocin-induced

T Triawanti[1*], F D Alexandra[2], A Frethernity[2] and S Mahmud[3]

[1]Department of Biochemistry of Medical Faculty of Lambung Mangkurat University, Banjarmasin, Indonesia
[2]Department of Pharmacology of Medical Faculty of Palangka Raya University, Palangka Raya, Indonesia
[3]Medical Faculty of Palangka Raya University, Palangka Raya, Indonesia

*Corresponding author's email: tria_fkunlam@yahoo.co.id

Abstract. Hyperglycemia that occurs in diabetic mellitus leads to glycation reactions in protein molecules and oxidative stress resulting in damage to cells and organs. Swallow nest believed society can lower blood glucose. The objective of the study was to analyze the potency of water swallow nest extract to prevent nephrophaty diabetic. The study used Post test-Only with Control Group Design, which consisted of 1 control group (K = aquadest) and 3 treatment groups (nest swallow nest dose 1, 10 and 100 mg / kg BW) each group consisted of 6 *Rattus Norvegicius*. Before treatment, means of glucose level rats have been 68 mg/DL, then rats were induced Streptozotocin in a dose of 40 mg / kg BW intraperitoneally. Day 7th after induction, rats had elevated glucose ± 102 - 108 mg / dL. Then the rats were given water extract nest swallow white for 28 days orally. All data were analyzed by Kruskall-Wallis test followed by Mann-Whitney test with 95% confidence level. The results of blood glucose levels in each group (K, P1, P2 and P3) were 111.0 vs 88.5 vs 86 vs 83 mg / dL (p = 0.004), pancreas H_2O_2 levels were 10.16 vs 9.20 vs 8.81 vs 7.27 (p=0,000), serum H_2O_2 levels were 17.35 vs 16.39 vs 15.01 vs 11.96 (p=0.044), renal methylglioxal levels were 242.55 vs 134.34 vs 67.12 vs 50.48% (p = 0,000) and renal AOPP levels were 1.17 vs 1.09 vs 1.05 vs 1.00 (p=0.000) respectively. The Mann-Whitney test showed a dose of 100 mg / kgBW has the greatest potential. In conclusion the swallow nest water extract has potential as an antidiabetic and prevent of renal's damage.
Keywords: Swallow Nest (*Collocalia fuciphaga*) Water Extract, hyperglycemia, oxidative stress, nephropathy diabetic

1. Introduction

Diabetic Nepropathy is kidney damage that occurs due to the accumulation of protein in the glomerular basal mesangial membrane in the condition of diabetes mellitus. Kidney damage can occur through a mechanism of oxidative stress and an inflammatory response. Chronic hyperglycemia will trigger the formation of reactive oxygen species (ROS). The combination of the glycation and oxidation reactions of glucose forms compounds of AGEs [1]. Previous studies have shown an increase in the levels of AGEs in streptozotocin-induced mice (STZ) [2]. The accumulation of AGEs compounds will aggravate diabetes mellitus because it causes fibrosis formation, atherosclerotic plaques, increased systolic and diastolic blood pressure [3]. One of the potent AGEs compounds, methylglioxal (MG) is formed from the degradation of glycolytic intermediates, protein glycation and lipid peroxidation [4, 5].

Advanced oxidation protein products (AOPPs) are protein products that contain dityrosin and cross-linking and are formed during oxidative stress. AOPP is a marker of oxidative stress and inflammatory mediators. AOPP in the diabetic group of patients is significantly higher than the control [6]. The chronic accumulation of AOPPs can trigger renal inflammation in diabetes presumably through activation of renal NADP oxidase. Increased AOPP triggers an increase in the expression of monocyte chemoatractan protein 1 (MCP-1) and infiltration of macrophages as markers of inflammation in diabetic kidneys [7].

DM therapy still uses pharmacotherapy drugs. However, it is not uncommon for people to choose traditional treatments that are believed to be able to treat diabetes mellitus, including white swallow nests (Collocalia fuciphagus). The wallet bird nest in Indonesia contains amino acid histidine, leucine, threonine, valine, methionine, isoleucine, phenylalanine, serine, aspartate, arginine, lysine, proline, glutamate acid, glycine, alanine, and tyrosine [8]. Administration of amino acids can increase insulin secretion in type 2 DM patients [9]. Lysine, arginine, alanine, aspartate and glutamate amino acids showed antiglycating effects by means of competitive inhibition. Amino acids can also increase tissue sensitivity to insulin and reduce oxidative stress [10]. Administration of the alanine to obese mice can improve systemic glucose tolerance by activating the AMP kinase system and modulating systemic glucose metabolism [11]. So far, not many studies have evaluated the potential extract of swallow nest water as antidiabetic while preventing kidney damage. This study aims to analyze the potential of swallow nest extract in preventing diabetic nephropathy.

2. Material and method

2.1. Animal study

This study used male white rats *(Rattus norvegicus)* as the models. This work was approved by the Animal Care and Experimentation Committee (Ethical Committee], Faculty of Medicine Palangka Raya University) by letter of number 839/UN24.9/LL/2018. This research is a true experimental design that uses Posttest Control Group Design. The treatment group was divided into 4 groups: (K) rats induced Streptozotocin + aquadest; (P1) rats induced strptozotocin + swallow's nest water extract 1 mg/kgBW; (P2) rats induced streptozotocin + swallow's nest water extract 10 mg/kgBW; (P3) rats induced strptozotocin + swallow's nest water extract 100 mg/kgBW.

2.2. Preparation of swallow's nest water extract

Swallow nest swiftlet cleaned of feathers attached using tweezers. Then swallow nest is cleaned under running water for ± 5 minutes, then dried at room temperature. Once clean, the sample is smoothed using a blender. 300 grams swallow's nest was smoothed using of mortar and dissolved in 4.5L aquades and homogenized for 30 minutes. Then in centrifuge at 10000 rpm for 10 minutes. The dosage of swallow's nest water extract in rats was divided into three doses: 1 mg/kgBW, 10 mg/kgBW, and 100 mg/kgBW.

2.3. Induction of streptozotocin in experimental animals

Rats underwent acclimatization weighed, then fasted for 10 hours. After being fasted for 10 hours, blood samples were taken from the vein of the tail at minute 0 to determine baseline blood glucose levels. Furthermore, white rats were induced streptozotocin (Bioworld®) with a dose of 40 mg/kgBW of intra-peritoneal[12]. After induced rats continue to be given food and drink, wait in 4 days, then measure blood glucose levels. Mice are considered hyperglycemia when blood glucose levels >100 mg/dL. On the seventh day, blood glucose levels were measured to determine whether there was an increase.

2.4. Administration of swallow's nest water extract

The swallow nest water extract is made into a solution and administered daily using gastric tube, 2 hours after meals with a dose of 1mg/kgBW, 10 mg/kgBW, and 100 mg/kgBW. Glucose levels were measured at weeks 4th after administration of the test solution using a glucometer.

2.5. Measurement of hydrogen peroxide (H_2O_2) levels

Measurement of H_2O_2 was using a spectrophotometer. At first, making a standard curve. A total of 20 μmol H_2O_2 (Merck®) was added with 2 ml of dichromate:glacial acetic acid (Merck®) (1:3) mixture. Then the mixture was heated in boiling water for 10 minutes. Then the cooled mixture was measured for absorbance at a wavelength of 570 nm. The same procedure was done for 40, 60, 80, 100, 120, 140, 160 and 180 μmol H_2O_2. A graph was made between the absorbance on the Y axis with levels of H_2O_2 on the X axis to obtain a linear equation.

Preparation of test solution was made with a total of 1 ml of serum homogenate was added 5 ml of PBS (Merck®) pH 7.4. A mixture of 1 ml was taken and added to 2 ml of dichromate:acetate (Merck®) (1:3) mixture and then wrapped in aluminum foil for 30 minutes. The mixed solution was heated using a water bath for 10 minutes at 100°C. The solution was cooled to room temperature. The solution was then transferred into the cuvette and measured its absorbance using UV-VIS at a wavelength of 570 nm [13].

2.6. Measurement of methylglioxal levels

On the 29th day the rats were sacrificed and the kidneys were taken to be homogenized. A total of 25 μL homogenate was added 350 μL DNPH (Merck®) (0.1% DNPH in 2 N HCl) and then added 2.125 mL of aquadest. After that incubated for 15 minutes at 37°C, then add 1.5 mL 10% NaOH (Merck®). Read the absorbance at $\lambda = 576$ nm (A1). Furthermore, as much as 0.25 mL of renal homogenate was added DNPH of 1 mL. Then the solutions were incubated for 45 min at room temperature and shielded from light, and shaken with vortex every 15 minutes. The next step adds 1 mL of TCA (Merck®) 20% to then incubated in ice for 5 minutes. Centrifuged for 5 minutes at 1400 rpm, and discarded the supernatant. After washing by adding 1 mL of ethanol-ethyl acetate into each tube, centrifugation for 5 minutes at a speed of 1400 rpm, then discard the supernatant. Washing is repeated 3 times. The end stage of measuring this carbonyl compound is to add 1 mL of urea 9 M, and incubate the solution at 37°C for 10 min while shaken. After that the solution was centrifuged at 1400 rpm for 5 minutes. Furthermore, the resulting color measured uptake at a wavelength of 390 nm. (A2) [14].

$$\text{Level of methylglioxal} = \frac{A1}{A2} \text{ x } 100\% \quad (1)$$

2.7. Measurement of AOPP levels

The test solution was prepared by mixing 200 μL homogenate, 600 L PBS, and 100 L KI 1.16 M (Merck®). Then the blank solution was prepared by mixing 800 L of PBS, 100 L KI 1.16 M. Test and blank solution were placed for 2 min and then add 200 μL of acetic acid. Absorbance was measured at = 340 nm. The concentration of AOPPs were expressed through A = ε b C with $\varepsilon = 26\ mM^{-1}cm^{-1}$ and b = 1 cm [15,16].

2.8. Statistical analysis

Data were analyzed by normality and homogenity test. Therefore, the data is not normal then the non parametric test is Kruskall Wallis followed by Mann Whitney test with 95% confidence level.

3. Results

In this study, the protein content of swallow's nests was tested using spectrophotometric examination methods with a wavelength of 546 nm and with repetitions three times, the average yield obtained was 38.33%.

3.1. Blood Glucose Level

STZ-induced rats at a dose of 40 mg / kgBW intraperitoneally, after the seventh day obtained an increase in glucose levels of 36.3 mg / dL (Table 1). Then followed by giving swallow nest water extract at a dose of 1 mg / kgBW, 10 mg / kg BW and 100 mg / kgBW for 28 days (Figure 1).

Table 1. Blood glucose level of rats before and after Streptozotocin-induced.

Streptozotocin-induced	Glucose level	p value
Before	68 mg/dl	0.000
after	104.2 mg/dl	

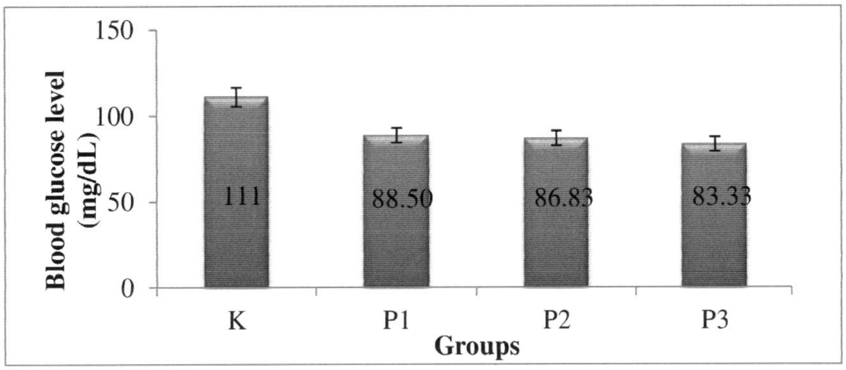

Figure 1. Mean of blood glucose level after Swallow nest water extract treatment ($p<0.05$) K = Control (aquadest); P1 = 1 mg/kg BW; P2 = 10 mg/kgBW; P3 = 100 mg/kgBW.

Kruskal Wallis test obtained $p <0.05$, then continued by post hoc test by using Mann Whitney test. The results of the Mann Whitney test showed differences between treatment groups and negative control groups. This proves that the water extract of bird nest can lower blood glucose levels in STZ-induced mice. Between dose of 1 mg/kgBW vs. 10 mg/kgBW vs. 100 mg/kgBW did not significant difference, meaning dose of 1 mg/kgBW had the same ability with dose of 100 mg/kgBW in lowering blood glucose level.

3.2. Pancreas Hydrogen peroxide (H_2O_2) levels

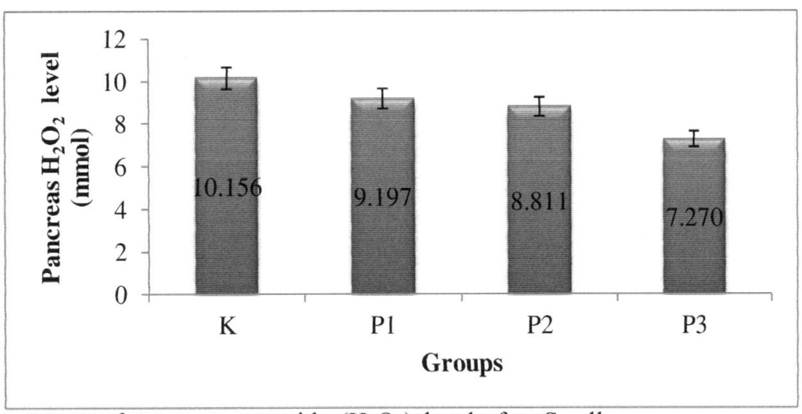

Figure 2. Mean of pancreas peroxide (H_2O_2) level after Swallow nest water extract treatment K = Control (aquadest); P1=1 mg/kgBW; P2=10 mg/kgBW; P3=100 mg/kgBW ($p=0.000$).

The results of the Kruskal Wallis standard test obtained a value of p = 0,000. Then followed by post hoc test using the Mann Whitney test. The results of the Mann Whitney test found differences between all treatment groups and negative control groups, P1 and P2 ($p = 0.002$), P1 and P3 ($p = 0.002$) and P2 and

P3 ($p = 0.002$). This proves that the administration of swallow's nest water extract can reduce levels of H_2O_2 in STZ-induced rat pancreas. The higher the dose of swallow nest extract, the lower the H_2O_2 content is formed.

3.3. Serum H_2O_2 levels

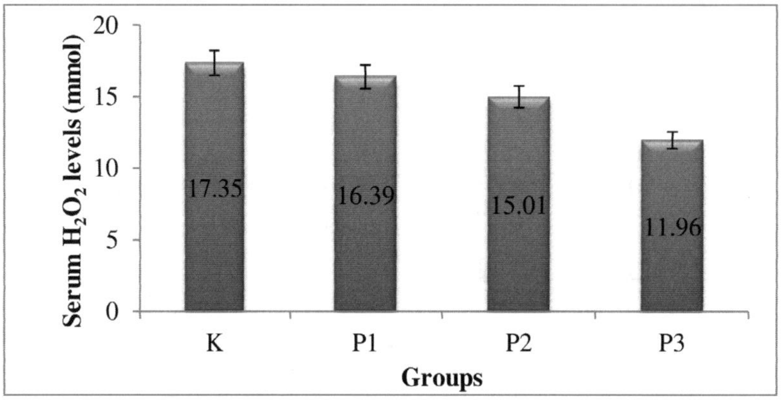

Figure 3. Mean of serum peroxide (H_2O_2) level after Swallow nest water extract treatment K = Control (aquadest); P1=1 mg/kgBW; P2=10 mg/kgBW; P3=100 mg/kgBW (p=0.044).

The results of the statistical test with nonparametric analysis, namely the Kruskal Wallis test showed that $p = 0.044$ ($p < 0.05$), which means that there are minimal differences in the 2 treatment groups. Then followed by post hoc test using the Mann Whitney test. The results of the Mann Whitney test found that there were differences between all treatment groups and negative control groups, but there was no difference between P1, P2, and P3. This proves that the administration of swallow's nest water extract can reduce serum H_2O_2 levels of hyperglycemic mice at a dose of 1 mg / kgBW.

3.4. Renal methylglioxal (MG) levels

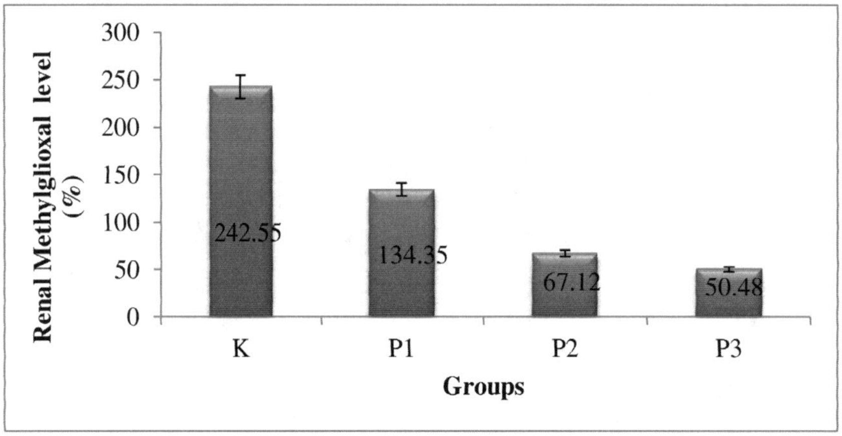

Figure 4. Mean of Methylglioxal (MG) level after Swallow nest water extract treatment K = Control (aquadest); P1=1 mg/kgBW; P2=10 mg/kgBW; P3=100 mg/kgBW (p=0.002).

Data was analyzed used Kruskal Wallis test followed by Mann Whitney test with 95% confidence level. The result of statistical test showed that there was significant difference between the negative control group and the group given the extract of swallow nest (p = 0.002), and between group of dose 1 mg/kgBW vs. 10 mg/kgBW vs. 100 mg/kgBW obtained significant difference (p = 0.002). This suggests that the swallow's nest water extract can reduce the formation of MG compounds where the higher the dose given the less the resulting MG will be.

3.5. Renal Advanced oxidation protein products (AOPPs) level

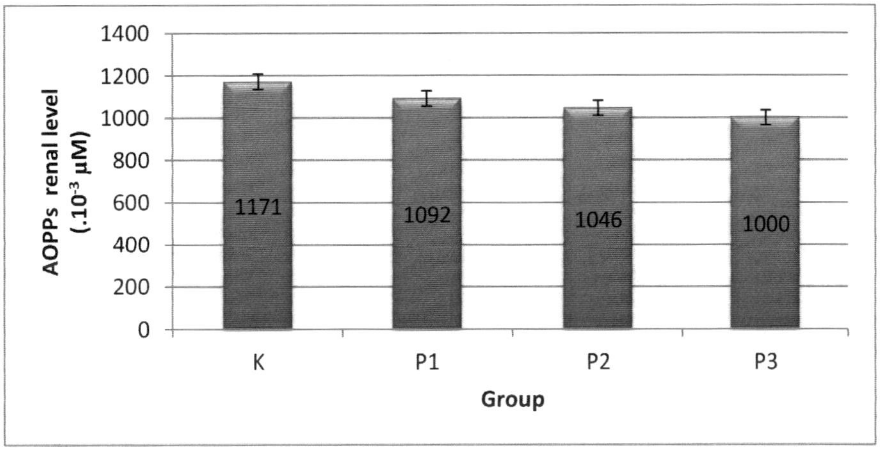

Figure 5. Mean of renal AOPPs level after Swallow nest water extract treatment K = Control (aquadest); P1=1 mg/kg BW; P2=10 mg/kgBW; P3=100 mg/kgBW (p=0.000).

The results of the Kruskal Wallis statistical test p value = 0.000, followed by the Mann Whitney post hoc test. The results of the Mann Whitney test found differences between all treatment groups and negative control groups, P1 and P3 (p = 0.009), and between groups P2 and P3 (p = 0.041). This proves that the administration of swallow's nest water extract can reduce AOPPs levels in hyperglycemic mice. The higher the dose of swallow nest extract, the lower the level of AOPPs formed.

4. Discussion

In this study, intraperitoneal low doses of Streptozotocin 40 mg / KgBB were used to induce rats of Type 2 diabetes mellitus, because high doses were reported to cause many deaths in the first week because of extensive damage to the pancreatic beta cells which caused an increase in blood glucose levels that were not under control. In small doses, pancreatic beta cell partial damage occurs. The increase in fasting blood glucose levels in STZ administration is caused by pancreatic beta cell damage and death. Initial damage to pancreatic beta cells occurs within 2-4 days after STZ administration, which is characterized by pancreatic swelling and beta cell degeneration in Langerhans island [12].

Based on table 1 the results of the average glucose level of the rat before it was induced by streptozotocin 68 mg/dL. After streptozotocin was induced, the rat's blood glucose levels increased to 104.2 mg/dL. Based on the results obtained, normal fasting blood glucose levels are in the range between 56 - 80 mg/dL. After being induced by STZ, blood glucose levels experienced a significant increase, which was an average of 104.2 mg/dL, so that it could be said that rats had experienced hyperglycemia.

The formation of the hormone insulin requires essential amino acids obtained from outside intake. In this study the administration of swallow's nest water extract which contains many essential amino acids can increase the synthesis of the hormone insulin. With the increase in the production of the hormone insulin, blood glucose levels can be lowered, as evidenced from this study where the blood glucose levels of rats given swallow nest extract with various doses were lower than that of the aquadest group (Figure 1). Previous study which provided a carbohydrate + free amino acid diet in patients with

Type 2 diabetes showed an increase in insulin secretion. Leucine amino acids, arginine and phenylalanine and their derivatives stimulate pancreatic Beta cell function and insulin secretion [9].

Lower pancreatic peroxide levels and serum in the group given swallow nest extract proved that swallow nest extract at a dose of 1 mg/kgBB was able to reduce free radicals formed in the pancreas due to STZ induction (Figures 2 and 3). STZ cytotoxicity causes the release of free radicals which trigger intracellular oxidative stress. Streptozotocin selectively tends to enter and accumulate in pancreatic beta cells, which are mediated by glucose 2 transporter bonds (GLUT2) in the plasma membrane. Other organs expressing GLUT2 such as the liver and kidneys also suffer damage from STZ induction [12].

The decrease in serum and pancreatic peroxide levels is thought to be due to the influence of the amino acid alanine contained in swallow nests. Among the amino acids that play a role in increasing insulin secretion, alanine has a long action in regulation of the function and integrity of pancreatic beta cells. Alanine stimulates the beta cell anti-apoptotic mechanism which involves decreasing levels of nitrogen species (RNS) and ROS. This is related to changes in gene expression involved in cellular signaling, metabolism, gene regulation, protein synthesis, apoptosis and cellular stress response. The greatest protective effect of Alanine does not depend on the decrease in NO production but is related to upregulation of catalase. This speeds up the conversion of H_2O_2 and its implications, decreasing levels of O_2- and peroxynitrite [10].

High level of blood glucose is chronic hyperglycemic state form advanced glycation products (AGEs) because glucose which binds to proteins (glycated protein) through the Maillard reaction can be oxidized and produce reactive Oxygen Species (ROS). The combination of glycation and glucose oxidation results in the formation of AGEs [1]. Methylglyoxal is one of the AGEs compounds. Glioksal, methylglioxal (MG) and 3-deoxyglucosone (3-DG) are powerful glycation compounds formed from the degradation of the compounds of glycolytic intermediates, protein glycation and lipid peroxidation. They react quickly with proteins to form AGEs, especially in arginine residues. The most important quantitative AGEs are arginine-derived hydroimidazolones formed by glyoxal, methylglioxal and 3-DG [4]. Physiologically, MG could detoxified by the enzyme glyoxalase 1 (Glo1) , using reduced glutathione as a co-factor. In the diabetic mellitus MG has also been shown to cause pathologic structural alterations and impair kidney function. Conversely, MG scavengers (such as N-acetylcysteine, aminoguanidine or metformin) or Nrf2/Glo1 activators (such as trans-resveratrol / hesperetin) are shown to be useful in preventing MG-induced cardiovascular and renal complications in diabetes [17].

In the study of Triawanti et al [2] it was reported that hyperglycemic rats given caramunting fruit juice at a dose of 1 mg / gBB had a low level of methylglioxal compared to those given aquadest significantly. When compared with the group given metformin, the levels were not significantly different. This is presumably because of the antioxidant content in caramunting fruit juices, namely flavonoids. Antioxidants function to reduce free radicals (ROS) that are formed due to glycation reactions.

In this study, swallow's nests have been shown to reduce levels of methylglyoxal kidney, where at a dose of 1 mg / kgBB it has shown significantly lower levels compared to those given only aquadest. Meanwhile at higher doses of 10 and 100 mg / kgBW the decrease in methylglioxal levels was greater (Figure 4). Swallow nests also contain the amino acid glycine. Previous studies have shown that diabetic rats given glycine showed less enlargement of the glomerular basal membrane than controls. This is because the secondary effect of glycine is its high solubility which prevents the formation and precipitation of AGEs products [10].

Advanced oxidation protein products (AOPPs) are a family of oxidized protein compounds that appear as new inflammatory mediators. AOPPs are protein products that contain dityrosin and cross-linking, formed during oxidative stress. Plasma AOPPs are mainly carried by albumin. The accumulation of AOPPs was first found in dialysis patients and is clearly shown in diabetic patients. Shi et al [7] reported that administration of AOPP-albumin infusion significantly increased macrophage infiltration in rat kidneys. Increased macrophage influx was also shown in the aorta of hypercholesterolemic mice stimulated by AOPPs. Apriani et al [15] study reported that in diabetic model rats, bone AOPP levels were positively correlated with increased blood glucose levels.

In this study it was seen that mice with hyperglycemia showed a high increase in AOPP levels. However, after administration of swallow nest extract at a dose of 1 mg/kgBW, 10 mg/kgBW and 100 mg/kgBW there was a significant decrease. The dose that shows the best decrease is 100 mg/kgBW. If it is associated with a decrease in glucose levels, the effective dose of swallow nest extract can reduce glucose levels as well as AOPPs levels starting at a dose of 1 mg/kgBW. Meanwhile at higher doses of 10 and 100 mg/kgBW the decrease in AOPPs levels looks bigger.

Swallow nests contain amino acids glycine, glutamic acid, and cysteine which are the substrate of the gluthathion enzyme which is an enzymatic antioxidant that plays a role in converting hydrogen peroxide compounds to H_2O and O_2. Previous studies have reported that the administration of seluang fish containing cysteine can reduce oxidative stress in the brains of malnourished mice. In this study the H_2O_2 levels of the groups given Seluang fish proved to be lower compared to those given standard feed. It is suspected that the cysteine content found in Seluang fish can increase gluthathion production [18]. Glutathione is a three-amino acid antioxidant tripeptide, cysteine, glycine, and glutamate, which catalyzes the reaction of H_2O_2 to H_2O and O_2. Glutathione is an important antioxidant contained in vital organs and bone marrow of life [19].

In this study it was proven that H_2O_2 levels in the pancreas and blood were significantly lower after administration of swallow nest extract. This indicates that the amino acids contained in swallow nests can increase insulin secretion, and reduce oxidative stress that occurs in the pancreas, so as to reduce glucose levels and prevent kidney damage due to glycation and inflammatory process.

5. Conclusion

Swallow's nest water extract had potency as an antidiabetic and prevent to renal's damage through the mechanism of (1) increased production of insulin so that glucose can be lowered; (2) the reduction of free radicals by enzymes gluthathion peroxidase thus lowering production methylglioxal and advanced oxidation protein products as a marker oxidative stress and inflamation.

References

[1] Novotny K T, Jung A, Hohn D, Weber and T Grune 2015 Advanced glycation end products and oxidative stress in type 2 diabetes mellitus *Biomolecules* **5** 194-222

[2] Triawanti, Suhartono E, Ramadhan Y, Zakina OF, Nurikhwan PW, Sanjaya A and Alam G A A 2017 Effect of karamunting fruit juice (Melastoma malabathricum L) to advanceds glycation end products (AGEs) and lipid profile as advanced complications of diabetes mellitus *Journal of engineering and Applied Sciences* **12** (2) 186-194

[3] Sadowska-Bartoz I and Bartosz G 2015 Prevention of protein glycation by natural compounds *Molecules* **20** 3309-3334

[4] Karachalias N, Babaei-Jadidi R, Rabbani N and Thornalley P J 2010 Increased protein damage in renal glomeruli, retina, nerve, plasma and urine and its prevention by thiamine and benfotiamine therapy in a rat model of diabetes *Diabetologia* **53** (7) 1506–1516

[5] Turk Z 2010 Glycoptoxine, carbonyl stress and relevance to diabetes and its complications *Physiol. Res.* **59** 147-156

[6] Jakus V, Sandorova E, Kalninova and Krahulec B 2014 Monitoring of glycation, oxidative stress and inflammation in relation to the occurence of vascular complications in patients with type 2 diabetes mellitus *Physiol. Res.* **63** 297-309

[7] Shi X Y, Hou F F, Niu H X, Wang G B, Xie D, Guo Z J, Zhou Z M, Yang F, Tian J W and Zhang X 2008 Advanced oxidation protein products promote inflammation in diabetic kidney through activation of renal nicotinamide adenine dinucleotide phosphate oxidase *Endocrinology* **149** (4) 1829-1839

[8] Elfita L 2014 Analisis profil protein dan asam amino sarang burung walet (*Collocalia fuchiphaga*) asal Painan *Jurnal Sains dan Farmasi Klinis* **1** (1) 27-37

Journal of Physics: Conference Series **1374** (2019) 012050 doi:10.1088/1742-6596/1374/1/012050

[9] Van Loon L J C, Kruijshoop M, Menheere P P C A, Wagenmakers A J M, Saris W H M and Krezer H A 2003 Amino acid ingestion strongly enhances insulin secretion in patients with long-term type 2 diabetes *Diabetes Care* **26** (3) 625-630
[10] Anuradha C V 2009 Aminoacid support in the prevention of diabetes and diabetic complications *Current protein and Peptide Science* **10** 8-17
[11] Adachi Y, Sousa-Coelho A L D, Harata I, Aoun C, Weimer S, Shi X, Herrera K N G, Takahashi H, Doherty C, Noguchi Y, Goodyear L J, Haigis M C, Gerszten R E and Patti M E 2018 L-alanine activates hepatic AMP-activated protein kinase and modulates systemic glucose metabolism *Molecular Metabolism* **17** 61-70
[12] Nugroho A E 2006 Hewan percobaan diabetes mellitus: patologi dan mekanisme aksi diabetogenik *Biodiversitas* **7** (4) 378-382
[13] Suhartono E and Setiawan B 2010 Model Indeks Peroksidatif dan Indeks Protein Teroksidasi Saliva Penderita Tuberkulosis Paru Berdasarkan Lama Pengobatan *JKM* **2** (9) 118-123
[14] Biswas U K, Banerjee S, Das A and Kumar A 2011 Elevation of serum methyglyoxal may be used as a screening marker in oral premalignant lesions *Biomedical Research Journal* **22** (3) 273-278
[15] Apriani N, Suhartono E and Akbar I Z 2011 Korelasi kadar glukosa darah dengan kadar advanced oxidation protein products (AOPP) tulang pada tikus putih model hiperglikemia *Jurnal Kedokteran Maranatha* **11** (1) 15-18
[16] Setiawan B, Adenan and Suhartono E 2012 Model indeks peroksidatif dan kadar advanced oxidation protein products (AOPPs) saliva penderita tuberkulosis paru berdasarkan lama pengobatan *J. Respir. Indo.* **32** (2) 72-77
[17] Sankaralingam S, Ibrahim A, Rahman M, Eid A H and Munusamy S 2018 Role of methylgyoxal in diabetic cardiovascular and kidney diseases: insight from basic science for application into clinical practice *Current Pharmaceutical design* **24** (26)
[18] Triawanti, Sanyoto D D and Nur'amin H W 2017 Reduction of oxidative stress by seluang fish (Rasbora Spp.) in brain of malnourished rats (Rattus norvegicus) *International Journal of Food Enginering* **3** (2) 107-111
[19] Piste P 2013 Cysteine-master antioxidant *Int. J. Pharm. Chem. Biol. Sci.* **3** (1) 143–149

Effects of Probiotic Treatment on Histopathology of Koi Carp (*Cyprinus carpio*) Infected by *Myxobolus* sp.

U Yanuhar[1*], N S Junirahma[1], K Susilowati[1], N R Caesar[1] and M Musa[1]

[1]Faculty of Fisheries and Marine Science, Brawijaya University, Jl. Veteran, Malang – East Java, Indonesia, 65145

*Corresponding author's email: doktoruun@ub.ac.id

Abstract. Koi fish (*Cyprinus carpio*) is one of the ornamental fish of a high economic value. However, raising koi can be troublesome due to poor water quality and high levels of stress that they are attacked easily by parasites. One of the parasites often attacking the gill of koi is *Myxobolus* sp. The presence of *Myxobolus* sp. can be observed through a histopathological observation of gill tissues. The purpose of this study was to determine the histological effect of probiotic administration to the damage on the gill tissue of Koi fish (*Cyprinus carpio*) infected with *Myxobolus* sp. The method used in this study was an explorative experimental method using factorial RAL. The observation of gill tissue was carried out on samples of koi fish consisting of two (2) control treatments, namely positive control (K+) and negative control (K-). Then, six (6) treatments, consisting of fish infected with *Myxobolus* and healthy fish, were done by applying a dose treatment of 1.1 ml/30 litters of water, 0.55 ml/30 litters of water, and 1.65 ml/30 litters of water. Treatments on control groups were carried out in a controlled tank by maintaining water quality parameters, which include temperature, pH, DO, and CO_2. The level of gill tissue damage treated with probiotic formulations was determined by scoring analysis. The results of histopathological observations commonly showed tissue damage, namely damage to Edema, Hyperplasia, and Lamela Fusion. The treatment having the best influence according to the 5% BNT test was the PB treatment (infected fish with a dose of probiotic formulation of 0.55 ml/30l).

1. Introduction

Koi fish (*Cyprinus carpio*) is one of the ornamental fish of a high economic value. However, raising koi can be troublesome due to extreme weather, changes in aquatic environment like poor water quality, and high stress levels in fish; these problems increase the number of cases of disease attacks on koi fish. Parasites are a disease that often attacks fish. Unstable environmental conditions can cause fish to have stress. So that its self-defense mechanism becomes weak and ultimately vulnerable to parasites [1]. Myxobolus has been known to be one of the parasites that often attacks the gills and sheets of koi.

Myxosporeasis or Myxobolusis is a parasitic disease caused by *Myxobolus* sp. The life cycle of this parasite is horizontal (indirect) involving invertebrates (Oligochaeta) and a number of vertebrates (fish or amphibians). The species has developed quite rapidly and need to be aware because it can cause economic losses. Clinical symptoms caused by the parasite *Myxobolus* sp. depending on the type of fish attacked, generally they have white nodul in their gills (operculum). These nodules are collection of thousand spores [2]. Diagnosis of the presence of *Myxobolus* sp. can be done through visual observations of behavior and clinical symptoms. Further observations can be carried out microscopically through histopathological observation of tissues of organs susceptible to *Myxobolus* sp.

Appropriate management of water quality and observation of tissues are the preventive measures to do to decrease the possibility of deaths. Water quality plays an important role in raising koi. In addition, to support the survival of fish, probiotic formulations are generally possible as an effort to manage water quality and as anti-parasites. The purpose of this study was to examine the histopathological damage in gill tissues of koi fish (*Cyprinus carpio*) infected with *Myxobolus* sp. and to determine the effect of probiotic formulations toward the damage. By knowing the histopathological damage and the effect of probiotic formulations, it is expected that we can provide better water quality for koi fish so the death rate due to parasites, especially *Myxobolus* sp., can be reduced.

2. Methods

The study was experimental. Samples were koi fish showing symptoms of being infected with *Myxobolus* sp. The samples were 7-12 cm long. As many as 288 koi were used in the experiment and were put into 24 treatment tanks, so there were 12 fish on each tank. The koi fish were soaked in the tanks and were given a probiotic formulation.

The doses of probiotic formulation administered to each healthy and infected fish to each tank were as follows: PA (infected fish with a dose of probiotic formulation of 1.1 ml/30 liters), PB (infected fish with a dose of probiotic formulation of 0.55 ml/30 liters), PC (infected fish with a dose of probiotic formulation of 1.65 ml/30 liters), PD (healthy fish with a dose of probiotic formulation of 1.1 ml/30 liters), PE (healthy fish with a dose of probiotic formulation of 0.55 ml/30 liter), and PF (healthy fish with a dose of probiotic formulation of 1.65 ml/30 liters). The treatment was carried out for 6 hours in water and probiotic formulations were given every other day for two weeks.

Koi fish were observed to determine the clinical symptoms. Infected fish that showed the most severe clinical symptoms were dissected and gills (lamella) were used for histopathological analysis. Then, the observation using microscope to analyse gill tissue damage. Histopathological analysis using the scoring method. Scoring data that obtained from observation were processed and the results were analysed by using ANOVA. Sampling was done from January to March 2019.

3. Results and Discussion

3.1. Clinical Symptoms of Koi Fish Infected with Myxobolus sp.

Koi fish infected with *Myxobolus* sp. showed clinical symptoms that could be observed directly, i.e. the existence of a reddish white cyst or bulge. The cyst contained thousands of spores covering the gills of the fish that the gill cover appeared open and made it difficult for fish to breathe. Infected fish were found to swim slowly, to stay more at the bottom of the water, and to show low feeding responses; many were dead [3]. Koi fish have clinical symptoms in the form of open or unable to close the operculum completely caused by swelling of the *Myxobolus* sp. reddish white color making it difficult for fish to do the process of respiration as in general. Furthermore, the health status of fish can be known from the body and behavior—healthy fish have no wound on their body, have bright color, are actively moving, and do not swim on the surface of the water [4].

3.2. Measuring the Damage of Koi Fish Gills

3.2.1. Edema. Based on observations of gill tissue (lamella) of koi fish infected with *Myxobolus* sp., the damage to the gill tissue was different in each treatment (Figure 1). The damage to gill (lamella) of koi fish infected with *Myxobolus* sp. can be seen from the presence of edema, which is indicated by an arrow in the observation Figure 1. Edema is cell swelling or excessive accumulation of fluid in tissues. The accumulation of fluid causes the cells to be irritative, which causes the cells to swell [5]. Edema can be characterized by the loosing of the basement membrane and the narrow of lacuna cells leading to deficiencies in the function of the gills and difficulties in the respiratory process, and then the metabolism begins to be disturbed [6].

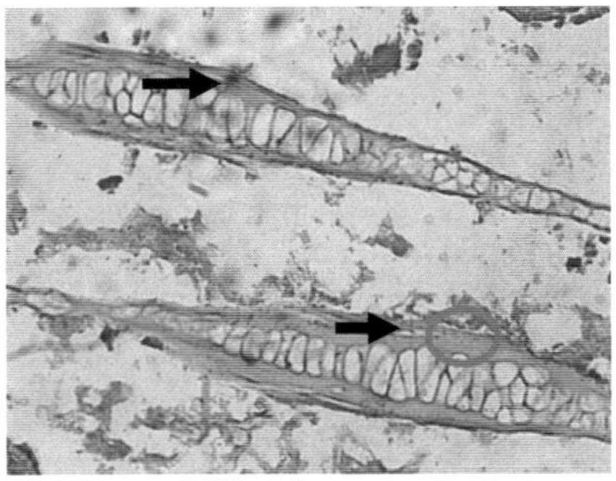

Figure 1. The damage of edema.

Based on the findings, the highest average score of edema damage was found in the PC treatment (infected fish with a dose of probiotic formulation of 1.65 ml/30 L) and the lowest was in the PB treatment (infected fish with a dose of probiotic formulations of 0.55 ml/30 L) and PE (healthy fish with a dose of probiotic formulation of 0.55 ml/30 L). Based on the 5% BNT test, the administration of the probiotic formulation of 0.55 ml/30 L showed the best effect to edema damage in the gill tissue of koi fish.

Table 1. Analysis of Variance Analysis (ANOVA) Edema Damage.

Sources of Diversity	Db	JK	KT	F Count	F 5%	F 1%
Treatment	7	14,58	2,083	63,314**	2,66	4,03
A	1	1,71	1,707	68,267**	4,49	8,53
B	3	2,43	0,811	32,444**	3,24	5,29
AB	3	10,44	3,480	139,2**		
Error	16	0,4	0,025			
Total	23	14,98				

Information : (**) = Very Real Difference, (TN) = Not Real

3.2.2. Hyperplasia. Based on observations of gill tissue (lamella) of koi fish infected with Myxobolus sp., the damage to hyperplasia is presented in Figure 2, as indicated by the arrow [7] hyperplasia is excessive tissue formation due to the increase in the number of cells. This is a physiological response to protect tissues from toxic substances. Hyperplasia of gill lamella begins with several case including edema, death cells and release of epithelium cells in gill lamellae [8]. Lamella with hyperplasia results in thickening of the epithelial tissue at the end of the filament or thickening of the epithelium tissue located near the base of the lamella. Hyperplasia occurs with the addition of a number of secondary lamella epithelium, so the secondary lamella gets bigger and coincides; consequently, the secondary lamella sticks together and converges [9]. This makes the gill lamella appears larger than normal and differences between the primary and secondary lamella cannot be seen clearly.

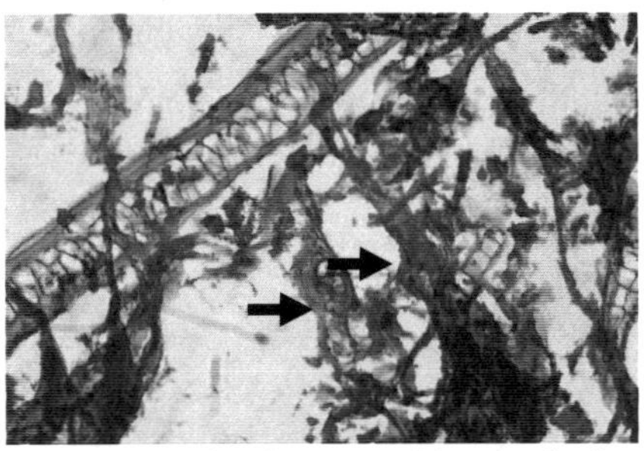

Figure 2. The damage of hyperplasia.

Based on the findings, the highest average score of hyperplasia damage was found in the PA treatment (infected fish with a dose of probiotic formulation of 1.11 ml/30 L) and PC treatment (infected fish with a dose of probiotic formulation of 1.65 ml/30 L) and the lowest was in the PB treatment (infected fish with a dose of probiotic formulation of 0.55 ml/30 L) and PE (healthy fish with a dose of probiotic formulation of 0.55 ml/30 L). Based on the 5% BNT test, the administration of probiotic formulations of 0.55 ml/30 L showed the best effect to hyperplasia damage in the gill tissue of koi fish.

Table 2. Analysis of Variance Analysis (ANOVA) of Hyperplasia Damage.

Sources of Diversity	Db	JK	KT	F Count	F 5%	F 1%
Treatment	7	17.23	2.461	67.117**	2,66	4,03
A	1	0.67	0.667	18.182**	4,49	8,53
B	3	3.05	1.018	27.758**	3,24	5,29
AB	3	13.51	4.502	122.788**		
Error	16	0.59	0.037			
Total	23	17.81				

Information : (**) = Very Real Difference, (TN) = Not Real

3.2.3. Lamella Fusion. Based on observations of gill tissue (lamella) of koi fish infected with *Myxobolus* sp., the damage to lamella fusion is presented in Figure 3, as indicated by the arrow. The fusion of two secondary lamellae can be caused by a parasitic wound on the secondary lamella forcing the organ to release much mucus to cover the wound. Lamella fusion occurs by hyperplasia that extends to basal cells and epithelium so secondary lamella will coalesce. This event results in the obstruction of the respiration and expiration processes [10]. Lamella fusion represents severe damage because it is an advanced stage of hyperplasia damage.

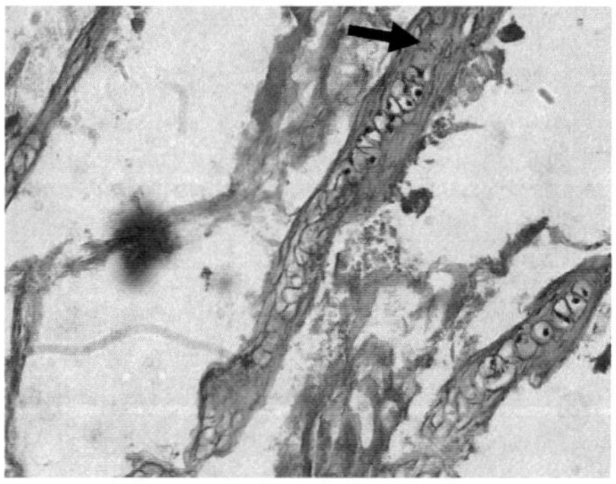

Figure 3. Lamella Fusion.

Based on the findings, the highest average score of lamella fusion damage was found in the PC treatment (infected fish by administering a dose of probiotic formulation of 1.65 ml/30 L) and the lowest was in the PB treatment (infected fish with a dose of probiotic formulation of 0.55 ml/30 L). Based on the 5% BNT test, the administration of probiotic formulations of 0.55 ml/30 L showed the best effect to lamella fusion damage in the gill tissue of koi fish.

Table 3. Analysis of Variance Analysis (ANOVA) of Lamela Fusion Damage.

Sources of Diversity	Db	JK	KT	F Count	F 5%	F 1%
Treatment	7	15,06	2,151	56.124**	2,66	4,03
A	1	2,16	2,160	56.348**	4,49	8,53
B	3	2,73	0,909	23.710**	3,24	5,29
AB	3	10,17	3,391	88.464**		

Error	16	0,61	0,038	
Total	23	15,67		

Information : (**) = Very Real Difference, (TN) = Not Real

3.3. Molecular Response of Probiotic Formulations on the Survival of the Body of Koi Infected with Myxobolus sp. This study used *Bacillus* spp., *Lactobacillus* sp., and *Nitrosomonas* sp. as the starters of probiotic formulation, with a quantity of 108 cfu/ml each. The use of Bacillus spp. with the right concentration and dosage as an ingredient in probiotic formulations will able to increase the number of red blood cells and hemoglobin levels—these are believed to be one indicator of an increase in the ability of fish to supply nutrients to the entire body and tissue repair to increase fish growth [11].

The effect on fish related to the administration of probiotic formulations is the formation of an adaptive immune response. Fish have an innate and adaptive immune defense system. Congenital parameters are at the forefront of immune defense and are an important factor in disease resistance. Adaptive responses are usually delayed but are very important for long-term immunity and are a key factor in successful treatment [12]. This response is characterized by the formation of IGM in koi fish.

Fish only synthesizes one class of immunoglobulin (IgM). IgM serum is tetrameric in teleosts fish and is pentametric in cartilaginous fish. IgM is more efficient than IgG in complement activation, opsonization, viral neutralization, and agglutination. IgM is found in fish mucus and is cell-mediated immunity. T-cytotoxic cells help kill viral-infected cells and abnormal cells [13].

4. Conclusion

Histopathological observations on gill tissue of koi fish not infected with *Myxobolus* sp. shows damage to different gill tissue (lamella). The damage to the gill tissue of koi fish infected with *Myxobolus* sp. is edema, hyperplasia, and lamella fusion. The treatment administered to tackle *Myxobolus* sp. with probiotic formulations at a dose of 1.1 ml/30l, 0.55 ml/30l, and 1.65 ml/30l of water in koi fish of different sizes give different effects on the level of gill tissue damage. The best effect of probiotic formulations on the level of histopathological damage to gill tissue, i.e. damage to edema, hyperplasia, and lamella fusion, is the one with the dose of probiotic formulation of 0.55 ml/30l according to 5% BNT.

References

[1] Pujiastuti N and Setiati N 2015 Identifikasi dan Prevalensi Ektoparasit Pada Ikan Konsumsi di Balai Benih Ikan Siwarak *UNNES Journal of Life Science* **4** (1) 9-15
[2] Mahasri G and Kismiyati 2011 *Buku Ajar Parasit dan Penyakit Ikan I (Ilmu Penyakit Protozoa pada Ikan dan Udang)* (Surabaya: Universitas Airlangga) hal. 3-4
[3] Deriyanti A 2016 Korelasi Kualitas Air dengan Prevalensi Myxobolus Pada Ikan Koi (Cyprinus carpio) di Sentra Budidaya Ikan Koi Kabupaten Blitar, Jawa Timur *Skripsi* Fakultas Perikanan dan Kelautan Universitas Airlangga Surabaya
[4] Rahmaningsih S 2016 *Hama & Penyakit Ikan* (Yogyakarta: Deepublish 352 hal
[5] Tandjung S 1982 The Toxicity of Alumunium for Organs of Salvalinus Fontanalis Mitchill In Acid Water Jakarta
[6] Fitriawan F Sutarno and Sunarto 2011 Perubahan mikroanatomi pada insang dan ginjal kerang air tawar (Anodonta woodiana) terhadap paparan kadmium *Bioteknologi* **8** (1) 42-52
[7] Setyawan N 2013 Gambaran Mikroanatomi Pada Insang Ikan Sebagai Indikator Pencemaran Logam Berat di Perairan Kaligarang Semarang *Skripsi* Jurusan Biologi Fakultas Matematika dan Ilmu Pengetahuan Alam Universitas Negeri Semarang
[8] Widayati E D 2008 Studi Histopatologi Insang Ikan Mujair (*Oreochromis mossambicus*) Pada Konsentrasi Sublethal Air Lumpur Sidoarjo. *Skripsi.* Institut Teknologi Sepuluh Nopember Surabaya
[9] Yuniar V 2009 Toksisitas Merkuri (Hg) Terhadap Tingkat Kelangsungan Hidup Pertumbuhan, Gambaran Darah dan Kerusakan Organ Pada Ikan Nila *Skripsi* Institut Pertanian Bogor, Bogor

[10] Suparjo M N 2010 Kerusakan Jaringan Insang Ikan Nila (Oreochromis niloticus L) Akibat Deterjen *Jurnal Saintek Perikanan* **5** (2) 1-7

[11] Rajikkannu M, Natarajan N, Santhanam P, Deivasigamani B, Ilamathi J and Janani S 2015 Effect of probiotics on the haematological parameters of Indian major carp (*Labeo rohita*) *International Journal of Fisheries and Aquatic Studies* **2** (5) 105-109

[12] Secombes C J and Wang T 2012 The innate and adaptive immune system of fish Infectious disease in aquaculture Woodhead Publishing Limited 2012

[13] Ode I 2013 Kajian Sistem Imunitas Untuk Pengendalian Penyakit Pada Ikan dan Udang *Jurnal Ilmiah agribisnis dan Perikanan (agrikan UMMU-Ternate)* **6** (2) 41-43

Anti-Inflammation Assay of Black Soybean Extract and Its Compounds on Lipopolysaccharide-Induced RAW 264.7 Cell

W Widowati[1]*, S Prahastuti[1], N L W Ekayanti[2], U Z Munshy[2], H S W Kusuma[2], S H B Wibowo [2], A Amalia[2], W S Widodo[2] and R Rizal[2]

[1]Faculty of Medicine, Maranatha Christian University, Jl Surya Sumantri No. 65, Bandung 40164, West Java, Indonesia
[2]Biomolecular and Biomedical Research Center, Aretha Medika Utama, Jalan Babakan Jeruk II No. 9, Bandung 40163, West Java, Indonesia

*Corresponding author's email: wahyu_w60@yahoo.com

Abstract. Inflammation response is related with various diseases. One of the useful therapeutic method to suppress inflammatory mediator synthesis is by application of compounds isolated from herbal medicine as treatment for inflammatory diseases. The aim of this study was to analyse the anti-inflammatory activity of black soybean extract (BSE), daidzein, and genistein trough *in vitro* analysis of inflammatory mediators such as prostaglandin 2 (PGE2) and cytokines interleukin 1β (IL-1β), and tumor necrosis factor α (TNF-α). Safety of samples was determined by viability test using MTS (3-(4,5-dimethylthiazol-2yl)-5-(3-carboxymethoxyphenyl)-2-(4-sulfophenyl)-2H-tetrazolium). Concentration tested for viability assay were 40, 200, 1000 μg/mL for BSE, daidzein, and genistein. Anti-inflammation activity of samples was determined by ELISA quantification of PGE-2, TNF-α, and IL-1β in conditioned medium (CM) of supplemented pro-inflammatory activated RAW 264.7 cell. Inflammation on cells were induced by Lipopolysaccharide (LPS). BSE 1000 ug/ml, daidzein 1000 ug/ml, and genistein 1000 ug/ml treatments shows <80% cell viability average compared to control cell, indicating the treatments have cytotoxicity effect on RAW 264.7 cells. Hence, concentration used for treatments are 40 and 200 μg/mL for each sample. Genistein with concentration of 40 μg/ml treatment result shows highest anti-inflammatory activity which indicated from PGE-2, TNF-α, and IL-1β concentration. This study suggests that BSE, daidzein, and genistein with concentration of 40 and 200 μg/ml were safe to use for RAW 264.7 cell and genistein with concentration of 40 μg/ml have the best anti-inflammatory activity compared to daidzein and BSE.
Keywords : inflammation, black soybean, RAW 264.7 Cell, PGE-2, TNF-α, IL-1β

1. Introduction

Inflammation is a biological response of body tissues to perilous stimuli, such as pathogens, irritants, and damaged self-cell. This response relates with various diseases such as rheumatoid arthritis, Alzheimer's, inflammatory bowel disease, atherosclerosis, and has a role in various cancer developments [1]. Many cells are involved in this response including macrophage which is the primary cells of chronic inflammation. During inflammation, macrophage is activated by exposure to interferon-ɣ, pro-inflammatory cytokines, or bacterial lipopolysaccharide (LPS) [2]. Activated macrophage releases amounts of inflammatory mediators such as nitric oxide (NO), reactive nitrogen species (RNS), inflammatory mediator nitric oxide (NO) and prostaglandin 2 (PGE2), and also pro-inflammatory cytokines such as interleukin (IL-12, IL-1β, IL-6) and tumor necrosis factor alpha (TNF-α) [2]. Bacterial endotoxin lipopolysaccharides (LPS) exposure can increase mRNA expression of those inflammatory cytokines and mediators which can stimulates innate immunity [3–5].

One of the useful therapeutic method in the treatment of inflammatory diseases is suppression of inflammatory mediator synthesis. Recently, application of compounds isolated from herbal medicine for the treatment of inflammatory diseases has been gaining interest. Herbal medicine also had several other properties such as non-toxic, Antioxidant (scavenge free radicals) and pharmacologically safe to use due to its pleiotropic immune modulatory properties [4,6,7].

Black soybean (*Glycine max* L. Merr) is not only used as protein source, but also functional food due to its isoflavone content that could prevent degenerative disease. Isoflavone in black soybean consist of four forms which are malonyl-glicoside, acetyl-glicoside, glicoside, and aglicon. The most significant biological activity is shown by aglicon form of isoflavone, genistein (5,7,4'-tryhidroxy isoflavone), daidzein (7,4'-dihydroxy isoflavone) and glycitein (6-metoxy-7,4' dihydroxy isoflavone) [8,9]. Daidzein is an isoflavone mainly extracted from soy plants. Daidzein can be used as anti-inflammation by inhibition of signal transducer and activator of transcription 1 (STAT-1) and nuclear factor kappa-light-chain-enhancer of activated B cells (NF-kB) activations [10], meanwhile genistein has numerous anti-oxidative and anti-cancer effects and is known to restrict tyrosine specific protein kinases [11].

The aim of this study was to analyze the anti-inflammatory activity of black soybean extract (BSE), daidzein, and genistein on the in vitro production of inflammatory mediators such as, PGE2 and cytokines IL-1β and TNF-α.

2. Materials and Methods

2.1. Extract preparation

Black soybeans were obtained from Unit Pengelolaan Benih Sumber (UPBS) Balai Penelitian Tanaman Aneka Kacang dan Umbi, Malang, East Java. The plants were identified by herbarium staff, Department of Biology, School of Life Science and Technology, Bandung Institute of Technology, Bandung, West Java, Indonesia. A total of 250 g of black soybean simplicia was added to the maserator for extraction. The solvent used was 70% ethanol. The filtrate was collected every 24 h and ethanol was added until the resulting filtrate was colorless. The filtrate was then evaporated using a *rotary evaporator* (Zhengzhou Well-known, RE-201D) at 50 °C until a paste extract was formed. The extract from black soybeans was then stored at -20 °C [12–14].

2.2. Cell culture

RAW 264.7 macrophages cell line (ATCC® TIB- 71™) was obtained from Aretha Medika Utama, Biomolecular and Biomedical Research Center, Bandung, Indonesia. Cells were grown in Dulbecco's Modified Eagle Medium (DMEM, Biowest L0060) supplemented with 10% fetal bovine serum (FBS) (Gibco, 10082147) and 1% penicillin-streptomycin (Biowest L0022-100), and then incubated at 37 °C and 5% CO_2 until the cells were confluent. The cells then washed using Phosphate Buffer Saline (PBS) and harvested using trypsin-EDTA (Biowest, L0931-500) [12,15,16].

2.3. Pro-inflammatory activation of RAW 264.7 cells

The activation of inflammatory condition of the macrophage cells was performed according to previous study [12]. The cells with density of $5x10^5$ cells/well were plated in 6-well plate, incubated for 24 hr at 37 °C in a humidified atmosphere and CO_2 5%. After the culture medium was discharged, 1,600 µM of fresh medium were supplemented to each well along with 200 µl of sample with various concentration (40 and 200 µg/ml of each sample: BSE, daidzein (Chengdu Biopurify Phytochemical Ltd, BP0445), genistein (Chengdu Biopurify Phytochemical Ltd, BP0635). Based on the viability assay. 200 µl of LPS from *Escherichia coli* (Sigma Aldrich, L2880) (1 µg/ml) was added into each well two hours following the addition of extract or compound, and the plate was incubated for 24 h at 37 °C, humidified atmosphere, 5% CO_2. The medium then was collected and centrifuged at 2000xg for 10 min. The supernatant was collected and stored at -80 °C for further assay [12,15,16].

2.4. *RAW 264.7 cell viability assay*
The assay was conducted using Proliferation Assay Kit (Abcam, ab197010). This assay is based on the conversion of yellow tetrazolium salt to form a purple formazan product. Cell with density 5 x 10^3 cell/well is seeded and incubated for 24 h (37°C, 5% CO_2). Briefly 20 µL of sample with various concentration (40, 200, 1000 µg/mL of each sample: BSE, daidzein, genistein) was added into each well and incubated for 24 h (37°C, 5% CO_2). After incubation, 20 µL of MTS 3 - (4,5-dimethylthiazol – 2 – yl) – 5 - (3 - carboxymethoxyphenyl) – 2 - (4-sulfophenyl) - 2H tetrazolium) Cell Proliferation Assay Kit (Abcam, ab197010) was added into each well. After 3 h of incubation (37 °C, 5% CO_2), absorbance was measured by Multiskan-GO Microplate Reader at 490 nm [12,15,16].

2.5. *Total protein assay*
Bovine Standard Albumin (BSA) standard solution was made from dilution series of BSA stock. The stock was obtained by dissolving 2 mg of BSA (Sigma, A9576, Lot. SLB2412) in 1000 µL ddH_2O, briefly 20 µL of standard solutions and 200 µL Quick Start Dye Reagen 1X (Biorad, 5000205) was added into each well plate. After 5 min of incubation in room temperature, absorbance was measured by microplate reader at 595 nm [17].

2.6. *PGE-2, IL-1β, and TNF-α assay*
PGE-2, IL-1β, and TNF-α concentration was measured using Mouse ELISA Kit (E-EL-M0052/0049/0037). Briefly 100 µl of each standard solution was added into well. The plate was sealed and incubated for 90 min 37 °C. The solution in well was discarded and 100 µL/well of *Biotinylated Detection Ab* was added. The plate was closed and incubated for 1 hr 37 °C. The solution was discarded and each well was washed with *Wash Buffer* and settled for 1-2 min. The washing process was repeated for three times. After that on each well was added 100 µL/well of *HRP Conjugate*. The plate was sealed incubated in 37 °C for 30 min and followed by five times washing. 90 µL/well of substrate was added and incubated for 15 -30 min 37 °C. After the color change 50 µL/well of *stop solution* was added and *Optical Density* (OD) was measured by Multiskan GO Microplate Reader at 450 nm [12,15,16].

2.7. *Statistical Analysis*
All the experiment was done in triplicate. Statistical analysis was conducted using SPSS software (version 20.0). Data were presented as mean ± standard deviation. The Significant differences within the groups were determined using One-Way ANOVA followed by Tukey's HSD Post-hoc Test.

3. Results and Discussion
Inflammation is a dynamic process involving pro-inflammatory cytokines, and it acts as an important biological response towards perilous stimuli [18,19]. In this study, we examined the anti-inflammatory properties of daidzein, genistein, and black soybean (*Glycine max*) extract on LPS-induced RAW 264.7 murine macrophage cell line which has been widely used as an inflammatory model *in vitro* [12,15,16,18,20,21]. During inflammation induced by LPS, RAW 264.7 cell release PGE2 and pro-inflammatory cytokines such as IL-1β and TNF-α [12,16,20,21,25].

3.1. *RAW 264.7 cell viability*
Cell viability was measured by MTS assay based on the conversion of yellow tetrazolium salt to form a purple formazan product. The viability of the RAW 264.7 cells can be seen on table 1. BSE 1000 ug/ml, daidzein 1000 µg/mL, and genistein 1000 µg/mL treatments shows <80% cell viability compared to control cell, indicating the treatments have cytotoxicity effect on RAW 264.7 cells. Meanwhile, other treatment used were nontoxic and safe to the cells with cell viability 90% [12], this concentration to be

applied for the next assay. Hence, concentration used for treatment are 40 and 200 μg/mL for each sample.

Table 1. RAW 264.7 cell viability assay of BSE, daidzein and genistein.

Concentration (μg/mL)	Cells viability (%)		
	BSE	Daidzein	Genistein
0 (Control cell)	100.00 ± 3.54	100.00 ± 3.54	100.00 ± 3.54
40	98.86 ± 7.63^{b}	104.31 ± 10.56^{b}	89.94 ± 4.87^{b}
200	87.76 ± 1.75^{b}	99.95 ± 13.20^{ab}	87.56 ± 3.40^{b}
1000	67.20 ± 0.60^{a}	78.61 ±2.12^{a}	71.92 ± 4.63^{a}

Data are presented as mean ± SD. Different superscript letters a,b for BSE, a,ab,b for daidzein, a,b for genistein show significant differences among concentrations based on Tukey HSD post hoc test (*P*-value < 0.05)

3.2. PGE-2, TNF-α, and IL-1β assay

Measurement of standard curve for total protein determination is presented. The linearity of the curve is in acceptable range. Total protein measured is 125.30 ± 4.79 (μg/mL). The results of PGE-2, TNF-α, IL-1β quantification which is shown in figure 1, 2 and 3 reveal that BSE had concentration dependent manner inhibition activity towards mentioned inflammatory mediators production in LPS induced RAW 264.7. Meanwhile, figure 1, 2 and 3 also reveal that daidzein and genistein also had inhibition activity towards mentioned inflammatory mediators but treatment with lower concentration which is 40 μg/ml resulted in better activity compared to dose of 200 μg/mL. PGE-2, TNF-α, and IL-1β level in positive control are significantly higher compared to the level of negative control which prove that LPS can increase mentioned Inflammatory mediators production in RAW 264.7 cell.

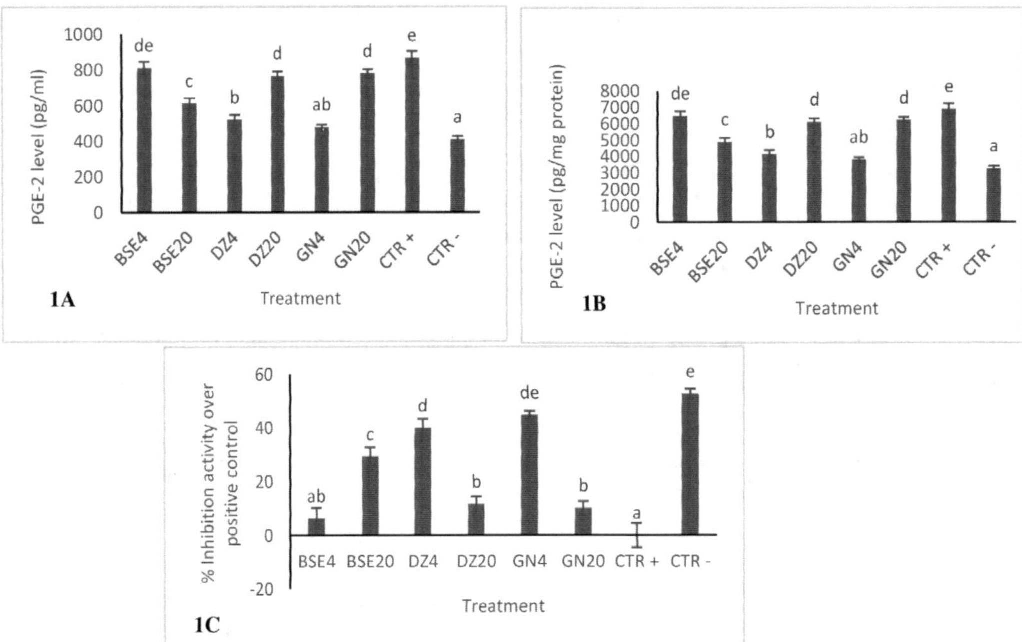

Figure 1. Effect BSE, daidzein, and genistein toward PGE-2 level, inhibition activity on LPS-induced RAW 264.7 cells.

The data are presented as histogram, of mean±standard deviation.

Journal of Physics: Conference Series **1374** (2019) 012052 doi:10.1088/1742-6596/1374/1/012052

BSE 4 = Black Soybean Extract 40 µg/mL, BSE 20 : Black Soybean Extract 200 µg/mL, DZ4 : Daidzein 40 µg/mL, DZ20 : Daidzein 200 µg/mL, GN4 : Genistein 40 µg/mL, G20 : Genistein Daidzein 200 µg/mL, CTR+ : Positive control, CTR- : Negative control

1A. Effect BSE, daidzein, and genistein toward PGE-2 level (pg/mL), different superscript letters (a,ab,c,d,de,e) show significant differences among treatments based on Tukey HSD post hoc test (*P*-value < 0.05).

1B. Effect BSE, daidzein, and genistein toward PGE-2 level (pg/mg protein), different superscript letters (a,ab,c,d,de,e) show significant differences among treatments based on Tukey HSD post hoc test (*P*-value < 0.05).

1C. Effect BSE, daidzein, and genistein toward inhibiton activity over positive control (%), different superscript letters (a,ab,b,c,d,de,e) show significant differences among treatments based on Tukey HSD post hoc test (*P*-value < 0.05).

PGE2 the most abundant PGs produced in the body, has a central role in inflammation and both cyclooxygenase-2 (COX-2). In inflammation, the expression of COX-2 and inducible microsomal prostaglandin E synthase-1 (mPGES-1) are enhanced by bacterial products and cytokines, resulted in increased PGE2 formation in macrophages and tissue cells [10]. PGE2 involves important mediators of many biological functions, such as regulation of immune responses, blood pressure, gastrointestinal integrity, and fertility in physiological regulation. PGE2 Deregulated synthesis or degradation has been associated with a wide range of pathological conditions [32]. PGE2 indicating process of inflammation, leading to the classic signs of inflammation such as redness, swelling, and pain [33]). PGE2 has a role and its receptors in modulating the inflammatory response has been observed In atherosclerosis [34]. In this study, daidzein, genistein, and black soybean could inhibit PGE2, TNF-α, and IL- 1β production in RAW 264.7 cell lines which suggest that they have anti-inflammatory effect through down regulation of those pro-inflammatory cytokines. Daidzein plays a vital role in the regulation of mammary tumor cell invasion induced by TNF-α, related to nuclear factor-kappaB (NF-κB) signaling pathway. Daidzein treatment can inhibit breast cancer cells MDA-MB-231 migration trough suppression of TNF-α induced NF-κB and AP-1, followed by a reduction in the secretion of uPA from breast cancer cells [35]. Estrogen receptor-mediated response of Daidzein-treatment on Alzheimer's disease can decrease mRNA expression of IL-1 which is a key inducer of the expression inflammatory process related molecule in astrocytes cultured in the presence of LPS. These results further increase the possibility that daidzein may have potential to ameliorate the inflammatory process and also alleviate the risk of Alzheimer's disease progression [36].

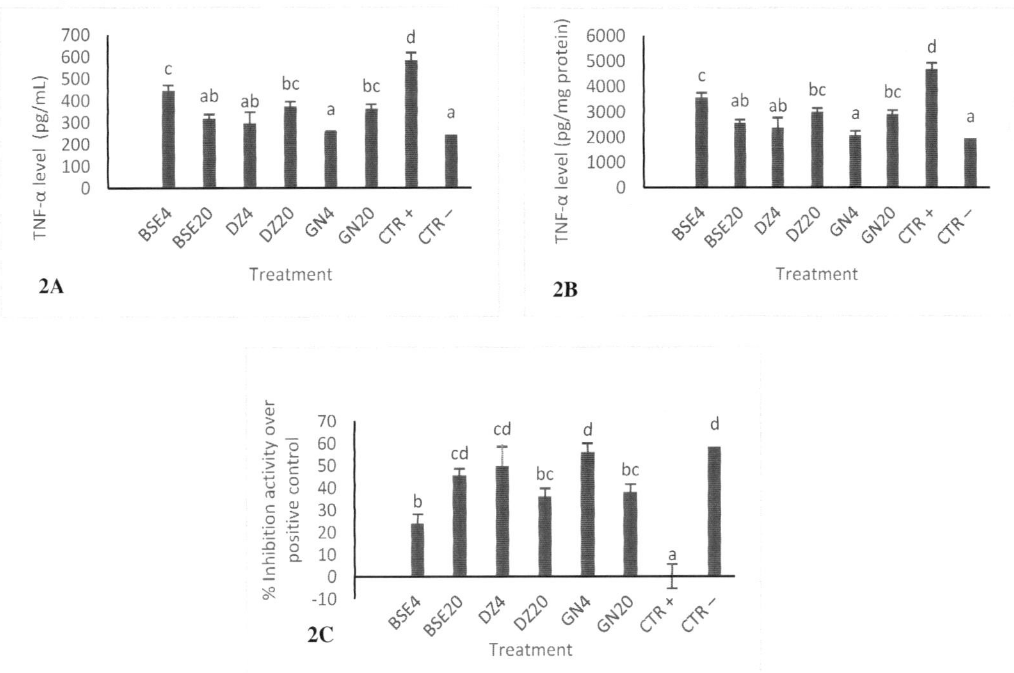

Figure 2. Effect BSE, daidzein, and genistein toward TNF-α level, inhibition activity on LPS-induced RAW 264.7 cells.

The data are presented as histogram, of mean±standard deviation.
BSE 4 = Black Soybean Extract 40 μg/mL, BSE 20 : Black Soybean Extract 200 μg/mL, DZ4 : Daidzein 40 μg/mL, DZ20 : Daidzein 200 μg/mL, GN4 : Genistein 40 μg/mL, G20 : Genistein Daidzein 200 μg/mL, CTR+ : Positive control, CTR- : Negative control
2A. Effect BSE, daidzein, and genistein toward TNF-α level (pg/mL), different superscript letters (a,ab,bc,c,d) show significant differences among treatments based on Tukey HSD post hoc test (*P*-value < 0.05).
2B. Effect BSE, daidzein, and genistein toward TNF-α level (pg/mg protein), different superscript letters (a,ab,bc,c,d) show significant differences among treatments based on Tukey HSD post hoc test (*P*-value < 0.05).
2C. Effect BSE, daidzein, and genistein toward inhibiton activity over positive control (%), different superscript letters (a,b,bc,cd,d) show significant differences among treatments based on Tukey HSD post hoc test (*P*-value < 0.05).

TNF-α is also a cytokine that plays a significant role in inflammation. This cytokine is produced chiefly by activated macrophages and play role during inflammatory response activating adhesion molecule inducer and nuclear factor kappa-light-chain-enhancer of activated cells (NF-κB) (25–27). TNF, being an endogenous pyrogenic, is able to induce fever, apoptotic cell death, cachexia, inhibit tumor genesis and viral replication, respond to sepsis via IL-1 and IL-6 producing cells. Deregulation of TNF production has been implicated in a variety of human diseases including Alzheimer's disease, cancer, major depression, psoriasis and inflammatory bowel disease (IBD) (28–30). With all effects caused, TNF-α inhibitory activity measurement is important in anti-inflammatory potential agent screening since this cytokine is an important mediator of inflammation (31).

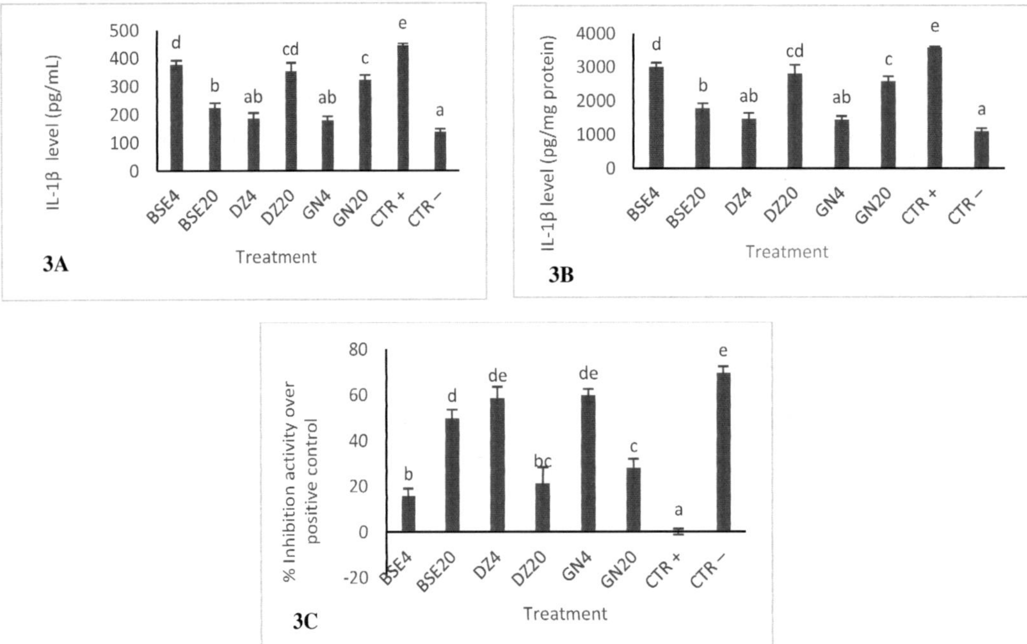

Figure 3. Effect BSE, daidzein, and genistein toward IL-1β level, inhibition activity on LPS-induced RAW 264.7 cells.
The data are presented as histogram, of mean±standard deviation.
BSE 4 = Black Soybean Extract 40 μg/mL, BSE 20 : Black Soybean Extract 200 μg/mL, DZ4 : Daidzein 40 μg/mL, DZ20 : Daidzein 200 μg/mL, GN4 : Genistein 40 μg/mL, G20 : Genistein Daidzein 200 μg/mL, CTR+ : Positive control, CTR- : Negative control
2A. Effect BSE, daidzein, and genistein toward IL-1β level (pg/mL), different superscript letters (a,ab,b,c,cd,d,e) show significant differences among treatments based on Tukey HSD post hoc test (*P*-value < 0.05).
2B. Effect BSE, daidzein, and genistein toward IL-1β level (pg/mg protein), different superscript letters (a,ab,c,cd,d,e) show significant differences among treatments based on Tukey HSD post hoc test (*P*-value < 0.05).
2C. Effect BSE, daidzein, and genistein toward inhibiton activity over positive control (%), different superscript letters (a,b,bc,c,d,de) show significant differences among treatments based on Tukey HSD post hoc test (*P*-value < 0.05).

IL-1β is important for the initiation and increase the inflammatory response to microbial infection during inflammation process due to its role to induces secretion of proinflammatory cytokines such as IL-6 and IL-8 [19,22]. IL-6 has a wide range effect on immune system cells resulted in the acute inflammation response. Increasing of IL-6 level are reported in rheumatoid arthritis, psoriasis, and encephalomyelitis individuals [22–24]. Therefore, inhibition of IL-1β synthesis would indirectly be useful for autoimmune disease and inflammation treatment [12,15,16,20,21,25].

Genistein as the major isoflavone in soybean has advantageous effects in metabolic disorders and inflammatory diseases. Genistein in dietary concentrations of (1, 5, and 10 μM) can inhibit LPS-induced IL-6 and TNF-α overproduction in plasma and liver of NASH model rats [36,37]. This isoflavone also significantly reduced LPS-induced NO production, increased antioxidant enzyme activity, and suppressed NF-κB activation in RAW264.7 macrophages. Major isoflavones contained in black soybean (genistein, daidzein, and glycitein) reported to have inhibition activity of LPS-induced NO production and decrease iNOS activity and gene expression in RAW264.7 macrophages The administration of genistein significantly decreased the IL-β expression in the dry-eye rat model group due to ovariectomy IL-1β involved conjunctiva inflammation which occurs in more than 80% of patients with dry eye syndrome [31-33, 40-42]. IL-1β are fundamental in the acute phase proteins synthesis and are produced in large quantities [43]. Genistein could suppress NF-κB activation in LPS-treated macrophages, which is in agreement with observations that genistein inhibits NF-κB that play role as pleiotropic regulator of many proinflammatory cytokines and has been found to be activated by a variety of stimuli. [42,44,45]. The results suggest that genistein's ability to inhibit LPS-induced TNF-α and IL-6 release may be explained in part by blocking NF-κB activation [46]. By more than 50% COX 2 mRNA expression is inhibited in Murine J774 macrophages by genistein. As measured by quantitative RT PCR. All three isoflavones studied, daidzein, genistein, and genistein, inhibited PGE2 production and COX 2 mRNA accumulation [10]. Genistein has earlier been shown to down regulate PGE2 production and COX 2 expression in macrophages [47,48]. In breast cancer cell line MCF-7, both daidzein and genistein inhibited phorbol 12-myristate 13-acetate (PMA)-induced COX 2 expression [49].

The inhibition activity of BSE is due to its major isoflavone content, daidzein and genistein. Other than those two substance, BSE also contain other isoflavone such as glycitein which is the least studied isoflavone, due to the fact that it is the least abundant isoflavone found in soy. Glycitein has been studied for some therapeutic effects including anti- oxidant, hypocholesterolemic, antitumor activity and was also tested for its anti-inflammatory activity both *in vitro* and *in vivo* [8,9]. Based on the results of this study and that has been reported, daidzein, genistein, and BSE may have potential to be used as safe anti-inflammatory agent to prevent chronic disease related to inflammation according to the result of viability assay.

4. Conclusion

This study suggests that BSE, daidzein, and genistein with concentration of 40 and 200 μg/mL were safe on RAW 264.7 cell and genistein with concentration of 40 μg/mL have the best anti-inflammatory activity compared to daidzein and BSE by inhibiting PGE-2, TNF-α, IL-1β.

5. Conflict of interest

All contributing authors declare no conflicts of interest.

6. Acknowledgement

The authors gratefully acknowledge to Biomolecular and Biomedical Research Center, Aretha Medika Utama, Bandung, West Java, Indonesia for research grant, research methodology and laboratory facilities. We are thankful to Ika Adhani Sholihah, Dewani Tediana Yusepany, Dwi Surya Artie, Rizki Amalia from Biomolecular and Biomedical Research Center, Aretha Medika Utama, Bandung, West Java, Indonesia for their valuable assistance.

References

[1] Ferrero-Miliani L, Nielsen O H, Andersen P S and Girardin S E 2007 Chronic inflammation: importance of NOD 2 and NALP 3 in interleukin-1b generation *Clin. Exp. Immunol.* **142** 227-35

[2] Duque G A and Descouteaux A 2014 Macrophage cytokines : involvement in immunity and infectious diseases *Front. Immunol.* **5** 1-12

[3] Kim K M, Kwon Y G, Chung H T, Yun Y G, Pae H O, Han J A, Ha K S, Kim T W and Kim Y M 2003 Methanol extract of Cordyceps pruinosa inhibits in vitro and in vivo inflammatory mediators by suppressing NF- kB activation *Toxicol. Appl. Pharmacol.* **190** 1-8

[4] Kim A R, Cho J Y, Zou Y, Chof J S and Chung H Y 2005 Flavonoids differentially modulate nitric oxide production pathways in lipopolysaccharide-activated RAW264.7 cells *Arch. Pharm. Res.* **28** 297-304

[5] Chang H Y, Sheu M J, Yang C H, Lu T C, Chang Y S, Peng W H, Huang S S and Huang G J 2011 Effects and the mechanisms of anti-inflammation of hispolon in mice *Evid. Based. Complement. Alternat. Med.* 1-8

[6] Yoon S B, Lee Y J, Park S K, Kim H C, Bae H, Kim H M, Ko S G, Choi H Y, Oh M S and Park W 2009 Anti-inflammatory effects of Scutellaria baicalensis water extract on LPS-activated RAW 264.7 macrophages *J. Ethnopharmacol.* **125** 286-90

[7] Krenn L and Paper D H 2009 Phytomedicine Inhibition of angiogenesis and inflammation by an extract of red clover (*Trifolium pratense* L.) *Phytomedicine.* **16** 1083-8

[8] Sulistyowati E, Martono S, Riyanto S and Lukitaningsih E 2018 Analisis daidzein dan genistein pada kedelai (*Glycine* Max L. Merril) varietas Anjasmoro, Argomulyo dan dena 2 menggunakan metode KCKT *Media. Farm. Indones.* **13** 1299-304

[9] Danciu C, Avram S, Pavel I Z, Ghiulai R, Dehelean C A, Ersilia A, Minda D, Petrescu C, Moaca E A and Soica C 2018 Main isoflavones found in dietary sources as natural anti-inflammatory agents. *Curr. Drug. Targets.* **19** 841-53

[10] Hämäläinen M, Nieminen R, Vuorela P, Heinonen M and Moilanen E 2007 Anti-inflammatory effects of flavonoids: genistein, kaempferol, quercetin, and daidzein inhibit STAT-1 and NF-kappaB activations, whereas flavone, isorhamnetin, naringenin, and pelargonidin inhibit only NF-kappaB activation along with their inhibitory effect on iNOS expression and NO production in activated macrophages *Mediators Inflamma.* 1-10

[11] Vitale D C, Piazza C, Melilli B, Drago F and Salomone S 2013 Isoflavones : estrogenic activity, biological effect and bioavailability *Eur. J. Drug. Metab. Pharmacokinet.* **38** 15-25

[12] Widowati W, Darsono L, Suherman J, Fauziah N, Maesaroh M and Erawijantari P P 2016 Anti-inflammatory effect of mangosteen (*Garcinia mangostana* L.) peel extract and its compounds in LPS-induced RAW264.7 cells *Nat. Prod. Sci.* **22** 147-53

[13] Widowati W, Janeva B W, Nadya S, Amalia A, Arumwardana S, Kusuma H S and Arinta Y 2018 Antioxidant and antiaging activities of *Jasminum sambac* extract, and its compounds *J. Reports. Pharm. Sci.* **7** 270-285

[14] Widowati W, Rani A P, Hamzah R A, Arumwardana S, Afifah E, Kusuma H S and Rihibiha D D 2017 Antioxidant and antiaging assays of *Hibiscus sabdariffa* extract and its compounds *Nat. Prod. Sci.* **23** 192-200

[15] Laksmitawati D R, Prasanti A P, Larasinta N, Syauta G A, Hilda R and Ramadaniati H U 2016 Anti-Inflammatory potential of gandarusa (*Gendarussa vulgaris* Nees) and soursoup (*Annona muricata* L) extracts in LPS stimulated-macrophage cell (RAW264.7) *J. Nat Remedies.* **16** 73-81

[16] Laksmitawati D R, Widyastuti A, Karami N, Afifah E, Rihibiha D D, Nufus H and Widowati W 2017 Anti-inflammatory effects of Anredera cordifolia and Piper crocatum extracts on lipopolysaccharide-stimulated macrophage cell line *Bangladesh J. Pharmacol.* **12** 35-40

[17] Pluemsamran T, Onkoksoong T and Panich U 2012 Caffeic acid and ferulic acid inhibit uva-induced matrix metalloproteinase-1 through regulation of antioxidant defense system in keratinocyte HaCaT cells *Photochem. Photobiol.* **88** 961-8.

[18] Blonska M and Czuba Z P 2003 Effect of flavone derivatives on interleukin-1beta (IL-1 beta) MRNA expression and IL-1 beta protein synthesis in stimulated RAW 264.7 macrophages *Scand. J. Immunol* **57** 162-6

[19] Dinarello C A 2010 Anti-inflam matory agents : present and future *Cell.* **140** 935-50

[20] Dewi K, Widyarto B, Erawijantari P P and Widowati W 2015 In vitro study of *Myristica fragrans* seed (Nutmeg) ethanolic extract and quercetin compound as anti-inflammatory agent *Int. J. Res. Med. Sci.* **3** 2303-10

[21] Sandhiutami N D M, Moordiani M, Laksmitawati D R, Maesaroh M and Widowati W 2017 *In vitro* assesment of anti-inflammatory activities of coumarin and Indonesian cassia extract in RAW264.7 murine macrophage cell line *Iran. J. Basic. Med. Sci.* **20** 99-106

[22] Brøchner A C and Toft P 2009 Pathophysiology of the systemic inflammatory response after major accidental trauma *Scand J Trauma Resusc Emerg Med* **17** 1-10

[23] Simpson R J, Hammacher A, Smith D K, Matthews J M and Ward L D 1997 Interleukin-6 : structure-function relationships *Protein. Sci.* **6** 929-55

[24] Ishihara K and Hirano T 2002 L-6 in autoimmune disease and chronic inflammatory proliferative disease *Cytokine Growth Factor Rev.* **13** 357-68

[25] Novilla A, Djamhuri D S, Nurhayati B, Rihibiha D D, Afifah E and Widowati W 2017 Anti-inflammatory properties of oolong tea (Camellia sinensis) ethanol extract and epigallocatechin gallate in LPS-induced RAW 264.7 cells *Asian Pac. J. Tro.p Biomed.* **7** 1005–1009

[26] Libby P 2002 Inflammation in atherosclerosis *Arterioscler. Thromb. Vasc. Biol.* **32**(9)2045-51

[27] de Cássia D S, Andrade L N, Dos Reis B D and de Sousa D P 2014 A review on anti-inflammatory activity of phenylpropanoids found in essential oils *Molecules* **19** 1459-80

[28] Swardfager W, Lanctôt K, Rothenburg L, Wong A, Cappell J and Herrmann N 2010 A meta-analysis of cytokines in Alzheimer ' s disease *Biol. Psychiatry* **68** 930-41

[29] Locksley R M, Killeen N and Lenardo M J 2001 The TNF and TNF receptor superfamilies : integrating mammalian biology *Cell* **104** 487-501

[30] Dowlati Y, Herrmann N, Swardfager W, Liu H, Sham L, Reim E K and Lanctôt K L 2010 Meta-analysis of cytokines in major depression *Biol. Psychiatry* **67** 446-57

[31] Boots A W, Wilms L C, Swennen E L, Kleinjans J C, Bast A and Haenen G R 2008 In vitro and ex vivo anti-inflammatory activity of quercetin in healthy volunteers *Nutrition* **24** 703-10

[32] Legler D F, Bruckner M, Uetz-von Allmen E and Krause P 2010 Prostaglandin E 2 at new glance : Novel insights in functional diversity offer therapeutic chances *Int. J. Biochem. Cell Biol.* **42** 198-201

[33] Funk C D 2001 Prostaglandins and leukotrienes : advances in eicosanoid biology *Sci* **294** 1871-5

[34] Babaev V R, Chew J D, Ding L, Davis S, Breyer M D, Breyer R M, Oates J A, Fazio S and Linton M F 2008 Macrophage EP4 deficiency increases apoptosis and suppresses early atherosclerosis *Cell. Metab.* **8** 492-501

[35] Valachovicova T, Slivova V, Bergman H, Shuherk J and Sliva D 2004 Soy isoflavones suppress invasiveness of breast cancer cells by the inhibition of NF-kappaB/AP-1-dependent and -independent pathways *Int. J. Oncol.* **25** 1389-95

[36] Liu M, Lin Y, Sheu S and Sun J S 2009 Anti-inflammatory effects of daidzein on primary astroglial cell culture *Nutr Neurosci.* **12** 123-34

[37] Ji G, Yang Q, Hao J, Guo L, Chen X, Hu J, Leng L and Jiang Z 2011 Anti-inflammatory effect of genistein on non-alcoholic steatohepatitis rats induced by high fat diet and its potential mechanisms *Int. Immunopharmacol.* **11** 762-8

[38] Sheu F, Lai H H and Yen G C 2001 Suppression Effect of soy isoflavones on nitric oxide production in RAW 264.7. Macrophages *J. Agric. Food Chem.* **49** 1767-72

[39] Aldina R, Sujuti H, Permatasari N and Widodo M A 2018 The effects of genistein on estrogen receptor-β, IL-1β levels, and MUC5AC expression in ovariectomized rats with dry eye *Nutr. Exp*

[40] Stahl J L, Cook E B, Graziano F M and Barney N P 2010 Differential and cooperative effects of TNFa, IL-1beta, and IFNgamma on human conjunctival epithelial cell receptor expression and chemokine release *Invest. Ophthalmol. Vis. Sci.* **44** 2010-5

[41] Solomon A, Dursun D, Liu Z, Xie Y, Macri A and Pflugfelder S C 2001 Pro- and anti-inflammatory forms of interleukin-1 in the tear fluid and conjunctiva of patients with dry-eye disease *Invest. Ophthalmol. Vis. Sci.* **42** 2283-92

[42] Raffoul J J, Wang Y, Kucuk O, Forman J D, Sarkar F H and Hillman G G 2006 Genistein inhibits radiation-induced activation of NF-kappaB in prostate cancer cells promoting apoptosis and G2/M cell cycle arrest *BMC Cancer.* **6** 1-10

[43] Sanchez-Garcia L, Martín L, Mangues R, Ferrer-Miralles N, Vázquez E and Villaverde A 2016 Recombinant pharmaceuticals from microbial cells : a 2015 update *Microb. Cell. Fact.* **15** 1-7

[44] Xu X, Harris K S, Wang H J, Murphy P A and Hendrich S 1995 Bioavailability of soybean isoflavones depends upon gut microflora in women *J. Nut.* **125** 2307-15

[45] Kang J L, Lee H W, Lee H S, Pack I S, Chong Y, Castranova V and Koh Y 2001 Genistein prevents nuclear factor-kappaB activation and acute lung injury induced by lipopolysaccharide *Am. J. Respir. Crit. Care Med.* **164** 2206-12

[46] Ji G, Zhang Y, Yang Q, Cheng S, Hao J, Zhao X and Jiang Z 2012 Genistein suppresses LPS-induced inflammatory response through inhibiting NF-kB following AMP kinase activation in RAW 264.7. Macrophages *PLoS One* **7** 1-7

[47] Akarasereenont P, Mitchell J A, Appleton I, Thiemermann C and Vane J R 1994 Involvement of tyrosine kinase in the induction of cyclo-oxygenase and nitric oxide synthase by endotoxin in cultured cells *Br. J. Pharmacol.* **113** 1522-8

[48] Liang Y C, Huang Y T, Tsai S H, Lin-Shiau S Y, Chen C F and Lin J K 1999 Suppression of inducible cyclooxygenase and inducible nitric oxide synthase by apigenin and related flavonoids in mouse macrophages *Carcinogenesis.* **20** 1945-52

[49] Lau T Y and Leung L K 2006 Soya isoflavones suppress phorbol 12-myristate 13-acetate-induced COX-2 expression in MCF-7 cells *Br. J. Nut.* **96**:169-76

Journal of Physics: Conference Series **1374** (2019) 012053 doi:10.1088/1742-6596/1374/1/012053

The decrease in bacterialcount in adult chronic rhinosinusitis patients after nasal irrigation therapy with NaCl 0,9% Solution

N Qamariah[1], N Muthmainah[2*] and R B B Tanjoto[3]

[1]Department of Ear, Neck, and Throat - Head and Neck Surgery, Faculty of Medicine, Lambung Mangkurat University, Ulin General Hospital Banjarmasin
[2]Department of Microbiology, Faculty of Medicine, Lambung Mangkurat University
[3]Medical Study Program, Faculty of Medicine, Lambung Mangkurat University

*Corresponding author's email: aizmother@yahoo.co.id

Abstract. Rhinosinusitis is a disease in which there is inflammation of the nasal mucosa and paranasal sinuses. One of the treatments suggested by several ENT specialists for chronic rhinosinusitis is nasal irrigation. In nasal irrigation therapy, 0.9% NaCl solution has been shown to accelerate the time of mucociliary transport which has an indirect effect on bacterial growth, and to remove inflammatory mediators leading to a decrease in inflammation. This study aimed to provide evidence of a decrease in bacterial count in chronic rhinosinusitis patients after nasal irrigation therap. The research method was descriptive analytical with a cross sectional approach. The study sample originated from adult chronic rhinosinusitis who were treated at the ENT Polyclinic of RSUD Ulin Banjarmasin, TK III Hospital, dr. Soeharsono Banjarmasin, and Banjarmasin harmony clinic in the period December 2018. Sampling was done by purposive sampling technique. The results showed that of the 7 samples tested there was a decrease in the number of bacteria after nasal irrigation therapy, showed a decrease of 38.08% and statistical tests showed a significant decrease in the number of bacteria after nasal irrigation therapy 0.9% NaCl solution (p=0,0001.

1. Introduction

Rhinosinusitis is a disease where inflammation of the nasal mucosa and paranasal sinuses occurs. Some of the causes of inflammation in rhinosinusitis are allergens, virus or bacterial infections, irritant substances, air pollution, extreme temperature changes, drought or high humidity.Inflammation of the mucosa causes stenoticssticking to the two opposite sides of the mucosa and cause mucociliary transport disorders, resulting in mucus retention and increasing bacterial colonization [1,2]

Chronic rhinosinusitis in adult patients is defined as an inflammatory disease involving inflammation of the nasal mucosa and paranasal sinuses with two or more symptoms more than 12 weeks, such as facial pain, facial congestion, nasal congestion, mucopurulent secretions, decreased olfactory ability and endoscopic or CT Scan examination findings.The incidence of rhinosinusitis in the United States based on National Health Interview Surveillance is estimated to be around 12.5% of the entire population of the United States or 31 million patients each year [3,4]

The rinosinusitis case in Indonesia based on data from the Ministry of Health of the Republic of Indonesia in 2003 stated that nasal and sinus disease was ranked 25th of 50 major ranking diseases or around 102,817 outpatients in hospitals. Data from the Department of Health, Ear, Nose, Throat,Head and Neck Surgery of Cipto Mangunkusumo Hospital showed a high incidence of rhinosinusitis, there were 300 (69%) out of 435outpatients in rhinology polyclinic who came during the period of January-August 2005.In Banjarmasin the incidence of rhinosinusitis was quite high; in 2015 there were 303 sufferers.Based on the results of the medical record, the Ulin General Hospital in Banjarmasin shows that there are still many chronic rhinosinusitis patients with an average of 349 patients each year [3-5]

The bacteria that often cause rhinosinusitis are *Streptococcus pneumonia* (30-50%), *Haemophilus influenza* (20-40%), *Moraxella catarralis* (4%), and others such as *Staphylococcus aureus*, *Streptococcus haemoliticus*, and *Pseudomonas sp.* The results of the study by Kamath et al in 2013 were conducted on 100 people with common symptoms of headaches, other symptoms such as nasal congestion, repeated sneezing, and olfactory disorders, resulted 51.24% of anaerobic bacteria with the most bacteria were *Staphylococcus aureus* and *Klebsiella*.The results of the 2010 Mantovani et al study in patients with chronic rhinosinusitis were found to be 27.6% of *Pseudomonas aeruginosa*, 13.9% of *Staphylococcus aureus* and 10.4% of *Streptococcus pneumonia* [6]

The treatment of chronic rhinosinusitis has the aim of restoring the physiological function of the nose that is disrupted and overcoming the symptoms.The principle of managing chronic rhinosinusitis is to avoid, eliminate the causes of irritation and inflammation, and improve the defence function of nasal mucosilia.Standard therapy to relieve complaints of chronic rhinosinusitis is by giving antibiotics, decongestants, and anti-inflammatory; while the additional therapy is nasal irrigation.Nasal irrigation helps prevent intranasal stasis and reduces crusting, while increasing the efficacy of intranasal topical drugs and improving ciliary function. Increased cilia function can inhibit the growth of bacteria that cause chronic rhinosinusitis [3,7]

Some ENT experts recommend nasal irrigation as being treated for patients with acute and chronic rhinosinusitis, post nose and sinus surgery.But there is no consensus on uniformity regarding the protocols for nasal irrigation and how to use them, including the tools used for irrigation such as nasal spray, bulbsyringe, snuffing, and cupped hand and other commercial tools [8,9]

It has been reported that saline solutions with various tonicity have an influence on the time of mucociliary transport.The results of the study suggesting that irrigation using soluble hypertonic saline improves the time of mucociliary transport, but that hypertonic solutions cause side effects in their use, including nasal burning, nasal irritation and irritation of the pharyngeal wall [10,11]

Min et al. (2001) conducted a study using saline solutions with various tonicity ranging from 0.06% - 7% to cilliary beat frequency (CBF) in vitro to the human nasal mucosa.It is reported that isotonic and hypotonic solutions such as 0.9% NaCl do not slow down cilia while hypertonic solutions (3% and 7%) result in ciliostatic. The use of 0.9% of NaCl solution will improve mucosilier clearance function which causes a decrease in bacterial colonization [10-13].

This study was conducted to prove a decrease in the number of bacteria in adult chronic rhinosinusitis patients after nasal irrigation therapy of 0.9% of NaCl solution. There has never been a previous study that examined the number of bacteria after the nasal irigassi therapy with 0.9% of NaCl solution, therefore this study was conducted as primary research.

2. Subject and methods

The method of this research is descriptive analytic with paired T-test and cross-sectional approach, where the study was conducted with the intention to determine the decrease in the number of bacteria in adult chronic rhinosinusitis patients after nasal irrigation therapy of 0.9% of NaCl solution.Sample collecting was done by doing a nose swab by an ENT specialist using a sterile cotton stick. After that, the cotton stick was put into the transport medium of the container and wrapped in aluminum foil, and put in an ice flask to be taken to the Microbiology Laboratory of the Medical Faculty, Lambung Mangkurat University, Banjarmasin.The samples were then planted on nutrient agar media and incubated for 24 hours at room temperature (27°C)

The calculation of the number of bacteria was conducted by counting the amount of bacterial colony that grew on nutrient agar media using colony counter. The next sample collecting was after the the nasal irigation therapy with 0.9% of NaCl solution, in every day for a week.

3. Result and discussion

Research on decreasing the number of bacteria in adult chronic rhinosinusitis patients before and after nasal irrigation therapy has been carried out at ENT Polyclinics of Ulin General Hospital, TK III dr. Soeharsono Hospital, Harmoni Medical Clinic Banjarmasin in December 2018. The results of the research had seven samples of chronic rhinosinusitis patients that were given 0.9% of NaCl solutions therapy, with the calculation of bacteria's number shown in table 1. below.

Table 1. The amount of bacterial colony in adult chronic rhinosinusitis patients before and after the nasal irigation therapy with 0.9% of NaCl solution.

Patient	The amount of bacterial colony		Decreasing percentage (%)
	Before nasal irigation therapy	After nasal irigation therapy	
1	68	44	35,29
2	42	29	30,95
3	31	17	45,16
4	47	28	40,43
5	52	31	40,38
6	65	36	44,62
7	37	26	29,73
	Decreasing average		38,08

Table 1 above shows that there was decreasing of the amount of bacterial colony in adult chronic rhinosinusitis patients after the nasal irigation therapy with 0.9% of NaCl solution. The decreasing was at the average 38.08% of the total amount of the bacteri before given by nasal irrigation therapy, with the highest decreasing was at 45.16% and the lowest was at 29.73%.

Based on a research by Sofyan and Tabary, the decreasing happened because nasal irrigation therapy can fix the nasal mucocilier function by clearance irrigation, eliminate the inflammation mediator, moisturizing cavum nasi, and fixing mucocilier function that shown by increasing cilia's movement frequency. The fixing of mucocilier will prevent mucus retention and the forming of bacterial colony, as of there will be a decreasing of the amount of bacteria in that area. This is in accordance with the clinical conditions of these patients after being given irrigation therapy by using 0.9% of NaCL, the patient's complaints are reduced.[13-17]

The average decrease of the amount of bacterial colony in adult chronic rhinosinusitis patients after the nasal irigation therapy with 0.9% of NaCl solution can be seen in Table 2 below.

Table 2. The average decrease of the amount of bacterial colony in adult chronic rhinosinusitis patients after the nasal irigation therapy with 0.9% of NaCl solution.

	Before (Mean+SD)	After (Mean+SD)	P
N=7	48,86+13,825	30,14+8,395	0.0001

Statistical analysis result using paired T-test shows significant differences between before and after nasal irrigation therapy in adult chronic rhinosinusitis patients with value p=0,0001. This result showed that there was significant differences after being given by irrigation therapy of 0,9% of NaCl solution.

This result has a resemblance as Ferryan Sofyan research in 2017 that studied the effect of nasal irrigation therapy to the pH ofnasal discharge; and also Hauptman research in 2007 of nasal irrigation to mucocilliary clearance in chronic rhinosinusitis patients. That research resulted significant effect from nasal irrigation therapy to mucocilliary clearance and the pH of nasal discharge. For normal people, nasal ciliary always should be covered by some mucus layer to keep it active, and the ciliary only works optimally in normal pH, 7 – 9. Outside that pH rage, there will be decreasing of frequency and dryness that can harm ciliary. People with inflammation in nasal mucose will have pH decreasing to rage 5,5 – 6,5. Thus cause a decrease in the fuunction of mucociliary transport. Within the increasing of nasal discharge pH after the therapy, ciliary in nasal can work effectively again. It also make the function of mucociliary improved and prevent mucuse retention that can make the bacterial growth inoptimally.[17]

A research by Georitis et al reported that nasal irrigation therapy with normal saline solution can reduce the amount of histamine and leukotriene nasal significantly ($p<0,5$) measured in rhinitis allergy patient at the 30 minutes, 4 and 6 hours after nasal irrigation. While a research by Tabary et al had result nasala irrigation with sea water physiologically can reduce kemokin IL-8 in order to reduce the inplammation in respiratory mucosa. Nasal irrigation with0,9% NaClsolutionhas not been proved having effect in decreasing the amount of eosinofil. This happened because 0,9% NaCl solution only consists of natrium 3500 mg/L ion and cloride 5500 mg/L, while reducing inflamation mediator secretion and cell degranulation needs magnesium ion which are only contained in physiological seawater. The decrease of eosinofil amount in nasal irrigation therapy might happen because of the mechanical clearance by NaCl in nasal cavity. By the reduce of inflamation mediator in nasal irrigation therapy, the inflamation in nasal mucosa and paranasal sinus will be reduced, and prevent mucosa retention that leads to the reducing of bacterial colony and the decreasing of bacteria's amount.[18-20]

4. Conclusion

Based on the results, it can be concluded that there was a decreasing in the amount of bacteria in the samples from adult chorinc rhinosinusitis after having nasal irrigation therapy of 0,9% NaCl solution, with the average decreasing was 38,09% and there was also significant decreasing of the amount of bacteria from statistic test results, p=0,0001

References

[1] W A Newman D 2011 *Kamus Saku Kedokteran Dorland Ed. 28* EGC Medical Publisher **28** pp 244
[2] Effiaty A S, Iskandar N, Bashiruddin J D 2015 Buku ajar ilmu kesehatan telinga hidung tenggorok kepala dan leher Jakarta **1** pp 1689–99
[3] Bachert C, Pawankar R, Zhang L, Bunnag C, Fokkens W J, Hamilos D L et al 2014 ICON: *Chronic rinosinusitis, World Allergy Organization Journal* 7
[4] Desrosiers M, Evans G A, Keith P K, Wright E D, Kaplan A, Bouchard J et al 2011 Canadian clinical practice guidelines for acute and chronic rinosinusitis *Allergy, Asthma Clin. Immunol.* [Internet] **7** (1) 2
[5] Perhimpunan dokter spesialis Indonesia THT K-L 2013 *Panduan Praktik Klinis PanduanPraktik Klinis Prosedur Tindakan Clinical Pathways* Pengurus Pus Perhati-KL **1** pp 4–8
[6] Mantovani K, Bisanha A A, Demarco R C, Tamashiro E, Martinez R, Anselmo-Lima W T 2010 Maxillary sinuses microbiology from patients with chronic rinosinusitis *Braz. J. Otorhinolaryngol.* **76** (5) 48–51
[7] Fokkens W J, Lund V J, Mullol J, Bachert C, Alobid I, Baroody F et al 2012 European position paper on rinosinusitis and nasal polyps *Rhinology* **50 (SUPPL. 23)** 4–305
[8] Taccariello M, Parikh A, Darby Y, Scadding G 1999 Nasal douching as a valuable adjunct in the management of chronic rinosinusitis. *Rhinology* **37** (1) 29–32

[9] Rabago D, Barrett B, Marchand L, Maberry R and Mundt M 2006 Qualitative aspects of nasal irrigation use by patients with chronic sinus disease in a multimethod study *Ann. Fam. Med.* **4** (4) 295–301
[10] Adam P, Stiffman M, Blake R L 1998 A Clinical Trial of Hypertonic Saline Nasal Spray in Subjects with the Common Cold or Rinosinusitis *Arch. Fam. Med.* **7** (1) 39–43
[11] Rabago D, Zgierska A, Mundt M, Barrett B, Bobula J, Maberry R 2002 Efficacy of daily hypertonic saline nasal irrigation among patients with sinusitis: a randomized controlled trial *J. Fam.* **51** (12) 1049–55
[12] Min Y G, Lee K S, Yun J B et al 2001 Hypertonic saline decreases ciliary movement in human nasal epithelium in vitro *Otolaryngol. Head Neck Surg.* **12** (3) 313-6
[13] Mustafa M 2015 Acute and Chronic Rinosinusitis, Pathophysiology and Treatment *Int. J. Pharm. Sci. Invent. ISSN.* **4** (2) 2319–6718
[14] Leung R S and Katial R 2008 *The Diagnosis and Management of Acute and Chronic Sinusitis*, Primary Care - Clinics in Office Practice **35** pp 11–24
[15] Chow A W, Benninger M S, Brook I, Brozek J L, Goldstein E J C, Hicks L A et al 2012 IDSA clinical practice guideline for acute bacterial rinosinusitis in children and adults *Clin. Infect. Dis.* **54** 8
[16] Singh V and Tiwari K 2014 *An Update of Rinosinusitis* ISSN **4** pp 1
[17] Sofyan F and Tami DRI 2017 Pengaruh Cuci Hidung dengan NaCl 0,9% terhadap Peningkatan Rata-Rata Kadar pH Cairan Hidung Departemen Telinga Hidung Tenggorok - Bedah Kepala Leher Fakultas Kedokteran Universitas Sumatera Utara/ Rumah Sakit Haji Adam Malik Medan *ORLI.* **47** pp 1
[18] Keojampa B K, Nguyen M H, Ryan M W 2004 Effect of buffered saline solution on nasal mucociliary clearance and nasal airway patency *Otolaryngol – Head and Neck Surgery* **131** 679-82
[19] Georgitis J W 1994 Nasal hyperthermia and simple irrigation for perennial rhinitis Changes in inflammatory mediators *Chest.* **106** 1487-92
[20] Tabary O, Muselet C, Yvin JC, Halley-Vanhove B, Puchelle E and Jacquot J 2001 Physiomer reduces the chemokine interleukine-8 production by activated human respiratory ephithelial cells *Eur. Respir. J.* **18** 661-6
[21] Zahra S S, Mailasari A and Marliyawati D 2016 Pengaruh Irigasi Hidung terhadap Derajat Sumbatan Hidung pada Perokok *JKD.* **5** 4
[22] Hauptman G and Ryan M W 2007 The Effect of Saline Solutions on Nasal Patency and Mucociliary Clearance in Rhinosinusitis Patience *American Academy of Otolaryngology–Head and Neck Surgery Foundation.* **137** 5

Molecular docking of the active compound *Garcinia mangostana* on the RANKL/RANK/OPG system

B Setiawan [1]*, E Suhartono[1], S Kaidah[2], I Z Akbar[3] and Z Noor[3]

[1] Department of Biochemistry and Biomolecular Sciences, Faculty of Medicine, Lambung Mangkurat University, Banjarmasin, South Kalimantan, Indonesia
[2] Department of Physiology, Faculty of Medicine, Lambung Mangkurat University, Banjarmasin, South Kalimantan, Indonesia
[3] Department of Orthopaedics and Traumatology, Ulin General Hospital, Faculty of Medicine, Lambung Mangkurat University, Banjarmasin, South Kalimantan, Indonesia

*Corresponding author's email: ganesh79setiawan@gmail.com

Abstract. This study aims to analyze the molecular docking between the active compounds of the *Garcinia mangostana* against the RANKL/RANK/OPG system and its potential as an antiosteoporosis. The research protocol includes the search and modeling of protein and ligand structures and their docking. Software used includes OpenBabel, HEX 8.0, Chimera 1.6.2, Discovery Studio 4.1, LigPlot + and LigandScout 3.1. Tovophillin has the most negative interaction energy with RANKL-OPG (–332.8 Kj/mol) and RANKL-RANK (–298.1 Kj/mol). It was concluded that fourteen active compounds of *Garcinia mangostana* did not interfere with the physiological function of RANKL against RANK. In addition, the active compound will not affect the RANKL-OPG complex. The antiosteoporosis mechanism of *Garcinia mangostana* does not by inhibiting RANKL-RANK interactions.

1. Introduction

Osteoporosis is a systemic skeletal disease characterized by decreased bone mass density and worsening microarchitecture, thereby increasing bone fragility and fracture. The pathophysiology of osteoporosis is very complex, a combination of genetic factors, nutritional factors, and endocrine abnormalities [1-3]. Diabetes mellitus is a metabolic disease characterized by hyperglycemia and has the potential to disrupt bone metabolism and trigger osteoporosis. Specifically, diabetes causes bone microangiopathy and decreased bone formation markers [4]. Diabetes decreases osteoclast activity and interferes with osteoblast activity, resulting in a low bone turnover. This will inhibit osteoclasts and osteoblasts through the induction of oxidative stress, hyperglycemia, and weight loss [5, 6]. Hyperglycemia triggers osteoblast lineage suppression [7], osteoblast apoptosis [8], inhibition of osteoblast-specific factors, and decreased in vitro mineralization [9].

Management of osteoporosis in diabetes mellitus includes lifestyle interventions, optimization of lifestyle control, analysis of the impact of diabetes mellitus treatment on fracture risk, and treatment with antiosteoporosis [10]. Until now, as far as researchers know, the application of herbs in the treatment of osteoporosis due to diabetes mellitus is still rarely done. Hesperidin increases bone turnover markers in serum, including osteopontin, osteocalcin, and decreases alkaline phosphatase in type 1 diabetes mellitus rats [11]. Green tea has not been able to improve bone mineral content and bone mass in individuals with diabetes mellitus [12]. Anhydroicaritin suppresses osteoclast differentiation and improves bone loss in diabetic rats [13].

The RANKL/RANK/OPG system plays an important role in bone remodeling. RANK is a receptor located on the surface of osteoclasts. RANKL is a RANK ligand that is synthesized and secreted by osteoblasts and bone marrow stromal cells. When the RANKL bond to RANK forms, osteoclast differentiation and bone resorption occur. OPG as a receptor trap against RANKL will block this activity [14, 15]. In type 1 diabetes mellitus found a significant decrease in OPG compared to controls [16]. In ovariectomy model mice suffering from type 1 diabetes mellitus, there was a significant increase in the RANKL / OPG ratio compared to controls [17].

Garcinia mangostana is a fruit plant native to Indonesia. *Garcinia mangostana* contains prenylated and oxygenated xanthone compounds [18]. Various pharmacological actions of *Garcinia mangostana* have been revealed, including as an antioxidant, antimalarial, antiinflammatory, and anticancer [19-22]. Until now, the potential of *Garcinia mangostana* as an antiosteoporosis has not been revealed. Specifically, the effect of *Garcinia mangostana* on the RANKL/RANK/OPG system has also not been revealed. Therefore, this study aims to analyze the interaction in silico between the active compounds of the *Garcinia mangostana* against the RANKL/RANK/OPG system and its potential as an antiosteoporosis.

2. Material and Methods

2.1. Determination of amino acids making up RANKL, RANK, and OPG

The composition of amino acids RANKL (GI: 2612922) RANK (GI: 19924309), and OPG protein (GI: 2072185) was downloaded from the National Center for Biotechnology Information (NCBI), United States National Library of Medicine (NLM), National Institute of Health (NLM) NIH) (http://www.ncbi.nlm.nih.gov). The 3D structure of RANKL, RANKL and OPG will be converted from * .sdf file format, to * .pdb file using OpenBabel software [23].

2.2. Searching and engineering the structure of active components of Garcinia mangostana

The 3D structure of the active compound component *Garcinia mangostana* was obtained from the PubChem Open Chemistry Database. Eleven active compounds were obtained, namely cudraxanthone G (CID426453), 8-deoxygartanin (CID392450), garcimangosone B (CID11143989) (3), garcinone D (CID5495926), garcinone E (CID102985a), gartanin (CID5281633), 1-isomangostin (CID5281641), α-mangostin (CID5281650), γ-mangostin (CID5464078), mangostinone (CID6478778), smeathxanthone A (CID1150950)). For 8-hydroxycudraxanthone G, mangostingone, tovophyllin A was obtained by structural engineering using Discovery Studio 3.5 software [24].

2.3. Structural modeling and molecular docking

The 3D structure of the target proteins is predicted using the SWISS-MODEL webserver with the homology modeling method. The 3D structure of the protein was then validated using Ramachandran plot analysis [24, 25]. HEX 8.0 software is a device for simulating docking between active compounds of *Garcinia mangostana* and target proteins [26]. The docking protocol consists of three stages of visualization, namely minimization of rigid-body energy, semi-flexible repairs, and finishing refinement in explicit solvents. The docking results are then visualized with Chimera 1.6.2 software and Discovery Studio 4.1. The docking analysis results are visualized using Discovery Studio 4.1, LigPlot + and LigandScout 3.1 software [27, 28]. Analysis of the interaction between proteins and ligands is done to see the number and type of bonds formed.

3. Results

The interaction energy between the RANKL-OP complex, tovophyllin A has the most negative interaction energy (−332.8 Kj/mol). The interaction energies of other active compounds in sequence include mangostingone (−331.2 Kj/mol), garcinone E (−309.8 Kj/mol), cudraxanthone G (−306.2

Kj/mol), 8-hydroxycudraxanthone G (–305.7 Kj/mol) , 1-isomangostin (–290.7 Kj/mol), α-mangostin (–285.1 Kj /mol), γ-mangostin (–284.6 Kj/mol), garcinone D (–282.6 Kj/mol), mangostinone (–277.4 Kj/mol) mol), 8-deoxygartanin (–274.0 Kj/mol), smeathxanthone A (–273.1 Kj/mol), gartanin (–268.8 Kj/mol), and garcimangosone B (–264.6 Kj/mol).

The interaction energy between the RANKL-RANK complex, of the fourteen active compounds *Garcinia mangostana*, tovophyllin A has the most negative interaction energy (–298.1 Kj/mol). Other interaction energy of active compounds include mangostingone (–283.8 Kj/mol), garcinone E (–274.7 Kj/mol), smeathxanthone A (–272.5 Kj/mol), α-mangostin (–265.4 Kj/mol), γ-mangostin (– 258.2 Kj/mol), gartanin (–254.7 Kj/mol), cudraxanthone G (–250.4 Kj/mol), 8-hydroxycudraxanthone G (–249.6 Kj/mol), 1-isomangostin (–248.9 Kj/mol), garcimangosone B (–248.4 Kj/mol), garcinone D (–246.1 Kj/mol), 8-deoxygartanin (–245.5 Kj/mol), and mangostinone (–243.7 Kj/mol).

4. Discussion

First, we simulate an interaction model between fourteen active compounds of *Garcinia mangostana* and the RANKL-OPG complex. Tovophyllin A is an active compound that is the most easily interacted with this complex compared to other compounds. However, the active compound will not affect the RANKL-OPG complex. Our previous study analyzed the interaction between RANKL-OPG and obtained an interaction energy of –584.74 Kj/mol [29]. This indicates that OPG will continue to work as a decoy against RANKL in the context of regulating the balance of bone formation and resorption.

For the second model, we simulate the interaction between the active ingredient *Garcinia mangostana* and the RANKL-RANK complex. The interaction energy of the RANKL-RANK complex is –660.95 Kj/mol [29]. All active compounds of *Garcinia mangostana* have more positive energy interactions with RANKL-RANK. That is, the fourteenth active compound *Garcinia mangostana* has no potential to inhibit RANKL-RANK signals. This has two meanings. First, in physiological views, this indicates that the active compound will not interfere with RANKL-RANK interactions. RANKL is expressed in various tissues, including mammary gland epithelial cells, prostate, pancreas, skeletal muscle, thymus, liver, colon, small intestine, adrenal gland, and osteoblasts [30]. RANKL–RANK interactions are involved in the formation of mammary and lymph node tissue, and temperature regulation [31]. Second, in the osteoporosis pathway, the RANKL-RANK complex is a signal of osteoclast activation and bone resorption. The Antiosteoporosis mechanism of these compounds does not go through the anti - resorption activity. This study extends previous finding that *Garcinia mangostana* can reduce bone resorption activity, through decreasing TRAP markers [32]. In addition, garcinol is able to inhibit bone resorption induced by RANKL [33].

It was concluded that fourteen active compounds of *Garcinia mangostana* did not interfere with the physiological function of RANKL against RANK. In addition, the active compound will not affect the RANKL-OPG complex. The antiosteoporosis mechanism of *Garcinia mangostana* does not by inhibiting RANKL-RANK interactions.

References

[1] Liu T, Ding S, Yin D, Cuan X, Xie C, Xu H, Wang X and Sheng J 2017 Pu-erh tea extract ameliorates ovariectomy-induced osteoporosis in rats and suppresses osteoclastogenesis in *vitro Front Pharmacol* **8** 324

[2] Tabatabaei-Malazy O, Salari P, Khashayar P and Larijani B 2017 New horizons in treatment of osteoporosis *Daru* **25** (1) 2

[3] Tai T W, Chen C Y, Su F C, Tu Y K, Tsai T T, Lin C F and Jou I M 2017 Reactive oxygen species are required for zoledronic acid-induced apoptosis in osteoclast precursors and mature osteoclast-like cells *Sci. Rep.* **7** 44245

[4] Presinger E 2009 Physiotherapy and exercises in osteoporosis and its complications *Z. Rheumatol.* **538** 534-536

[5] Abbassy M A, Watari I and Soma K 2010 The effect of diabetes mellitus on rat mandibular bone formation and microarchitecture *Eur. J. Oral Sci.* **118** 364–9
[6] Hamada Y, Kitazawa S, Kitazawa R, Fujii H, Kasuga M and Fukagawa M 2007 Histomorphometric analysis of diabetic osteopenia in streptozotocin- induced diabetic mice: a possible role of oxidative stress *Bone* **40** 1408–14
[7] Okazaki K, Yamaguchi T, Tanaka K, Notsu M, Ogawa N, Yano S and Sugimoto T 2012 Advanced glycation end products (AGEs), but not high glucose, inhibit the osteoblastic differentiation of mouse stromal ST2 cells through the suppression of osterix expression, and inhibit cell growth and increasing cell apoptosis *Calcif. Tissue Int.* **91** 286–296
[8] Alikhani M, Alikhani Z, Boyd C, MacLellan C M, Raptis M, Liu R, Pischon N, Trackman P C, Gerstenfeld L and Graves D T 2007 Advanced glycation end products stimulate osteoblast apoptosis via the MAP kinase and cytosolic apoptotic pathways *Bone* **40** 345–353
[9] Sanguineti R, Storace D, Monacelli F, Federici A and Odetti P 2008 Pentosidine effects on human osteoblasts in vitro *Ann. NY Acad. Sci.* **1126** 166–172
[10] Chandran M 2017 Clinical aspects and management of osteoporosis and fragility fractures in patients with diabetes *Osteoporosis & Sarcopenia* **3** 123-127
[11] Shehata A S, Amer M G, El-Haleem M R A and Karam R A 2017 The ability of hesperidin compared to that of insulin for preventing osteoporosis induced by type I diabetes in young male Albino rats: A histological and biochemical study *Expe. Toxicol. Patho.* **69** 203-212
[12] de Amorim L M N, Vaz S R, Cesário G, Coelho A S G and Botelho P B 2018 Effect of green tea extract on bone mass and body composition in individuals with diabetes *J. Funct. Foods* **40** 589-594
[13] Zheng Z G, Zhang X, Zhou Y P, Lu C, Thu P M, Qian C, Zhang M, Li P, Li H J and Xu X 2017 Anhydroicaritin, a SREBPs inhibitor, inhibits RANKL-induced osteoclastic differentiation and improves diabetic osteoporosis in STZ-induced mice *Eur. J. Pharmacol.* **809** 156-162
[14] Kohli S S and Kohli V S 2011 Role of RANKL-RANK/osteoprotegerin molecular complex in bone remodeling and its immunopathologic implications *Indian J. Endocrinol. Metab.* **15** 175–181
[15] Liu J Z, Ji Z L and Chen S M 2003 The OPG/RANKL/RANK system and bone resorptive disease *Sheng Wu Gong Cheng Xue Bao* **19** 655–660
[16] Starup-Linde J 2013 Diabetes, biochemical markers of bone turnover, diabetes control, and bone *Front. Endocrinol. (Lausanne)* **4** 21
[17] Ferreira E C S, Bortolin R H, Freire-Neto F P, Souza K S C, Bezerra J F, Ururahy M A G *et al* 2017 Zinc supplementation reduces RANKL/OPG ratio and prevents bone architecture alterations in ovariectomized and type 1 diabetic rats *Nutrition Res.* **40** 48-56
[18] Mohamed G A, Ibrahim S R, Shaaban M I and Ross S A 2014 Mangostanaxanthones I and II, new xanthones from the pericarp of *Garcinia mangostana Fitoterapia* **98** 215-221
[19] Bunyong R, Chaijaronkuel W, Plengsuriyakarn T and Na-Bangchang K 2014 Antimalarial activity and toxicity of *Garcinia mangostana* Linn *Asian Pacific J. Trop. Med.* **7** (9) 693-698
[20] Tjahjani S, Widowati W, Khiong K, Suhendra A and Tjokropranoto R 2014 Antioxidant properties of *Garcinia Mangostana* L (Mangosteen) rind *Procedia Chem.* **13** 198-203
[21] Li G, Thomas S and Johnson J J 2013 Polyphenols from the mangosteen (*Garcinia mangostana*) fruit for breast and prostate cancer *Front. Pharmacol.* **4** 80
[22] Tewtrakul S, Wattanapiromsakul C and Mahabusarakam W 2009 Effects of compounds from *Garcinia mangostana* on inflammatory mediators in RAW264.7 macrophage cells *J. Ethnopharmacol.* **121** (3) 379-382
[23] O'Boyle N, Banck M, James C A, Morley C, Vandermeersch T and Hutchison G R 2011 Open Babel: An open chemical toolbox *J. Cheminform.* **3** 33
[24] Arnold K, Bordoli L, Kopp J and Schwede T 2006 The SWISS-MODEL workspace: a web-based environment for protein structure homology modelling *Bioinformatics* **22** (2) 195-201

[25] Kiefer F, Arnold K, Kunzli M, Bordoli L and Schwede T 2009 The SWISS-MODEL repository and associated resources *Nucleic Acids Res.* **37** 387-392

[26] Macindoe G, Mavridis L, Venkatraman V, Devignes M D and Ritchie D W 2010 HexServer: an FFT-based protein docking server powered by graphics processors *Nucleic Acids Res.* **38** 445-449

[27] Laskowski R A and Swindells M B 2011 LigPlot+: multiple ligand-protein interaction diagrams for drug discovery *J. Chem. Inf. Model* **51** (10) 2778-2786

[28] Wolber G and Langer T 2005 LigandScout: 3-D pharmacophores derived from protein-bound ligands and their use as virtual screening filters *J. Chem. Inf. Model* **45** (1) 160-169

[29] Akbar I Z, Primula Dewi F R and Setiawan B 2019 In silico interaction of the active compounds of *Scurrula atropurpurea* with the RANK/RANKL/OPG system in diabetoporosis *Acta Informatia Med.* **27** (1) 8-11

[30] Wada T, Nakashima T, Hiroshi N and Penninger J M 2006 RANKL-RANK signaling in osteoclastogenesis and bone disease *Trends Mol. Med.* **12** (1) 17-25

[31] Liu W and Zhang X 2015 Receptor activator of nuclear factor-κB ligand (RANKL)/RANK/osteoprotegerin system in bone and other tissues *Mol. Med. Rep.* **11** 3212-3218

[32] Jia Y, Jiang J, Lu X, Zhang T, Zhao K, Han W, Yang W and Qian Y 2019 Garcinol suppresses RANKL-induced osteoclastogenesis and its underlying mechanism *J. Cell Physiol.* **234** (5) 7498-7509

[33] Vellapandian C, Sukumaran E S, Sivasubramanian L R and Vetrivelan V R 2017 A comparative study of *Actinidia deliciosa* and *Garcinia mangostana* in ovariectomy-induced osteoporosis in female wistar rats *BioMed. Res. Int.* **2017** 1-10

Journal of Physics: Conference Series **1374** (2019) 012055 doi:10.1088/1742-6596/1374/1/012055

Urological cancer: molecular docking of the active compound *Scurrula atropurpurea* against nuclear factor erythroid2-related factor2 (Nrf2)

H Prasetya[1]*, F R P Dewi[2] and B Setiawan[3]

[1]Department of Urology, Ulin General Hospital, Faculty of Medicine, Lambung Mangkurat University, Banjarmasin, South Kalimantan, Indonesia. Email: herup_uro@yahoo.com
[2]Malang In Silico Club, Malang, East Java, Indonesia. Email: firlyrahmah@gmail.com
[3]Department of Biochemistry and Biomolecular Sciences, Faculty of Medicine, Lambung Mangkurat University, Banjarmasin, South Kalimantan, Indonesia. Email: ganesh79setiawan@gmail.com

*Corresponding author's email: herup_uro@yahoo.com

Abstract. The nuclear factor erythroid2-related factor2 (Nrf2) is a transcription factor for redox homeostasis involved in antioxidant genes and detoxification enzymes. This transcription factor provides protection against organs and is involved in urological cancer progression. This study aims to investigate the interaction between the active compounds of *Scurrula atropurpurea* against the Nrf2 signal. This study was an in silico study. The research protocol consisted of searching for amino acids making up the Nrf2-Keap1 system, searching for the structure of the active component of *Scurrula atropurpurea*, modeling 3D protein structures, docking and visualization between protein-ligand, and analyzing bond interactions between proteins and ligands. The active compounds of *Scurrula atropurpurea* which are molecularly docking include aviculin, caffeine, catechin, epicatechin, kaempferol, quercetin, quercitrin, rutin, and theobromine. For interactions with Nrf2, rutin was easier to interact compared to other compounds. Energy interactions between caffeine, catechin, kaempferol, quercetin, quercitrin, and rutin were lower than the energy of interaction between Nrf2 and Keap-1. It was concluded that some of the active compounds of *Scurrula atropurpurea* can modulate the Nrf2 signal. Thus, there is an active compound from *Scurrula atropurpurea* which can be an anticancer urological candidate via an Nrf2 signal.

1. Introduction

Cancer is a group of diseases in which uncontrolled cell growth occurs and has the potential to spread to other parts of the body. Tumor suppressing mechanisms present in normal cells always distinguish between abnormal growing cancer cells and normal cells, but the problem arises when the function of these tumors suppressing genes is limited by different environmental factors (such as pollution, radiation, certain infections, etc.) or human habits (such as tobacco, bad diet, alcohol, etc.) [1-5].

Malignancy of urology, including prostate cancer (PCa), bladder (BCa) and kidney cancer (renal cell carcinoma, RCC) is a major cause of morbidity and mortality worldwide. About 326,000 new diagnoses of urological cancer and more than 63,000 deaths are estimated to occur in the United States in 2018, which contributes around 20% and 10% to total estimates of new diagnoses and deaths of each type of cancer [6]. Because of the high incidence, diversity in biology, and especially direct interaction with urine, urological cancer is an important resource for scientists and doctors for diagnostic discoveries and new therapies [7, 8].

Most aspects of cancer biology show a level of redox regulation. Carcinogenesis, cancer cell proliferation, migration, invasion, metastasis, and vascularization all appear to be under redox control [9]. Nuclear factor erythroid2-related factor2 (Nrf2) is the main regulator of redox homoeostasis, is the main transcription factor that regulates various genes for antioxidant enzymes and detoxification. It protects the organs from various types of toxic insults. On the other hand, activation of Nrf2 also correlates with cancer projection and chemoresistance [10] Downregulation of Nrf2 activity has attracted increasing attention because it may provide alternative cancer therapy.

Natural products isolated from medicinal plants have been used for the treatment of various diseases from ancient times. The first use of natural products as medicine dates to 2600 BC in Mesopotamia [11]. *Scurulla atropurpurea* or known by the Javanese as parasite tea is a parasitic plant for tea (Thea sinensis). This downgraded plant has been used by the inhabitants of Java as a cancer drug [12, 13]. *Scurulla atropurpurea* inhibits cervical cancer cell growth through an intrinsic pathway apoptosis mechanism [14]. *Scurulla atropurpurea* also acts as an antioxidant. Some of the active components of this plant are antioxidants of quercetin, quercitrin, kaempferol [15-19]. Until now, the potential of *Scurulla atropurpurea* for the treatment of urological cancer through Nrf2-Keap1 modulation has not been revealed. Therefore, this study aims to analyze the effects of modulation of Nrf2-Keap1 in the context of urological cancer therapy.

2. Material and methods

2.1. Search for amino acids making up Nrf2 and Keap1

The amino acid sequences of Nrf2 (GI: 693842) and Keap1 proteins (GI: 22027642) were obtained from the National Center for Biotechnology Information (NCBI) database, the United States National Library of Medicine (NLM), National Institute of Health (NIH) (http: //www.ncbi.nlm.nih.gov). The 3D structure of Nrf2 and Keap1 in the form of * .sdf file format, will be converted to * .pdb file using the OpenBabel software [20].

2.2. Search for the structure of the active compound Scurrula atropurpurea

The 3D structure of the active compound of *Scurrula atropurpurea* was obtained from the PubChem Open Chemistry Database. Nine active compounds were obtained, namely aviculin (CID 10391477), caffeine (CID 2519), catechin (CID: 9064), epicatechin (CID: 72276), kaempferol (CID 5280863), quercetin (CID 5280343), quercitrin (CID 5280459), routine (CID 5280805), and theobromine (CID 5429). The 3D structure of various compounds in the form of * .sdf file format, will be converted to * .pdb files using the OpenBabel software [20].

2.3. 3D protein structure modeling

The 3D structure of target proteins was predicted using the SWISS-MODEL webserver with the modeling homology method. The structure then validated by using Ramachandran plot analysis [21, 22].

2.4. Docking and visualization between protein-ligand

Docking simulations between active compounds *Scurrula atropurpurea* with target proteins were carried out using HEX 8.0 software [23]. The docking protocol consists of three stages of visualization, namely minimization of rigid-body energy, semi-flexible repairs, and finishing refinement in explicit solvents. The docking results are then visualized with Chimera 1.6.2 software and Discovery Studio 4.1.

2.5. Analysis of bond interactions between proteins and ligands

The results of the next docking analysis will be visualized using Discovery Studio 4.1, LigPlot + and LigandScout 3.1 [24, 25] software. Analysis of interactions between proteins and ligands was done to find the type of bond.

3. Results

Table 1 shows the interactions between various active compounds from *Scurulla atropurpurea* to Nrf2. The smallest bond energy for interaction with Nrf2 is found in rutin (-277.62 kJ/ mol). Sequentially the energy of the quercitrin bond (-268.08 kJ/mol), aviculin (-261.81 kJ/mol), catechin (-194.92 kJ/mol), quercetin (-192.86 kJ/mol), epicatechin (-191.44 kJ/mol), kaempferol (-189.96 kJ/mol), caffeine (-161.98 kJ/mol), and theobromine (-142.86 kJ/mol).

Table 1. Interaction between *Scurulla atropurpurea* and Nrf2 active compounds.

Interaction	Point Interaction	Category	Binding energy
Nrf2 – aviculin	Aviculin – Cys514	Hydrophobic Bond	-261.81 kJ/mol
	Aviculin – Ala511	Hydrophobic Bond	
	Aviculin – Arg449	Hydrophobic Bond	
	Aviculin – Pro469	Hydrophobic Bond	
	Aviculin – Lys472	Hydrophobic Bond	
	Aviculin – Arg517	Hydrophobic Bond	
	Aviculin – Gln512	Hydrophobic Bond	
	Aviculin – Asn513	Hydrophobic Bond	
	Aviculin – Lys516	Hydrophobic Bond	
Nrf2 - caffeine	Caffeine – Leu497	Hydrophobic Bond	-161.98 kJ/mol
	Caffeine – Ile498	Hydrophobic Bond	
	Caffeine – Phe468	Hydrophobic Bond	
	Caffeine – Ile466	Hydrophobic Bond	
	Caffeine – Pro469	Hydrophobic Bond	
	Caffeine – Ile473	Hydrophobic Bond	
	Caffeine – Ile501	Hydrophobic Bond	
	Caffeine – Ala461	Hydrophobic Bond	
	Caffeine – Val470	Hydrophobic Bond	
Nrf2 - catechin	Catechin – Val478	Hydrogen Bond	-194.82 kJ/mol
	Catechin – Pro477	Hydrophobic Bond	
	Catechin – Lys506	Hydrophobic Bond	
	Catechin – Arg503	Hydrophobic Bond	
	Catechin – Arg502	Hydrophobic Bond	
Nrf2 - epicatechin	Epicatechin – Cys514	Hydrophobic Bond	-191.44 kJ/mol
	Epicatechin – Lys516	Hydrophobic Bond	
	Epicatechin – Asn513	Hydrophobic Bond	
	Epicatechin – Gln512	Hydrophobic Bond	
	Epicatechin – Arg517	Hydrophobic Bond	
Nrf2 - kaempferol	Kaempferol – Asn521	Hydrophobic Bond	-189.96 kJ/mol
	Kaempferol – Ile522	Hydrophobic Bond	
	Kaempferol – Lys462	Hydrophobic Bond	
	Kaempferol – Glu520	Hydrophobic Bond	
Nrf2 - quercetin	Quercetin – Leu519	Hydrophobic Bond	-192.86 kJ/mol
	Quercetin – Lys516	Hydrophobic Bond	
	Quercetin – Asn513	Hydrophobic Bond	

	Quercetin – Arg517	Hydrogen Bond	
	Quercetin – Cys514	Hydrogen Bond	
Nrf2 - quercitrin	Quercitrin – Arg517	Hydrogen Bond	-268.08 kJ/mol
	Quercitrin – Pro469	Hydrophobic Bond	
	Quercitrin – Val523	Hydrophobic Bond	
	Quercitrin – Lys516	Hydrophobic Bond	
	Quercitrin – Leu519	Hydrogen Bond	
	Quercitrin – Asn521	Hydrophobic Bond	
Nrf2 - rutin	Rutin – Lys472	Hydrogen bond	-277.62 kJ/mol
	Rutin – Asn513	Hydrophobic Bond	
	Rutin – Lys516	Hydrophobic Bond	
	Rutin – Gln512	Hydrophobic Bond	
	Rutin – Cys514	Hydrophobic Bond	
	Rutin – Ala511	Hydrophobic Bond	
	Rutin – Arg517	Hydrophobic Bond	
	Rutin – Pro469	Hydrophobic Bond	
Nrf2 - theobromine	Theobromine – Ile474	Hydrogen bond	-142.86 kJ/mol
	Theobromine – Leu476	Hydrogen bond	
	Theobromine – Arg502	Hydrophobic Bond	
	Theobromine – Lys506	Hydrophobic Bond	
	Theobromine – Lys508	Hydrophobic Bond	

Table 2 shows the comparison of the interaction energy between the Nrf2-Keap1 complex with the Nrf2-Keap1 complex when there is an active compound from *Scurulla atropurpurea*. The bond energy of the formation of the Nrf2-Keap1 complex is -678.33 kJ/mol. The presence of an active compound *Scurulla atropurpurea* changes the energy of the Nrf2-Keap1 bond. Compounds that cause more negative bond energy are caffeine (-703.03 kJ/mol), catechins (-710.72 kJ/mol), kaempferol (-708.53 kJ/mol), quercetin (-722.01 kJ/mol), quercitrine (-749.01 kJ/mol), and rutin (-753.54 kJ/mol).

Table 2. Interaction between Keap1 and Nrf2 with or without the active compound *Scurulla atropurpurea*.

Molecule	Binding energy
Nrf2 – Keap1	-678.33 kJ/mol
Nrf2, aviculin – Keap1	-632.96 kJ/mol
Nrf2, caffeine - Keap1	-703.73 kJ/mol
Nrf2, catechin - Keap1	-710.72 kJ/mol
Nrf2, epicatechin - Keap1	-655.82 kJ/mol
Nrf2, kaempferol - Keap1	-708.53 kJ/mol
Nrf2, quercetin - Keap1	-722.01 kJ/mol
Nrf2, quercitrin - Keap1	-749.84 kJ/mol
Nrf2, rutin - Keap1	-753.54 kJ/mol
Nrf2, theobromine – Keap1	-652.39 kJ/mol

4. Discussion

The docking method considers the interaction between small molecules and activity on the target, and predicts the degree of attachment of their binding interactions from the docking orientation and the strength of the interaction between them [26]. Nrf2 is protective of normal cells against stress conditions. While in tumor cells, Nrf2 can also be protective due to its activation which is able to win stress or redox regulation based therapy. Thus, there is still controversy over whether activation or inhibition of Nrf2 in tumor cells is a useful therapeutic strategy [27]. In this study, we first analyzed the interaction of nine active compounds from *Scurulla atropurpurea* against Nrf2. The interaction between rutin and Nrf2 has the most negative bond energy. This indicates that rutin interactions with Nrf2 are easiest to occur among other active compounds.This study extends previous findings that interactions between routine and Keap can inhibit Keap-Nrf2 interactions [28].

In the second model we simulate the change in the interaction energy of Nrf2 with Keap1 without the presence of active compounds compared to the active compound is available. Active compounds that facilitate interaction between Nrf2 and Keap1 include caffeine, catechin, kaempferol, quercetin, quercitrin, and rutin. If Nrf2 with Keap1 forms stronger complexes, the more difficult Nrf2 translocation is for modulation of endogenous antioxidants. In cancer cells this will be destructive. In contrast, compounds that inhibit the interaction of Nrf2 with Keap1 include aviculin, epicatehin, and theobromine. This inhibition will make it easier for Nrf2 translocation into the nucleus for antioxidant modulation. In cancer cells this will be protective against redox regulation based therapy.

It was concluded that some of the active compounds of *Scurrula atropurpurea* can modulate the Nrf2 signal. Thus, there is an active compound from *Scurrula atropurpurea* which can be an anticancer urological candidate via an Nrf2 signal.

References

[1] Bagnardi V, Rota M, Botteri E, Tramacere I, Islami F, Fedirko V, Scotti L, Jenab M, Turati F, Pasquali E, Pelucchi C, Galeone C, Bellocco R, Negri E, Corrao G, Boffetta P and La Vecchia C 2014 Alcohol consumption and site-specific cance rrisk: a comprehensive dose–response meta-analysis *Br J Canc* **112** 580

[2] Gallagher E J, Neel B A, Antoniou I M, Yakar S and LeRoith D 2017 The increased risk of cancer in obesity and type 2 diabetes: potential mechanisms. In: Poretsky, L. (Ed.), *Principles of Diabetes Mellitus* Springer International Publishing, Cham, pp.731–753

[3] Leon M E, Peruga A, McNeill A, Kralikova E, Guha N, Minozzi S, Espina C and Schüz J 2015 European code against cancer. In: fourth ed. Tobacco and Cancer *Cancer Epidemiology* **39** S20–S33

[4] Massarweh N N and El-Serag H B 2017 Epidemiology of hepatocellular carcinoma and intrahepatic cholangiocarcinoma *Cancer Contr.* **24** 1073274817729245

[5] Taylor C, Correa C, Duane F K, Aznar M C, Anderson S J, Bergh J, Dodwell D, Ewertz M, Gray R, Jagsi R, Pierce L, Pritchard K I, Swain S, Wang Z, Wang Y, Whelan T, Peto R, McGale P and for the Early Breast Cancer Trialists'Collaborative Group 2017 Estimating the risks of breast cancer radiotherapy: evidence from modern radiation doses to the lungs and heart and from previous randomized trials *J. Clin. Oncol.* **35** 1641–1649

[6] Siegel R L, Miller K D and Jemal A 2018 Cancer statistics, 2018 *CA Cancer J. Clin.* **68** 7–30

[7] Di Meo A, Bartlett J, Cheng Y, Pasic M D and Yousef G M 2017 Liquid biopsy: a stepforward towards precision medicine in urologic malignancies *Mol. Cancer* **16** 80

[8] Dhondt B, Van Deun J, Vermaerke S, de Marco A, Lumen N, De Wever O and Hendrix A 2018 Urinary extracellular vesicle biomarkers in urological cancers: from discovery towards clinical implementation *Int. J. Biochem. Cell Biol.* **99** 236–256

[9] Hegedűsa C, Kovács K, Polgára Z, Regdon Z, Szabó E, Robaszkiewicz A, Forman H J, Martner A and Virág L 2018 Redox control of cancer cell destruction *Redox Biol.* **16** 59-74

[10] Basak P, Sadhukhan P, Sarkar P and Sil P C 2017 Perspectives of the Nrf-2 signaling pathway in cancer progression and therapy *Toxicol. Rep.* **4** 306-318

[11] Borchardt J K 2002 The beginnings of drug therapy: ancient mesopotamian medicine *Drug News Perspect* **15** 187–192

[12] Ohashi K, Winarno H, Mukai M and Shibuya H 2003 Preparation and cancer cell invasion inhibitory effects of C16-alkynic fatty acids *Chem. Pharm. Bull.* **51** (4) 463-466

[13] Ohashi K, Winarno H, Mukai M, Inoue M, Prana M S, Simanjuntak P and Shibuya H 2003 Indonesian Medicinal Plants. XXV. Cancer cell invasion inhibitory effects of chemical constituents in the parasitic plant *Scurrula atropurpurea* (Loranthaceae) *Chem. Pharm. Bull.* **51** (3) 343-345

[14] Parwati N W M, Lindayani I K, Ratnawati R, Winarsih S and Nurseta T 2015 Possible effect of tea plant parasite, *Scurrula atropurpurea* (Blume) Danser, on growth inhibition of culture HeLa cells *in vitro* through DNA repair and apoptosis intrinsic pathways mechanism *Asian Pacific J. Trop. Dis.* **5** (9) 743-746

[15] Athiroh N, Permatasari N, Sargowo D and Widodo M A 2014 Antioxidative and blood pressure-lowering effects of *Scurrula atropurpurea* on deoxycorticosterone acetate–salt hypertensive rats *Biomarkers Genom Med.* **6** (1) 32-36

[16] Afanasev I B, Dorozhko A I, Brodskii A V, Kostyuk V A and Potapovitch A I 1989 Chelating and free radical scavenging mechanisms of inhibitory action of rutin and quercetin in lipid peroxidation *Biochem. Pharmacol.* **38** 1763–1769

[17] Van Acker S A, van Balen G P, van den Berg D J, Bast A and van der Vijgh W J 1998 Influence of iron chelation on the antioxidant activity of flavonoids *Biochem. Pharmacol.* **56** 935–943

[18] Saija A, Scalese M, Lanza M, Marzullo D, Bonina F and Castelli F 1995 Flavonoids as antioxidant agents: Importance of their interaction with biomembranes *Free Radic. Biol. Med.* **19** 481–486

[19] Jung H A, Jung M J, Kim J Y, Chung H Y and Choi J S 2003 Inhibitory activity of flavonoids from Prunus davidiana and other flavonoids on total ROS and hydroxyl radical generation *Arch. Pharm. Res.* **26** 809–815

[20] O'Boyle N, Banck M, James C A, Morley C, Vandermeersch T and Hutchison G R 2011 Open Babel: An open chemical toolbox *J. Cheminform.* **3** 33

[21] Arnold K, Bordoli L, Kopp J and Schwede T 2006 The SWISS-MODEL workspace: a web-based environment for protein structure homology modelling *Bioinformatics* **22** (2) 195-201

[22] Kiefer F, Arnold K, Kunzli M, Bordoli L and Schwede T 2009 The SWISS-MODEL repository and associated resources *Nucleic Acids Res.* **37** 387-392

[23] Macindoe G, Mavridis L, Venkatraman V, Devignes M D and Ritchie D W 2010 HexServer: an FFT-based protein docking server powered by graphics processors *Nucleic Acids Res.* **38** 445-449

[24] Laskowski R A and Swindells M B 2011 LigPlot+: multiple ligand-protein interaction diagrams for drug discovery *J. Chem. Inf. Model* **51** (10) 2778-2786

[25] Wolber G and Langer T 2005 LigandScout: 3-D pharmacophores derived from protein-bound ligands and their use as virtual screening filters *J. Chem. Inf. Model* **45** (1) 160-169

[26] Kazmi S R, Jun R, Yu M S, Jung C and Na D 2019 Insilico approaches and tools for the prediction of drug metabolism and fate: A review *Computers in Biol. Med.* **106** 54-64

[27] Sporn M B and Liby K T 2012 NRF2 and cancer: the good, the bad and the importance of context *Nat. Rev. Cancer* **12** (8) 564–57

[28] Li M, Huang W, Jie F, Wang M, Zhong Y, Chen Q and Lu B 2019 Discovery of Keap1–Nrf2 small–molecule inhibitors from phytochemicals based on molecular docking *Food Chem. Toxicol.* **133** 1-8

Journal of Physics: Conference Series **1374** (2019) 012056 doi:10.1088/1742-6596/1374/1/012056

Effect Of Coal Dust Exposure To Serum Interleukin 6 (IL6) And Vascular Endothelial Growth Factor (VEGF) Level As Preeclampsia Induction In Pregnant Rattus Novergicus

S Wasilah[1], F Rachman[2], D Rosadi[3], A O Puteri[4], N Laily[2], F Yulidasari[5], L Anggraini[5], B Setiawan[6] and M S Noor[4,7*]

[1]Department of Biology, Faculty of Medicine, Lambung Mangkurat University, Banjarmasin, Indonesia; Email: wasilah77@gmail.com
[2]Department of Health Management, Public Health Study Program, Faculty of Medicine, Lambung Mangkurat University, Banjarbaru, Indonesia; Email: fauzie21@ulm.ac.id
[3]Department of Health Promotion, Public Health Study Program, Faculty of Medicine, Lambung Mangkurat University, Banjarbaru, Indonesia; Email: dian_rosadi@ymail.com
[4]Department of Reproductive Health, Public Health Study Program, Faculty of Medicine, Lambung Mangkurat University, Banjarbaru, Indonesia; Email: oputriandini@gmail.com
[5]Department of Nutrition, Public Health Study Program, Faculty of Medicine, Lambung Mangkurat University, Banjarbaru, Indonesia; Email: riniyulidasari@gmail.com
[6]Department of Biochemistry, Faculty of Medicine, Lambung Mangkurat University, Banjarmasin, Indonesia; Email: ganesh79setiawan@gmail.com
[7]Department of Public Health, Faculty of Medicine, Lambung Mangkurat University, Banjarmasin, Indonesia; Email: drmeitria@yahoo.com

*Corresponding author's email: drmeitria@yahoo.com

Abstract. South Kalimantan is one of provinces in Indonesia that has a lot of coal minings. south kalimantan living environtment committee showed that there are 49 coal minings in 13 regions. the most coal minings are in Kotabaru, Tanah Laut and Balangan, each has 10 coal minings. Air pollution because of coal dust can effect endothelial function, such as hypertension in pregnant or preeclampsia. One of its mechanisms is coal dust can induce inflammation. One of inflammation's indicator is high level of IL6. Inflammation disturb angiogenesis which is showed by low level of VEGF. High inflammation and low angiogenesis can induce endothelial dysfunction which is found in preeclampsia. This research's goal was to analyze the effect of coal dust exposure to serum IL6 and VEGF level as preeclampsia induction in pregnant *rattus novergicus*. The design was experimental with *post test only with control group design, consisted of 7 groups: control/K0* (group without coal dust exosure*),* group with coal dust exposure 6,25 mg/m^3 for 1 hour/day (K1-1), group with coal dust exposure 12,5 mg/m^3 for 1 hour/day (K1-2), group with coal dust exposure 25 mg/m^3 for 1 hour/day (K1-3), group with coal dust exposure 6,25 mg/m^3 for 2 hour/day (K2-1), group with coal dust exposure 12,5 mg/m^3 for 2 hour/day (K2-2), and group with coal dust exposure 25 mg/m^3 for 2 hour/day (K2-3).

Exposure started since first day of pregnant untill 19th day of pregnant, and termination was in the 19th of pregnant after exposure. Serum IL6 level in all of groups was in normal distribution (p value > 0,05) but not homogen (p value < 0,05), So IL6 was analyzed by nonparametric Kruskal Wallis test with p value = 0,000. It meant that there was effect of coal dust exposure to serum IL6 level as preeclampsia induction in pregnant *Rattus novergicus*. Serum VEGF level was not in normal distribution (p value < 0,05), so it was analyzed by nonparametric Kruskal Wallis test with p value = 0,002. It meant that there was effect of coal dust exposure to serum VEGF level as preeclampsia induction in pregnant *rattus novergicus*. Coal dust exposure caused high inflammation and disturb angiogenesis. So it could induce endothelial dysfunction in preeclampsia process. The conclusion was there was effect of coal dust exposure to serum IL6 and VEGF level as preeclampsia induction in pregnant *Rattus novergicus*.
Keywords: Coal dust exposure, IL6, VEGF, preeclampsia.

1. Introduction

South kalimantan is one of provinces in Indonesia that has a lot of coal minings. South Kalimantan Living Environtment Committee showed that there are 49 coal minings in 13 regions. The most coal minings are in Kotabaru, Tanah Laut and Balangan, each has 10 coal minings. Pollutant evidence in those 3 regions in 2017 were Kotabaru Region with sulfur dioxide 403 ug/Nm^3, Nitrogen Dioxide 78,5 ug/Nm^3 and Carbon Monoxide 5.032 ug/Nm^3; Tanah Laut Region with Sulfur Dioxide 132,34 ug/Nm^3, Nitrogen Dioxide 95,35 ug/Nm^3 and Carbon Monoxide 3.393 ug/Nm^3; Balangan Region with Sulfur Dioxide 58,92 ug/Nm^3, Nitrogen Dioxide 34,65 ug/Nm^3 and Carbon Monoxide 1.199 ug/Nm^3 [1].

Pollutant exposure to pregnant women can cause negative effects into mother and foetus. It also increases risk for preeclampsia and neonatal condition [2]. Air pollution also causes hypertention in pregnant such as preeclampsia that can induce high risk of premature delivery, low birthweight, small of gestation, stillbirth, and neonatal mortality [3].

Coal dust in epithel of alveolare will be fagocyted by macrofag to produce potential factors. Active macrofag will produce high level of inflammation [4]. One of inflammation cytocin that is involved in this process is interleukin 6 (IL6) [5]. Inflammation process can effect pregnant.

The result of normal process in early pregnant is placentation. It is signed by spiralis artery remodelling. This remodelling supplies oxygen and nutrition for foetus. Spiralis artery remodelling is a description of angiogenesis. Angiogenesis can be shown by vascular endothelial growth factor (VEGF) [6].

Some researches had shown that there was effect of oxidative stress and inflammation to pregnant women. One of the effect was preeclampsia. That effect's mechanism happened because of placental angiogenesis [7]. Untill now, there were not a lot of researches that had evaluated effect of coal dust to serum IL6 and VEGF level as the induction of preeclampsia. The goal of this research was to analyse effect of coal dust to serum IL6 and VEGF level as the induction of preeclampsia.

2. Materials and Methods

2.1. Research Design

This research design was experimental with *post test only with control group design*. The subjects were *Rattus novergicus* in 12 weeks. Female *Rattus novergicus* will be mated by male *Rattus novergicus*. Sample size was calculated by Frederer formula, with minimal sample was 5/group. Total samples were 35 *Rattus novergicus*.

Samples were divided into 7 groups: *control/K0* (group without coal dust exosure*)*, group with coal dust exposure 6,25 mg/m^3 for 1 hour/day (K1-1), group with coal dust exposure 12,5 mg/m^3 for 1 hour/day (K1-2), group with coal dust exposure 25 mg/m^3 for 1 hour/day (K1-3), group with coal dust exposure 6,25 mg/m^3 for 2 hour/day (K2-1), group with coal dust exposure 12,5 mg/m^3 for 2 hour/day (K2-2), and group with coal dust exposure 25 mg/m^3 for 2 hour/day (K2-3). Variables that were measured were serum IL6 and VEGF serum

2.2. Research Location
Research was done in Biochemistry Laboratory in Faculty of Medicine, Lambung Mangkurat University.

2.3. Research Procedure
2.3.1. Mating of Rattus novergicus. Female Rattus novergicus were injected by pregnant mare serum gonadotropine (PMSG) 10 IU 10 IU intraperitoneal. 48 hours after that, they were injected by human chorionic gonadotropin (HCG) 10 IU intraperitoneal. After HCG injection, they were mated by male Rattus novergicus with 1:1. 17 hours after mating, vaginal plug was observed in female. If it was positive, it was called as pregnant day 0.

2.3.2. Coal dust exposure. Coal dust exposure was done since in the first day of pregnant. Exposure used specific tools with dose of coal dust in each groups. Exposure finished in the day of 19.

2.3.3. Termination and sample taking. In the day of 19 of exposure, all of Rattus novergicus were terminated and heart blood was taken. Blood sample was made into serum and then checked for IL6 and VEGF.

2.4. Data analyzes
All of data was tabulated and analysed by homogeneity and normality test. If distribution data was normal and homogen, the test would use Anova test with 95% significant level. If data was not in normal distribution and not homogn, the test would be Kruskal Wallis test with 95% significant level.

3. Results and Discussion
This research was done in 7 groups: *control/K0* (group without coal dust exosure*),* group with coal dust exposure 6,25 mg/m^3 for 1 hour/day (K1-1), group with coal dust exposure 12,5 mg/m^3 for 1 hour/day (K1-2), group with coal dust exposure 25 mg/m^3 for 1 hour/day (K1-3), group with coal dust exposure 6,25 mg/m^3 for 2 hour/day (K2-1), group with coal dust exposure 12,5 mg/m^3 for 2 hour/day (K2-2), and group with coal dust exposure 25 mg/m^3 for 2 hour/day (K2-3). Variables that were observed were serum IL6 dan VEGF level.

Mean of serum IL6 level was in normal distribution (p value > 0.05) but not homogen (p value < 0.05). Statistic analyzes used Kruskal Wallis in Table 1.

Table 1. Analyzes of serum IL6 level.

GROUP	MEAN (pg/ml)	DEVIATION STANDARD	KRUSKAL WALLIS TEST
K0	5.26700	.994593	p = 0.000
K1-1	6.44983	.169871	
K1-2	7.15233	.514288	
K1-3	7.25033	.948923	
K2-1	6.17200	.105119	
K2-2	11.87650	4.123804	
K2-3	13.11650	6.010984	

IL6 is one of inflammation cytocin. It is included in preeclampsia process. Table 1 showed that mean of serum IL6 level in control group (without coal dust exposure) was lower than in coal dust exposure groups. Longer exposure and higher dose tend to cause higher serum IL6 level. Kruskal Wallis test showed that there was significant different between research's groups. So, inflammation happened in coal dust exposure with longer and higher doses.

The other variables was serum VEGF level. Mean of serum VEGF level was not in normal distribution (p value < 0.05), so the analyses test used Kruskal Wallis test in Table 2.

Table 2. Analyzes of serum VEGF level.

GROUP	MEAN (pg/ml)	DEVIATION STANDARD	KRUSKAL WALLIS TEST
K0	714.550	52.2709	p = 0.002
K1-1	717.133	.169871	
K1-2	937.133	148.5728	
K1-3	816.133	144.5831	
K2-1	700.133	224.0310	
K2-2	665.133	180.8963	
K2-3	571.633	48.2718	

VEGF is one of factors that is involved in angiogenesis to make spiralis artery remodeling. Spiralis artery remodeling in preeclampsia decreases. Table 2 showed the mean of serum VEGF level in control group (without coal dust exposure) was higher than in coal dust exposure groups. Longer and higher dose of exposure tend to make serum VEGF level became lower. Kruskal Wallis test showed that there was significant different between research groups. That data showed there was disturbances of angiogenesis and spiralis artery remodeling processbecause of coal dust exposure. This process started early clinical manifestation of preeclampsia.

Placentation in human was signed by wide utery spiralis artery remodelling. It also need the role of natural killer cell (NK cell) and invasive trophoblast. Normal pregnant has extravillous cytotrophoblast in foetus that invate spiralis artery in decidua and myometrium. Invation of cytotrophoblast will replace mother's spiralis artery in endothelial layer and makes it becomes wider. In preeclampsia, the change is not complete. Invation of cytotrophoblast in spiralis artery is only in the surface of decidua and myometrium [8]. Abnormality of this spiralis artery remodelling can caused by inflammation.

Small coal dust can reach alveolar and make inflammation and fibrogenesis. Components of coal dust that can cause inflammation are Fe dan Si [9]. IL6 will cause inflammation in the body. Inflammation inhibits trophoblast invation and disturbs utery spiralis artery remodeling. Inhibition of placentation will cause hypoxia and induce preeclampsia [10].

Serum IL6 level in coal dust exposure groups was tend to increase, so high level of IL6 could cause wide inflammation in longer and higher dose of exposure. Inflammation was followed by decreasing of serum VEGF level. Serum VEGF level tend to decreased in longer and higer dose of exposure. It showed spiralis artery remodelling also decreased, and it could induce preeclampsia.

4. Conclusion
The conclusion was there was effect of coal dust exposure to serum il6 and vegf level as preeclampsia induction in pregnant *Rattus novergicus*.

5. Acknowledgement
We would like to thanks to Laboratory of Biochemistry Lambung Mangkurat University as the location of research. We also would like to show our thank for Faculty of Medicine, Lambung Mangkurat University for giving funding to this research.

References
[1] Balai Lingkungan Hidup Daerah Provinsi Kalimantan Selatan 2017 *Laporan Data Pencemaran* (Banjarmasin: BLHD Prov. Kalsel)

[2] Lu X, Lin C, Li Y, Yao T, Fung J C and Lau A K 2017 Assesment of Health Burden Caused by Particulate Matter in Southern China Using High-Resolution Satellite Observation *Environt. Int.* **98** 160-170

[3] Lee H, Hyo J K, Yoo J L, Min Y L, Hyeonjin C, Hyemin L *et al* 2011 Kruppel-like factor KLF8 plays a critical role in adipocyte differentiation *PLoS ONE* **7** (12)
[4] Armutcu F, Gun B D, Altin R, Gurel A 2007 Examination of lung toxicity, oxidant/antoxidant status and effect of erdosteine in rats kept in coal mine ambience *Environmental Toxicology and Pharmacology* **24** (2) 106-113
[5] Perrone S, Maria L T, Simona N, Mariangela L, Maria S, Maria G A *et al* 2016 Placental histological examination and the relationship with oxidative stress in preterm infants *Placenta* 72-78
[6] Burton G J J A 2009 Oxygen, the Janus gas; its effects on human placental development and function *Epub* **215** (1) 27-35
[7] Pereira R D, De Long N E, Wang R C, Fereshteh T, Yazdi, Alison C *et al* 2015 *Angiogenesis in the Placenta: The Role of Reactive Oxygen Species Signaling. Biomed Research Internasional* Hindawi Publishing Corporation
[8] Kim Y J 2013 Pathogenesis dan promises non invasive markers for preeklmasia *Obstet. Gynecol. Sci.* **56** (1) 2-7
[9] Errickson A and Arbour L 2014 The Shared Pathoetiological Effects of Particulate Air Pollution and the Social Environment on Fetal-Placental Development *Journal of Environmental and Public Health*
[10] Laresgoiti-servitje E, Lopez N G and Olson D M 2010 An immunological insight into the origins of preeclampsia *Human Reproduction Update* **16** (5) 510-524

Estimation of leaf antioxidant activity using image processing

E Suhartono[1]*, B Setiawan[1], P B Santosa[2], R Idroes[3] and M S Indrawan[4]

[1] Department of Medical Chemistry/ Biochemistry, Faculty of Medicine, University of LambungMangkurat, Banjarbaru, South Kalimantan, Indonesia
[2] Banjarbaru Enviroment and Forestry Research Development Institute, Research Development and Innovation Agency, Ministry of Environment and Forestry, Banjarbaru, South Kalimantan, Indonesia
[3] Department of Chemistry/Pharmacy, Faculty of Mathematics and Natural Sciences, Syiah Kuala University, Aceh 24415, Aceh, Indonesia
[4] Telkom School Banjarbaru, South Kalimantan, Indonesia

*Corresponding author's email: esuhartono@ulm.ac.id

Abstract. Physiologically, the influence of inundated conditions on many plants varies. The eco-physiological response of most plantations to flooding is wilt and stomata closure, root damage, reduction of root biomass[2.] This results in a decrease in the production of bioactive compounds, which have an effect on decreasing antioxidant activity.Various methods of measuring antioxidant activity still use chemicals through extraction processes that require a long time. So, it is necessary to examine alternative methods that are cheap, easy and fast. This method is digital processing.The leaf samples were then photographed for image analysis in the Telkom School Banjarbaru Multimedia Laboratory. Antioxidant activity was analyzed through the hydroxyl scavenging method at the laboratory of Biochemistry and Biomolecular, Faculty of Medicine, Lambung Mangkurat University, Banjarbaru.The results of image analysis showed that the antioxidant activity model of RGB values was AA= -54.9592 + 1.0295 R + 0.3757 G + 2,006 B -0,0004 R*G - 0.0012 R*B + 0.004 G*B (AA = antioxidant activity; R = red; G = green; B = blue) with $p = 0.001$ ($p < 0.005$) and $r = 0.76$. In this model, the R value has a significant effect on antioxidant activity.Image analysis using RGB can be used to measure leaf antioxidant activity. However, further research is needed.

1. Introduction

Indonesia's peat swamp forests have been degraded due to more drainage and conversion of forests to agricultural land, and irreversible environments, and large-scale fires [1]. The condition of post-combustion peat swamp forests is not beneficial for plants. Site conditions that have become open areas, physically and chemically damaged peat, burning peat forms a lower surface which is flooded in the rainy season. This condition is a barrier to the success of restoration.Physiologically, the influence of inundated conditions on many plants varies. The eco-physiological response of most plantations to flooding is wilt and stomata closure, root damage, reduction of root biomass [2].

Research using the image method continues to increase in development. In classifying plant diseases, the last 25 years continue to improve the ability to detect and classify approval, accuracy and accuracy of image analysis [3]. Some reports show increasing use of image processing in plant research. Study on plant recognition based on image processing has been rising, and more and more popular because this method is rapid and objective, is not dependent on the person's subjective judgment, effectively and improves work efficiency [4].

Digital images processing techniques to analyze and detect, classify, quantify plant leaf diseases [5] and help to support in the identification of diseases at an early or initial stage and provide useful information for its control [6]. Various methods of measuring antioxidant activity still use chemicals through extraction processes that require a long time. So, it is necessary to examine alternative methods that are cheap, easy and fast. This method is digital processing. Therefore, the digital image processing should be useful as a simplified method for estimating antioxidant activity of plantations.

2. Materials and method

2.1. Study design

This research is an experimental in-vitro study for one week in May 2019.

2.2. Experimental section

The taking of leaves samples was done on *Nothaphoebe*sp.Fresh *Nothaphoebe*sp leaves were collected from plants in the Tumbang Nusa Research Station, Central Kalimantan, Indonesia, at 39 months age, during rainy season in April 2019.The taking of leaves samples was carried out under conditions of raised and inundated plants. Extraction of leaeves samples was carried out on plants randomly. All leaves sample then out in paper bag and taken to the laboratory. Handle samples musch reach the laboratory in good condition and clean [7].

2.3. Measurement of RGB

The leaves samples were then photographed for image analysis in the Telkom School Banjarbaru Multimedia Laboratory, Banjarbaru, South Kalimantan. Fresh leaves are then cleaned. To get a picture, the leaves are scanned with Canon LiDE 110. Nikai RGB is determined with imageJ the free software

2.4. Measurement of antioxidant activity

Gemor leaf obtained in maceration for 3 days with methanol as a solvent. The extract obtained was determined antioxidant activity using the method by Purwanto et al [8]. Antioxidant activity was analyzed through the hydroxyl scavenging method at the laboratory of Biochemistry and Biomolecular, Faculty of Medicine, Lambung Mangkurat University, Banjarbaru, South Kalimantan.

3. Results and discussion

Wet and dry leaves obtained are scanned. The scanned image is seen in Figure 1.

(a) (b)

Figure 1. Scanned image of wet (a) and dry (b) leaves.

Measurement of antioxidant activity with RGB technique is based on phytochemical compound producing a certain color spectrum. Changes in the color spectrum reflect changes in phytochemical structure and antioxidant activity. RGB measurement results for plants are presented in figure 1.Looking at figure 2, theres is a change in the RGB value in (a) and (b).

The results of image analysis showed that the antioxidant activity model of RGB values was AA= -54.9592 + 1.0295 R + 0.3757 G + 2,006 B -0,0004 R*G - 0.0012 R*B + 0.004 G*B (AA =

antioxidant activity; R = red; G = green; B = blue) with p = 0.001 ($p < 0.005$) and r = 0.76. In this model, the R value has a significant effect on antioxidant activity.

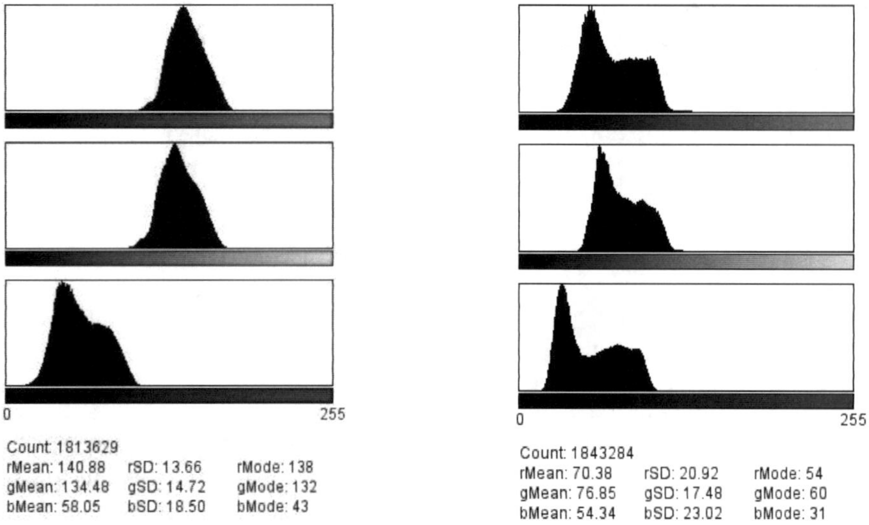

(a) (b)

Figure 2. RGB value of wet (a) and dry (b) leaves.

Figures 1 and 2 show that in wet leaves the value of Red (R) and Green (G) is higher than that of dry leaves. Changes in the value of RGB indicate changes in chemical structure such as covalent bonds or groups of molecules that are bound. This structural change causes the molecule to release H^+ ions to be donated to other molecules.This molecule is called an antioxidant.

Antioxidants in plants have long been used by people for herbal medicine. Antioxidants are believed to inhibit aging, prevent coronary heart disease, hypercholesterolemia, enhance immunity and others.Antioxidants work by hydroxyl radical scavenging, peroxide scavenging, and metal chlelating. A high green color is assumed to be a leaf green or chlorophyll or anthocianin compound which gives green pigment (figure 3).

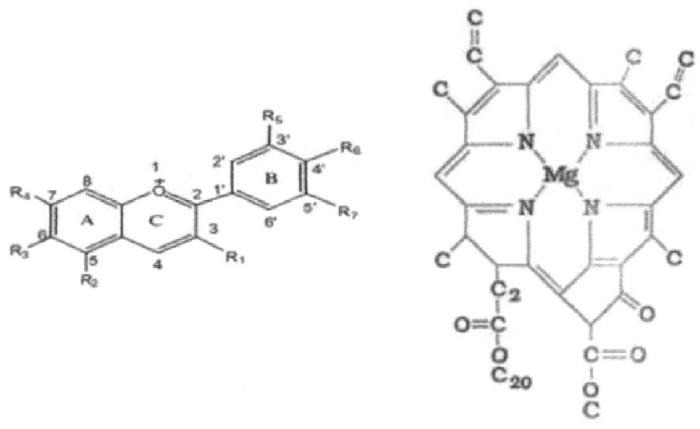

(a) (b)

Figure 3. Chlorophyll or anthocianin structure.

Journal of Physics: Conference Series **1374** (2019) 012057 doi:10.1088/1742-6596/1374/1/012057

Many studies reveal the antioxidant activity of anthocinins. In that study it was found that anthocinin is also an anti-inflammatory [9-10].

4. Conclusion

Image analysis using RGB can be used to measure leaf activity. However, further research is needed.

Reference

[1] K Hergoualc'h, R Carmenta, S Atmadja, C Martius, D Murdiyarso and H Purnomo 2018 Managing peatlands in Indonesia challenges and opportunities for local and global communities *Info. Brief*

[2] H Chen, R G Qualls and G C Miller 2002 Adaptive responses of Lepidiumlatifoliumto soil flooding: biomass allocation, adventitious rooting, aerenchyma formation and ethylene production *Environmental and Experimental Botany* **48** 119-128

[3] G D Sandhu, V Kumar and H D Joshi 2017 Study of digital image processing techniques for leaf disease detection and classification *Multimed. Tools Appl.* **77** 19951-20000

[4] Z Wang, H Li, Y Zhu and T F Xu 2017 Review of plant identification based on image processing *Arch. Computat. Methods Eng.* **24** 637-654

[5] P K Sethy, Negi, B Negi, S K Behera, N K Barpanda and A K Rath 2017 An image processing approach for detection, quantification, and identification of plant leaf diseases-a review *International Journal of Engineering and Technology (IJET)* 635-648

[6] D S Prabha and J S Kumar 2014 Study on banana leaf disease identification using image processing methods *IJRCSIT* **2** (A) 89-94

[7] D M Hassan and M Mankilik 2014 Phytochemical screening of aqueous extract of Luffaaegyptaca(spone gourd) leave sample from northern Nigeria: a short cominucation *International Journal of Pharma Sciences and Research* **5** 344- 345

[8] P B Santosa, I Thalib, E Suhartono and M Turjaman 2016 Antioxidant and anti-lipid peroxidation activities of leaves and seed extracts of gemor (Nothaphoebecoriacea) *International Journal of Pharmacognosy and Phytochemical Research* **8** (7) 1149-1153

[9] M G Miguel 2011 Anthocyanins: Antioxidant and/or anti-inflammatory activities *Journal of Applied Pharmaceutical Science* **01** (06) 07-15

[10] H E Khoo, A Azlan, S T Tang and S M Lim 2017 Anthocyanidins and anthocyanins: colored pigments as food, pharmaceutical ingredients, and the potential health benefits *Food Nutr. Res.* **61** (1) 1361779

Abdominal Circumference, Body Fat Percent, and VO_2 Max in Pilgrims of Hulu Sungai Tengah Regency

Huldani[1*], Asnawati[1], D Auliadina[2], F R Amilia[2], N Nuarti[2] and R Jayanti[2]

[1]Department of Physiology, University of Lambung Mangkurat, Banjarmasin, Indonesia
[2]Student of Medicine, Faculty of Medicine, University of Lambung Mangkurat, Banjarmasin, Indonesia

*Corresponding author's email: huldani@gmail.com

Abstract. Maximum oxygen capacity (VO_2 max) is the maximum volume of oxygen that the body processes when carrying out intensive activities. VO_2 max units is milliliters per minute per kilogram of body weight. Factors that determine VO_2 max include age, gender, heart lung function, aerobic muscle metabolism, exercise, genetics, multivitamins, and nutritional status such as BMI, abdominal circumference, and percent body fat. The study is an analytic descriptive. The number of samples is 50 people. Selected using purposive sampling technique. The VO_2 max measurement is done by Rockport test. Data analysis uses frequency distribution. The results showed that the majority of the research subjects were 26-45 years old (33%), female (40%), normal BMI (32%), central obesity (44%), normal body fat percent (28%). Data on abdominal circumference relationships and body fat percent with VO_2 max were analyzed using the chi square test with a confidence level of 95%, $p = 0.04$ and $p = 0.013$. The conclusions of this study were that the majority of Hulu Sungai Tengah pilgrims who had VO_2 max enough. There is a significant relationship between abdominal circumference and body fat percent with VO_2 max.

Keywords: pilgrims, VO_2 max, age, gender, BMI, abdominal circumference, body fat percent.

1. Introduction

Indonesia is a country whose population is predominantly Muslim, the Hajj activity which is 70% in the form of physical activity and the remaining 30% is non-physical spiritual worship, and the majority of pilgrims range from young adults to old age. As it is known that the cells of the body begin to degenerate function which result in a decrease in the endurance of the lung heart, flexibility, muscle strength, muscle endurance. So that it requires a good physical condition in order to be able to carry out all the series of activities in the Hajj properly and become a hajj mabrur. To be able to achieve good physical condition means that you must have good physical fitness [1].

Physical fitness has 4 basic components, namely endurance of the heart and lungs (cardiopulmonary), strength and endurance of muscles, flexibility and body composition. Fitness for heart and lung endurance is defined as the maximum capacity for breathing oxygen or abbreviated as VO_2 max. Factors that determine VO_2 max include age, gender, heart lung function, aerobic muscle metabolism, exercise, genetics, multivitamins, and nutritional status such as BMI, abdominal circumference, and percent body fat [2,3].

The peak age of the endurance of the heart and lungs is in the age of 20-30 years old, then it will decrease. This is because the more a person ages, the various sports activities will decrease because someone tends to choose to work, besides the decrease in heart contraction and capacity total, and heart muscle mass also affects [4,5].

VO_2 max of men and women is different, the difference in VO_2 max is not apparent at a young age and is most evident during adulthood or middle age. The difference between VO_2 max male and female is related to differences in body composition and size because the physiological bodies of men and women are different. In addition to the body composition the difference in VO_2 max is also due to the different hemoglobin concentrations of men and women. The composition of the female body is more fat than muscle compared to men which causes women has smaller VO_2 max. The male hemoglobin concentration is also higher than women. The hemoglobin concentration is closely linked to the binding of oxygen that the body needs for the process of burning energy, so that men can optimally produce energy [6].

Nutritional status can be assessed based on body mass index, abdominal circumference and body fat percent. BMI is obtained by dividing body weight (kg) by the square of height (m). Body weight tends to be inversely proportional to maximum oxygen capacity. This means that the greater the weight (kg), the lower the maximum oxygen capacity. Being overweight is associated with an increase the amount of fat tissue. The abdominal circumference is the amount of length around a person's body parallel to the navel. The abdominal circumference is generally used as a measure of central obesity. Central obesity is defined as excess fat in the body accompanied by a partial buildup of fat in the visceral part of the stomach. Central obesity is considered a risk factor associated with several chronic diseases [7,8,9].

The lifestyle of people who prefer to eat fast food compared to more nutritious home-made foods causes high rates of obesity [10]. Body fat percent is related to obesity because it is a constituent component of the body besides bone mass, muscle mass, and body water content. describes the condition of the weight or fat mass that is in a person's body in general, both subcutaneous fat and visceral fat (fat contained in organs). At fitness centers, before starting an exercise program, it is usually done checking body composition, one component of which is body fat percent.

2. Method

2.1. Study design

This study uses a analytic descriptive method to describe VO_2 max and the factors that influence it and to know the relationship between abdominal circumference and body fat percent with VO_2 max in pilgrims. The study was conducted at the Hulu Sungai Tengah District Health Office in May-June 2018.

2.2. Participants

The population of this study included all prospective pilgrims of Hulu Sungai Tengah regency 2018. Of the 319 pilgrims there were 50 people who met inclusion criteria that are physical health, did not take stamina multivitamins one day before the study was held, had no history of chronic heart and lung disease, mild to moderate physical activity, and cooperative.

2.3. Physical test

Before measuring VO_2 max, data collection and informed consent were carried out. Informed consent included the approval of the research subjects to take measurements of VO_2 max. The measurement of VO_2 max uses the rockport method by means of the study subjects warming up and stretching the muscles first in the form of walking slowly 5-10 minutes, then walking fast constantly along the 1.6 km track. Travel time (in minutes and seconds) is measured using a stopwatch and the results are matched with the available VO_2 max tables.

Calculation of body mass index by measuring height and weight. Height measurement is carried out with the subject standing against the wall and below the measuring instrument, the measuring instrument is pulled up to the top of the head, the number seen above that is indicated by the measuring line is the subject's height. Measurements of body weight were carried out by means of subjects being asked to rise above the weighing instrument with the position of the feet right in the middle of the weighing device but not covering the reading window, calmness, and the head not looking down, the numbers on the weigher will appear and wait until the numbers do not change.

Measurement of abdominal circumference is performed on subjects with a standing position and breathing normally (normal expiration) using a measuring tape. Examination is done by feeling the last rib of the subject to set the measurement point of the lower edge of the rib cage. Next, the end point of the pelvis, and the midpoint of measurement are between the last rib point and the end point of the hip bone. Measurements of the abdominal circumference are taken from the midpoint and then parallel horizontally around the waist and abdomen back towards the midpoint.

Measuring body fat percent using carada scan body fat gauge by entering the identity of the research subjects such as age, gender and height. Then the research subjects put their feet on the sensor until the measurement results appear.

3. Result and Discussion

The characteristics of the study subjects consisted of age, gender, body mass index, abdominal circumference and body fat percent.

Table 1. The characteristic study subject.

Characteristic	n	%
Age (years)		
1. 26-45	19	38
2. 46-65	31	62
Gander		
1. Men	19	38
2. Women	31	62
BMI		
1. Underweight	2	4
2. Normal	24	48
3. Overweight	9	18
4. Obesity	15	30
Abdominal circumference		
1. Normal	16	32
2. Obesity	34	68
Persen Lemak Tubuh		
1. Normal	13	26
2. Obesity	37	74

Table 2. VO_2 max distribution of pilgrims.

VO_2 maks	n	%
Low fit	12	24
Moderate fit	31	62
High fit	7	14

Table 2 shows that the majority of prospective pilgrims have enough VO2 max.

Table 3. VO_2 max distribution of pilgrims based on age.

Age (years)	VO_2 maks					
	Low fit		Moderate fit		High fit	
	n	%	n	%	n	%
26-45	6	12	12	24	1	2
46-65	1	2	19	38	11	22

Based on Firman's research on Nganjuk district soccer athletes in 2016 stated that the risk of physical fitness of athletes aged 18-35 years had physical fitness 42 times higher than athletes aged over 45 years. The difference in VO_2 max between adult and elderly in the study subjects was caused by a decrease in the function of the heart, lung and blood vessels. The decrease in the average VO_2 max per year is 0.46 ml / kg / minute for men, and 0.54 ml / kg / minute for women. So, at the age of 55 years VO_2 max is 27% lower than the age of 25 years. In addition, heart muscle mass factors, and decreased heart contractions also have an effect [13,14].

Table 4. VO_2 max distribution of pilgrims based on gender.

Gender	VO_2 maks					
	Low fit		Moderate fit		High fit	
	n	%	n	%	n	%
Men	4	8	11	22	4	8
Women	8	16	20	40	3	6

Differences in male and female maximal VO_2 are related to differences in body size and body composition because the physiological bodies of men and women are different. The composition of the female body is more fat than muscle compared to men which causes women to have smaller VO_2 max. The difference between VO_2 max between men and women apart from body composition is one of them also because of the different hemoglobin levels of men and women. Higher hemoglobin levels in men than women cause differences in blood transport capacity which causes men to get far more oxygen during the training process so that their aerobic capacity is better, due to lower oxygen transport and greater percentage of body fat then women tend to show lower VO_2 max values than men [15,16].

Table 5. VO_2 max distribution of pilgrims based on BMI.

BMI	VO_2 maks					
	Low fit		Moderate fit		High fit	
	n	%	N	%	n	%
Underweight	0	0	2	4	0	0
Normal	4	8	16	32	4	8
Overweight	1	2	6	12	2	4
Obesity	8	16	6	12	1	2

Journal of Physics: Conference Series **1374** (2019) 012058 doi:10.1088/1742-6596/1374/1/012058

The VO_2 max level in obese people is always lower than the normal BMI because obese individuals will find it harder to move which will result in an increase in the amount of fat tissue in the body. Excess fat tissue will inhibit the function of the cardio respiration system to take enough oxygen. Excessive amount of body fat provides a burden that does not benefit and inhibits heart function, especially when on the move. In obese people there is an increase in type II muscle fiber and a decrease in type I muscle fiber which has the effect of reducing oxygen absorption [17].

Table 6. VO_2 max distribution of pilgrims based on abdominal circumference.

Abdominal circumference	VO_2 maks					
	Low fit		Moderate fit		High fit	
	n	%	N	%	n	%
Normal	2	4	9	18	5	10
Obesity	10	20	22	44	2	4

The greater the waist hip ratio, the lower the level of VO_2 max. This is related to more fat tissue not only as a place to store triglycerides, but as a producer of bio- active substances called adipokines namely leptin which can affect the cardiovascular system [18].

Table 7. VO_2 max distribution of pilgrims based on body fat percent.

Body fat percent	VO_2 maks					
	Low fit		Moderate fit		High fit	
	n	%	N	%	n	%
Normal	2	4	6	12	5	10
Obesity	10	20	25	50	2	4

Based on table 7 shows the research subjects with percent of normal body fat and VO_2 max which is less or less than the study subjects who have percent of obesity body fat. Based on Susilo's research, people who have percent of normal body fat will have a high cardiorespiratory endurance and conversely people who have a percentage of obesity body fat will have low cardiorespiratory endurance [11].

According to Chatterjee et al, the percentage of excessive body fat will have an impact on decreasing cardiac output while doing activities. The excessive amount of fat tissue will cause narrowing of the arteries and peripheral resistance. As a result, the pumped blood becomes less so that the maximum oxygen capacity also decreases and causes a decrease in cardiovascular ability [19].

Table 8. The relationship between VO_2 max with abdominal circumference of pilgrims.

Abdominal circumference	VO_2 maks						Total		
	Low fit		Moderate fit		High fit				
	n	%	n	%	n	%	n	%	*p Value*
Normal	2	4	9	18	5	10	16	100	0,04
Obesity	10	20	22	44	2	4	34	100	

Based on the statistical test obtained $p = 0.04$ this states that there is a significant relationship between abdominal circumference and VO_2 max. The results of this study are in line with Nurlim's research in 2012 Physiotherapy students at Hasanuddin University medical school, stating that there was a significant relationship between abdominal circumference and VO_2 max ($p = 0.002$) [20]. The 2012 Ramba study showed a significant relationship between central obesity and VO_2 max to

Makassar Polytechnic Physiotherapy students (p = 0.002). In students, central obesity is generally caused due to obesity in childhood which persists into adolescence. Incorrect eating patterns and low physical activity in childhood cause obesity in adolescence. If this condition persists, it will increase the risk of obesity in advanced adulthood, where there is a decrease in metabolism and disposition of fat tissue [21].

Table 9. The relationship between VO_2 max with body fat percent of pilgrims.

Body fat percent	VO_2 maks						Total		p Value
	Low fit		Moderate fit		High fit				
	n	%	N	%	n	%	n	%	
Normal	2	4	6	12	5	10	13	100	0,013
Obesity	10	20	25	50	2	4	37	100	

Based on statistical tests, it was found that p = 0.013, this stated that there was a significant relationship between percent body fat and VO_2 max. These results are in line with previous research which states that there is a relationship between physical fitness and body fat in the gymnastics group (p = 0.02) and non-gymnastic groups (p = 0.41) [22] and there is a strong correlation (r = -0, 75) between percent body fat and VO_2 max, in this study it was concluded that an increase in percent body fat was associated with a decrease in maximal oxygen capacity [23].

4. Conclusion

Based on the description obtained in this study, it can be concluded that most of the prospective pilgrims of Hulu Sungai Tengah regency in 2018 are aged 26-45 years (33%), female sex (40%), normal BMI (32%), central obesity (44%), percent of normal body fat (28%) has enough VO_2 max. There was a significant relationship between abdominal circumference with VO_2 max p = 0.04 and a significant relationship between body fat percent with VO2 max p = 0.013. Pilgrims should prepare their physical 1 to 2 years before departure and the government should hold regular physical training for pilgrims.

5. Acknowledgments

A lot of peoples have contributed by giving advices and supports to the researchers. Our gratiful and appreciation for all of them, especially for the regend of Hulu Sungai Tengah dan the head of religion ministry who have allowed us to conduct the research there. We also say thanks to all participants that have would be participant of this research. May God bless all the kindness.

References

[1] Ningsih I S and Junaidi 2005 Pengaruh latihan senam haji terhadap peningkatan daya tahan jantung paru pada calon jamaah haji non resiko tinggi *Jurnal Fisiologi Indonesia* **5** (2) 45-9

[2] Depkes RI 2005 *Petunjuk teknis pengukuran kebugaran jasmani* (Jakarta: Direktorat Jenderal Bina Kesehatan Masyarakat)

[3] Sugiarto 2012 Hubungan asupan energi, protein, dan konsumsi suplemen dengan tingkat kebugaran. *Jurnal Media Ilmu Keolahragaan Indonesia* **2** 94-101

[4] Cengiz A, Robert A R and Len K 2008 Prediction of VO_2 max from an individualized submaximal cycle ergometer protocol *JEP online* **11**(3).

[5] Hodges A N H, Sheel A W, Mayo Jr and McKenzien D C 2007 Human lung density is not altered following normoxic and hipoxic moderate-intensity exercise: implications for transient edema *J. appl. physiol.* **103** 111-8

[6] Katch 2011 *Essentials of exercise physiology* Edition 4 (Amerika Serikat: Lippincott)

[7] Puskesjasrek 2000 *Pedoman dan modul pelatihan kesehatan olahraga bagi pelatih olahragawan pelajar* (Jakarta: Depdiknas)

[8] Guyton A C and Hall J E 2008 *Buku Ajar Fisiologi Kedokteran* Edisi 11 (Jakarta: EGC)

[9] Capodaglio P, Castelnuovo G, Brunani A, Vismara L, Villa L and Capodaglio E M 2010 Functional limitations and occupational issues in obesity: A review *International Journal of Occupational Safety and Ergonomics* **16** (4) 507-23

[10] Vinet A, Mandigouz S and Nottin S 2003 Influence of body composition, hemoglobin concentration, cardiac size and function of gender differences in maximal oxygen uptake in prepubertal children *Chest.* **124** 1494-9

[11] Susilo E A 2013 Hubungan persentase lemak tubuh terhadap daya tahan kardiorespi atlet pencak silat di klub SMP Negeri 01 Ngunut Tulungagung [skripsi] (Surabaya: Fakultas Ilmu Olahraga Universitas Negeri Surabaya)

[12] Purwati S 2001 *Perencanaan menu untuk penderita kegemukan* (Jakarta: Penebar Swadaya)

[13] Firman F B 2016 Faktor yang berhubungan dengan kebugaran jasmani (VO2 maks) atlet sepakbola (Surabaya: Unair)

[14] Retnosari D 2016 Hubungan antara nilai volume oksigen maksimum dengan nilai panjang napas pada anggota paduan suara mahasiswa universitas Hasanuddin [Karya Tulis Ilmiah] (Makassar: Universitas Hasanuddin)

[15] Hoeger W W K and Hoeger S A 2010 *Principles and labs for physical fitness* (7th ed) (USA: Wadsworth)

[16] Yunus F 2011 Ambilan oksigen maksimal dan faal paru laki laki sehat penyelam dan bukan penyelam *Jurnal Respirologi Indonesia* **31** (2) 61-71

[17] Setty P, Padmanabha B V and Doddamani B R 2013 Correlation between obesity and cardio respiratory fitness *International Journal of Medical Science and Public Health* **2** 300-4

[18] Sharma M, Kamal R B and Chawla K 2016 Correlation of body composition to aerobic capacity; a cross sectional study *International Journal of Applied Research* **2** (1) 38-42

[19] Chatterjee S, Chatterjee P and Bundyopudhyay A 2005 Cardiorespiratory fitness of obese boys *Indian J. Physiol. Pharmacol.* **49** (3) 353-7

[20] Nurlim I 2012 Hubungan obesitas sentral dengan VO2 maks pada mahasiswa fisioterapi fakultas kedokteran universitas Hasanuddin tahun 2012 (Makassar: Fakultas Kedokteran Universitas Hasanuddin)

[21] Ramba Y H 2013 Hubungan obesitas sentral dengan VO_2 max mahasiswa fisioterapi Poltekkes Makassar *Media Kesehatan* **8** (2) 41-54

[22] Dewi P K 2015 Hubungan kebugaran jasmani dan lemak tubuh pada kelompok senam dan kelompok tidak senam [skripsi] (Bandung: Fakultas Kedokteran Universitas Islam Bandung)

[23] Mondal H and Mishra S P 2017 Effect of BMI, body fat percentage and fat free mass on maximal oxygen consumption in healthy young adults *Journal of Clinical and Diagnostic Research* **11** (6) 17-20

AUTHOR INDEX

Adriani 175
Aflanie, I 227
Aisiyah, M M 147
Akbar, I Z 364
Alexandra, F D 334
Alfaqihah, S 94
Ali, A 170
Ali, M 22
Ali, M M 111
Amalia, A 129, 159, 220, 349
Amansyah, A 58
Amilia, F R 384
Andrijono, A 300
Anggraini, L 375
Anita, K W 275
Anita, T Z 327
Annisa, D F 15
Apriasari, M L 15, 187
Ardyati, T 175
Arif, M F 48
Arif, M H 265
Arimaswati 170
Aristya, G R 48
Arrochmah, N 201
Asnawati 384
Asnawati, A 238
Aulanni'am, A 1, 116, 327
Auliadina, D 384
Azhar, A Z 100
Bangkit, I 122
Banthani, G 122
Barlianto, W 111
Baskara, I K A 79
Beltran, M A G 1, 116, 327
Caesar, N R 343
Cahyanto, M N 28
Carabelly, A N 187
Chandradikusuma, D 265
Chiuman, L 58, 159
Devi, W R 232
Dewanti, L P 122
Dewi, F R P 369
Dewi, S S 141
Edyson, E 94
Ekayanti, N L W 349
Endharti, A T 269
Engel, M M S 275
Fatchiyah, F 71, 111, 288, 311
Febrianto, A 100
Firdaus, I W A K 187
Fitri, L E 265
Fitriyani, N 79
Frethernity, A 334
Gaffar, S 316
Ginting, C N 34, 58, 220
Girsang, E 34, 58, 220
Gunanegara, R F 159
Hadi, S 79
Handayani, D 275
Handono, K 159
Hasianna, S T 129
Haskito, A E P 213
Herawati, T 122
Herenda, H 196
Heriansyah, T 322
Hermana, W 256
Hidayat, M 129
Huldani 384
Idroes, R 380
Indahsari, L I N 71
Indarto, D 8, 88, 250
Indrawan, M S 380
Iskandar 122
Ito, M 79
Jayanti, R 384
Junirahma, N S 343
Kaidah, S 364
Kania, N 22
Kasiamdari, R S 48
Khatimah, H 65
Kurnia, A 79
Kurnianingsih, N 111
Kurniati, M 300
Kurniawan, N 71
Kusuma, H S W 129, 349
Kusumastuty, I 275
Laily, N 375
Lestari, S D 141
Lisnawati, L 269
Lister, I N E 34, 58, 159, 220
Lubis, Y E P 220
Lusiana, S E 147, 154
Mahdi, C 42, 213, 327
Mahmud, S 334
Maksum, I P 316
Mardhiyyah, K 265
Marisa, D 227
Marlina, M 159
Mashuri, M 94
Masruri, M 100, 147, 154
Massi, N 170
Maulina, N T A 48
Mawarti, H 292
Meidinna, H N 311
Melati, D 8
Mu'jizah, E Y 288
Mulyaningsih, T 79
Mulyati, M 288

AUTHOR INDEX

Munshy, U Z 34, 349
Musa, M 343
Muthmainah, N 227, 359
Mutia, M S 220
Nahari, D S 1
Nainggolan, B 106
Nasihin, Z D 147
Nasution, S L 34
Noor, M S 375
Noor, Z 364
Noorhamdani 175
Nuarti, N 384
Nugraha, J 292
Nugraheni, M A 88
Nur'Adya, S S 201
Nurhayati, A P D 207
Nurikhwan, P W 22, 232
Nurmardina, P C 187
Nurseta, T 265
Nurwidyaningtyas, W 322
Octavianty, T D 181
Pamungkasari, E P 88
Panghiyangani, R 300
Poeranto, S 269
Prahastuti, S 129, 349
Pramitha, S R 15
Prasetya, H 369
Prasetyawan, S 1
Prastiyanto, M E 141
Pratiwi, D I N 232
Pratiwi, R 207
Purnamasari, Y 170
Purnomo, B B 22
Purwanto, D A 292
Puspitasari, D 15
Puteri, A O 375
Putri, A S 181
Putri, D A 187
Qamariah, N 359
Rachman, F 375
Rahman, E Y 22
Rahman, M F 147
Rahmaniah, G 42
Raissa, R 116
Ratnawati, R 111
Retnani, Y 256
Ristanti, I K 250
Rizal, R 34, 58, 129, 159, 220, 349
Robbie, M H 213
Rohmah, R N 311
Romdhini, M U 79
Roosdiana, A 196, 201
Rosadi, D 375
Rosida, L 65
Rostika, R 122
Rosyada, I 100
Sadiq, A L O 141
Safitri, A 42, 111, 116, 181, 201
Safitri, W A 141
Santosa, P B 380
Santoso, M I E 269
Sanyoto, D D 238
Setiawan, A 15
Setiawan, B 364, 369, 375, 380
Sihombing, L D M 94
Silaban, S 106, 316
Simorangkir, M 106, 316
Skripsiana, N S 65
Soeharso, P 300
Soekardiman 207
Soeroso, J 292
Soetarto, E S 28
Srihardyastutie, A 147, 154
Subagiyo, G J R 243
Subroto, T 316
Suhartina, S 34
Suhartono, E 65, 94, 227, 364, 380
Suryandari, D A 300
Susilowati, K 343
Suyanto, E 28, 111
Syarif, I 256
Syarifin, A N K 170
Tambunan, S H 311
Tamtomo, D G 8, 250
Tanjoto, R B B 359
Tien 170
Tjahjono, C T 322
Triawanti, T 238, 334
Utomo, D H 22
Uyunin, U Z 147
Vidiastuti, D 196
Wahyudi, N T 181
Wardhani, S O 243
Wasilah, S 375
Wibowo, S H B 159, 349
Widiastuti, A 48
Widodo, W S 159, 349
Widowati, W 34, 58, 129, 159, 220, 349
Wihastuti, T A 322
Winarsih, S 175, 265
Wiweko, B 300
Yamada, I 79
Yanti, N L W E 58
Yanuhar, U 343
Yuliana, I 65
Yulidasari, F 375
Yunita, E P 243
Yusepany, D T 129

Institute of Physics
Dirac House, Temple Back
Bristol BS1 6BE UK

ISSN: 1742-6588
ISBN 978-1-7138-1131-2

CURRAN ASSOCIATES INC.
proceedings.com